DYNAMICS OF TECHNOLOGICAL CHANGE

DYNAMICS OF TECHNOLOGICAL CHANGE

Louis A. Girifalco
The University of Pennsylvania

VNR VAN NOSTRAND REINHOLD
New York

Library of Congress Catalog Card Number 90-45717
ISBN 0-442-00563-6

Printed in the United States of America.

Van Nostrand Reinhold
115 Fifth Avenue
New York, New York 10003

Chapman and Hall
2-6 Boundary Row
London, SE1 8HN, England

Thomas Nelson Australia
102 Dodds Street
South Melbourne 3205
Victoria, Australia

Nelson Canada
1120 Birchmount Road
Scarborough, Ontario MIK 5G4, Canada

16 15 14 13 12 11 10 9 8 7 6 5 4 3 2 1

Library of Congress Cataloging-in-Publication Data

Girifalco, L. A. (Louis A.)
Dynamics of technological change / Louis A. Girifalco.
p. cm.
Includes bibliographical reference and index.
1. Technological innovations. I. Title. II. Title.
Technological change.
T173.9.G57 1991 90-45717
600—dc CIP

To the memory of my father
who lived his life for family, honor, and truth

and to my advisor, Robert J. Good,
who transformed me from a student
into a colleague

Contents

List of Figures in Text x
List of Figures in Appendix 1 xiii
List of Tables in Text xiv
Preface xvii

Chapter 1. Basic Concepts of Technological Change 1
Defining Technological Change 1
Measuring Technological Change 8
Theories of Technological Change 12
Determinants of Technological Change 20
Enterprises, Systems, and Hierarchies 28
Five Patterns 30
Interactions Among Systemic Trends 40
A Systemic Model of Technological Change 40

Chapter 2. Technical Progress and Performance 46
Aspects of Technological Progress 46
Measurement of Technical Performance 49
Case Study: Aviation 55
The Learning Curve 60
Production Costs and the Learning Curve 63
Case Study: Farm Tractors 67
Case Study: Underground Coal Mining 73

Chapter 3. Innovation, Research, and Development 80
Clusters and Waves 80
Frequency of Innovation 81
The Origin of Clustering 86
Inventions and Time Lags 88
Innovation and Business Cycles 91
The Origins of Innovations 95
The Effect of Corporate Size 99
Functions of R&D 101
Definitions of Success in R&D 104
Technological Cycles and R&D 106
The Technology-Driven Corporate Cycle 109
Technological Discontinuities 112

The R&D Portfolio 115
Characteristics of Industrial R&D 119
Determinants of Successful R&D 123

Chapter 4. Technological Diffusion and Substitution 130
Technological Diffusion 130
Diffusion Analysis 136
Substitution Analysis 143
Multilevel Substitution 145
The Diesel Locomotive: An Example of Binary Substitution 148
The Diffusion of Radio Broadcasting 154
Steelmaking: An Example of Multilevel Substitution 160
Results of Diffusion and Substitution Studies 165
Nuclear Power: A Case of Interrupted Substitution 169
Diffusion and the Clustering of Technologies 170
Some Comments on Causal Factors 173

Chapter 5. The Dynamics of Energy Usage 176
The Energy Transition 176
Energy Sources 183
Energy in the Economy 185
The Efficiency of Energy Utilization 195
Patterns of Consumption 205
The Energy Inventory 215
Patterns of Production: The Hewett-Hubbert Model 225

Chapter 6. The Electronic Digital Computer 230
Significance of the Computer 230
The Precursors 233
COLOSSUS 241
The ENIAC 242
Turing and von Neumann 247
Downfall of the Moore School 251
Post-ENIAC Machines 253
Computer Companies 258
Evolution of Modern Computer Technology 266
Computer Performance 269
Diffusion of Computers 279
Summary 282

Chapter 7. The Industrial Transition 285
Measures of Industrialization 285
World Industrialization 288
Economic Growth and Industrialization 295
Postwar World Economic Growth 298

Catch-Up 302
The First Transition 305
The American Transition 309
Material Resources 321
Materials and World Industrialization 326

Appendix 1. Mathematics of Technological Growth 335
Forms of Growth Laws 339
Properties of the Exponential 343
Properties of the Logistic 348
Properties of the Decaying Exponential Rise 351
The Gompertz Equation 353
Superacceleration 354
Competitive Growth 356
Multilevel Substitution 359
The Lognormal Distribution 360
Horizontal Analysis 366
Learning Curves and Technological Progress 366
Data Smoothing 370

Appendix 2. Chronology of Innovation 372

Appendix 3. Data Tables 411
List of Tables 416

References 490

Additional Readings 495

Index 515

List of Figures in Text

1-1 Long Cycles of Innovation According to Mensch **32**
1-2 Efficiency of Incandescent Lamps **33**
1-3 Efficiency of Fluorescent Lamps **34**
1-4 Efficiency of Artificial Light Sources **35**
1-5 Percent of Households with Television **36**
1-6 Learning Curve for Pressman Productivity **38**
1-7 Frequency Distribution of Research Funds at Ninety-One Universities **39**
1-8 Schematic of the Systemic Model of Technological Change **44**
2-1 Horsepower to Engine Weight Ratio of Reciprocating Aircraft Engines **56**
2-2 Consecutive Maximum Cruising Speed of U.S. Commercial Aircraft **58**
2-3 Passenger Miles per Hour of Commercial Aircraft (Consecutive Values) **59**
2-4 Maximum U.S. Military Aircraft Speeds **60**
2-5 Cost Curves for Learning and Prorating of Initial Costs: Low Production Volumes **65**
2-6 Cost Curves for Learning and Prorating of Initial Costs: High Production Volumes **66**
2-7 Successive Maximum Tractor Fuel Efficiencies **70**
2-8 Mechanical Efficiency of Tractors **71**
2-9 Total Efficiency of Tractors **72**
2-10 Learning Curve for Fuel Efficiency of Tractors **72**
2-11 Learning Curve for Total Efficiency of Tractors **73**
2-12 Cumulative Number of Tractors Produced Worldwide **74**
2-13 Labor Productivity in Underground Coal Mining **77**
2-14 Test of Learning Curve Equation for the Productivity of Underground Coal Mining **78**
3-1 Nine-Year Running Sum of Major Innovations from Data of van Duijn **83**
3-2 Nine-Year Running Sum of Innovations from Data of Girifalco **84**
3-3 Innovations in the Farm Equipment Industry: Five-Year Running Sum **84**
3-4 Major Semiconductor Product Innovations: Five-Year Running Sum **85**
3-5 Major Semiconductor Process Innovations: Five-Year Running Sum **85**
3-6 Number of Television Innovations per Decade **86**
3-7 Number of Major Inventions: Nine-Year Running Sum **89**
3-8 Time Lags Between Inventions and Innovations **89**
3-9 Mean Time Lags Between Inventions and Innovations **90**
3-10 Ideal Technological Growth Curve **108**

3-11 Ideal Technological Productivity Curve **108**
3-12 Productivity Advantage of a New Technology **109**
3-13 Model Corporate Growth Curve **110**
3-14 Market Shares of Piston, Turboprop, and Jet Engine Aircraft **113**
3-15 The Technological Discontinuity **114**
3-16 Technology Portfolio of Petrov **117**
3-17 The Dynamic Technology Portfolio **118**
4-1 The Generalized Logistic with Different Delay Coefficients **142**
4-2 Fraction of New and Old Technologies During a Substitution **145**
4-3 Three-Level Technological Substitution **148**
4-4 First Adoptions of the Diesel Locomotive **150**
4-5 The Number of Diesel Locomotives in Service in the United States **152**
4-6 Fraction of Diesel and of Steam Locomotives in Service in the United States **152**
4-7 Number of Commercial Radio Stations in the United States **160**
4-8 Percent of Households with Radios in the United States: Curve Computed from $P = 1 - \exp(-0.09938*\text{YR} + 1923.4)$ **161**
4-9 Multilevel Substitution of Steelmaking Technologies **164**
4-10 Fraction of Electricity Generated by Nuclear Energy in the United States: Curve Computed from $F = 1/[1 + \exp(-0.27519*\text{YR} + 546.065)]$ **170**
4-11 Annual Telegraph Messages per Capita for Western Union **171**
4-12 Incandescent Lamps per Capita Produced in the United States **172**
5-1 Long-Term Energy Usage versus Time **177**
5-2 Annual Energy Consumption in the U.S. per Capita **177**
5-3 Fraction of Fuel Use from Mineral Sources in the United States **179**
5-4 Fraction of Energy Consumption by Source **179**
5-5 World Energy Consumption per Capita **182**
5-6 Integral Distribution of per Capita World Energy Consumption **182**
5-7 United States Fossil Fuel Costs as Fraction of GNP **187**
5-8 Expenditures on Fuel and Electricity in the United States as a Fraction of GNP **188**
5-9 Relation Between GNP and Energy Consumption in the United States **189**
5-10 Energy–GNP Ratio for the United States **191**
5-11 Relation Between Energy Consumption and GDP for 102 Countries in 1960 **193**
5-12 Relation Between Energy Consumption and GDP for 106 Countries in 1979 **193**
5-13 Steam Engine Efficiencies **202**
5-14 Thermal Efficiency of Electricity Generation **203**
5-15 Rail Freight Transport Efficiency in Ton-Miles Per 1000 Btu's **203**
5-16 Miles per Gallon of U.S. Passenger Vehicles **204**
5-17 Flow of Energy from Source to End Use **206**
5-18 Fraction of Energy Use in Economic Sectors **207**
5-19 Electricity Production in the United States **211**
5-20 Percent of United States Dwellings with Electric Service: Curve Computed from $P = \exp(0.96558*\text{YR} - 181.9165)$ **212**

5-21 Fraction of Primary Energy Consumption Used for Electricity Generation in the United States **213**
5-22 Total United States Annual Oil Consumption **214**
5-23 Auto Registrations per Capita in United States **215**
5-24 Classification of Mineral Resources **217**
5-25 Cumulative Oil Production in the United States **227**
5-26 Annual Production of Oil in the United States **228**
6-1 Maximum Scientific Computing Power **275**
6-2 Maximum Commercial Computing Power **275**
6-3 Computer Power per unit Cost in Operation per Second per Penny **276**
6-4 Maximum Addition Rate for Computers with Higher Addition Rates than any Previous Computer **276**
6-5 Successive Maxima of Computer Power: NSF Index **277**
6-6 Successive Maxima of Computer Power: L.A.G. Index Using NSF Data: $ln(\text{index}) = [2ln(\text{speed}) + ln(\text{mem}]/3$ **278**
6-7 NSF Data for Memory and Speed of Computers **278**
6-8 NSF Data on Computer Cost for Computers with Maximum Performance in KOPS Per Dollar **279**
6-9 The Diffusion of Computer Capacity in the United States in OPS per Capita **282**
7-1 Ratio of Industrial Level of U.S. and U.K. to the Rest of the World **289**
7-2 Ratio of Industrial Level of Developed to Underdeveloped Countries **290**
7-3 Industrial Level of the United States and the United Kingdom as a Function of Time **291**
7-4 Industrial Level of Japan as a Function of Time **292**
7-5 Industrial Level of the World as a Function of Time **292**
7-6 Growth Rate of Industrial Level Versus Date of Industrial Takeoff **294**
7-7 Consumption of Steel per Capita in the United States: Five-Year Moving Average **313**
7-8 Installed Nonautomotive Horsepower in the United States **314**
7-9 Installed Horsepower in Factories and Mines in the United States **315**
7-10 Installed Horsepower on United States Farms **316**
7-11 Installed Horsepower in Electric Generating Plants **317**
7-12 Automotive Horsepower per Capita in the United States **317**
7-13 GNP per Capita From Berry/Rostow Data: Five-Year Moving Average **319**
7-14 GNP per Capita From Government Data **320**
7-15 Silver Production Cycles of the Comstock Lode **324**
7-16 The Demographic Transition **331**

List of Figures in Appendix 1

A1-1 The Exponential Curve **344**
A1-2 The Logistic Curve **349**
A1-3 The Rate Curve for the Logistic **350**
A1-4 The Decaying Exponential Rise **352**
A1-5 The Rate for the Decaying Exponential Rise **353**
A1-6 The Gompertz Curve **355**
A1-7 The Rate for the Gompertz Curve **355**
A1-8 The Super Acceleration Curve **356**
A1-9 The Normal Distribution **361**
A1-10 The Lognormal Distribution **364**
A1-11 The Lognormal Integral **365**
A1-12 The Learning Curve **368**
A1-13 The Cost Reduction Curve **369**

List of Tables in Text

2-1 Learning Rate Parameters for the Production of Machine Tools **62**
2-2 Learning Rate Parameters for Products in Different Industires **62**
3-1 Percent Share of Industry and of Innovation for the Four Largest Firms **100**
3-2 Characteristics of the Technology-driven Corporate Cycle **111**
4-1 Time Constants for the Spread of Diesel Locomotives **150**
4-2 Multilevel Substitution Parameters for Steelmaking **164**
4-3 Data From Mansfield's Study of Adoptions **166**
4-4 Fisher-Pry Results for 17 Substitutions **167**
4-5 Diffusion of Appliances in American Households **168**
4-6 Periods of Highest Growth Rates for the Diffusion of Technologies in the United States **173**
5-1 Properties of Various Coals **185**
5-2 U.S. Energy Costs and Expenditures in 1985 **186**
5-3 Efficiencies in the Steps of Converting Fuel to Work in an Average Automobile: 1978 **199**
5-4 Possible Efficiency Improvements for Automobiles **200**
5-5 Efficiency of Use of Electricity for Various Devices **201**
5-6 Percentage Consumption of Fossil Fuel Minerals by End Use for 1982 **208**
5-7 Estimates of World Fossil Fuel Resources **220**
5-8 Estimates of Nuclear Fuel Resources **221**
5-9 Estimates of World Renewable Resources **225**
5-10 Half Life in the Production Cycle of Fossil Fuels as Computed from the Hubbert-Hewett Model **228**
6-1 Values of the Technological Progress Parameter From 1952 to 1962 **271**
6-2 Values of the Technological Progress Parameter From 1962 to 1966 **272**
6-3 Annual Percentage Growth Rates in Average Computer Power at Constant Cost for Scientific and Commercial Computing **273**
6-4 Percentage Share of Computer Installations in the U.S. Economy **280**
7-1 Systemic Growth Rates of GNP Per Capita From Data of Rostow **297**
7-2 Growth Rates of GDP Per Capita for Selected Industrialized Countries in Percent Per Annum **301**
7-3 Distribution of Growth Rates per Annum of GDP per Capita for Countries of the World and for World Population **301**
7-4 Catch-up of Some Industrial Countries to the United States: Real GDP Per Capita; U.S. = 100; (Lipsey and Kravis 1987) **303**

7-5 Ratio of per Capita GDP in 1984 to That in 1950 Computed From Data of Lipsey and Kravis (1987) for National Groupings **304**

7-6 Ratio of per Capita GDP in 1984 to That in 1950 **304**

7-7 Curve Fitting Parameters for Real GNP per Capita in the United States **320**

7-8 Ore Grade, Copper Content and Energy Costs at Cuajone **324**

7-9 Resource Inventory of the Most Commonly Used Elements Referred to World Consumption in 1986 **329**

Preface

Technology is not an end in itself, but a way of satisfying human wants. It shows us how to solve the age-old economic problem of surviving and prospering in a hard world. But to optimize the benefits of technological advance requires an understanding of how it happens. The purpose of this book is to provide some of that understanding. The subject is so enormous and so intertwined with every human activity that a small selection of it, and that from a special viewpoint, is inevitable.

The selection of subject matter has been, of course, conditioned by what interests me and is somewhat heterogeneous. However, it is connected by two major themes. The first is that it emphasizes the dynamic nature of technology, in the sense that it must be approached as a process evolving in time that can often be described in quantitative terms. The second is that I have chosen topics that I believe are essential for a strategic sense of how to plan for, execute, and respond to technological change. These two themes complement each other because the strategic sense requires an appreciation of the dynamics and the dynamics naturally lead to a consideration of how to deal with technology so that it can be used to achieve human objectives.

The unifying thought behind the book is that technological change has a systemic as well as an idiosyncratic aspect. While much of change is the result of unique circumstances and events, even more can be described as a regular evolution in time. It is these systemic regularities that hold the key to understanding technological change in the large and that provide us with strategic tools. One objective of this book is to show that this approach provides a coherent framework that illuminates many aspects of the technological process.

I have adopted quantitative methods, historical desciption, and case studies in developing the subject matter, but this is not a book of case studies, a history, or a text on mathematical methods. I have brought in engineering, social, and economic considerations as needed, but this is a text on none of these subjects. Much of the material is new and originates with me; I have treated existing data in new ways, but this book is not intended to be a research monograph. The purpose of this text is to help its readers learn about technological change from a particular viewpoint that is not only academically sound but also useful.

Technological change occurs as a group of evolutionary organic processes punctuated by revolutionary changes that constitute new beginnings. These processes range from general industrialization to the improvement of a specific industrial process, with time scales ranging from centuries to months. A long-range historical perspective is therefore necessary, as well as an understanding of short-term change. Topics have been chosen to illustrate this dichotomy and to display the interrelations among long- and short-term processes, the hierarchical relations among technologies and subtechnologies, and the critical importance of nested time scales in technological progress.

The fundamental concepts of technological change and its treatment as a systemic phenomenon are presented in Chapters 1 to 4. These chapters describe the tools needed for an organized understanding of technological dynamics. The first chapter deals with the necessary definitions of terms and provides an overview of the systemic model, with introductory examples of the quantitative measures on which the book is based. The next three chapters elaborate on these ideas in detail and present specific examples of the quantitative description of the origin, development, and diffusion of technology. While some discussion of specific technologies is included in the first chapters, these are given in the context of the methodology being presented. The systemic nature of technological change is best understood from an extensive consideration of specific technologies. I have chosen energy and computers for this purpose. Chapter 5 deals with energy usage, encompassing some of the most pervasive and critical technology in existence. The availability of energy in large amounts is essential for a technological society, and the dynamics of energy usage is an excellent field for the application of the systemic understanding of technological change. Chapter 6 focuses on the electronic digital computer which is the technology that is characteristic of postindustrial technology as energy is for industrial technology. It is the basic technology on which much else depends. The transition from an agricultural to an industrial society is discussed in Chapter 7. Some historical sense of how far the modern industrial world differs from its predecessors is essential for anyone who wants to understand the processes of technological change and their implications.

Some comment about the use of mathematics in this book is necessary. First, the required mathematics is rather simple. Elementary methods suffice for a remarkably large part of the quanitative analysis of technological change. To a large degree, this reflects the fact that such theory is still in an early stage of development. Second, the important mathematical results can all be displayed graphically, and I have made extensive use of graphs. The major mathematical considerations have been collected into an Appendix so that the reader can choose how deeply to go into mathematics. The object in doing this was to make the material accessible to as wide an audience as possible.

The readers envisioned for this book are those who wish to learn about technology from a dynamic, analytical perspective and those who, now or in the future, must take part in technological decision making. The latter include engineering and business school students at the senior or graduate level. I have, in fact, been teaching courses for the past five years based on the material in this book to students from Engineering, Business, and the Arts and Sciences. I owe a great deal to my students, whose critical faculties combined with a receptivity to new ideas have helped me to refine many aspects of this book. Engineering and R&D personnel, project managers, corporate planners, and indeed anyone who must have some strategic sense in dealing with technological change should find this book of value. Since it represents a systemic approach to technological change, it provides a framework within which specific problems in the innovation process for specific technologies can be addressed. It is my belief that the quality of decision making for technological change can only be enhanced by an appreciation of its systemic nature. The book therefore has a practical utility as well as the academic function of increasing understanding.

1

Basic Concepts of Technological Change

DEFINING TECHNOLOGICAL CHANGE

Technology is the corpus of knowledge and the set of techniques by which we manipulate and control the physical world to satisfy human wants. *Technological change* generally focuses on the techniques, their attendant devices, products, and processes, and the effects of these on individuals and society. In this chapter, we briefly introduce some of the basic concepts and vocabulary that are useful in describing and analyzing the dynamics of technological change.

The study of how technology changes is called *technological dynamics.* It is concerned both with the rates of change in time and the forces that impel these changes. It asks two kinds of important questions. The first questions are of a descriptive nature; they simply ask what is going on and how fast it is happening. It isn't always easy to extract the information and data needed to answer these questions, but at least they are objective and can be treated by analytical and historical methods. The second kind of questions are more profound and ask for the reasons the changes take place and for the identification of the forces controlling the rates. That is, the second set of questions tries to identify the nature and magnitude of the forces that drive the kinetics described by the first set. Of course, more progress can be made at the descriptive level, but the answers to the really critical issues lie at the second, explanatory level.

Technological change is a dynamic process that encompasses an enormous array of events, influences, motivations, individuals, and institutions. It is a process in the sense that it takes place in time as a series of linked events. While it is often easy to recognize technological change, it is often not easy to define it with sufficient generality and precision to support a detailed analysis of how and why it occurs. For our purposes, we start with an obvious observation: if a device, product, or process, which did not exist at some given time is found to exist at a later time, then technological change has taken place. Also, if a device, product, or process is found to be different at two different times, then technological change is said to have taken place over the intervening time period. These definitions require qualifying comments to make them useful. In the first place, for the second definition to apply, the device, product, or process must be a recognizable entity with a sufficient number of constant characteristics so that it preserves its identity over time in spite of the changes it exhibits. In the second place, the two definitions overlap, and which one to use is often arbitrary. For example, audio amplifiers have undergone very great improvements since they were first introduced, but they are all recognized as closely related products, and it is legitimate to compare two different vintages to determine the extent of technological change. We then regard the process of change as an evolutionary development of a given functional device and the second definition is appropriate. However, at another level, there is a profound difference between solid state and vacuum tube amplifiers so that technological change should be considered separately for the two different types. The solid state amplifier is then regarded as a new entity which undergoes its own evolution and the first definition applies. Clearly, technological change takes place at a variety of interrelated levels so that the task of defining what is undergoing change is not trivial.

For purposes of analysis it is often useful to describe technological change in terms of function. That is, change is defined in terms of the means of performing a given technical task, fulfilling a certain need, or achieving a specific objective. This definition has the merit of being able to describe technological change at all levels from large systems to individual device components, while including new as well as existing functions. Also, the functional description often provides quantitative measures for changes in technical performance. As an example, consider the function of transportation, under which we include all means of moving people and goods. This description includes major systems such as airlines and highways, major devices such as trains and automobiles, components such as engines, and subcomponents such as wheels and spark plugs. There are a large number of parts, consisting of devices, assemblies of devices, subsystems, and systems all interconnected to fulfill the overall function of transportation. Each of these parts has a function of its own to support that of transportation. Any one of them can be taken as an object of study, thereby identifying the entities to be examined with respect to technological change.

Describing technological change as a process implies that it is a time series of interconnected events. While this is certainly true, it does not mean that the series is a linear sequence of easily recognizable parts with characteristic properties. Quite often, in fact, the sequence is confusing and complex. Nevertheless, different kinds of activities and motivations can usually be discerned at different points in time, and labeling the different parts facilitates an understanding of the process. There is a generally accepted classification of the sequence of technological change, although it must be emphasized that the classification is somewhat formal and the real situation has more complexities than can be captured by any classification scheme.

With this proviso in mind, the time sequence of technological change can be divided into three broad eras: invention, innovation, and diffusion. *Invention* is the process of arriving at an idea for a device, product, or process and demonstrating its feasibility. It arises from a massive substructure of experience, imagination, and scientific and engineering knowledge; but its end result is always a demonstrated concept that has a practical utility for some human need or desire. *Innovation* is the process by which the invention is first brought into use. It involves the improvement or refinement of the invention, the initial design and production of prototypes, pilot plant testing, and construction of production facilities. *Diffusion* is the process of the spread of the innovation into general use as it is adopted by more and more users. These eras may extend over considerable periods of time and can overlap each other, even though they generally represent distinctly different activities.

Although each of these three eras represents a continuum with stages that sometimes meld into each other, it is often important to attach dates to them. Accordingly, the date of an invention is defined as the time at which it is first made public through an accessible medium such as a technical journal, a patent disclosure, a professional society conference, or a press announcement. The date of an innovation is defined as the first time it is available for use. For a commercial product, this would be the time of first sale or the beginning of its production. For an industrial process, it would be the first time the process is used in a regular production run. For a military device or weapon, it would be the time of its first deployment. The diffusion process defines the spread of an innovation after its first introduction and normally begins at or near the date of innovation and can continue for many years afterwards. It is often temporally characterized by the time it takes for the innovation to be adopted by some given fraction of all its potential users. The half-time and 90 percent time are useful for this purpose, these being the times at which one-half or nine-tenths of potential users respectively have adopted the innovation. A proper designation of each of these dates requires some overall understanding of the specific technology under consideration. There can be, for example, some ambiguity in the date of innovation if a company introduces a product, sells a few copies, and then goes out of business. If some other company takes up the product and sells it successfully, then extensive diffusion takes place. Is the date of innovation that at which the first company introduced the product, or is it the date at which the second company started its successful marketing? The answer depends on the relation between the efforts of the two companies. If the two dates are reasonably close, and if the success of the second company was assisted by the fact that the first company introduced the product, then the earlier date should be chosen. If the second company's success was completely independent of the other company's activity, and if there was an appreciable difference between the two dates, then the later date is the date of innovation. This ambiguity is more frequent than is commonly supposed, but it can usually be resolved by applying the principle that the innovation is the event that initiates diffusion. Similar considerations apply to deciding on the dates of invention. The important invention for technological change is that on which the innovation is directly based.

Any case study of a specific sequence of technological change shows that the division of the process into invention, innovation, and diffusion must be interpreted with care. During the innovation process, while the invention is being brought to commercialization, improvements may be made that must be classified as inventions. Also, a product or industrial process is seldom in its final form when it is first introduced. It undergoes many changes, involving both inventions and innovations. A little thought about any of the major modern technologies reveals the complex interaction of the different kinds of change. Automobiles and television sets, for example, are very different products today than when they were first introduced commercially. Their evolution included a host of inventions, innovations, and diffusion processes, including such things as the all-metal body, new engines, power brakes, new phosphors, larger picture tubes, and solid state circuits.

The danger of labeling different eras of the process of change is that labeling prejudges the nature of that process and therefore constrains the way it is to be interpreted. On the other hand, such labeling is very useful if it is regarded as a model to be tested by the analysis. Of course, a variety of constructs can be used to describe the sequence of change, each useful for different purposes. For example, to describe the activity of a development from initial conception to a final product or process in a corporation, the sequence is defined to consist of applied research, project preparation, pilot plant phase, preparation of manufacturing facilities, and manufacturing startup. In many corporations, the management of innovation is actually organized along lines similar to this. The point is that the framework and language for describing the process of change will depend on the purposes of the classification as well as on the particular process itself.

Because of the enormous size and complexity of the subject, there can be no simple, precise definition of the process of technological change that applies universally. However, in trying to understand particular technological changes, it is important to clarify and define our terms. That this is not just an academic exercise is clear by considering the meaning of "technological innovation." (Unfortunately, this term is commonly used to describe the entire process of change as well as the activities between invention and commercialization, but its meaning is usually clear from the context.) We all have an intuitive sense of technological innovation as a new combination of materials, devices, and/or processes based on scientific and engineering principles that results in a new product or process that performs an old function in a better way, or performs a function that could not be performed before. After all, it is not hard to recognize something new. But this is not enough; innovations are extremely variable events. Photography, television, and semiconductors were innovations with very far-reaching consequences, based on scientific principles and resulting in great new industries that changed the lives of entire populations. Power steering, on the other hand, was not based on radically new science or engineering and was more incremental in both its nature and its effects. It is clear that innovations range from small, modest improvements to revolutionary changes of enormous impact. It would be desirable to have a quantitative measure of the importance of innovations, perhaps in terms of their economic consequences, perhaps in terms of how radical they are, perhaps in terms of their scientific and engineering content. Construction of such a scale would be an enormous task, but at least we can subjectively classify innovations as radical, substantial, and incremental. Even this broad classification can be misleading, since a series of incremental innovations can have large-scale important consequences. Nevertheless, the classification is useful and illuminates both the descriptive and explanatory levels of technological change.

Incremental technology is the continual improvement of existing products and processes by small steps. An example would be the continual increase of automobile mileage by greater use of lightweight materials, changes in engine design, and more efficient combustion. Incremental technology is essentially conservative. It is not approached with the idea of developing whole new systems, or looking for a major breakthrough, but with the intention of improving what already exists. It is the easiest kind of innovation to deal with since it is a natural extension of current activities. It carries the lowest risk and the highest probability of success. The entire system, from corporate organization and personnel to users, is geared to accept and take advantage of incremental technology. Of course, there are times when incremental technology leads to changes that are not merely quantitative, but also dramatic and far-reaching. An example of this is the development of integrated chips for computers. Incremental advances in optical lithography permitted more and more functions to be placed on a chip, leading to microprocessors and computers of such power and small size that they are creating a revolution in information processing and process control. The next step is to go to submicron lithography. This is not achievable with optical methods. Electron beam methods or synchrotron radiation are likely paths. This step will not be an example of incremental technology because new radiation technology for the litho process will have to be perfected. However, it is still a lithographic process and therefore represents a substantial but not radical change for computer technology. A radical change would be the development of optical computers in which the basic logical and arithmetic operations would be done by optical rather than electronic elements.

Radical technological change is that which can change an industry or create new industries. It results in rapid jumps into new systems, products, or processes. It is characterized by a willingness to take risks for the possibility of high returns, by measuring markets in terms of future needs and desires, by seeing great possibilities at early development stages, and above all by insight. The risks are great for radical technology and it takes a long time to come to fruition, but when it is successful the returns are very high. Radical innovations are based on newly emerging scientific results that are incomplete and qualitatively different from what has gone before and for which the nature and profitability of the applications are uncertain. A current example of the pursuit of radical technology is molecular genetics.

Compared to all other technological work, the amount of activity in radical technology is not large, but it is the aggressive, future-oriented approach to technological change that gives us our modern civilization. Electric power systems, telecommunications, computers, transistors, integrated circuits, polymers, and nuclear energy all started with an aggressive, radical view of technology. Its hallmark is the attitude that asks what are the inherent possibilities in technology and science, rather than asking how can we maintain what we have and perhaps make it a little better.

Clearly, aggressive, substantial, and incremental innovations are but three points in a continuous spectrum. These terms are relative but useful. Another useful term to describe technological change is that of defensive technology.

Defensive technology, as its name implies, is designed to maintain a corporation's present position. The most obvious example is that in which antipollution devices are developed to satisfy Federal regulations. There is, in fact, widespread and intense technological activity throughout industry designed to satisfy environmental and safety regulations. This activity is obviously defensive in nature. Defensive technology also arises in response to changing markets and competition. The American automobile industry responded to foreign competition by developing smaller, high mileage automobiles because of declining markets for traditional large cars. In this case, the incremental technological advances of foreign manufacturers forced American companies to change the direction of their own incremental technology. In a sense, incremental technology is also defensive, but we will take it to mean the technology that changes in small steps as a result of the inner logic of the technology itself as it interacts with the day-to-day corporate operations and as dictated by the standard, usually conservative, mind set of the people involved. Defensive technological changes will be defined as arising from challenges and changing conditions external to the normal technological routine of the corporation.

Another type of defensive technological activity is that in which a technology is developed that may or may not be used. In fact, much technology is developed without any clear intention of putting it in place. An example might be the development of a hydroextractive process to replace pyrometallurgy for the extraction of copper. The corporation may feel that it must have this technology to protect its current business and will decide to work it out, with appropriate patents, without deciding that it will be used and even if it believes that it will not be used.

Both incremental and defensive technology are introduced in response to perceived needs. Radical technology, however, is seeking opportunities, creating functions, and opening possibilities that did not previously exist. It embodies a quite different psychology and an unusual degree of self-confident commitment.

Incremental and defensive technology both may be called conservative. They fit well in existing corporate and social structures with their networks of groups and organizations all established for doing things by current methods and running the current business, rather than inventing a new one. Aggressive or radical technology, on the other hand, often does not find a home in the established structures, and when it does, it creates a considerable disturbance. To deal with it, people have to alter their ways of thinking. It often poses a threat to established norms, but it is the kind of technology that has altered our lives. Understanding and utilizing its dynamics properly is a major problem, as well as an opportunity, for our technological society.

The components of modern technology are very varied and extremely numerous. Furthermore, they are interdependent to such a degree that it is hard to imagine any one major portion of technology being able to exist without the others. The development of modern computers, for example, depends not only on advances in microchips but also on communications, information theory, materials, cathode ray tubes, and magnetics. The jet airplane depends on combustion chemistry, materials, computers, electronics, communications, and even food processing. A little thought reveals that most of our modern systems and products draw on a large array of technological resources. In spite of this, we often call out a portion of the technological universe and label it "a technology." This does no harm if the strong interdependence of all parts of technology is kept in mind and if a sufficiently precise meaning of the term "a technology" is constructed for particular cases of analysis or study. The phrase is widely used in the literature without any real attempt at definition, its meaning being determined by the content in which it is used. We will take the term *a technology* to mean some subset of the technological corpus that performs a particular function, or is embodied in a particular product or process, or that arises from closely related scientific and engineering principles. These are three definitions for the same term, but all three are in use. For clarity of thought these definitions should be made explicit in any discussion of technological change.

Again consider the example of transportation. This includes automobiles, trains, aircraft, ships, trucks, motorcycles, carts, and so on. From the point of view of function, transportation technology is the complete set of technological factors that make the transport of people or goods from one point to another possible. This is such a large set that we divide it into subgroups such as automotive technology and air travel technology, still using function for the classification scheme. The division of a technology into subgroups can be carried to many levels, all based on function. From the point of view of the product, we should have a somewhat different orientation. In this case, we could define an *automobile technology* as those technological factors that are necessary to create the device we call an automobile. Again, since the automobile consists of many parts, subclassification to many levels is possible. The internal combustion engine, for example, is a subtechnology for automobiles, while the diesel–electric is a subtechnology for railroad locomotives. From the point of view of scientific or engineering principles, transportation is the result of the implementation of mechanical engineering, thermodynamics, polymer science, materials engineering, aerodynamics, and so on.

Operationally, the reason for defining a technology in the first place is to limit our field of inquiry in particular instances so that questions small enough to be answered can be asked, and so the answers will respond to the objectives of our study. But whatever the basis of definition, it is obvious that any classification scheme for technologies will result in a branching structure and that a truly complete description of any technology would ultimately encompass a major portion of all technology.

MEASURING TECHNOLOGICAL CHANGE

A primary concern of technological dynamics is the quantitative analysis of technological change, so it is important to identify parameters that can be expressed numerically and provide a measure of the degree of change from one time to another. These parameters are essentially of four kinds: those that describe technical performance, those that describe the process of technological change, those that describe its effects, and those that describe the extent of use of a technology.

Technical performance can be described at a variety of levels. The performance of artificial light sources, for example, can be described in terms of their efficiency in converting chemical or electrical energy into visible light. This is a broad, macro measure and can be used to follow the evolution of artificial lighting from the candle to the electrical resistance lamp to the fluorescent lamp to the sodium vapor lamp. It can also be used to track the progress of any one of these devices. Of course, other parameters such as size, cost, and operating lifetime can be used to measure changes in lighting technology. Usually, there are a number of parameters that all contribute to the technical performance of a device or process. For an amplifier, these would include power, harmonic distortion, bandwidth, intermodulation distortion, and dynamic headroom. For an automobile, they would include cruising speed, gasoline mileage, turning radius, size, interior noise level, braking distance, and so on. In general, an adequate description of technological change requires the analysis of a number of parameters for any particular device or technology. It is sometimes possible to derive a single measure of technical performance by constructing a linear and multiplicative combination of the set of important parameters for a particular device. This has the virtue of reducing a multidimensional description to a single number, but has the disadvantage of introducing the subjective element of weighting the parameters that constitute that number.

Technical performance parameters provide the simplest way to follow technological change and are relatively easy to interpret because they deal directly with the most immediate product of technology. They provide the technical description of technological function.

Parameters that represent the process of technological change itself include such measures as the frequency of innovations, the rate of patent disclosures, the size of research and development budgets, numbers of scientists or engineers working in particular fields, numbers of articles published in technical journals, the time lag between inventions and innovations, numbers of engineering graduates, and the rate of formation of new technology based corporations. Such measures yield information on the magnitude of the enterprise that is directly involved in producing technological change, on the structure of that enterprise, and on relations among its parts.

The widest array of measures for the quantitative description of technological change are those that describe its effects. These include effects on the physical infrastructure of society, such as miles of railroad track, communication networks, or numbers of airplanes; effects on health, such as life expectancy or infant mortality; effects on personal life styles, such as length of the work week, leisure and entertainment, extent of formal education, housing and travel; effects on the environment, including pollution and risk of accidents. The list of measurable effects of technology is very long because of the profound and widespread integration of technology in all aspects of modern life. While this permits us to approach the study of technological change from a variety of perspectives, it presents us with the need to determine the degree to which the parameters being used really represent technological factors rather than changes in such things as tax structure, organization, capital investment, or social preferences. Nevertheless these measures are of great interest because they reflect the integration of technology into the social fabric and address the most important aspect of technological change: its impact on individuals, groups, and institutions in society. A major class of technological effects is economic. The Gross National Product, personal incomes, corporate sales and profits, productivity, and foreign trade are all strongly influenced by technology. The relation between economics and technology is complex and pervasive.

Of course, a technology has important effects only if it is adopted. An important technological measure is therefore the diffusion of technology as a function of time. Technological diffusion starts when a technology first becomes available; it is the spread of a technology into the population which can use it. It is usually measured by determining the number of users that have adopted the technology.

The very idea of measuring technological change over time raises the question of causality and determinism. That is, to what extent is technological change an inevitable phenomenon of society that follows regular laws and to what extent is it the result of random processes dependent on unpredictable events such as individual acts of creativity? If we see a regular succession of points through which a smooth curve can be drawn for some parameter, we are inclined to think that some deterministic law is operating that is responsible for the observed regularity. This is the case, for example, for a plot of the efficiency of light sources as a function of time and for the fraction of electrically wired households having certain appliances as a function of time. These sets of data exhibit remarkably smooth curves. In the first case we are led to look for factors that determine a regular progression in technological advance; in the second case we seek social and economic factors that are constant enough to cause a continual increase in the adoption of home appliances. Smooth, regular data attract us to a search for causality, but any case study of technological change shows the existence of a large number of acausal, unpredictable events. Flashes of insight, happy accidents, vagaries of financing, chance meetings, and a host of other events of a nonsystematic nature enter into most innovative developments. Somehow, the well-known nondeterminism of innovative events exists alongside the causality implied by the regular progression of many technological measures. One possible rationalization of this is that while the process is nondeterministic at the microlevel of individual human actions and choices, these microevents integrate into causal progressions at the macro level. This is analogous to statistical mechanics, in which the chaotic motion of atoms and molecules leads to well-defined thermodynamic laws. Another possibility is that the state of technology at any given time, along with the social and cultural ambience, constrains the choices that can be made. This means that while individual events may seem random, they actually reflect the inertial movement along a limited set of technical and social possibilities.

I want to avoid a simple Newtonian interpretation of causality for the social and economic systems involved in technological change. Such an interpretation would assign a linear sequence of causes to a series of linked events. This is not appropriate for social systems in which there are a large number of strongly interacting factors and varying degrees of purposefulness and free choice in operation. Accordingly, I will use the terms *systemic* and *idiosyncratic* to describe the factors influencing technological change. The systemic factors are those that are inherent in the technoeconomic system, or some part of it, while idiosyncratic factors are those that depend on special circumstances, unpredictable events, or free human choice. An important task of technological dynamics is to sort out the systemic from the idiosyncratic factors.

It is best to take the clues from the data. The greater the degree of regularity, the more important the systemic factors. It is important to extract as much regularity from the data as they allow, but it is also important to recognize the inherent irregularities and not to accept systemic interpretations that go beyond those of the data. An interesting example is that of the evolution of steelmaking processes. Over the past hundred years, the four major processes for making steel have been competing with each other with the better processes gradually replacing the others. These processes are the Bessemer, the open hearth, the basic oxygen, and the electric furnace. If the fraction of steel production by each process is plotted as a function of time, four surprisingly smooth curves are obtained that span over one hundred years. It is hard to escape the conclusion that certain economic and technical considerations have been operating constantly for more than a century to give such regularity. However, it would be a mistake to conclude from this that the evolution of the steel industry was completely regular. In fact, if total production of steel, either as a whole or for each process, is plotted instead of fraction of production, the plots are far from smooth. They give the appearance of erratic swings superimposed on a long-range rising trend. The influence of wars, booms, and depressions is clearly evident in such plots. The point of all this is that quantitative measurement of technological change is very useful, but its interpretation must be done with care and take into account many factors in addition to the data.

With this in mind, I have adopted the following posture towards the question of systemic versus idiosyncratic factors with respect to technological change. Systemic factors certainly exist. They arise from the nature of our culture and society, the commonality of many of our aspirations, and the structure of our institutions, as well as from the natural momentum of technological progress. At the same time, free choice and unpredictable factors also exist, as must be the case for human beings interacting with an uncertain world. This choice, of course, is not totally free; it is constrained by cultural factors, social attitudes and structures, and limitations on the possible alternatives inherent in any real-world situation. Nevertheless, enough freedom of action exists to inject a strong random element in many processes of change. One task for the study of technological change is to sort out the two kinds of influences and to give them their proper weight in specific cases. A critical element in this task is the construction of a time series, which is simply a sequence of values of a parameter for a series of times. The degree of regularity of the time series is a measure of the extent of systemic influence in the parameter and, by implication, in the underlying factors that give rise to it. Standard techniques of mathematical statistics are used to measure the degree to which the parameter can be described by deterministic curves.

In a rigorous sense, the only information given by measurement of a changing parameter is that on the parameter itself. Anything more depends on inference. In this regard, the level of specificity of the parameter must always be kept in mind. That is, there are macroparameters which are broad measures of overall technological evolution and microparameters which describe detailed situations, and there is an entire spectrum between these two extremes. As an example, the total number of scientists and engineers is a macroparameter with respect to the stock of technical expertise. At a somewhat finer level, the numbers of chemists, biologists, electrical engineers, and so on refer to subsets of the overall technical manpower pool. The subdivision can be continued down to ever-increasing degrees of specialization. Each of these gives increasingly specific information as the subdivision is extended. Another example would be the amount of steel produced relative to all materials production. This is a macroparameter with respect to the materials industries. The fraction of steel produced by the various steelmaking processes, described above, is a macroparameter with respect to the steel industry. The fraction of steel produced by a specific configuration of all components of a basic oxygen plant linked to a specific continuous casting configuration would be closer to a microparameter. Different levels of detail give different levels of information and often exhibit different amounts of systemic regularity. It often happens that the time series for macroparameters are more regular than those for microparameters.

This leads us to the following interesting hypothesis: at a micro level, there is a strong element of indeterminism; while there is a general social and economic framework within which people make decisions, there is no way of telling which decision will be made among the possible alternatives by particular individuals or firms. It is not possible, for example, to predict precisely when a particular firm will adopt a particular innovation. The total sum of individual actions, however, is more regular. The total number of firms adopting the innovation as a function of time can show a considerable degree of regularity. The individual, indeterminate actions get averaged into a smooth progression in a manner similar to that in the kinetic theory of gases in which the chaotic motion of molecules leads to deterministic variables such as pressure and viscosity.

Clearly, quantitative measurement cannot replace careful historical analysis. The latter is critically important for an understanding of the individual motivations, psychology, and social forces attending technological change. But the quantitative approach is important in revealing the overall results of the flow of the historical events and investigating general influences to which they contribute. Quantitative analysis can also guide and illuminate historical analysis. An important example of the danger of ignoring quantitative considerations is that of the role of large corporations in the innovation process. In earlier years, distinguished scholars believed that only the largest corporations could create significant innovations and that the role of the individual and of small companies could no longer be important. This was widely believed until quantitative studies showed it to be wrong.

THEORIES OF TECHNOLOGICAL CHANGE

The origins of change in human societies have been the subject of a great deal of study and speculation, some of which has been specific to technological change and some of which has been more general. We will abstract from these studies those features that are most pertinent to technology.

One class of theories of change seeks to determine long-range regularities in the entire sweep of human history and to understand the basic forces that shape such regularities. These are large-scale theories that try to put all of history in a single framework. It is not necessary for us to decide whether or not such theories are valid, or even if it is possible to capture the enormous scope of human history in a finite theoretical structure. In fact, no single theory of history can be universally valid. Nevertheless, some theories contain elements that can elucidate the causes of technological change, at least in specific cases. The three such theories we will be concerned with are those of Oswald Spengler (1926), Pitirim A. Sorokin (1937–41), and Arnold Toynbee (1946).

Oswald Spengler

Spengler interpreted history as the evolution of a culture as it worked out the consequences of its world view. He believed that, while each culture was unique, they all went through similar organic cycles of birth, growth, and decline. In fact, he constructed a comparative morphology of cultures that purported to show that all cultures exhibited similar stages during their development and that if the time axes were shifted, the general outlines of the evolution of all cultures could be superimposed on each other. History, therefore, is cyclic. While each culture differs in detail, the stages of development are similar for all, and the underlying cause of change is the basic ethos of the culture.

Spengler's work had a strong mystic coloration. For our purposes, the important point is the concept of the life cycle. This concept can be expressed in analytic language as follows:

A coherent society (or culture) is defined and held together by a set of implicit assumptions about itself, its members, and the rest of the world that is shared by all the individuals. These assumptions constitute the axiomatic core of the society, which determines its most fundamental values and even how it interprets space and time. In its youth, the axiomatic core is not even recognized as such; the world and people are simply seen to be built in a certain way that is obvious and unquestioned. The axiomatic core has potentialities within it that are realized in the forms of art, literature, mathematics, science, rituals, religion, politics, family, and social relations. These elements are the direct results of the basic axioms and derive their vitality from them. As long as the axioms are implicitly accepted, the society is dynamic and sure of itself. It is discovering and working out the consequences of its world view, which appears to be the only one that can be true. But sooner or later there comes a time when little can be done that is new. The potential development of the axiomatic core is finite; inconsistencies and contradictions appear; the smooth, easy flow of creativity that marked the early stages of development slows down; people start to question fundamentals; in short, the limits of what can be done with the basic axioms are reached, and the society goes into decline. Either it must find a new axiomatic base or it will be displaced by a younger, more vital system.

The application of these ideas to technological change is obvious. A technological organization such as a corporation or an R&D laboratory is constructed around a given set of missions and purposes, has a certain strategy, and sees itself in relation to the world in a particular way. It has an axiomatic core which is a driving force for change as its consequences unfold. But if the core stays constant, then growth must inevitably slow down as its limits are approached. The idea of organizational life cycle is inherent in the Spenglerian concept of birth, growth, and decline as the fundamental ethos realizes its possibilities.

The same idea can be applied to technologies. A technology is based on a limited set of scientific and engineering principles. These constitute the "axiomatic core" of the technology. As the inherent capabilities of the basic principles are worked out, the technology develops and is improved, and in the early stages this is done easily and rapidly. But there is a limit to how far the improvement can go. As this limit is approached, further improvement becomes more and more difficult until the technology becomes static. A recent example is the long-playing phonograph record which experienced significant improvement for years but could not be developed further because of the inherent limitations of materials, the nature of mechanical contact, and the difficulties of analog processes. Further advance came only with the advent of the optical laser disc, which was based on a different scientific and engineering core.

Pitirim A. Sorokin

Sorokin was also interested in large-scale history, but his work was much more analytic than that of Spengler. With the aid of specialist scholars in a variety of fields, Sorokin studied developments of the past 2,500 years in the arts, literature, science, and other aspects of Western culture. The thesis he developed was that there were three general ways in which a society can view the world that determine that society's broad characteristics. He called these views the ideational, the sensate, and the idealistic. The ideational type emphasizes spiritual factors, the sensate sees the world in material, physical terms, and the idealistic is a mixture of the two. Sorokin's analytical work consisted largely of assigning numerical scales that measured the degree of ideational or sensate content of various ages of Western culture. He identifies the rise of rationalism and modern science and technology with the replacement of an ideational by a sensate world view.

For our purposes, the most interesting of Sorokin's theses is that change is immanent; that is, the inherent nature of any society is that it changes because of its own internal characteristics. Change occurs as a result of the actions and choices of the members of society. Whenever an act is performed, it automatically alters something; a relation between people, the way of performing a ritual, a technique, a mode of expression, and so forth. Since time flows and circumstances are unique, there is no way of doing the same thing twice in exactly the same way. Change is therefore part of the natural character of any society.

The application to technology is straightforward. In the ordinary course of living in a technological society, people will encounter a variety of circumstances to which they will respond in different ways. Some of these will lead to changes in the technology. But the principle of immanent change must mean more than this for a technological society. Change is immanent directly and by design; it is part of the process of technology and the basis of the economy. The reinvestment of a portion of the national product and the large expenditures on innovation are only two aspects of this.

The idea of immanent change simply means that a system changes because of its own properties that determine its own course of evolution. Clearly, this is applicable to technological systems whose inner characteristics of performance, ease of use, maintenance requirements, size, cost, and so forth all affect the way in which such systems change. Often, the system calls forth its own changes as it evolves to correct its deficiencies or fulfill its technical potential.

Arnold Toynbee

The important idea for technological change that results from Toynbee's work is that of challenge and response. In a competitive technological society the existence of this factor is obvious in the competition between individuals, corporations, industries, and nations. Toynbee used the concept of challenge and response as an organizing principle for understanding human history. The challenge might be environmental such as a weather change, from an alien population as in migration or war, from a depletion of resources, or from internal disarray. Any condition that threatens the society will elicit a response to mitigate and rise above the threat. It is important that the threat be of the right degree of severity. If it is trivial, the response will be mild or nonexistent and change will not occur. If it is overwhelming, the society will simply succumb, since it sees no way to mount a successful response. Thus, challenges of the right kind and degree are agents of progress in human history. A great deal of technological progress must be based on the challenge and response syndrome. Every new technology challenges someone that must respond to it to succeed, and new technologies are often responses to challenges posed by the market or by competitors.

Karl Marx

We now turn to theories of change that are more specific to our time; that is, to the time since the Industrial Revolution. These theories are largely economic in nature and they start with Karl Marx (1848, 1867). Marx was a keen student of technology; his observations on this score have a value that is independent of his ideas on the class struggle and communism. Marx's basic assumption is that economics determines history, and that the mode of production is what determines change. For Marx, the key element of the modern age was the substitution of machine labor for manual labor. The specialization of labor in factories was already a reality before the advent of mechanical power sources. This specialization made it possible to mechanize labor. Mechanization was a major event because it eliminated the variability and unreliability that accompanies all human work, especially after it has been rendered boringly repetitious by specialization. Mechanization of production was a great agent of change of the Industrial Revolution.

According to Marx, the process by which change takes place is that of dialectic materialism. That is, all human activity reflects the material conditions of human existence, which are characterized by continual change and development. Processes of gradual development eventually expose inner conflicts which can only be resolved by a qualitative shift in the material conditions of life. Thus, an event occurs or something is introduced and a line of development begins; this inevitably produces a negative reaction, which grows in strength until there is a significant tension between the new introduction and its reaction, which exposes the contradictions inherent in the situation. The conflict between the two results in modifications that resolve the contradictions. An important example of this would be that of the environmental effects of electric power generation. The benefits of electrical power were widely and enthusiastically adopted, and its availability was one of the important measures of technological progress. But with the growth of electric power, it became apparent that the negative polluting effects were serious. Our desire for electricity was in conflict with our need for clean air. This conflict is still in the process of resolution; the end result will most probably be that electricity will continue to be generated in increasing amounts, but its price will include the cost of keeping the pollution it generates to acceptable levels. Thus, a new approach to generating electricity is arising from two contradictory human desires.

Many of Marx's insights into technology and the economy are still of value. He gave a careful analysis of the idea of economies of scale, recognized that innovations often bear the burden of concepts of previous technology, and that innovations undergo improvement as they are used. He recognized that the transition from hand manufacture to mechanization was a critical technological event, since it took control of production away from variable human hands and placed it with machines that could be standardized and indefinitely improved. He regarded innovation in capital goods as being particularly important. Improvement in capital goods means that the rate of return on investment increases since more can be done with the same expenditure.

Marx's great failure was that he did not look beyond the industrial technology of his day and could not predict its evolution. With hindsight, it is easy to see that technological advance empowers individuals and is a powerful force for individual well-being. Technology must have markets, it must give people what they want, and large numbers of people must have enough money to buy it so that its costs can be prorated over large numbers of users. The wealth that technology creates must be distributed among a large number of people. Otherwise there will be no way to effect technological change, and there will be no reason for it. The technology itself liberates people. Greater mobility, ease of communication, better health, and more leisure time are all results of technological change that work against Marx's thesis that revolutionary class struggle is inevitable.

Nikolai Kondratieff

Technology and the economy are intertwined, and a great deal of effort has gone into trying to understand their relationship. Kondratieff was a Soviet economist who analyzed movements in commodity prices, coal, iron, and lead production, foreign trade turnover, and other economic indicators back to the latter part of the eighteenth century (1922–28; translated by Daniels and Synder, 1984). In this search for long-term regularities he performed a statistical analysis of the data and concluded that capitalist economies rise and fall in cycles of 50 years duration. He also noted a remarkable correlation between innovation and the long wave cycle, the degree of innovation being very high for about 20 years before the beginning of the rising wave of the long cycle. However, Kondratieff refused to draw a causal connection between technical innovation and the long wave and wrote, ". . . I am absolutely disinclined to believe that it offers any explanation of the causes of the long cycles." Instead, he asserted that the causal factor was investment in new large-scale capital goods, such as railroads, canals, and major construction. When the upswing in a cycle has gone far enough, there is an excess of capital available which makes it profitable to invest in large-scale projects. But this cannot continue indefinitely. Capital then becomes scarce because it is tied up in the large projects, and this results in a downturn in the economy. This is the same reasoning that Marx applied to ordinary business cycles. It is sad to note that Kondratieff was later imprisoned by Soviet authorities, presumably because they could not accept a theory that described the continual renewal of capitalist economies.

It was Kuznets (1930) and Schumpeter (1934, 1939, 1942) who ascribed a central role to technical innovation as an explanation of business cycles. They maintained that new products are the major cause of massive changes in the economy. This idea was taken up by Mensch (1979) and combined with that of Kondratieff's long waves. Mensch constructed a data base of important innovations from the Industrial Revolution to the present and counted the number of innovations as a function of time. He concluded that there were 50-year cycles during which the number of innovations rose and fell and that these innovation cycles were out of phase with the long wave business cycles. The Kuznets–Schumpeter–Mensch picture is as follows: During a rising business long wave, the economy is exploiting a cluster of innovations that were introduced earlier. These innovations offer new products and new opportunities. But as time goes by, nearly everyone acquires these products; they become part of the normal scene and represent replacement industries rather than growth industries. Competition results in lower profits in a static market, increasing automation contributes to fewer jobs, and the opportunities associated with the products decrease. The result is recession and depression. But this releases both capital and labor. In the search for new opportunities, innovations are introduced and the resulting new products start the cycle anew. There is an inherent discontinuity in technological progress as the old is periodically swept away by the new.

The evidence for the relationship between innovation and economic growth is strong, but that for the regularity of waves of innovation is not. Analysis of the data for the clustering of innovations does not support the idea of 50-year cycles. However, there is no doubt that clustering does occur, which has a powerful effect on the economy.

Usher (1954), on the other hand, has emphasized the continuous nature of technological change by pointing out that the accumulation of small, incremental improvements over time often results in dramatic changes when viewed on a long time scale. It is unfortunate that this has been often regarded to be in opposition to the idea that technological progress takes place in large, discontinuous steps. It is clear that both kinds of changes take place. There certainly are major events that drastically alter the course of technical advance in a short time. At the same time, there certainly are long series of closely spaced incremental changes that result in great advances in the long run.

These theories leave a fundamental question unanswered that has engaged the interests of historians. Why did the Industrial Revolution take place in the West? Even after it occurred, it did not take root naturally in non-Western societies, but was imported as an alien thing, useful but not natural. In his study of the rise of modern technology in Europe, Landes (1969) identified a number of conditions that favored Europe, particularly England, for the development of industry. These included wealth, rising consumer demand, the rule of law and contracts, limits on arbitrary power, a relatively flexible class structure, well-developed transport and communication systems, a high level of technical development, a willingness to import technical ideas, and a proclivity for risk and innovation. While these are important historical conditions, they cannot be ultimate explanations of the uniqueness of Western society with respect to the growth of technology. At a more fundamental level, the differences must reside in the modes of thought and world views that define the possible paths a society can take. Rationalism and its application to nature already had an illustrious history, and the analytic mode of thought was well entrenched before the Industrial Revolution. Even the religion that permeated Europe for over a millennia was conducive to technological growth. Christianity had destroyed animism and regarded the function of nature to be to serve man. The concept of a personal God stressed a kind of duality and separateness that made it natural to break the world down into parts that could be studied and exploited. In short, there was an all-pervasive complex of ways of thinking that made the evolution of technology natural for the West. When the physical conditions permitted it, technology arose as a natural expression of the European mind.

The progress of technology is often described from two seemingly opposed viewpoints; one focuses on the individual major innovations, which are treated as discontinuous events that alter the direction of technology, while the other emphasizes the evolutionary, continuous character of technological change. This dichotomy is related to another in which one view is that innovation is the result of heroic individual achievement while the other view is that progress is determined by the systemic forces of society. A variation of this last idea is that what is critically important is the interaction of people involved in technology with the individuals, institutions, attitudes, and value systems of the larger social system.

Rather than adopting one or the other way of describing and trying to understand the totality of technological change, it is well to recognize that technological change is a set of diverse phenomena and that the appropriate description depends on what we wish to analyze as well as on the particular technological phenomenon under discussion. In this regard, a consideration of the relationship between discontinuities and continuous progress is of value.

With hindsight, the steam engine, the Bessemer steelmaking process, the radio, and the transistor were major "breakthroughs" that dramatically altered the course of technology in relatively short times. They can be regarded as discontinuous events associated with heroic invention and, indeed, we associate specific dates and specific names with them. But the discontinuity is apparent only if the innovations are examined on a long time scale. A detailed historical analysis shows that on shorter time scales, the steps leading to these innovation were many, were often close to each other in both time and conception, and involved many more people than just the ones we labeled as inventors. The situation seems to be that each dramatic innovation is the culmination of a long line of incremental progress; on a fine enough scale of time and analysis, technological advance seems to be continuous. But this must not mislead us. There are aspects of the process that can be thought of as discontinuous. Even on a short time scale, there are individuals that have a key idea without which the previous incremental work would not bear fruit. They supply an insight that everyone else missed. On a longer time scale, there are innovations that start a new line of technology and are therefore discontinuous in the sense that there is a sharp change in the direction of technological change.

There is not much point in trying to decide between continuous and discontinuous or between social and individual factors as general frameworks for technological change. However, there are radical changes that induce a long period of incremental advance; there are long periods of incremental change whose end result is so great that it constitutes a technology that is qualitatively different from the starting point. This is not as chaotic as it sounds. A technological innovation either leads to further advances or it does not. If it does, it goes through a growth curve until technical limits bring it to a point of diminishing returns. The growth consists of other innovations, each of which can generate a growth curve of its own. The longer the growth and the more important its technical, economic, or social consequences, the more we are inclined to think of the original innovation as a radical, discontinuous event.

The foregoing is a sample of the attempts to explain technological change at several levels of aggregation. It is hard to believe that any of these theories has universal validity, and it is easy to criticize each of them. However, they all contain elements that undoubtedly influence the course of technological change, at least at some times and places. Such theories of change should be regarded as permissive rather than necessary in that they specify conditions and mechanisms that can lead to change but are not required for change to take place. The actual conditions and causes surrounding any particular change are multiple and complex and require the most sophisticated historical analysis for their elucidation in detail, but in many cases generalized causal factors of the type discussed here are useful in understanding technological change. These may be called macro and middle level approaches. At levels of analysis that are more fine grained, more specific models are needed, and the factors that enter into such models we will call determinants of technological change.

The primary fact of modern technological change is that it is self-sustaining and that it is coupled to a self-sustaining economy. Any theory of modern technology must therefore account for the self-sustaining growth of technology and the economy. Such a theory would start from a few premises rooted in the nature of people and their societies, their personal and social values, and the possibilities inherent in physical resources. The premises would then permit the possible courses of technological development to be derived, would account for the differences of development among various peoples, would expose the origins of the rates of technical progress, and would describe the evolution of modern technological societies in detail. No such theory exists, and it is unlikely that one can be constructed in the near future. But some of the important factors responsible for self-sustaining growth can be identified and some of the important consequences can be accounted for.

DETERMINANTS OF TECHNOLOGICAL CHANGE

The influences and causal factors affecting technological change are so various that they include every sphere of human activity and their interactions with the physical world. While this is not a very useful statement, it should remind us that it is easy to construct lists of the determinants of change, but it is also easy to leave out factors that may be important. We must recognize that any general statement of causal factors is bound to be incomplete and that different factors may be operative in different circumstances. Yet, it is necessary to have some schema in mind if some order is to be imposed on what would otherwise be an unmanageable subject. Accordingly, we classify the determinants of technological change to be of the following types: values and attitudes, physical resources, knowledge and skills, national objectives and policies, political factors, governmental regulatory requirements, world competition, macroeconomics, microeconomics, institutional and social structures, markets, and inherent possibilities in existing technology. These categories are neither inclusive nor unique. Also, factors of the various types interact strongly, and many factors can be defined as belonging to more than one type. But as a general framework for discussion, this classification is very useful.

Values and Attitudes

Values and attitudes include a wide range of beliefs and preferences, from the cluster of implicit assumptions that define a culture to localized individual preferences. History displays ample evidence that cultural values have a pronounced effect on the generation and implementation of new technology. A culture that believes its most important mission is spiritual and that the pursuit of wealth, power, or novelty is the expression of an undesirable materialism will experience a much slower rate of technological change than one which believes in the ability and desirability of controlling nature for secular ends. Also, the cultural view of the place of the individual in society has a considerable effect on the rates of change in that society. A statist culture, in which the individual is valued primarily for his contribution to the group and whose position and functions are determined by some centralized mechanism, will result in approaches to innovation very different from those in a culture in which the purpose of society is believed to be to support the individual aspirations of its members. The history of the Industrial Revolution suggests that an important contributing factor to its beginnings in England was the fact that the ideas of private property, contracts, freedom of individual action, and personal rights were more advanced there than anywhere else. Furthermore, commercial success was socially acceptable and even desirable. Conversely, the beginnings of industrialization were slowed down on the Continent by an attitude that valued social status in terms of hereditary privilege and government careers and looked down on commercial activities.

Even within a given cultural framework there is a great variety of values and preferences. Belief in the importance of education, the desire to travel, attitudes toward family size, whether or not medical care is regarded as a universal right, the definition of death, views on national defense, the perceived legitimacy of large corporations, positions on nuclear energy and even preferences in entertainment are just some of the examples of human values that can influence technological change. The point is that technology exists to satisfy human wants and that these are conditioned by what human beings need and what they hold to be important. This is the basic origin of change and constitutes one of the most fascinating aspects of the history of technology. For our purposes, we must recognize the critical role of values in technological change, the fact that values themselves change with time, and that there is a reflexive effect in which the evolution of technology modifies and even shapes values.

Physical Resources

It is obvious that physical resources are important determinants of technological activity, as is shown by the continuing concern about energy sources. The availability of large amounts of materials and energy is a necessity for a technological society and at least partially defines the technological possibilities. In a larger sense, physical resources include the entire physical environment, including land mass, climate, coastline, waterways, and biosphere. A long coastline and an extensive waterway system, for example, is an incentive to develop water transport, which in turn can provide impetus for other developments by providing cheap transportation; a large forest endowment is an incentive for development of forest products, while arid regions can either hinder agriculture or spur irrigation technology. The physical endowment certainly is an important part of the pattern of challenge and response postulated by Toynbee, in which progress is often made by people's reactions to both the opportunities and difficulties confronting them. However, the particular location of physical resources is not as important as it once was. With modern means of cheap transport and communications, raw materials can be brought from and to any part of the globe. The costs are low enough so that location alone does not form a barrier to technological progress. Japan has shown this to be true.

Knowledge and Skills

Technological change is implemented by people who know how to do it. The knowledge base from which they work and the level of their skill in using it are crucial elements. The success of a technological society depends on the existence of large numbers of people with the appropriate education and expertise to create, discover, invent, test, manufacture, and service sophisticated systems and products. Technological change is therefore strongly influenced by the availability of such people. The first group that comes to mind is the engineers, who indeed are usually the ones most intimately instrumental in technological change. In the modern world, they generally draw on a large base of science for their work. Scientific research tries to pursue its own ends, dictated by a desire to understand the world independently of possible applications, and it is difficult to trace a connected path from pure research to applied research to engineering to manufacture for specific technological advances. But science creates, replenishes, and continually enlarges the pool of knowledge that is essential for technological progress.

Engineers and scientists are not the only ones that affect the rate of change. Technicians, machinists, craftsworkers, pipefitters, and a host of skilled and semiskilled workers all contribute to the technological enterprise in important ways. In fact, without the human technological infrastructure represented by these people, technological change could not be implemented no matter how energetic and creative the engineers and scientists. And the personnel needs of a technological society are not restricted to those directly involved in the technical process. Financial analysts and accountants to estimate costs, risks, and returns, lawyers to construct contracts and interpret the law, market analysts, insurance underwriters, and venture capitalists all influence the rate of technological change. The particular mix of human talents and education largely defines how and to what extent a society can participate in the technological process.

National Objectives

Perhaps the atomic bomb and the space program are the most dramatic examples of the effects of national objectives on technological change. The resulting changes in the technology of war, energy production, and communications resulting from these programs are well known, and they have yielded a great many spinoffs in such areas as instrumentation and materials. Since the Federal government is such a large supporter of research and development, it supplies a major impetus in areas it deems to be in the national interest. Through such agencies as the Department of Defense, the National Institutes of Health, the National Science Foundation, and the Department of Energy, national objectives exert a continuing pressure on the course of technical developments. Somewhat more indirectly, monetary and fiscal policies, tax structures, investment credits, degree of support for education, and antitrust laws all embody national objectives that affect both the rate and the direction of technological change. Government has, in fact, become such a large presence in the technological enterprise that it is almost always important, often pervasive, and sometimes completely determining. The role of government in other countries is no less apparent. In Japan, the Fifth Generation Computer Project was started by a government initiative; France and other European countries have adopted excellence in the information technologies as national goals; in the Soviet Union, practically all technological progress is a government responsibility. Both in the past and the present, governments have recognized the great power of technology in the pursuit of national aims and have acted accordingly.

Political Factors

National objectives can certainly be described politically. However, we will take the term to mean that set of goals which has been adopted on behalf of the country as a whole through the instrumentality of the national government. The set of laws, actions, and procedures that furthers more limited ends for subgroups of the population will be labeled political factors. They often operate through the national government, but also through state or local governments, political action committees, labor unions, professional societies, and institutions or associations that represent the specific interests of their members. Examples that affect technological change include a Congressman's sponsorship of a hydroelectric facility, lobbying for a particular defense system, antinuclear organizations and demonstrations, municipal ordinances on waste disposal, regional tax incentives to attract industry, and local objections to the transport of hazardous materials. The list could be extended indefinitely because, in a technological society, much of its activity and aims are linked to technology directly or indirectly.

Governmental Regulations

Govermental regulations are such important determinants of technological change that they merit separate consideration from other political and policy factors. The entire nuclear industry, for example, is the subject of detailed and extensive government control; broadcasting and air transport are regulated; the Food and Drug Administration affects the time it takes to bring a new health product to market by defining the tests required for its approval; environmental and safety regulations strongly affect the costs and time of construction of electrical generation plants; mandated programs such as seat belts and antipollution devices for automobiles, environmental protection regulations, and industrial safety rules all have a direct bearing on the direction and rate of technological change.

World Competition

The days in which one or two countries could economically or technologically dominate the world are gone. Now more than ever before, national economies are becoming increasingly interdependent, innovations are globally mobile, and competition is intense. The great international struggle developing over the computer business, the positions of the steel and automobile industries, the never-ending debates on protectionism and industrial policy, the concern over foreign labor costs, and the foreign invasion of American market are all the gross manifestations of an overriding phenomenon. Industrialization has taken place in many countries and is well on the way in many more. Technological competition is now international in scope, and the achievements, strategies, and competence of each nation will have an effect on change in all other nations.

Macro- and Microeconomics

The close connection between technology and economics is obvious. Even the economy of primitive societies is based on the tools and techniques they use to make their living. In modern society, all aspects of the economy are intimately entwined with technology. It is also obvious that the implementation of technology requires capital, labor, and resources and is therefore an economic activity. Economic factors are crucial and often completely determining in the evolution of technological change. Overall macroeconomic conditions define the economic background against which technological change takes place. The money supply, interest rates, consumer expenditures, the unemployment rate and business cycles affect the degree to which new ventures can be implemented, whether or not possible innovations are brought to the market place, and the extent of risk people are willing to assume. Microeconomic factors determine the decisions on specific projects and innovations. If an innovation is perceived to be profitable relative to other possible investments, it is likely to be adopted. Technological innovation by corporations is essentially completely driven by such microeconomic considerations. Throughout the process of change, technical and economic factors interact and influence each other, and an innovation has little chance of adoption if it doesn't pay.

Institutional Structures

The corporation is the primary vehicle we have chosen to implement technological innovation. Applied research and development, process innovation, product selection and improvement, manufacturing, and marketing are mostly the province of corporations, at least in the Western world. The nature of corporations, their objectives, incentives, structure, and strategic posture are therefore important components of the determinants of technological change. The way corporations carry on their business, choose their products or services, interact with labor and government, and organize their activities has a great deal to do with the degree of success with which they initiate or accommodate change. Strategic decisions can determine whether or not a particular technology will be pursued at all; R&D budgets and projects define the future course of their innovations; their organization and internal culture affects their propensity and efficiency for innovation. The corporate factors are many, and their strong impact is apparent in any analysis of the evolution of modern technology. Of course, other institutions are also important. These include universities, with their responsibility for higher education and their enormous contribution to basic research, government, nonprofit research laboratories, Federal agencies, and public school systems. Again, technology is so pervasive that a great many of our institutions are continually interacting with it, being changed by it, and affecting it. An understanding of technological change must include the institutional and social setting in which it takes place.

Markets

In the final analysis, new technology will not take hold unless it satisfies a need that someone is willing and able to pay for. This is why the market place is such a major force in technological change. What people, groups, institutions, or governments want will always control what is actually done. The structure of markets, the needs and desires of their members, the degree of free competition or its absence, the extent of consumer affluence, and the level of marketing competence go a long way in deciding which innovations are successful, which are not, and who reaps the attendant benefits.

Inherent Possibilities in Technology

No matter how these various influences are arrayed, no innovation can occur if it is not possible within the context of knowledge that exists or that can be developed from the existing base. Innovation at any time is directed and limited by the inherent technical and scientific possibilities at that time. This is one reason why so much innovation is of an incremental nature; it is easier to do because the amount of new knowledge and techniques required is small. A radical innovation, on the other hand, generally requires a considerable extension of existing technical practice and even new scientific research. The rate of innovation, as well as its type, is thus a strong function of what is already known. This is also what contributes to the self-sustaining character of technical advance. The larger the knowledge base, the greater the future possibilities; and every new advance enhances the knowledge base. The impact of the existing technological base makes itself felt in a specific as well as a general sense. A particular firm, for example, pursues those innovations for which it possesses the necessary expertise, and when it is required to move too far from those areas in which it has competence it experiences great difficulties. A challenge in which an outside group invades a firm's markets using new technology produces crisis and turmoil since the firm is constrained by the technological resources it is accustomed to. Similarly, the work of a research and development laboratory is based on the specific expertise of its personnel, the kind of experimental facilities at its disposal, and the thrust of its recent efforts. For each component of the technological process, as well as for technology as a whole, the rate of change and the direction of change depend on what is already known and what has already been done.

This description of the various influences on technology is hardly complete.

Every sphere of human activity can be shown to bear on the nature of the innovations we adopt and how fast we adopt them. Society is not an entity that can be neatly separated into parts, each with an exclusive and well-defined function. Rather, it consists of a large number of people, organizations, institutions, and informal associations all interacting with each other, sometimes indirectly, sometimes very strongly. Sometimes they overlap, their actions and aims conflict or are parallel, their attitudes are affected by each other, and they are continually involved in a multiplicity of pursuits. Modern society is a dynamic, changing organism whose parts are closely linked to each other. Few are the groupings that are not tied into the technological process.

The complexity we have described might seem overwhelming, and it indeed is so, if a complete understanding from grand design to small detail is to be obtained. But this is the case with any aspect of modern knowledge. That is why we construct frameworks of thought that organize the complexity. The picture that emerges from the above description is that every technological change is the result of a number of forces acting over time. Some of these forces are positive and some negative, in the sense that they accelerate or retard change; some are relatively constant in time, giving rise to the observed regularities in technological parameters; some of them change rapidly with time, and some even appear spontaneously and then vanish. The net effect is that of a force vector whose fluctuating magnitude and direction drives a changing technology at varying rates and in different ways as time passes. Sometimes it is one component of the force that is predominant, sometimes another, or a set of forces might have comparable importance. That is why technological change is sometimes determined by investment policy, sometimes by foreign competition, sometimes by the vision of an individual, sometimes by government policies, and often by a group of influences acting concurrently. The task of technological dynamics is to describe this process and sort out the operative forces. In doing this, we usually search for a relatively small number of forces that seem to predominate in particular instances. However, it must always be kept in mind that the forces are not independent and can affect each other strongly. Also, the process of technological change itself interacts and alters the forces that determine it. Thus, tax structures can affect the conditions of investment, foreign competition, and even market acceptance; a successful innovation can alter corporate structure, spawn other innovations, and affect international competition; international competition can spur innovation or give rise to protective action.

One more comment on the complexity of this subject may be useful. The physical world is also a highly complicated system with many strongly interacting parts, and if we approached it with the idea of immediately comprehending all of nature, we would have an impossible task. But by generations of careful work and continual refinement, we have learned how to make some sense of it. The physical sciences are intellectually organized in such a way that much of our knowledge of nature is rather well defined. The study of technological change in a quantitative way is relatively new, but even now specific parts of it can be understood in relatively simple terms if we adopt a position of some modesty in our objectives in specific investigations. A number of examples of this understanding are given in this book.

ENTERPRISES, SYSTEMS, AND HIERARCHIES

It is clear that technological change can be described from a variety of viewpoints and that each viewpoint gives a different kind of information. To illustrate the richness of the subject and how it might be approached, let us consider a specific innovation, that of the electronic computer. This can be examined in many ways, but we will start from a direction of obvious interest, namely the increase of computer power over time, and branch out from there.

The first task is to define what we mean by computer power. This can be done by constructing measures based on the speed of standard numeric operations and memory size, or by measuring the speed of execution of standard benchmark programs. Next, we need to define the meaning of the evolution of computer power. A simple definition is that of the most powerful machine available at any given time, so that we would be tracking the best state of the art according to our measure of computer power based on speed and memory or benchmark programs. A curve could then be drawn of maximum available computer power as a function of time. When this is done, it is found that the ENIAC, constructed at the University of Pennsylvania in 1941, was almost one thousand times more powerful than any computation devices previously available, so that it qualifies as a radical innovation. For over twenty years after the birth of the ENIAC, maximum available computer power increased steadily at a rate of 50 to 75 percent per year, making it an extremely rapid technological change.

Other measures of growth can be used. Instead of finding the most powerful computer as a function of time, we could take the average computer power of all computers introduced in a given year, or we could multiply computer power for each computer by the number of such computers sold and determine the growth in aggregate computational resources. Also, the total number of computers sold can be measured along with the types of markets (government, business, universities). Even with such a gross measure as computer power, a variety of analyses can be performed, each giving different information and different perspectives on the growth of computation, markets, and the state of the art.

A somewhat different perspective can be developed by tracking the amount of money spent on electronic data processing, the number of firms adopting computers, the growth of data processing centers, and the number of firms in the computer industry. Or we can determine the growth in the number of computer science curricula at universities, the number of programmers and systems analysts, and the total number of computer-related jobs. Measures such as these give a picture of the evolving integration of computers into the economy.

The measures described above are of a macroscopic sort; they are macroparameters for computer technology as a whole, which is defined as the set of systems, devices, and processes that contribute to, and result in, the function of data and information processing.

Now let us start from a different direction by considering a specific computer system consisting of processing and core memory units, mass storage devices, input and output devices, and all the necessary interconnections. Any one of these parts represents a subtechnology whose evolution can be studied separately. The computer, for example, evolved from vacuum tubes to transistors to integrated circuits to large-scale and very large-scale integrated circuits. Each of these subtechnologies themselves consists of parts and is related to other technologies, such as materials preparation, high vacuum, and microlithography. A hierarchy of technologies exists, all with a history of change and all contributing to overall computer technology.

Yet another direction can be taken by examining the agents through which computer development has taken place. These include individuals, government agencies, university departments, research and development laboratories in industry, marketing groups, and business firms. Each of these has a history of change intimately connected to the growth of computers.

This discussion, along with that presented earlier, is sufficient to introduce the concepts of technological enterprises, systems, and hierarchies. A *technological enterprise* is any organization whose objectives include or require technological change, or whose functions depend on technology. These might be technology-based corporations, research and development groups, manufacturing plants, consulting firms, and government agencies. Some organizations may have multiple functions, and technology may have a greater direct importance for some than for others. In a corporation, for example, the highest priority may be marketing, while technological change is of urgent and immediate concern to its development laboratory. Technological enterprises are the institutional agents through which change takes place. Since technology is normally not pursued for its own sake, but for some desired human purpose, many organizations have basic purposes that are not technological. The degree to which technology is a component in achieving its objectives defines the extent to which an enterprise is technological. The growth and life cycles of technological enterprises are important aspects of the study of technological change.

A *system* is an entity consisting of interacting parts organized to perform a specific function or set of functions. A well-known example is the stereophonic audio system whose purpose is the accurate reproduction of sound. Such physical systems are relatively easy to understand since their behavior is determined by physical laws and the random element is minimal. The technological enterprise is an example of a social system, since it consists of people organized into groups held together by common objectives. Whether physical or social, systems have two closely related attributes that are important for understanding technological change. The first is that they exist in *hierarchies*; that is, a system is usually part of a larger system, and the parts of a system are themselves systems. The second is that systems can interact, communicate, and affect each other. In a computer system, for example, the state of the central processor is determined by data communicated by input devices, and the time it takes for a specific computation is determined by various parts and their architecture. Furthermore, the computer system is linked to other systems, such as the electric power system, other computers, user groups, and the data processing staff. These in turn have subsystems and are linked to other systems. In a stereo system, amplifier performance is affected by loudspeaker impedance, and overall sound quality is determined by the interaction of the components with each other and the coupling of the speaker output with the room acoustics.

The strength of the coupling among systems is important because it determines the extent to which a system can be treated as an independent entity. In studying technological change, we often take a system, mentally isolate it, and follow its evolution in time. This system may be physical, such as the airplane, or it may be an organization, such as a company, or it might be an industry, such as that for producing automobiles. In any case there is a built-in assumption that we have isolated something that retains its identity throughout the time period of the analysis. This means that we assume that the entity has some attributes of its own that it retains, and that the strength of its interactions with the outside world are not sufficient to destroy its identity. In abstracting a system from the complex of existing interrelationships, it is necessary to account for its coupling with other systems and how that coupling affects its evolution. Some attributes of a system evolve as a result of the properties of the system itself, while others are more strongly influenced by outside factors. As an example, a corporation's inner culture and organization may be committed to high quality, state-of-the-art electronic instrumentation and may continue to be a leader in this respect because of factors internal to it. However, the specific products it develops and markets may be strongly influenced by market demand and by competition. The nature of its growth, therefore, is determined by a mix of factors, some of which are intrinsic to it and others which result from its coupling to other systems.

FIVE PATTERNS

The recognition that there are both regularities and special individual events that determine technological change is an important insight and provides an organizing principle to describe the technological process. The idiosyncratic factors can be dealt with only by historical analysis and case studies that expose the social, economic, and political conditions that influence individual events. The regular trends, however, can be treated as quantitative measures of technological evolution, even though they are embedded in a historical and social context.

Analyses of time series have exposed five regularities that are particularly important for describing technological change. These are: the clustering of innovations, the diffusion of new technologies, the improvement of technical performance, productivity learning curves, and the skewed size distribution of growing systems.

The significance of these patterns goes beyond the fact that they illuminate specific aspects of technological change; they interact with, reinforce, and limit each other to show that a substantial part of the process of technological change has the characteristics of an evolving system. These regularities are called systemic trends, in recognition of the fact that they are properties of the techno-economic society in which they occur. Some of them will be treated in greater detail in later chapters. The purpose of the presentation here is to give a brief overview of the regularities, to describe the interactions among them, and to show how they lead to a useful model of technological change.

Clustering of Innovations

The clustering of innovations means that they do not take place at a constant rate. Rather, there are periods of time when the frequency of innovation is high, separated by periods when it is low. Often, a cluster of innovations includes a group of related technologies or technologies that complement and reinforce each other.

The evidence for clustering has been constructed from counts of the number of innovations as a function of time. The problems in this approach are that the dates of innovation are hard to identify, and the judgment as to what constitutes an important innovation is subjective. Different authors have drawn up different lists of innovations and have sometimes even used different dates for the same innovation. All of these lists, however, do exhibit clustering. There is no doubt that each of them shows time periods in which the number of innovations is high and other times when it is low. Clustering does exist. But the controversial point is whether or not the clusters occur at regular time intervals and whether or not they are tied to business cycles. For the purposes of this chapter the question as to whether or not there are regular cycles is not important; it is only necessary to note the existence of innovation clusters.

The most striking example of data that show clustering is that of Mensch (1979), who collected data on the dates of important technological innovations over several centuries. A plot of a five-year moving average of Mensch's data for 112 innovations dating back to 1800 is shown in Figure 1-1. (These data can be interpreted as exhibiting regular innovation cycles of half a century as well as clustering, although generalizing from three peaks and a limited data set is questionable.) The first cluster, centered on the decade

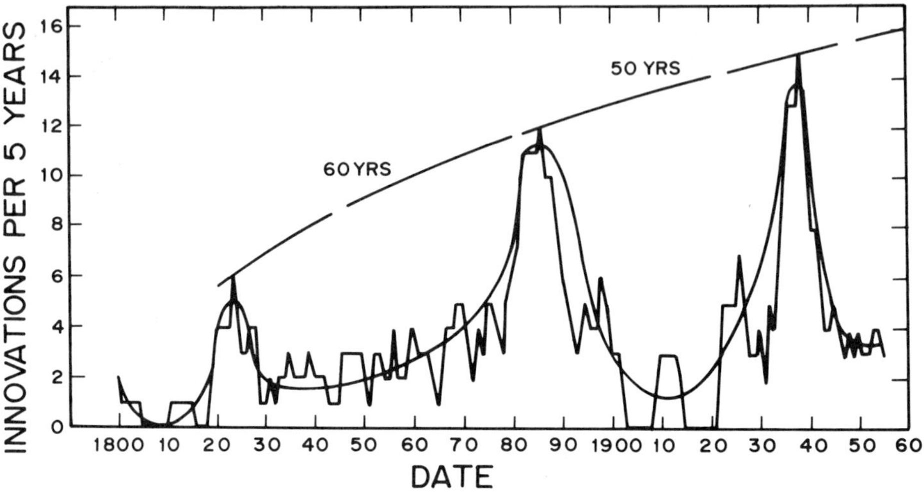

Figure 1-1 Long cycles of innovation according to Mensch.

of 1820, represents innovations in railroads, steel production, and pharmaceuticals. The second cluster, centered on the 1880s, is associated with a group of technologies including vulcanized rubber, photography, electric lighting, the telephone, and the steam turbine. The cluster centered on the 1930s includes the jet engine, plastics and polymers, petroleum technology, and electronics.

An important point is that the technologies in a given cluster often reinforce each other. Thus, railroads and steelmaking spurred each other's development, as did electric lighting and the telephone, the jet engine and electronics, and plastics and petroleum technology.

The clustering of technological innovations will be taken as a fact (but not necessarily its regular cyclic character). This fact has important implications for understanding the systemic evolution of technology.

An innovation seldom performs at a high level of technical capability. But as time passes, improvements are made and performance rises. At first this is a slow process, but as more is discovered and the knowledge base expands, the rate of improvement increases. But every technology has a maximum beyond which it cannot go because of inherent engineering or physical limits. The rate of improvement then starts to slow down as maximum performance is approached. An important concomitant of this process is that the cost per unit improvement increases rapidly as the limits of technical performance are approached. This occurs for all technologies, whether they are major or minor.

When a new technology takes root, it starts a line of technical development that may go on for years with ever-improving technical performance. An excellent example of this is the technology of artificial light sources. A good measure of the performance of artificial lighting is the efficiency with which

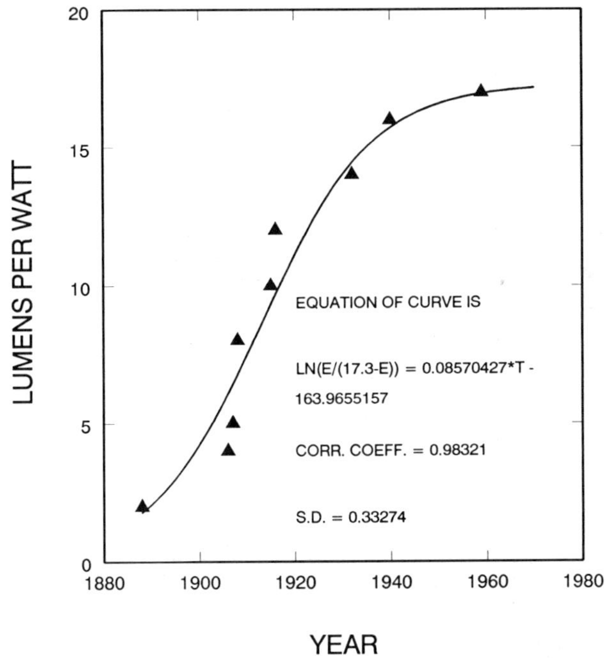

Figure 1-2 Efficiency of incandescent lamps.

electrical energy (or chemical energy, as in the case of candles) is converted into light. Figures 1-2, 1-3, and 1-4 show the time dependence of the efficiency of incandescent lamps, fluorescent lamps, and all artificial light sources respectively. The data for these plots were taken from Martino (1983).

Figure 1-2 refers to the evolution of the incandescent lamp from the carbon filament to the cellulose filament to the tungsten filament to the inert gas filled lamp and the incremental improvements attending these major changes. Starting at only 2 lumens per watt, the efficiency followed a sigmoid curve that is typical of many technical performance curves and ultimately increased by a factor of nearly eight. This is a large improvement, but it took three-quarters of a century to achieve it. The curve shows that the rate of technical improvement was much greater in the early years and ultimately became quite small. Clearly, by the 1960s the limits of the tungsten lamp were reached. These limits are determined by the temperatures at which the lamp can be operated and still have a reasonable life, by the presence of impurities, and by the fact that an incandescent lamp must generate heat. This heat is wasted energy which is not available for lighting.

Fluorescent lighting avoids many of the problems of incandescent lamps. A plot of the efficiency of fluorescent lamps as a function of time is shown in Figure 1-3. Here, too, the technology undergoes rapid advancement in the years right after its introduction and ultimately slows down and levels off. However, the curve differs from that for the incandescent lamp in several

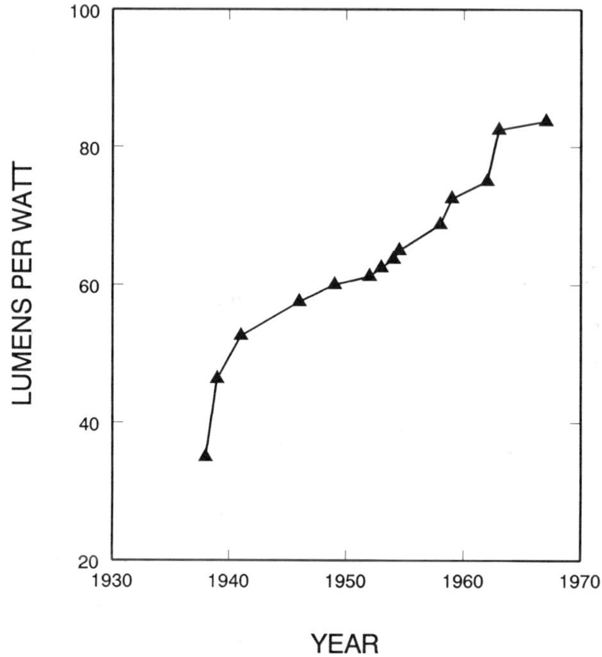

Figure 1-3 Efficiency of fluorescent lamps.

important respects. While its general shape is sigmoid, the initial rates of improvement for the fluorescent lamp are much higher than for the incandescent lamp. Also, the time taken to approach the technical efficiency limits was much shorter for the fluorescent lamp. The most important point is that the fluorescent lamp was capable of being developed to a much higher level, its final efficiency of over 80 watts per lumen being over four times that of the best incandescent lamp.

The progress of artificial lighting in general is shown in Figure 1-4, in which the efficiency of light sources from the candle to the Gallium Arsenide diode is given as a function of time. This plot is quite different from those for the incandescent or fluorescent lamps, in that it continually increases without any leveling off. In fact, it follows an exponential curve quite well. Of course this exponential rise cannot continue indefinitely. One lumen is equivalent to 0.001496 watts, so a perfectly efficient light source would put out 668.45 lumens per watt. The best incandescent lamps are therefore about 2.6 percent efficient, while fluorescent lamps have an efficiency of about 12 percent. Even the Gallium Arsenide diode has an efficiency of only about 27 percent. We can certainly expect that as better light sources are developed the curve for efficiency versus time will bend over and approach a limit.

The course of technical development shown by artificial lighting is typical. A cycle of performance improvement exists for each technology that performs a given function. As the cycle reaches its maximum, further improvement

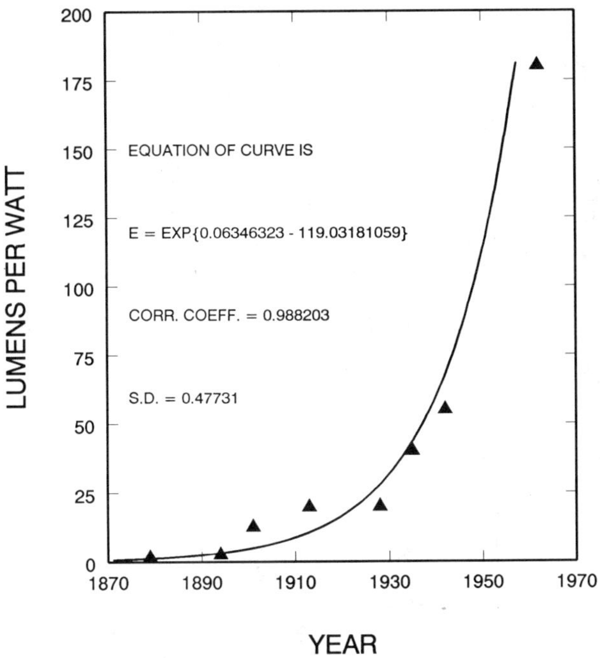

Figure 1-4 Efficiency of artificial light sources.

can be obtained only by developing another technology. The overall technical progress for a given functional requirement is therefore a series of overlapping sigmoidal performance cycles for specific technologies and subtechnologies.

The Diffusion of New Technologies

It takes time for an innovation to be adopted by all its potential users. At first, the new technology spreads slowly, but as it is improved and more people adopt it the rate of diffusion increases rapidly. Since the number of potential users is finite, the rate of diffusion must ultimately slow down. A variation of technological diffusion is the process in which a new technology is not only spreading but also replacing an old technology. This process is called *technological substitution.*

The typical course of diffusion of a new technology is sigmoidal. When a technology is first introduced it usually spreads into a particular market niche either because it satisfies an urgent need that overcomes the barriers of cost and novelty, or because only a segment of the market can afford its initially high cost. The initial market penetration is therefore small. As the technology proves itself and costs decrease, more and more people adopt it, and the rate of diffusion increases. At some point, the market starts to saturate, so the rate of diffusion levels off as penetration approaches com-

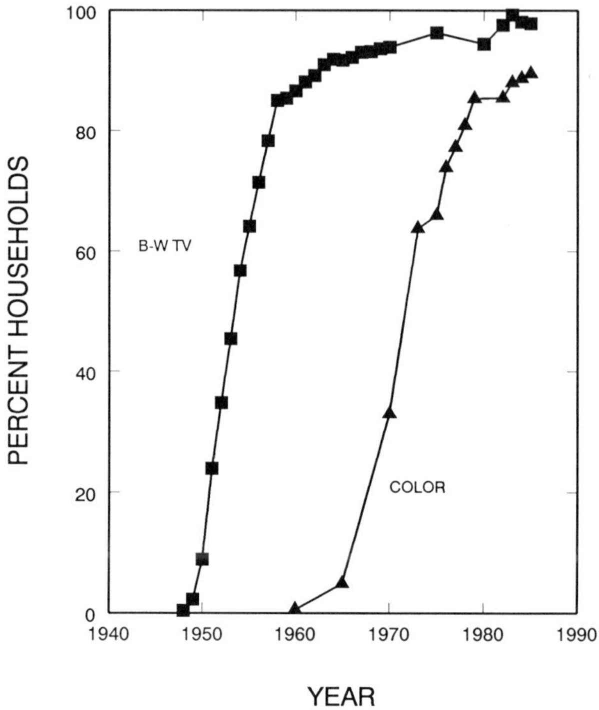

Figure 1-5 Percent of households with television.

pletion. This process leads to the sigmoidal or S-shaped curve for technological diffusion.

An illustration of technological diffusion is given in Figure 1-5, which shows the percentage of electrically wired households with television sets in the United States. (The data were taken from *Historical Statistics of the United States;* Department of Commerce 1970; p. 796, and *Statistical Abstract of the United States;* Department of Commerce 1986; p. 545.) After a slow start in the late forties, there was a rapid increase in the adoption of black and white television in the home until the fraction of homes with television exceeded 80 percent. The rate of diffusion then decreased as the market penetration approached saturation. The curve for color television followed a similar course, except that it started about fifteen years later.

Figure 1-5 shows that by 1960 the rapid growth of black and white television was over, and the market was limited by population growth and replacements. Fortunately for the television industry, color television was available by then and another growth curve was started. The major conclusion from technological diffusion curves is that a new technology leads to rapid growth only for a limited time, after which growth slows down dramatically unless another technology is ready.

Improvement of Technical Performance

The delivery of a new technology normally requires new methods for its production. As experience accumulates, the efficiency of the production, distribution, and servicing of the new technology increases and its unit cost decreases. This result is embodied in the productivity learning curve, which is often a power law.

Improvements in technological performance refer to an increase in the inherent capabilities of a technology. Learning to use a new technology has a similar effect, in that it increases efficiency and productivity. Learning, however, generally refers to the improvement in using a given level of technology as people become more familiar with it and, as a result of practice, avoid mistakes and increase their proficiency.

Productivity Learning Curves

The typical situation in which learning is important involves the adoption of a new production technology which is embodied in new equipment or a new manufacturing process. At first, the rate of production is much less than the inherent capabilities of the new installation. But, as experience accumulates, the error rate goes down and productivity goes up.

An example of a learning curve is shown in Figure 1-6, which shows the productivity of a pressman (with two assistants) on a new offset printing press as a function of the total number of impressions he accumulates as he works on the new machine over a period of a year. (Data for this plot were taken from Levy, 1965, p. B-152.) At first, with only 100,000 impressions behind him, he produced 2700 impressions per hour. But by the time he had made 1.12 million impressions his productivity went up to 3,560 impressions per hour. This was an increase of 32 percent. Many learning curves show considerably higher increases. While the overall trend of productivity versus experience is up, this obviously cannot continue indefinitely. Sooner or later the learning must reach a maximum; indeed the pressman of Figure 1-6 reached a limit that fluctuated around 3500 impressions per hour after about 800,000 impressions.

The general rise in productivity with experience occurs at a variety of levels, ranging from the performance of individual operators to the outputs of entire plants and industries. The results are similar for all levels. For a given level of technology, productivity increases regularly until a saturation point is reached. Along with the productivity increase there is a decrease in costs. After saturation, productivity and costs can be improved only by improving the technology.

The ability to learn is critically important in taking advantage of technical advances. Obviously, an intelligent and educated work force is a necessity for a technological society.

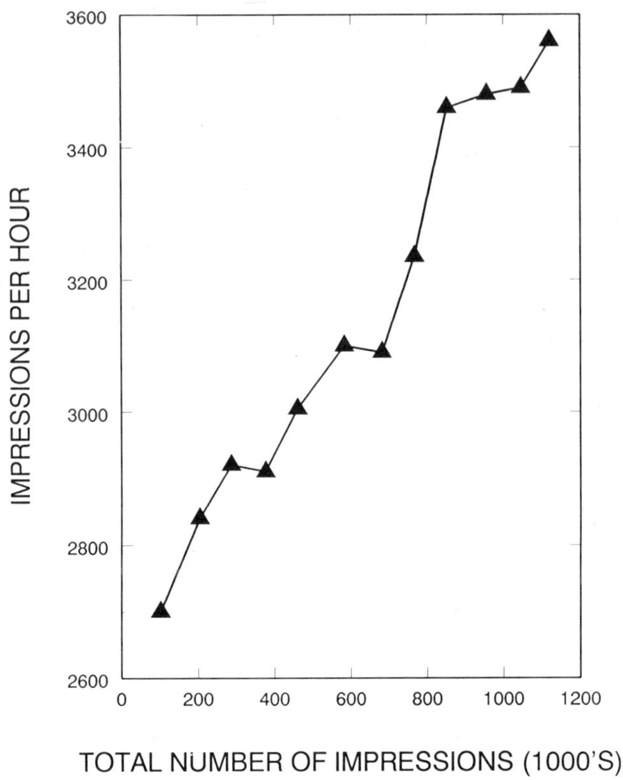

Figure 1-6 Learning curve for pressman productivity.

Skewed Size Distribution

The skewed size distribution is observed for a number of parameters that describe technology-based systems. R&D budgets for universities, the sizes of corporations, personal incomes, and even the sizes of cities display size distributions such that a few entities are very large while a great many are small. The lognormal distribution is often a reasonable representation of the sizes of sets of growing entities. It is easy to see that for any group whose members grow at different rates, their sizes will differ substantially after a time even if they all started out at the same size. If a group of employees all start out at the same salary, but a few get consistently higher percentage raises than the others, they will ultimately have salaries that are a large fraction of the total. In Appendix 1 it is shown that if growth rates of a large set of growing entities are distributed randomly according to the normal distribution, then their sizes will be distributed lognormally, which is a highly skewed distribution. But it is natural that the distribution of growth rates be normally distributed. The growth of a company in any given year depends on an enormous number of factors, including economic conditions, union

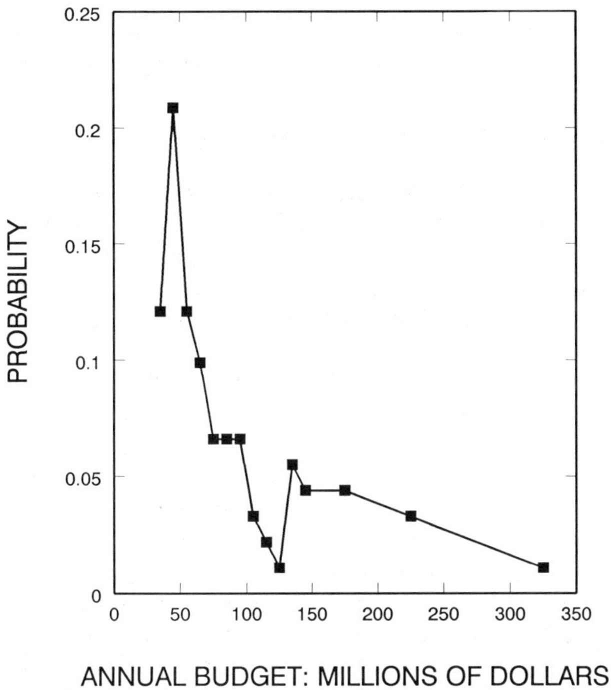

Figure 1-7 Frequency distribution of research funds at ninety-one universities.

contracts, marketing success, environmental accidents, competitive pressures, monetary exchange rates, costs of supplies, and so on. These factors all contribute something to the degree of corporate growth and are of differing importance for different companies. This situation automatically leads to a Gaussian distribution of growth rates. That is, when the influences on corporate growth are many and variable among companies and through time, the average growth rates of corporations must be distributed according to the laws of random numbers, which yields the bell-shaped curve. The highly skewed distribution of sizes is therefore an inevitable consequence of arithmetic.

Figure 1-7 shows the distribution of research expenditures at the ninety-one universities with the largest research budgets in 1985. The ordinate is the probability that a university has expenditures given on the abscissa. Clearly, most universities have small research funds, only about one-third of the major research universities having research budgets above 100 million dollars. This, of course, is a very small fraction of all universities.

The sizes of industrial research laboratories and the sizes of corporations also display the highly skewed form of Figure 1-7. This fact has important implications for technological innovation, in that a few large institutions have a large effect on R&D and technological innovation.

These five systemic patterns and their mutual interactions illustrate the regular evolutionary character of much of technological change.

INTERACTIONS AMONG SYSTEMIC TRENDS

The systemic patterns of technological change do not exist in isolation and are not independent of each other. In fact, they reinforce or inhibit each other to a great degree, as one would expect in a system of strongly interacting parts. These interactions give technological change its systemic character and are important in any model that seeks to describe the mode of advance of technology.

To explore the relations among the patterns of change, first consider the mutual effects of technical performance cycles and the clustering of innovations.

Innovation clusters have two important effects on the increase of technical performance. The first is that, to the extent that innovations start within the same time periods, different technologies will progress in parallel. The second is that it often happens that progress in one technology depends on advances in another technology. Thus, advances in the performance of steam engines would not have been possible without new materials and improved fabrication and machining methods. Bessemer steel made it possible to produce rails that could bear more weight and permitted steam locomotives to evolve in size and power. Technological performance curves are therefore interactive, in that the development and growth of one technology accelerates advances in another. The clustering of innovations strengthens these interactions. Innovations are extensions of current practices and draw on the technologies that are available. This truism makes it inevitable that clustering and technical performance curves reinforce each other.

The improvement of technical performance of a given technology is the result of a continuing series of advances, each of which is minor relative to the initial innovation. The end result of these incremental changes may bring about much greater technical progress and have much more economic impact than the original innovation. While improvements are still being made at a reasonable rate, there is no reason to seek a new technology, and indeed a new technology will find it difficult to take hold since the market is being served well by continually improving products, investors are still getting good returns, and demand is still high. R&D, production, and marketing are all expanding, and all energies are turned to the exploitation of the technology.

But when the technology approaches its limits, the cost of improvement is very high, the magnitude of possible improvement is low, products become static, diffusion becomes complete, demand goes down, profits fall, and significant progress in capabilities is possible only by developing a new technology. Thus the cycle of technical improvement acts as a deterrent to major innovation during its growth phase and a stimulant to major innovation during its stagnant phase. To the extent that major innovations occur in clusters and the times of their technical progress cycles are similar, the form of progress in technical performance reinforces the clustering of major innovations.

The relationship between learning curves and innovation is similar. The learning curve is steeper in its early stages and then tends to level off. That is, productivity gains for a given production process decrease as production continues. Cost savings from production learning are thereby harder to come by for a mature technology. Further progress depends on going to a new technology of greater potential and starting a new learning curve with higher limits than the old. Again, this has a negative effect on the introduction of major innovations during the growth of a technology and a positive effect when the technology is mature.

The diffusion cycle has a similar effect on innovation clustering to that of technical performance. When a technology is diffusing rapidly, the market is growing and profits are being made. There is no reason to do anything but improve the existing technology. But a saturated market is stagnant, and competitive advantage rests on such factors as advertising, service, reputation, and distribution. None of these has the force of increased functional capability. Thus, the growth phase of technological diffusion deters major innovations while they are stimulated as diffusion nears completion.

The relation between skewed size distributions and innovation clusters is somewhat different. Most incremental improvement of a technology is done in a relatively small number of large organizations. This process includes productivity learning, product and process improvement, and adaptations for market niches. Similarly, the majority of technological diffusion involves large companies either as users or vendors of the technology. This fact has significant implications for the introduction of major innovations. The life cycle of a major technology is long, and large corporations are well suited for the process of continual, gradual progress of a given technology. However, major changes are extremely troublesome for large organizations. All the know-how, organizational structure, production techniques, marketing methods, technical expertise, and customer expectations are well understood and well adapted for a technology that has existed in the company for a number of years. Switching to a new technology is a traumatic experience in which massive changes with risky investments must be made. Thus, while large companies are progressive within a given technology, they often tend to be conservative with respect to radically new technology. They will generally make a switch only after it is obvious that staying with the old technology poses a serious threat to the company. Large corporations do not have a natural affinity for major innovation. This is borne out by the fact that so many of the present high technology companies did not exist two or three decades ago, by statistical studies of the response of corporations to new technological threats, and by studies that show that a majority of innovations in an industry come from outside that industry.

Of course, large firms sometimes do make major innovations. Those that can do this successfully exhibit continuing long-range growth.

Large firms are important in the diffusion of a new technology. They have the development, production, and marketing resources required to improve performance, take advantage of economies of scale, and penetrate markets. But the early stages of diffusion are often driven by small firms that are flexible and more responsive to new developments. They often cater to market niches in which the advantages of the new technology are important even in the early stages of technological development. For large-scale diffusion, however, the small firms must either grow in size or leave large-scale marketing to others. In practice, a new technology is often the force that drives corporate growth. A major reason for the differences in firm sizes is the differential growth rates arising from the uneven success of firms in developing and adopting new technologies.

Superior technical performance of products and processes is a powerful force for corporate growth. It therefore has a strong effect on which firms will be large and which will be small. Also, since large corporations are much better at the incremental progress of a given technology than at introducing radical innovations, there is a continual change of firms with a given size in the skewed distribution as small firms successfully develop new technologies and become large.

Learning and the resulting increases in productivity are also important factors in corporate growth. Learning lowers costs and therefore contributes to increased market share. Not all firms have the same learning rates, which reinforces the differential growth that skews the distribution of size.

The diffusion cycle is closely linked to the continual improvement of a technology. A new technology usually starts out in a market niche in which its new functional capability is important and will be adopted regardless of its higher costs and crude state of development. As it spreads, areas of needed improvement become obvious because it is being used. Feedback from users is an important spur to technical improvement. Also, the profits from its growing use and the promise of further profits spur technical improvement. Competition among different firms exploiting the same or similar technology adds to this drive to increase functional capability and technical efficiency.

Conversely, the changing technical performance of an innovation has a strong effect on the rate of its diffusion, since improvements make it more useful and desirable to users. Even more importantly, improvements increase the total extent of diffusion, since they open the innovation to users who would not have considered it because of price or functional limitations. An example is the computer, which has expanded into new markets with every increase in capability and decrease in cost.

None of the systemic patterns of technological change exists in a vacuum. Each of them represents one facet of a network of interactions.

A SYSTEMIC MODEL OF TECHNOLOGICAL CHANGE

The regularities of time series certainly do not exhaust the description of technological change. But it is useful to think of innovation in terms of an evolutionary description that incorporates the features displayed by these regularities.

Let's start with the introduction of a major innovation, one that opens up new possibilities and has far-reaching effects, such as the steam engine or electric lighting or the airplane. Such innovations start in rather small market niches, but give rise to large-scale systems that permeate society. Often, the innovation is the result of several streams of progress coming together, each of which involved significant innovations of its own. Also, one major innovation often calls forth others, either directly, because of the needs of the new technology, or indirectly by exposing bottlenecks or creating opportunities in other technologies. The innovation is best thought of as a member of a cluster of new technologies that reinforce each other and more or less grow together; bottlenecks in one technology may call forth advances in another, advances in one may open up opportunities in another, or progress in one may have to await improvements in another.

A major innovation has a long history of precursors before it is successfully commercialized, but in spite of this the technology is in a rudimentary stage and is capable of extensive improvement.

In its early stages, the technology diffuses through its market niche, following a typical S-shaped curve. While this is happening, the technology is being improved, the technological performance describing an S curve of its own. These two developments reinforce each other; the diffusion provides the experience, feedback, and funds needed for technical improvements while the technical improvements assure the continual adoption of the technology by more and more users. As the number of users grows and performance improves, costs go down. This process is reinforced by the learning curve in the production process.

The increase in performance involves a large number of technical improvements, some of them of such a magnitude that they become major innovations in their own right. But most of the improvements are of an incremental nature and take place continually as the technology is used. While none of these minor improvements do much by themselves, their cumulative effect is enormous and accounts for most of the great increases in technical efficiencies and decrease in costs that are the hallmark of successful technologies.

Along with the improvements, there is differentiation of the technology as it is adapted to purposes different from those originally intended. These

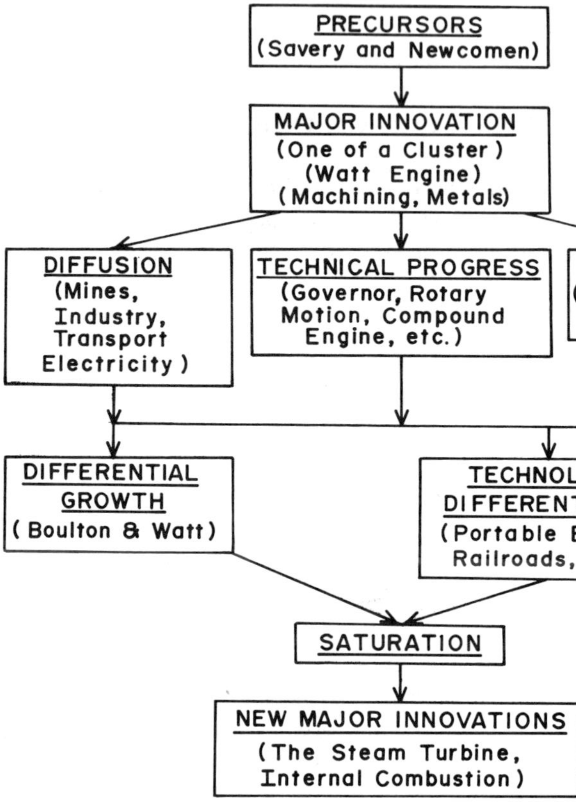

Figure 1-8 Schematic of the systemic model of technological change.

adaptations are coupled with other complementary technologies, resulting in new technologies which start out on growth curves of their own with new regimes of diffusion, technical progress, and learning. The result is a set of large-scale interlocking systems based on interrelated technologies.

Sooner or later markets are saturated, and the limits of the technology are reached. Further progress then depends on major innovations based on a new technology that can start the cycle again.

The evolution of steam engine technology provides an example of the application of the systemic model of technological change. The critical innovation for steam technology was the invention of the Watt engine with a separate condenser. Watt was fully aware of the precursors to his convention, which were the Savery and Newcomen engines. In fact, it was his efforts to get a model of a Newcomen engine to work that led to his own contributions to steam power. Once the major innovation of the separate condenser was introduced, a chain of subsidiary innovations was made that continually improved performance. These included rotary motion, the governor, the compound engine, the high pressure engine, and improvements in valves and boilers. Other technologies that enjoyed great improvements at the same time and made the technical advances in steam possible involved machining and improved metals, especially iron and steel.

Along with the technical progress, there was diffusion of the use of steam into successive markets. The first was to provide power for water pumps at mines. This was followed by the use of steam to pump water used for power in industry, particularly in textiles, and later for direct steam power in industry. The development of rotary motion led to portable engines and their use in land and sea transport as well as in agriculture. In the late nineteenth century, steam was coupled to another major technology when it was applied to the generation of electricity. There was a differentiation of the technology for different applications, which contributed to the growth of entire new technologies in transport and power generation. After a time, the full possibilities of the Watt engine had been exhausted, and further progress depended on the steam turbine and internal combustion engines.

A chart summarizing the model presented here as applied to steam is shown in Figure 1-8. It must be emphasized that this model provides only a skeletal framework within which to describe the process of technological change.

This concludes our basic introduction to some of the elementary concepts of technological change. These concepts will be used, elaborated, and applied to specific instances in later chapters.

2

Technical Progress and Performance

ASPECTS OF TECHNOLOGICAL PROGRESS

There is no doubt that the long-range trend of technological change has been toward ever-increasing complexity, sophistication, and greater technical performance. The analysis of technological change requires that some carefully defined measures be constructed that will permit the comparison of the state of technology at different times. Here, two different kinds of such measures will be considered. The first refers to the actual performance capabilities of a device, process, or system, while the second refers to the efficiency with which a technological system is utilized in practice.

Performance Capabilities

The concept of performance capabilities is a familiar one. The first step in defining a measure of performance is to define the entity whose performance is to be measured. These entities can range over a spectrum from large-scale systems to specific devices. At one end of the spectrum, for example, might be an electrical transformer whose performance is measured by the efficiency of energy transfer as the voltage of an electric current is changed from one value to another. This is a small-scale measure of a particular device. At the opposite end, a large-scale measure of an extensive and complex technological system would be the overall national efficiency with which chemical energy from burning fuels is converted into electrical energy that is used by consumers. In the first case, the energy going into the primary windings is compared to the energy coming out of the secondary windings. The ratio gives the efficiency of the transformer. In the second case, the national efficiency of electrical energy production is given by the ratio of the electrical energy consumed in the entire country to the energy content of the fuel used to generate that electricity. Both measures give information about the performance of energy conversion systems, but the information in the two cases is very different. The efficiency of national electricity production is an extreme example of a parameter resulting from the combination of a large number of very different factors. Each utility has a different generation efficiency, depending on the type and age of its equipment. Some generators may be coal-fired, while others may use oil, gas, or nuclear fuel. Different utilities have differing load demands and operate at differing percents of capacity. Also, the efficiency depends on losses that are incurred in the distribution of electricity from the generator to the point of end use. Clearly, a large number of systems and pieces of equipment are involved, each with different characteristics, and the national efficiency is a kind of average of all of these. The transformer efficiency, on the other hand, refers to one piece of equipment and does not depend on the efficiencies of a large number of other components.

There are, of course, systems of varying complexity between the overall system of national electrical production and the individual devices. Generating plants, generators, fireboxes, boilers, pumps, coolers, fly ash separators, motors, and rotors are components that have performance characteristics of their own. Each of these is composed of smaller subsystems. In each case, the state of the technology can be measured either by technical engineering specifications or by functional parameters that are closely related to the purposes of the technology.

Efficiency in Practice

Technical performance refers to the capabilities of the technological system being considered. But very often the performance depends on the abilities of the people operating the system. These abilities change as people become more familiar with technical devices or processes. Thus, when a new electrical generating plant is brought on line, operators, engineers, managers, and maintenance people need to get to know the characteristics of the new equipment. As time goes by and experience accumulates, each person learns how to do his job better and how to get better performance from the system. Whenever a new plant or a new process is introduced, the productivity increases with time as people learn how to deal with it. This phenomenon is most commonly studied in connection with the introduction of new industrial processes, the productivity increasing as the personnel go through a learning curve. A typical example of a learning curve would be the number of units produced per man-hour of labor as a function of the total number of units produced by some industrial process. As the total number of units increases, the number produced per unit of labor also increases. This is the most usual kind of learning curve. However, the learning phenomenon is a general one and is not restricted to industrial processes. New devices and new systems as well as new processes are all used more efficiently as they become more familiar.

While technological learning and technical performance are two different concepts, they are both measures that refer to the ability of a society to create and use technology. The two concepts are often related, as in the case of generating plants in which the actual technical performance achieved in practice is dependent on the learning curve.

The terms learning, technological progress, and technical performance are often used in a generic sense with overlapping meanings. For the purposes of clarity, it would be useful to give narrower, specific definitions to these terms. Accordingly, the following nomenclature will be adopted whenever possible: *learning* is the improvement in the efficiency of people as they gain experience with a new product or process, and the learning curve is the functional relation between human efficiency and experience. *Technological progress* is the improvement in the efficiency of the production or use of a product, device, or process as the result of increasing experience that includes organizational and technical changes as well as learning. It represents a generalized learning in which many people with different functions improve their efficiency by a variety of means. (Learning in the simpler sense is treated separately because it can then be applied to the specific issue of labor costs per unit of production in a particular process.) *Technical performance* refers to the functional technical parameters of a technological device or process, and the technical performance curve is the relation between these parameters and the amount of effort devoted to improve the technology.

Often, technological performance or technological progress is plotted as a function of time simply because data are not available on amount of effort. The regularity of these plots must not obscure the fact that the mere passage of time does not ensure improvements of any kind and that the important variable is the amount of effort expended to achieve improvement. This effort can be measured by the number of man-hours or the amount of money spent on its improvement if the number of people and the budgets for specific projects are known. But even these are not really adequate measures except in simple cases. They do not take into account the differing capabilities of people and the discontinuous advances of major technological changes. Nevertheless, such measures are preferable to the mere measurement of time.

Even with the definitions given here, the description of technological advance is complex. There are many situations which are a mix of the various phenomena that can be thought of as technological advance, and often they cannot be sorted out with any precision. Also, the use of these terms in the literature is such that a clean separation of these definitions is not possible. The concept of the learning curve, for example, is often used to describe improvements in productivity of a manufacturing process whether or not it results only from the direct learning of workers. However, the nomenclature adopted above at least identifies three important points in a spectrum and provides a vocabulary with which more complex cases can be described. The nature of the case is usually clear from the context.

MEASUREMENT OF TECHNICAL PERFORMANCE

The parameters that can be used to measure technical performance are determined by the functions and properties of the system being measured. Which parameters are appropriate is often evident from the intended use of the system and its engineering characteristics. In energy conversion systems, such as the transformers and generating plants mentioned above, for example, the efficiency of conversion is important, and its value as a function of time is a good measure of the temporal progress of the technology. For military aircraft, speed would be an important parameter, while for a chemical production process, percent yield might be the most important characteristic.

Sometimes a single parameter can be selected that gives a good measure of the performance of a technological system, but often more than one parameter is needed. For aircraft, for example, range, payload, and fuel efficiency as well as speed are needed to specify technical performance. Sometimes a number of parameters are combined to give a single measure, but sometimes several parameters need to be considered separately. Measures of technical performance are therefore classified into three types: single variable measures in which only one parameter is taken to represent performance, figures of merit in which one measure is constructed from several parameters, and multivariable measures.

Single Variable Measures

Single variable measures are most commonly used because of the ease with which one-parameter data can be acquired. The choice of a single parameter to describe technical performance must be made with care. In a chemical production system, for example, both temperature and yield may be important parameters. In the case where the reaction will go only at extremely high temperatures but yields are satisfactory, then the temperature is the critical technical parameter, because high temperatures pose severe materials problems and are expensive to generate and maintain. Advances in technical performance are then best measured by the decreases in temperatures that can be achieved by catalysis, increases in pressure, or other means. For cases in which the reaction temperatures are not too high, the important parameter may be yield, since the greater the yield the better the economic return on the process. In those cases for which both temperature and yield need to be improved, a single variable treatment is inadequate, and either a figure of merit must be constructed or a fuller two-variable analysis must be made.

Figures of Merit

There are many cases in which a single parameter measure of performance is inadequate. The performance of racing cars, for example, is well described by maximum speed, since this parameter reflects the purpose of racing vehicles. For passenger cars, however, speed is only one of a number of parameters that need to be considered. Braking distance, fuel mileage, and turning radius are quantifiable characteristics that are more important than speed. The speed that automobiles are capable of has exceeded the practical driving speed for many years, so it does not make sense to use speed as a measure of the performance of the modern auto. Another example is that of computers, in which the speed of an addition operation is only one of the important technical parameters. Others include the speed of logical operations, input–output times, and memory capacity. The figure of merit approach tries to reduce a number of technical parameters, P, Q, R, S, and so on, to a single one. The simplest way to do this would be to take a linear combination to construct a figure of merit of the form

$$\mathrm{FOM} = pP + qQ + rR + sS + \ldots \tag{2-1}$$

where the $p,q,r,s\ldots$ are weighting factors describing the relative importance of the different technical parameters.

This definition of figure of merit needs to be improved on two counts. The first is that each technical parameter is different and has different dimensions, and it would be better if all dimensions were the same. This can be accomplished by making each factor dimensionless by dividing it by one of its values. The figure of merit would then have the form

$$\mathrm{FOM} = p(P/P_0) + q(Q/Q_0) + r(R/R_0) + s(S/S_0) + \ldots \tag{2-2}$$

the subscript zero indicating some particular value of the parameter.

There is often a natural choice for the values of P_0, Q_0, and so forth, which arises from the fact that every technology has a limit. The efficiency of energy conversion cannot exceed unity, the yield of a chemical reaction cannot exceed one hundred percent, the strength of materials is limited by the strength of the chemical bond, no speed can exceed that of light, and electrical insulation is limited by dielectric breakdown. Generally, a limiting value of the technical parameter can be found that can be used as a divisor to make the parameter dimensionless and to restrict its range of values between zero and unity. Choosing such a limiting value is not always trivial. Consider, for example, aircraft speed as a technical parameter. If interest is centered on the development of piston engine aircraft, then the most informative way of presenting the data is to choose the maximum speed that piston engines can deliver as the limiting value. If an analysis is being made of the entire evolution of aircraft, regardless of type of engine, then a limiting value for the speed might be chosen to be that estimated for advanced rocket engines. The choice of limiting value is dictated by the nature of the technology being considered and the ideal value towards which the technology is evolving. Even so, the choice is not always easy, as a little reflection on what to choose for the maximum attainable speed using piston engines will reveal. This, of course, is a problem that can be avoided in single parameter studies since then there is no need to worry about dimensionality of technical parameters. However, a careful analysis of technical limits is of great value no matter how technical performance is described, since a clear picture of the capabilities of a particular technology can then be obtained.

There are cases in which the concept of technological limits cannot be easily used in defining a figure of merit. As an example, consider the total harmonic distortion of an amplifier. The ultimate limit of harmonic distortion in electronic amplifiers is zero, and modern amplifiers are actually approaching this limit. But using zero in the denominator to make harmonic distortion dimensionless gives infinity for the performance parameter, regardless of the value of total harmonic distortion. Actually, since the desirable technical situation is one in which there is a low harmonic distortion, we choose its reciprocal as a performance parameter, so the ultimate limit is infinity and the value of the performance parameter is zero, regardless of the value of the harmonic distortion. In any case, if the ultimate value of a performance parameter is either zero or infinity, then it cannot be used as a divisor, and the best that can be done is to choose some specific value of the parameter as a divisor. Another consideration is that there are cases in which improvement of a technical parameter beyond a certain point is not useful. Total harmonic distortion in stereo audio amplifiers is an example of this. Any improvement of this parameter below, say, 0.1 percent may have marketing value, but has no audible effect in a home sound system. In such cases, the limiting value can be chosen to be that which is of practical use, and all technically better values can be replaced by this definition of the limiting value. In the case of total harmonic distortion, any values less than 0.1 percent would be replaced by 0.1 percent.

The second modification of our definition of the figure of merit arises from the fact that a zero value for some technical parameters makes the technology totally useless. These are called essential parameters. If a power amplifier has zero power or an automobile gets zero miles to the gallon or a chemical reaction gives zero yield, then any figure of merit that includes these quantities must be zero when they are zero. Such parameters as the number of functions on a remote control for a television set or the accuracy of a self-timer for a camera, on the other hand, are not essential, since their absence does not totally incapacitate the device. They are nonessential parameters. The only way to treat essential parameters is to include them multiplicatively in the figure of merit, not additively. But even if different parameters are all essential, they may not all be equally important for the technological function. Thus, zero values for either braking ability or speed would make an automobile useless, but advances in braking ability are more important than advances in maximum speed. This means that multiplicative factors must also be weighted according to their importance in the technological system. This cannot be done by simply multiplying the factor by a constant, since this merely changes the scale of all factors. Instead, each multiplicative factor is raised to a power that is larger the more important the factor. The final form of the figure of merit is therefore

$$\text{FOM} = [(C/C_0)^c(D/D_0)^d \ldots][p(P/P_0) + q(Q/Q_0) + \ldots] \quad (2\text{-}3)$$

This definition of figure of merit is sometimes used in logarithmic form, which makes it clear that the numbers *c, d,* and so on are linear weighting factors for the logarithms of the essential parameters. Instead of Equation (2-3), the figure of merit would then be written as

$$\begin{aligned}\text{FOM} = {} & [c\ln(C/C_0) + d\ln(D/D_0) \ldots] + \\ & \ln[p(P/P_0) + q(Q/Q_0) \ldots]\end{aligned} \quad (2\text{-}4)$$

Equations (2-3) and (2-4) are equivalent definitions of the figure of merit.

The process of choosing a figure of merit is clearly not trivial. It can be summarized in the following steps:

1. Decide whether you wish to describe a particular technology, a set of technologies, or all technologies that perform the essential functions of the system, device, or process. For example, figures of merit for computers can be constructed for either electromechanical, vacuum tube, transistor, or integrated circuit computers or for all of these as a group. In other words, define the technological system, device, or process you wish to study.
2. Choose a set of essential parameters and nonessential parameters in such a way that the parameter increases as the technology improves. This may require that the reciprocal be taken of some parameters that are commonly used to describe a technical function. Braking distance and total harmonic distortion, for example, both decrease with improving technology, so their reciprocals should be used to describe technical advance.
3. Reduce all parameters to dimensionless form by dividing them by their limiting values if such limits exist. If limits do not exist, divide by some reasonable and convenient value.
4. Choose weighting factors for both the essential and nonessential parameters.
5. Put the parameters with their weighting factors in Equation (2-3).

Artificial lighting was presented in Chapter 1 as an example of technical evolution and the existence of technical limits. There was a definite limit to the efficiency that could be obtained from incandescent lamps because of the nature of electrical incandescence. The resistance of a wire to the flow of electricity heats the wire to such a high temperature that it emits light. The heat that must be generated is not available to produce light. The phenomenon of incandescence *requires* that a certain amount of energy is wasted in the form of heat so that there is a natural limit to the efficiency of incandescent lamps. This limit was exceeded only when a new method of the conversion of electrical to light energy was adopted; namely, fluorescence. The overall curve for the efficiency of artificial lighting sources shows how a continually increasing efficiency is obtained by moving from one technology to another. But sooner or later, an ultimate limit must be reached because, by definition, the efficiency cannot exceed unity. When all the electrical energy is converted to light, progress stops.

But even if the efficiency of the conversion of electrical energy to light is unity, there are still two ways in which artificial lighting can be improved. In the first place, it might be possible to convert chemical energy to light directly, rather than going through the transformation of chemical energy to heat, then to mechanical energy, and then to electricity. Measured relative to the chemical energy, electric lighting is much more inefficient than indicated by the curves in Chapter 1, in which the efficiency is computed relative to the electrical energy input.

The other way to improve lighting is to work on factors other than efficiency. These factors might include size, frequency distribution of the emitted light, and lamp lifetime. In fact, the concept of the figure of merit for a technology is designed to explicitly denote the multivaried nature of a technology. The point I am making here, however, is that the existence of technological limits does not imply an upper limit to overall technical progress. Technical limits exist with respect to particular technologies that are based on a given set of scientific and engineering principles, but they can be transcended by shifting to a new technology based on a different set of principles. This new technology might or might not require a differently defined figure of merit. For example, in incandescent lighting a technical figure of merit should include the filament temperature. Indeed, the success of incandescent lighting depended on finding a material that could be heated to high enough temperatures. In fluorescent lighting, however, filament temperatures do not exist. On the other hand, the functional parameters describing the performance of amplifiers are the same for transistor-based or tube-based amplifiers.

A distinction has been introduced between two kinds of figures of merit in the above considerations. A *functional figure of merit* is one in which only parameters describing the functional performance of a technology or a device appear. Examples would include airplane speed, computer memory, engine efficiency, or strengths of materials. A *technical figure of merit* is one in which the parameters describe those technical properties on which the useful performance of the technology is based. Such parameters might be electrical cable resistance, semiconductor purity, refractive index of glasses for lenses, or jet engine thrust. Of course, there are some parameters that are both functional and technical parameters, such as harmonic distortion, and figures of merit can be constructed that include both kinds of parameters.

The figure of merit not only yields a single parameter with which a technology can be measured, it also permits a quantitative analysis of technological trade-offs. Often, the state of a given technology permits a range of values of technical parameters to be achieved, but an improvement in one can be had only at the expense of another. In general, neither the desirable functions of a technology nor its technical properties are independent of each other. For example, the speed, capacity, and fuel mileage of trucks can not all be improved simultaneously. An improvement in any one of these parameters can only be achieved at the expense of one or both of the others. The figure of merit for a given state of technology can therefore be interpreted as a trade-off equation that defines the technical possibilities at a given time and shows how compromises among the desirable characteristics of the technology must be made.

Trade-off parameters can be defined that measure the degree to which one technical function must be given up in order to improve another. This is merely the differential coefficient between the two technical functions when the figure of merit is a constant. For example, if the parameter C in the figure of merit is related to two other parameters D and P, then a change in C can be described as

$$dC = \frac{\partial C}{\partial D}\, dD + \frac{\partial C}{\partial P}\, dP \tag{2-5}$$

The partial derivatives in Equation (2-5) are defined to be the trade-off parameters.

In applying the concept of the figure of merit to a device or a technology to describe the evolution of technical performance, two considerations must be kept in mind. The first is that the set of parameters chosen should adequately describe the technology; the second is that the functional figure of merit should be continuous, in the sense that the same parameters should be used throughout time to describe the technology. However, it is often useful to define figures of merit that give only partial information. For example, the product of speed and capacity certainly does not exhaust the description of aircraft, but passenger-miles per hour is of sufficient importance to be analyzed on its own.

A discontinuity in the technical figure of merit implies a fundamental change in the nature of the technology. For example, neutron cross sections and core temperatures are important for nuclear-powered electric generating plants but are not relevant to fossil fuel generators. However, a change in technology does not always require a change in figure of merit, particularly if the figure of merit is based on functional parameters. The overall progress of a technology can usually be described by functional performance parameters that are common to each of the specific subtechnologies.

The relationship between figures of merit and corporate strategy should be obvious. What customers are often buying is a set of technical parameters embodied in a product or a service. Keeping track of these as a function of time would inform the corporation of its progress in meeting customer needs, of how well it is meeting the competition, to what extent improvements can be made, and of the nature of trade-offs needed to improve any single characteristic.

Multivariable Measures

Multivariable measures are sets of single variable measures that are not combined into a single figure of merit. These are used, for example, when improvements in several different parameters are monitored simultaneously but independently. A case in point would be computer development, for which engineers should follow both speed and memory separately although a user might be more interested in a figure of merit for computer power that combines both speed and memory.

CASE STUDY: AVIATION

Two great technologies have dominated air transport; reciprocating engine propeller aircraft and jet engines. Each of these technologies has been the subject of extraordinary evolutionary improvement, and each has reached a stage of maturity, at least for commercial applications. Since the Wright brothers' first flight, planes have grown in size, speed, and range through the development of a series of innovations ranging from internal combustion engine improvements, structural and metallurgical advances, and the invention of the jet engine to the application of electronics for instrumentation and guidance systems.

Initially, the great barriers to powered flight were the high weight to power ratio of existing engines and the difficulty of controlling the vehicle during flight. Here, I will focus on the engine weight. Late in the nineteenth century it was clear that the reciprocating internal combustion engine was a prime candidate for powering airborne craft. But the heaviest component of an airplane was the engine, and high power to weight ratios were critical. The Wright brothers' first flight in 1903 used an engine with a ratio of 0.067 horsepower per pound of engine weight, which was barely adequate for the Kitty Hawk flight, although this was a great improvement over the first internal combustion engines of the 1880s, which generated only about 0.0023 hp/lb.

The evolution of the internal combustion, propeller-driven airplane depended on the development of engines with ever greater power to weight ratios. These ratios are shown in Figure 2-1 for a series of engines from 1902 to 1948, using data taken from page 367 of Martino's book (1983). At first, improvement was rapid, but by the mid-thirties the power to weight ratio reached a peak and actually declined in later years. The decrease after 1934 was the result of building larger engines in order to get greater aircraft speed and carrying capacity. The 1934 Rolls-Royce Merlin weighed 1,450 pounds and generated 1,730 horsepower, while the 1945 Pratt & Whitney

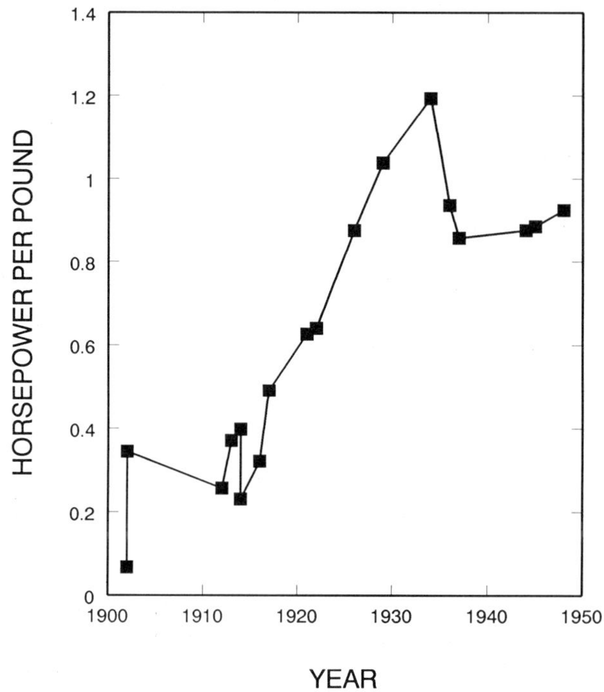

Figure 2-1 Horsepower to engine weight ratio of reciprocating aircraft engines.

weighed 3,670 pounds and generated 3,250 horsepower. The increased horsepower after 1934 could only be achieved at the expense of a greater than proportional increase in weight. This was a clear indication that reciprocating engine technology was approaching its limits.

Of course, engine developments were not the only important technical advances. Until the end of World War I, aircraft structures consisted of frameworks of wood and wire covered with painted fabric. In 1919, Junkers in Germany introduced a steel tube and metal skin construction. The skin was made of duralumin, an aluminum alloy containing small amounts of copper and magnesium that was developed in Germany. In the early 1920s the skin was made an integral part of the support structure by stressing it so that it lent strength to the wings and fuselage. The pressurized cabin, first used on the Lockheed XC-35 in 1937, and the tricycle landing gear, introduced with the DC-4E in 1938, were just some of the advances that contributed to the range, speed, comfort, and safety of air travel. Others included wing flaps, two-pitch propellers, and propeller reduction gearing, all introduced in the DC-2 and DC-3 along with retractable landing gear. The continuously variable pitch, the "feathering" propeller, and the reversible pitch propeller were introduced over a period from the mid-thirties to the late forties. These innovations were particularly important because they permitted the matching of propeller speed to plane speed and altitude (variable pitch), prevented the energy losses and excessive vibration (along with the danger of fire) that accompanied the propeller turning in an incapacitated engine (feathering), and provided a braking method that did not depend on wheels or flaps alone (reverse pitch).

But by the end of World War II it was evident that the propeller-driven plane could not be improved much more, at least within the economic limitations of commercial aviation, and it was generally recognized that the technical limitations of reciprocating engine propeller planes precluded attaining speeds of 500 mph or more, at least for commercial aviation.

Meanwhile, largely because of the initial efforts of von Ohain in Germany and Whittle in England, the jet engine was being developed during the 1930s and early 1940s.

The turbojet works by taking in air from the atmosphere, using a centrifugal or axial compressor to compress it to three to twelve times its original pressure, and adding fuel. The fuel–air mixture burns to reach temperatures of 1200 to 1700 degrees F, and the resulting hot gas drives a turbine which drives a compressor. The jet discharge at the compressor is at a pressure of about two atmospheres or more, which is high enough to drive the aircraft forward. The development of the jet engine removed the limitation of speed imposed by the maximum possible speeds of rotating propeller tips and reciprocating pistons.

The advantage of the jet engine is that it has a high power to weight ratio. Much more power can therefore be used for the aircraft. Also, the efficiency of the jet is high at high speeds, while that of the reciprocating engine is lower at high than at low speeds. Furthermore, the absence of a propeller permits the jet plane to be more streamlined and aerodynamically efficient than the propeller-drive plane.

Jet propulsion was adopted by the military fairly early, but commercial application began in 1959 with the introduction of the Douglas DC-8 and the Boeing 707. These turbojets exhibited greater speed, fuel economy, capacity, range, and passenger comfort than the propeller-driven airliners. They were also more reliable and easier to maintain. Of course, jet plane design did not stop with the DC-8 and the 707. In the early 1960s the turbofan engine was introduced, and in the beginning of the seventies a second generation turbofan yielded further advances, with a resulting improvement of specific fuel consumption of more than a factor of two over the first commercial turbojets. In the turbofan, a portion of the hot gases is diverted to a turbine that adds to the thrust by acting as a propeller. The turbofan is quieter than the turbojet and has greater efficiency, particularly at low speeds.

These technical factors lie behind the data represented in Figure 2-2, which displays the maximum consecutive values of cruising speeds of commercial aircraft. (Data for this figure were taken from Martino 1983, page 338.) For

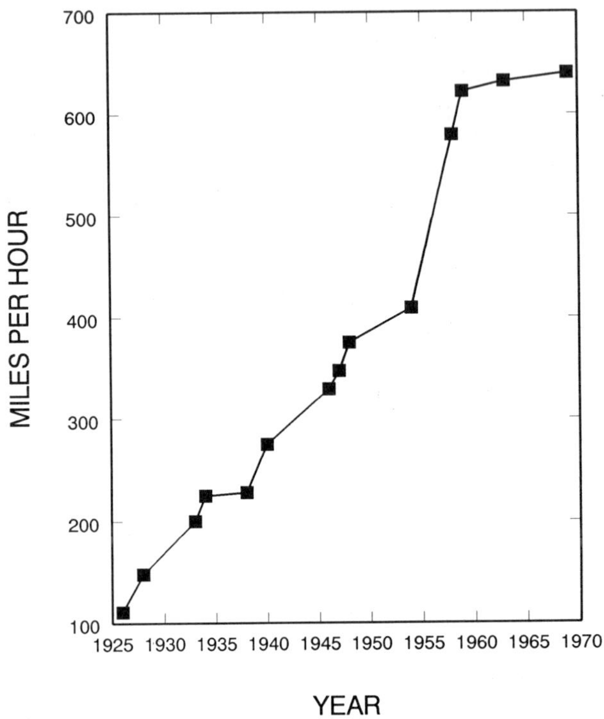

Figure 2-2 Consecutive maximum cruising speed of U.S. commercial aircraft.

reciprocating engine propeller-driven vehicles, there is a regular rise to a maximum of about 400 miles per hour up to 1954, which was not subsequently exceeded by commercial propeller-driven planes. With the introduction of the jet engine in 1959 there was a jump in speed which quickly leveled off to about 640 miles per hour. This represents the current limit of what can be done with subsonic jet technology.

In 1968 and 1969, two supersonic transports were put into service, the Tupolev TU-144 and the Concorde. These planes achieved speeds of 1,550 mph and 1,354 mph respectively. The supersonic transports represent an uneconomical extrapolation of technology and were built more as a matter of national pride than for economic reasons. In fact the SSTs have not been economically self-sustaining. The air fares cannot be pushed much higher, fuel consumption is large, R&D costs were enormous, and the passenger capacity of the SSTs is low. The TU-144 carries 140 passengers and the Concorde only 100, while the Boeing 747 carries 490 passengers. This illustrates an important point. Often, a technology can be pushed to extreme limits if cost is no object, but usually economic constraints are as important as technical ones.

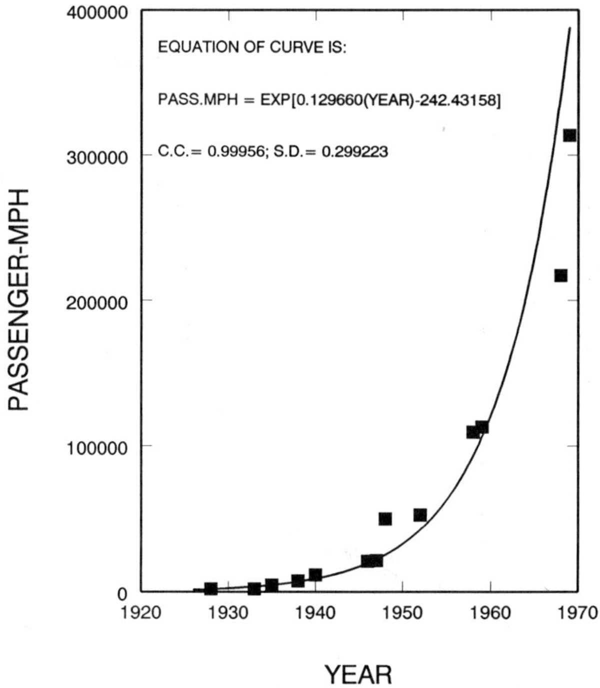

Figure 2-3 Passenger miles per hour of commercial aircraft (consecutive maximum values).

Both speed and passenger capacity are obviously important for commercial aircraft. Multiplying speed and capacity gives the combined measure of passenger miles per hour, a two-parameter figure of merit which is shown as a function of time in Figure 2-3. (The data in Martino 1983, was used to compute this figure of merit.) This measure had leveled off for propeller-drive planes by 1950 but started to rise again with the advent of the jets in the late fifties. This curve is for the successive maximum values of passenger miles per hour, and it is interesting to note that the Concorde does not appear on the curve. Even when it was built, the Concorde did not attain the values of this important parameter that had already been achieved by subsonic aircraft. The Tupolev represented an advance in passenger miles per hour in 1968 but was overshadowed by the 747 in 1969. Furthermore, the fuel cost per passenger mile for the Concorde is about fifteen times that for the 747. Thus, even though the Concorde can cross the Atlantic in half the time, it simply cannot compete economically. An interesting point that is illustrated by Figure 2-3 is that measures of technological performance do not always level off to their maximum values by following an S curve. The passenger miles per hour as a function of time followed an exponential curve right up to the introduction of the 747 in 1969 and has not been improved upon in

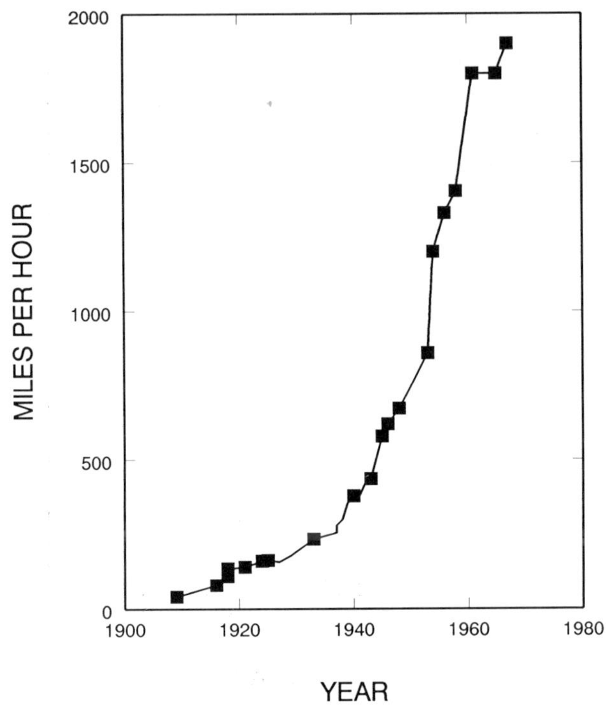

Figure 2-4 Maximum U.S. military aircraft speeds.

this respect to date. Rather than being described by an S curve in which the figure of merit approaches a maximum asymptotically, the maximum passenger miles per hour for commercial aircraft grew exponentially to 1969 and then stopped.

The military provides the best example of economic considerations being secondary to technical ones. The result is a steady increase of the maximum speed of military aircraft, as shown in Figure 2-4. Again, a faster rise is evident with the introduction of jet engines in the mid-1940s.

THE LEARNING CURVE

In its simplest form, the learning curve is a relationship between the efficiency of performing a task and the number of repetitions of that task when everything else is constant. That is, there are no changes in the tools, no improvements in the technology, and no improvements in the way the work is laid out or organized. Such stringent conditions usually are found only in artificial situations such as psychological laboratory tests. In actual practice, this simple kind of learning is often accompanied by other changes of a technical or organizational nature that are introduced to enhance productivity. These other changes are, of course, a kind of learning, but their origins and significance are much different from those of simple learning. The increases in productivity that result from a combination of learning, technical advance, and organizational change is called the *progress function.*

Examples of learning abound, and a number of them have been studied. Hirsch (1952) analyzed the relation between man-hours per unit and the total number of units produced for new products or new models of a large American machine tool manufacturer. The products varied widely in size and function, including semiautomatic and automatic machine tools, a textile machine, and a multipurpose construction machine. Hirsch used the usual power law to describe learning in the form

$$lnP = a - blnN \tag{2-6}$$

where P is the man-hours required to build the Nth unit and a and b are parameters that were determined from fitting observed data to Equation (2-6). (See Appendix 1.)

Production consisted of two major categories: machining of parts and assembly. Assembly was carried out by a crew of fixed size for a given product, and the rate of production was increased by having more crews assemble more machines rather than by increasing the size of the crews. Production was done in "lots" of a particular number of units, which were started at intervals that may or may not be regular. Thus, there is no sensible way to plot productivity versus time. Furthermore, a statistical analysis showed that there was no correlation between productivity and lot size. Productivity versus lot number, however, showed a good log–log plot. This result implies that learning took place on a lot-by-lot, rather than a unit-by-unit, basis.

There were no changes in plant, equipment, or management during the time of the analysis, which referred to data that extended from 1946 to 1950. The observed productivity improvements could therefore be ascribed to the increased efficiency of the crews as their members improved their skills with experience.

Hirsch's results are summarized in Table 2-1, where the learning rates are given in terms of the parameter b of Equation (2-6). The first set of b values is for the learning curves of the machining process, the second set is for the assembly, and the third set is for the combined machining and assembly processes. The entry N.A. means that data were not available, and the entry N.S. means that the correlation coefficient of the log–log plot is too small for the plot to be statistically significant.

These results show that learning for assembly is faster than for machining. This is to be expected because many machining operations are the same for any of a wide variety of parts, while the process of assembly is very different for different products. The opportunities to improve are therefore greater in the assembly process as workers become familiar with the techniques needed for the new product assembly.

Table 2-1 Learning Rate Parameters for the Production of Machine Tools

	b Values:		
Product	Machining	Assembly	Combined
Machine 1	N.S.	0.462	0.287
Machine 2	0.204	0.424	0.297
Machine 3	0.219	0.485	0.337
Machine 4	N.S.	0.392	0.260
Machine 5	0.209	0.378	0.287
Machine 6	N.A.	N.A.	0.277
Machine 7	0.324	0.303	0.323

Cole (1958) studied the learning curve in the form of cost versus number of units produced to determine if learning rates were similar for different industries. He reported his results in terms of the improvement percentage for each doubling of the number of units produced. The improvement ratio is related to the learning curve as follows: if the cost per unit is inversely proportional to the productivity, then it follows from Equation (2-6) that

$$lnC = A - blnN \tag{2-7}$$

Then for a doubling of the number of units produced to $2N$, we have

$$lnC_2 = a - bln(2N) \tag{2-8}$$

where C_2 is the cost when the number of units is $2N$. Subtracting Equation (2-8) from Equation (2-7) and solving for b gives

$$b = -ln(C_2/C)/ln2 \tag{2-9}$$

For a given value of b, the cost improvement ratio, C_2/C, is the same for every doubling of the number of units produced.

This relation was used to convert Cole's improvement ratios to the learning rate parameters b listed in Table 2-2.

From these data and that of other authors, Cole concluded that there was very little difference in the learning rate for different manufacturing pro-

Table 2-2 Learning Rate Parameters for Products in Different Industries

Product	*b* Value	Improvement Ratio
Electro-optical aircraft instrument	0.36	0.78
Electronic unit	0.34	0.79
Mechanical-hydraulic electrical unit	0.38	0.77
Laminated plastic aircraft assembly	0.38	0.77
EDP unit A	0.29	0.82
EDP unit B	0.36	0.78

cesses, and that a doubling of the number of units produced decreased the labor per unit by 20 percent. That is, a doubling of production decreases cost per unit to 80 percent of the original cost. This is known as the 80 percent rule for labor productivity or labor costs.

That the 80 percent rule is not always followed precisely is shown by the work of Conway and Schultz (1959), who found that for the assembly of an electronic product the b value was 0.456 and for the assembly of an electro-mechanical product the b value was 0.415. These results corresponded to a 72.9 percent and 75 percent decrease in labor per unit respectively for each doubling of the number of units produced. Yet, the results show that the 80 percent rule can be used at least as a first approximation.

The results presented so far were for labor-intensive processes where the opportunities for human learning are obvious. Baloff (1966) has studied the increase in production efficiency in machine-intensive processes by plotting production per unit time as a function of cumulative production and found that the learning curve also applies. After start-up, production and engineering personnel learn how to be more efficient so that there is less downtime because of maintenance problems, better production scheduling, improved inventory control, and better production standards. There was much less uniformity of b values for the machine-intensive learning curves, however. For eight different processes in a steel company, for example, the b values ranged from 0.174 to 0.714 (Baloff 1967). This finding corresponds to increases in the production rates of 61 percent and 88 percent respectively for each doubling of production volume. Actually, this case is somewhat closer to what we have termed technological progress than to learning, since the improvements reflect all changes, not only labor efficiency increases.

PRODUCTION COSTS AND THE LEARNING CURVE

The learning curve illustrates one of the economic advantages of volume production and can be an important tool for estimating the labor costs in repetitive manufacturing processes. Another advantage of high volume arises from the fact that all new products or processes require initial investments in such activities as research and development, capital equipment, marketing analysis, and planning. These investments must be paid for by revenues from sales of the product. The larger the number of units produced, the smaller the initial costs per unit. The decreasing labor costs because of learning and the decreasing prorated initial costs are two major contributors to the economies of scale that permit prices to go down.

To illustrate this, consider the production of a product which has three main cost elements, initial start-up costs, materials costs, and labor costs. The total cost of producing the Nth unit would therefore be

$$C(N) = F/N + m + AN^{-b} \tag{2-10}$$

where F is the start-up cost, m is the cost of materials per unit, and the last term is the cost of labor, which is assumed to decrease as a function of the number of units produced according to the learning curve, with a learning rate parameter b and the labor cost of the first unit being A.

The materials cost per unit is assumed to be constant, thereby ignoring the possibility of variable volume discounts. It is also assumed that the inverse of the labor cost follows the same learning curve as that for productivity (except for a constant factor). This is equivalent to assuming that wages stay constant over the period of application of the formula.

Usually, what is of interest is the total cost of production of a number of units, or equivalently, the average cost of production per unit as a function of production volume. This is the figure that determines the profitability of the product and the price at which it must be sold. Let us assume that the number of units produced is large, so that sums can be approximated by integrals. The average cost per unit is then obtained by integrating (2-10) from 0 to N and dividing by N (see Appendix 1). The result is

$$C(\text{ave}) = F/N + m + AN^{-b}/(1 - b) \qquad (2\text{-}11)$$

This equation displays the fact that costs per unit can decrease with volume of production in two major ways, as embodied in the first and last terms. The first is the decrease in start-up costs per unit produced. This decreases inversely as the number of units produced since each unit is defined to bear an equal amount of the initial costs. The second source of cost reduction is the learning curve. The relative importance of these two terms depends both on the magnitude of the fixed costs and the rate of learning. If the initial start-up costs are small, then they do not contribute much to the cost per unit, and the cost decreases are primarily the result of learning. If the initial costs are large, then the costs per unit go down rapidly as the initial costs are prorated over more and more units, and the relative savings from learning are not appreciable until many units are produced. To show the relationship between savings from learning and savings from prorating initial costs, we ignore the materials cost in Equation (2-11), which is assumed to be constant, and consider only the costs of labor and start-up. Also, let us define N^* to be the number of units produced at which the labor costs equal the prorated initial costs per unit. That is,

$$F/N^* = A(N^*)^{-b}/(1 - b) \qquad (2\text{-}12)$$

Now subtract the materials cost, m, from both sides of Equation (2-11) and divide by F/N^* to get

$$[C(\text{ave}) - m]/(F/N^*) = N^*/N + (N^*/N)^b$$

The left-hand side of this equation can be given a single symbol and written as

$$C(v) = N^*/N + (N^*/N)^b \qquad (2\text{-}13)$$

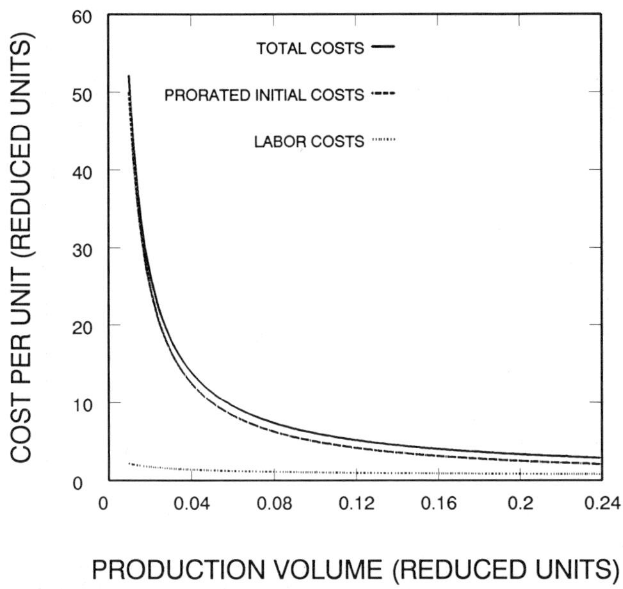

Figure 2-5 Cost curves for learning and prorating of initial costs: low production volumes.

where $C(v)$ is the average unit cost of labor plus start-up in terms of the cost of labor plus start-up when both these costs are equal. That is, $C(v)$ is given by

$$C(v) = [C(\text{ave}) - m]/(F/N^*) \tag{2-14}$$

Equation (2-13) was used to compute the cost curves for low and high production volumes shown in Figures 2–5 and 2–6, where it was assumed that the 80 percent rule holds so that $b = 0.32$. Low volumes and high volumes are defined by the relations $N < N^*$ and $N > N^*$ respectively.

For the low volume case shown in Figure 2-5, the decrease in cost with production volume is dominated by the effect of prorating the initial costs. By comparison, the cost of labor is low and the decrease in labor cost is slow. For high volumes, the situation is reversed, as shown in Figure 2-6. After the production volume is sufficient to make the prorated initial cost equal to the cost of labor, the major cost improvements come from the learning curve. Beyond a production volume of $N = 10N^*$ to $20N^*$, labor is the main cost element, and the important cost improvements are the result of the learning curve.

For low volumes, the decreases in cost per unit with increasing volume are very rapid and quite large. Because the prorated initial costs are inversely proportional to the production volume, a doubling of volume cuts the initial cost per unit in half, yielding decreases of more than a factor of ten as N goes to N^*. For high volumes, the cost decreases are much smaller, since

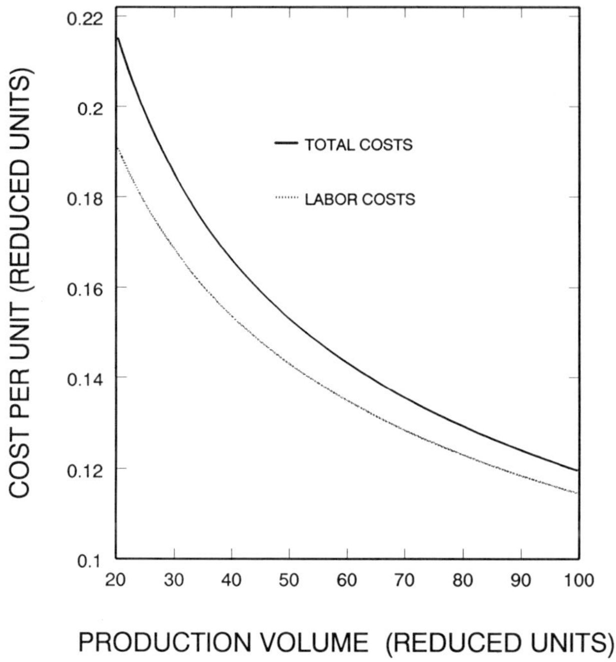

Figure 2-6 Cost curves for learning and prorating of initial costs: high production volumes.

they follow the learning curve for which it takes a doubling of volume to decrease costs by 20 percent.

The great importance of volume for decreasing costs is clear. If a sufficient number of units can be sold so that the volume of production reaches at least N^*, great cost savings are realized simply from prorating initial costs, while larger volumes result in significant savings in labor costs. Prices can therefore be reduced. Note that there is a feedback link here, in which larger market share leads to larger production volumes which permit lowering of prices which, in turn, leads to larger market share.

Obviously, there is a great benefit in being able to predict the future cost curve for a new product so that future prices can be set, profit margins established, marketing programs set up, and appropriate financing secured. The ability to estimate future costs has enormous strategic and tactical value. At start-up, the initial costs are known because they are in the past, but the labor cost curve has to be estimated. Several approaches to such an estimate can be taken. The first is simply to start production and collect productivity data as a function of units produced, construct a learning curve, and use the power law to extrapolate the results to higher production volumes. The second is to use learning curves from the start-ups of previous products to estimate the learning curve for the new product. The best way, of course, is to use earlier data for initial estimates of the learning curve, which is continually refined as data on the new production accumulate. In any case, the question arises of the degree of carryover of previous, similar learning curves to the new product. This question was addressed in some detail for airframe production by Alchian (1963), who examined existing learning data for the production of different types of airframes (bombers, fighters, trainers) at fifteen different manufacturing facilities. Twenty different model–facility combinations were included in the study.

The first question considered by Alchian was whether or not all combinations of models and facilities followed the same learning curve. A statistical analysis gave the result that different facilities and different models of aircraft gave different learning curves, so that no single master learning curve for the assembly of airframes throughout the industry could be constructed. Each specific combination of aircraft model and assembly facility had its own learning curve. Different facilities did not have the same learning curves even when assembling the same models, and a given facility did not have the same learning curve for all models.

Although each model–facility combination has its own learning curve, it is still possible to aggregate the data for different models and facilities and construct learning curves for the aggregates. The justification for doing this is the purely practical one of searching for a predictive tool that can be used to estimate future labor requirements. Accordingly, Alchian aggregated all data to get an industry-wide learning curve, and all data for a given airframe type (bomber, fighter, trainer) to get a learning curve that included assembly at all facilities of particular types of aircraft. He then used each of these curves to predict labor productivity for each model–facility combination and compared the prediction to the actual observed productivity. He also used specific learning curves computed for each model–facility combination from early production data to predict later labor productivity. According to his results, the overall errors were about the same whether the future productivity was predicted using learning curves computed from industry-wide data, from data for types of aircraft, or from initial data for specific models assembled at specific facilities. The average error of prediction was about 25 percent in the first two cases and about 22 percent for the specific model–facility case. Furthermore, the standard deviations were similar, being about 36 percent for the first two cases and about 32 percent for the third. These results imply that previous learning curves can be used as a rough guide, but that it is important to continue to monitor the production process and to continually revise predictive estimates in the light of accumulating data if serious errors are to be avoided.

CASE STUDY: FARM TRACTORS

Although the analysis of technical performance and of learning is best applied to specific devices and processes, it is often the case that they are also applied to aggregated data that represent complex systems. It is therefore important to illustrate the extent to which this can be done. Tractor technology is chosen for this purpose because it has been extensively studied, aggregated data are available, and it displays important features and limitations of the quantitative analysis of technology. A collection and treatment of the data, along with some description of the technology, is given by Sahal (1981), whose work has been drawn on heavily here. The analysis will address the change in technical parameters over time and the applicability of technological progress functions with the form of learning curves to the results.

The mechanization of agriculture began with the development of implements for such agricultural operations as plowing, reaping, and threshing. These devices were initially powered by horses, and it was only after the implements had reached a certain stage of development that horses were replaced by mechanical power. A number of developments in implement design occurred before the second half of the nineteenth century, but the commercialization and diffusion of the new farm implements began in earnest with the McCormick reapers and other machinery in the middle of the century.

The use of steam to drive farm machinery had been pioneered in England on model farms as early as the 1840s, and steam power, both stationary and mobile, was increasingly used until the early part of the twentieth century. But the development of the tractor with an internal combustion engine was the major contributor to the pervasive mechanization of agriculture. By 1900 the mechanization of the United States was well under way, with a total of nearly 64 million horsepower of inanimate prime movers installed. But agriculture was far behind industry in this respect. With nearly 37 percent of the work force, farms had only 6.3 percent of the mechanical power installed in the country. Most of the work was done by animals rather than by machines. The mechanization of the farm had to await the development of an efficient, mobile power source that could be used for the farm operations of plowing, seeding, reaping, threshing, and binding. This was the tractor, which provided mo*tor*-driven *trac*tion.

The tractor was largely an American development; the first known gasoline-powered tractor was built in 1892 by John Froelich, a farmer and blacksmith in Iowa. It had a 20-horsepower, one-cylinder engine, friction clutches, and reverse gearing. Over the next ten years at least eighteen serious attempts to construct viable, gasoline-driven tractors were made. In 1897, oil-burning tractors were being produced in England, but these were soon supplanted with gasoline-powered tractors of American design. The first factory devoted exclusively to tractor production was opened in 1905 by Hart and Paar, who had built and sold at least seventeen tractors by 1903. Steam-powered tractors were also developed and sold during this time, but the durability, economy, and convenience of the internal combustion tractor was so superior that no more steam tractors were produced after the mid-1920s.

The rapid diffusion of tractors in the United States began with the Fordson, which was first marketed in 1917. By 1920 this machine had taken the general form that was to dominate tractor technology right up to the present. The Fordson was of frameless, cast-iron construction and was the first tractor to be made on an assembly line.

At this time, the tractor was not much more than a pulling machine and was used primarily for plowing. Its gradual evolution into a versatile, multipurpose machine was the result of a number of innovations introduced over a period of time. A general purpose tractor, introduced in 1924, had high rear axles and adjustable front wheel spacing. The tractor could therefore move between rows for cultivation. In 1929 a mechanical lift was added so that implements could be lifted off the ground when the tractor was turned. These improvements made it possible to use the tractor for harvesting both row and non-row crops. Rubber tires began to replace steel tires in 1934, which greatly increased comfort, stability, and control. The introduction of the three-point hitch and control system in 1938 was the final innovation that made the attachment and use of a wide variety of implements simple and convenient. This, of course, was not the end of tractor innovation. The continuous running independent power takeoff and hydraulic remote control of drawn implements in 1947, power steering in 1953, on-the-go power shifting in 1954, front and four-wheel drive in 1967, and electric remote control of implements in 1970 were some of the more important innovations that contributed to the effectiveness of the modern farm tractor.

From the beginning, attempts were made to use oil-burning engines in tractors, but it was not until the middle of the century that gasoline began to be seriously displaced. In 1955, about 12.5 percent of the tractors produced in the United States were diesel-powered, and by 1970 this had increased to 71.5 percent, and now most of the tractors on farms are diesels.

The technical progress of tractor technology will be measured by changes in fuel efficiency (horsepower-hours per gallon of fuel consumed), mechanical efficiency (ratio of drawbar to belt or takeoff horsepower), and total efficiency, which is the product of fuel efficiency and mechanical efficiency in units of horsepower-hours per gallon.

Data for the analysis were taken from that collected by Sahal and given in the appendix of his book (1981). Sahal's data refer to tractors produced in each year from 1920 to 1970 in North America and Western Europe, so that it is essentially worldwide data. The data for the efficiencies given in Sahal's tables are averages for the worldwide production each year. To describe technical progress, I have treated only the data that give successive maximum improvement in efficiency. For many of the years, the efficiency did not improve over previous years, and sometimes it was even less than in previous years. This fact illustrates one of the problems in using aggregated data. The tractors produced in any given year include a variety of sizes, models, and manufacturers with varying markets and functional objectives. The average over such a group represents average current practice, not best

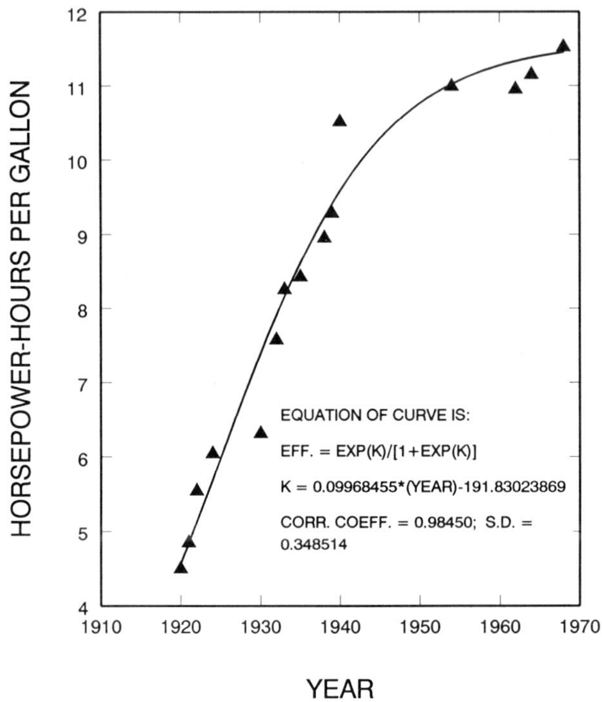

Figure 2-7 Successive maximum tractor fuel efficiencies.

practice. The closest that one can come to measuring advances in the state of the art under such circumstances is to take successive maxima as the technical parameters change in time. This then represents the successive improvements in best average practice.

Successive maximum tractor fuel efficiencies are shown in Figure 2-7. A logistic equation was used to fit the data, and the calculated curve describes the long-range improvement in efficiency reasonably well. It must be remembered, however, that the scatter of points from the curve is not experimental error but represents real deviations from the calculation. On a somewhat finer time scale, therefore, the points can be thought of as covering two regimes: the first being a rise and leveling off up to 1930, and the second being a rise and leveling off up to 1970. A variety of technical innovations contributed to increased fuel efficiency, including improvements in fuel quality, pneumatic rubber tires, the shift to diesel engines, better quality metals and alloys, and increased compression ratios. The introduction of pneumatic tires in 1934 and high compression engines that burned leaded gasoline were of major importance in the second wave of increased fuel efficiency that started in the mid-thirties.

The change in mechanical efficiency with time is shown in Figure 2-8.

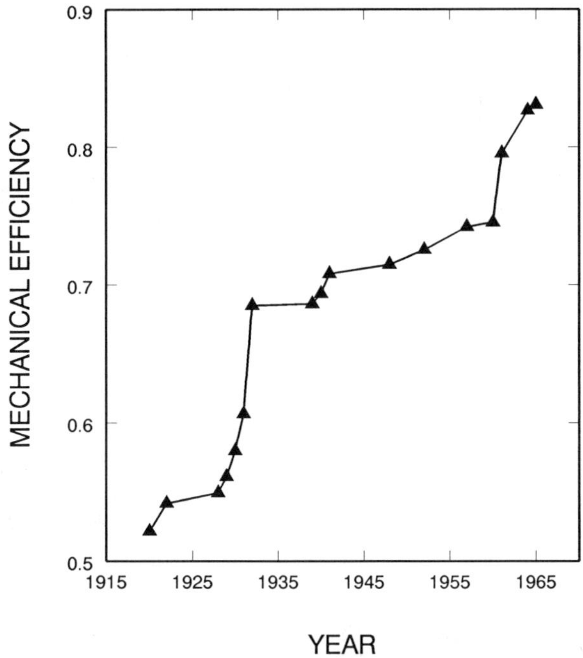

Figure 2-8 Mechanical efficiency of tractors.

Here the data fall into three regimes. The first is an exponentially increasing rise up to 1930, which was driven by developments in the power takeoff first introduced in 1918. The second is a relatively slow rise up to 1960, resulting from the use of pneumatic tires and the three-point hitch introduced in the thirties. A final spurt in mechanical efficiency during the sixties was made possible by the introduction of four-wheel drive.

Figure 2-9 shows the change in the total efficiency with time. Since this is the product of fuel efficiency and mechanical efficiency, its form is similar to that of the others.

To what extent can the improvements be regarded as learning curves? To answer this question, I have plotted the successive maxima in fuel efficiency and total efficiency as a function of worldwide production in Figures 2-10 and 2-11. The data for fuel efficiency were fitted to a *ln-ln* learning curve, using all but the last three points. The curve is drawn to give continual learning until the mid 1950s, after which learning is negligible. The learning curve for total efficiency, however, continues right up to the mid-sixties because of the improvements in mechanical efficiency. A closer examination of the data points shows that they fall into the same major regimes as when efficiencies are plotted against time. The conclusion to be drawn from this is that learning curves can be used to describe the general, long-term advances in tractor efficiencies, but the description of advances over shorter

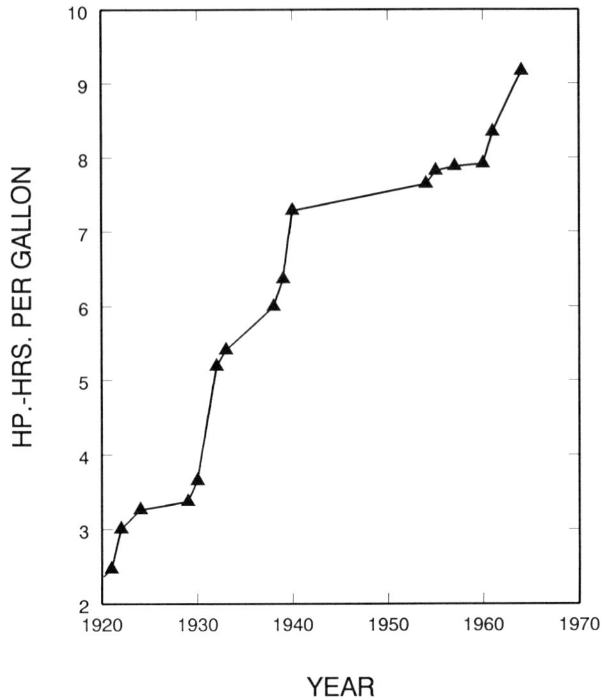

Figure 2-9 Total efficiency of tractors.

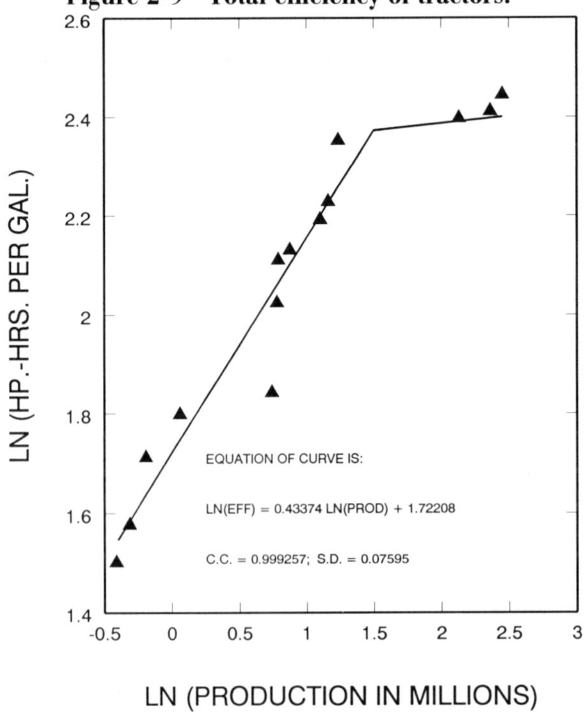

Figure 2-10 Learning curve for fuel efficiency of tractors.

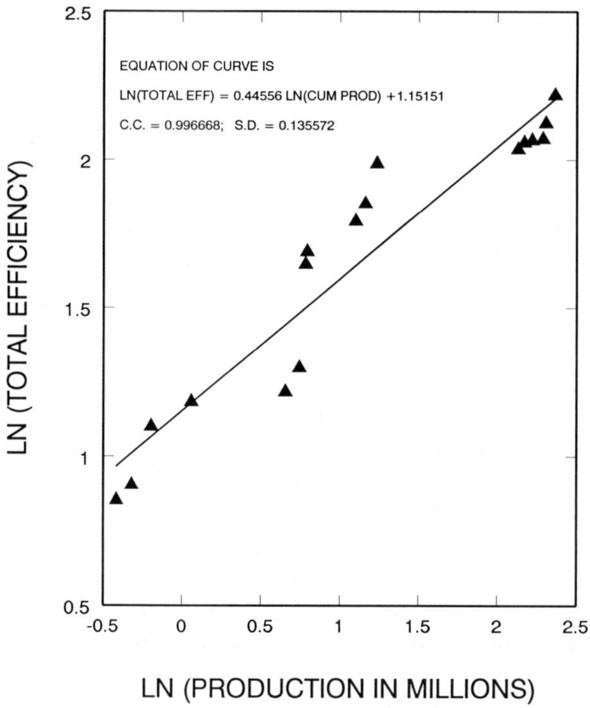

Figure 2-11 Learning curve for total efficiency of tractors.

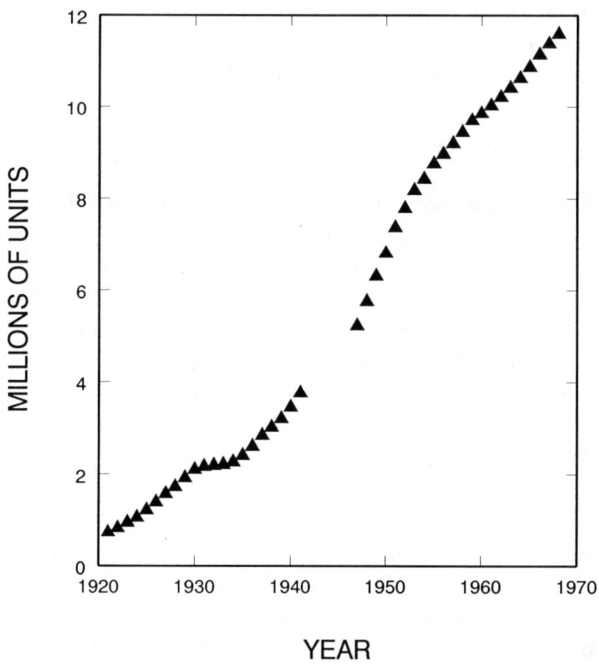

Figure 2-12 Cumulative number of tractors produced worldwide.

time periods requires a different kind of analysis. In actuality, technical advance depends on the degree of effort put into it, and a more appropriate independent variable would be the number of man-hours spent on improving the technology. Such data are seldom available, and cumulative production figures, as well as time, are proxies for the cumulative effort expended.

One more point to note is that cumulative production and time are obviously related, and if cumulative production is a smooth function of time, then technical parameters must be smooth functions of cumulative production to the same extent that they are smooth functions of time. This is certainly the case with tractor technology, as shown in Figure 2-12, which displays cumulated production as a very smooth function of time. By its nature, a cumulated variable is smooth, since it absorbs the annual fluctuations into an ever-increasing number so the relation between time plots and cumulated production plots is always close.

An important limitation of the type of quantitative analysis given here is that the numbers do not display some of the most important characteristics of technical change. Curves for the increase of efficiencies with time do not show the fact that the tractor continually improved in function in qualitative ways that were of critical importance for agricultural production. The changes that enabled the tractor to evolve from a pulling machine to a multiple purpose, mobile power source that operated a wide range of implements are not reflected in the efficiency numbers. Nevertheless, the efficiencies are important in their own right, since they embody the combined results of many technical improvements and contribute to the cost savings and increased productivity that tractors make possible.

CASE STUDY: UNDERGROUND COAL MINING

The examples of productivity analysis given above all followed the usual power law for simple learning. When applicable, this is a powerful tool for management and cost analysis purposes. However, not all productivity data can be represented by the simple learning curve. In general, the longer the time span and the larger the degree of technological change over time, the more likely it is that productivity data will deviate from the simple learning equation. An interesting example of such a case is that of underground coal mining in the United States, a major technology that was critical for industrialization and which illustrates some of the important issues of productivity analysis.

Technological progress has had an enormous effect on the productivity of underground coal mining. By the end of the nineteenth century, coal mining techniques had already come a long way from the early days of the Industrial Revolution, when all digging and breaking was done by men with simple hand tools, small boys opened and closed doors to allow access to traffic while controlling ventilation, and coal was transported to the surface in baskets on the backs of women. Even so, the technology was still primitive before the turn of the century. The evolution of coal mining technology has been well summarized by Gold, Pierce, Rosseger, and Perlman (1984), whose work provided the foundation for the analysis I am presenting here.

Underground mining was done in rooms carved out of a large coal seam, in which pillars of coal were left in place to support the roof. The miner removed coal from an ever-receding coal face by first cutting a groove or notch two or three feet into the bottom of the seam. The miner did this by using a pick while lying on the floor. It took one to three hours to cut ten feet of notch, depending on the conditions. Cutting was the most tedious of the mining operations and constituted the bottleneck to increased productivity.

After the initial bottom notching, holes were bored into the coal face, filled with explosive, and coal was blasted from the face into large lumps. The lumps were loaded into carts which were pulled away and out of the mine by animals. The sequence of mining operations was, therefore, cutting, drilling, blasting, loading, and hauling.

Mechanized cutting started in 1876 with a revolving bar cutter, which had teeth or bits, and a compressed air puncher patented by J. W. Harrison in 1877. In 1894, an electrically powered machine was introduced in which the cutting elements were steel teeth embedded in an endless chain. Modern cutters are descendants of this machine. In spite of these early technological developments, 75 percent of all bituminous coal mined underground was hand cut. However, the diffusion of machine cutting continued regularly, as cutting machines were improved so that cutting could be done along the entire width of the face in a continuous operation, cuts could be made at a variety of levels and angles, and cutting machines were mounted on tracks. By 1955 about 90 percent of the underground coal being mined was cut by machine.

Compressed air drills were available by 1890 and electric drills were introduced in 1911, but mechanical drilling started late, and in 1940 more than half the drilling was still done by hand. Power drilling peaked in 1955 when it accounted for drilling of 80 percent of the production of underground bituminous coal.

Once cutting was mechanized, loading was the chief bottleneck in the mining operations. Mechanized loading therefore started later than power cutting and diffused more rapidly. Mechanical loading started in the mid 1920s, and by 1960 accounted for 90 percent of all bituminous coal loaded underground. The machines that came to dominate mechanical loading consisted of rotating arms or other devices to push coal onto a conveyor which led to a car.

With the mechanization of cutting, drilling, and loading, the next bottleneck was haulage. In 1900, hauling was done primarily by animals pulling cars on wooden rails. In passageways that were too small for animals, the carts were pushed by men. The savings in time and money that resulted from the mechanization of the other operations, along with the increasing demand for coal and the opportunity to decrease costs, were the driving forces for the adoption of mechanical hauling.

The problem of hauling has two parts: transport of coal from the mine face to the mainline and transport of coal along the mainline and to the surface. The latter process was easier to mechanize because the mainline was more spacious and straighter than the often cramped quarters in the mining rooms. A variety of methods for mainline haulage were adopted, ranging from steam or gasoline engines on the surface which pulled cars along the mainline by rope or cable, to trolley locomotives electrically powered by overhead cables. The trolley locomotive dominates modern mainline haulage, although other means, including battery locomotives and rope haulage units, are also used. The trolley locomotives had actually been successfully tried out by 1890, and by the mid 1920s were in widespread use. Electrical motive power is used, since the exhaust from internal combustion engines would result in serious ventilation problems.

Battery-operated shuttle cars for haulage from the mine face became practical after 1911 when improved batteries were available. Cable reel cars were also used, which received their power from trailing cables rather than from overhead wires. These cars could enter passages that trolleys could not reach, but did not solve the entire problem because the very existence of the rails placed limitations on the transport. This limitation was overcome after 1935 with the introduction of rubber-tired vehicles powered either by batteries or by trailing cables.

The technology described so far was an evolutionary development that gave a continuous improvement in productivity. But around 1950 a new technology was introduced that had a much greater effect. This was the introduction of the continuous mining machine, which integrated the separate functions of cutting, drilling, blasting, and loading. The machine broke or cut coal from the face and dumped it directly into cars or onto a conveyor. The continuous miner was made possible by advances in metallurgy and the use of tungsten-carbide cutting elements. These materials advances also led to improvements in the conventional mining equipment, so that the result was a dramatic increase in productivity even though not all mines adopted the continuous miner.

Figure 2-13 shows the productivity of labor in American underground mining of bituminous coal as a function of time. There are four regions of this curve, each of which merits comment. From 1900 to 1949 the data are well described by an exponential curve with a regular annual productivity growth of 1.72 percent. From 1950 to 1968 another exponential, with an annual growth rate of 5.57 percent, gave a good representation of the data. The remarkable increase in productivity growth from 1.72 to 5.57 percent can be ascribed to the introduction and diffusion of the new materials and continuous mining technology.

Starting in 1969, the productivity growth slowed down, and in a few years productivity was actually decreasing rapidly. This productivity decrease has

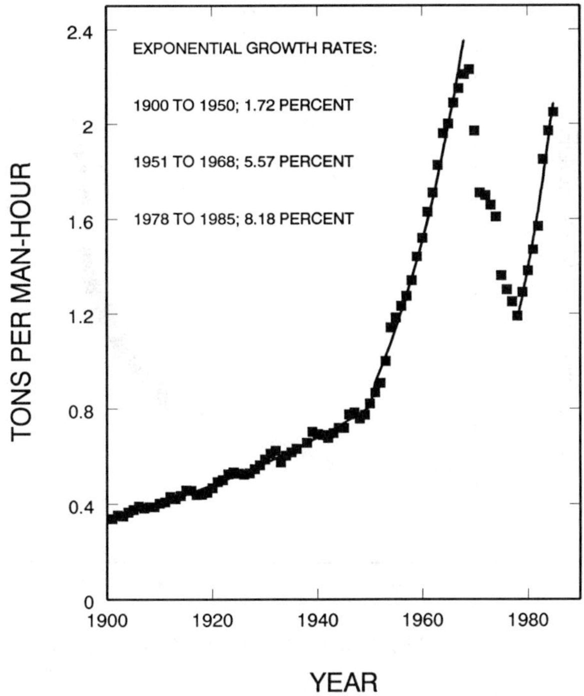

Figure 2-13 Labor productivity in underground coal mining.

been ascribed to two related factors. The first is the passage of the Federal Coal Mine Health and Safety Act of 1969, and the second is the rise of environmental concerns. Although both improvements in safety and decreases in environmental damage were overdue, they did have the effect of decreasing productivity. By 1978 the changes resulting from these two factors had been absorbed, and productivity growth was resumed at a rate of 8.18 percent per year.

The productivity data for underground coal mining are readily rationalized as a time series, but can they be described by a standard learning curve? The answer is no, as shown in Figure 2-14, which is a *ln-ln* plot of productivity versus cumulative amount of coal mined. If the simple learning curve applied, this plot should be linear. This is clearly not the case, either for the curve as a whole or for separate parts of it relating to different technological regimes.

There are several important points of this analysis which should be noted. The first, and the most obvious, is that not all productivity data fit the learning curve, and an empirical study is needed to determine the functional form that is most appropriate for particular cases, especially when large-scale data are considered. The learning rate of individuals and of groups using a constant technology is more likely to follow the power law, which is

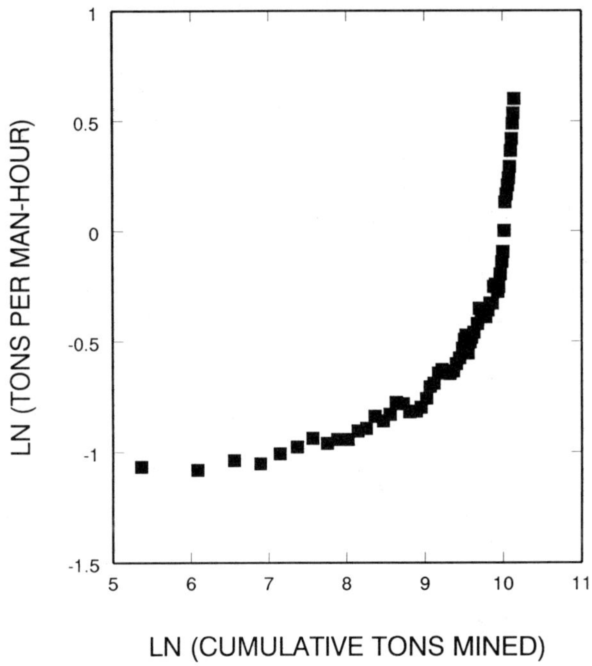

Figure 2-14 Test of learning curve equation for the productivity of underground coal mining.

often too simple to describe the overall progress of an industry. The second point is the profound effect of technology on productivity. Not only did technological advance improve productivity for three-quarters of a century, but when a major new technology was introduced, the growth rate of productivity was dramatically increased. The third point is that for most of the time the time series for productivity was regular, indicating a systemic cause which we maintain is the continual evolution of technology. However, a sharp change in political or social conditions constitutes a nonsystemic factor that usually introduces a dramatic interruption of systemic trends. After this change is absorbed, it becomes part of the system, and systemic trends once again take over. This was the case with the safety and environmental regulations of 1969.

It is important to note that the sequence of adoption and diffusion of technological advances was determined by the factors limiting productivity at the time. Thus, the mechanization of cutting and drilling was adopted first, because they were the limiting operations of the hand mining of coal. When these were mechanized, loading was the limiting factor. After the mechanization of loading, further savings were made by using mechanical haulage. Finally, the continuous mining machine integrated the mining into a connected sequence of mechanical operations.

A final point is that desirable social ends cost money, at least in the short run. The decade-long decrease in mining productivity meant that it was more expensive to mine coal, and it is actually possible to compute the increased cost over this decade. Later, however, the productivity started to grow at a greater rate than before, and at some point the productivity will be just what it would have been without the Federal Coal Mine Health and Safety Act and without environmental improvements. Ultimately, the costs of these social desiderata will be recovered, but this is far enough in the future so that the beneficiaries will not be those that currently bear the cost. This is often the case with technological change and the reactions to it. The long term is usually profitable and beneficial for society as a whole, while in the short term there are both winners and losers.

3

Innovation, Research, and Development

CLUSTERS AND WAVES

The frequency of innovations is a matter of great importance to a technological society. The rate at which innovations are adopted has a great effect on the economic position, defense posture, and general well-being of a modern nation. If that rate is too slow, the nation falls behind in international competition, its products and processes become obsolete, growth slows down, and jobs are lost. If it is too fast, the population cannot adjust to the rapid changes, industry finds it hard to recover investments from previous innovations, and confusion makes long-range planning impossible. A great deal of effort has therefore been expended on trying to understand the frequency of innovation and its relation to economic trends.

The first question to be considered is whether or not innovations do indeed appear in clusters, as stated in Chapter 1. This will be done by analyzing data on the dates of innovations. The answer will be found to be unambiguous: clusters of innovation exist.

But the existence of clusters has led to a speculation that is more dramatic and, if true, much more profound than the mere bunching of innovations in time. As illustrated by the data of Mensch in Chapter 1, it has been proposed that not only do innovations cluster, but the clusters appear at regular intervals, so that there is a repeating cycle in which a high innovation rate occurs about every half century. This cycle is presumed to have enormous economic significance in that it is bound up with long-wave fluctuations in economic growth. The innovation cycles were postulated to provide an explanatory framework for economic waves.

It is necessary to clarify and keep in mind the precise definitions of clusters, cycles, and waves. A *cluster* of innovations merely means that in some period of time the number of innovations is considerably larger than at other times. The time period used in defining clusters is short, never being more than a decade, but usually more than one or two years. *Cycles* are defined to be periodic in time, the phenomenon being measured having peaks and valleys at regular intervals. A *wave* is a continuous rise and fall that may or may not have any regular periodicity associated with it. The term "long wave" in economics is usually taken to mean a long-range rise and fall in economic activity in a regular periodic pattern, but this is a careless use of language. A better term would be long-wave cycle. It is important to remember that long waves without periodicity are possible.

Modern long-wave cycle theories of capitalistic economies had their origin in the work of a Soviet economist, Nikolai Kondratieff, who published a series of papers from 1922 to 1928. While others had proposed the existence of long-wave economic phenomena earlier in the century, Kondratieff's work was the most extensive and most widely disseminated. The most important of his papers appeared in 1925 and have recently been translated into English (Daniels 1984). On the basis of an analysis of various economic indicators in France, England, Germany, and the United States, Kondratieff concluded that the course of economic change in time consists of long-wave cycles with a periodicity of about half a century. He noted that the beginning of each new wave was accompanied by the introduction of new technical innovations, but explicitly denied that these were the causal factors driving the economic wave. For Kondratieff, as for some more modern long-wave theorists, capital investment plays the central role in driving the long waves. The detailed connection between technology and economics was first made by Schumpeter (1939), who developed a model in which major innovations started long-term trends, new industries grew, secondary improvement innovations followed in swarms, and the number of competitors increased. This is accompanied by a boom in capital investment, which is followed by a depression in reaction to the boom.

Schumpeter's theory has been recently extended by Mensch, whose data were noted in Chapter 1. Mensch contended that major innovations appeared in regularly spaced clusters, that they drove economic growth until their markets and technical potentialities were exhausted, and that growth then slowed down, with a resulting depression until a new set of innovations was introduced.

A theory that ascribes a long-term regularity to the appearance of innovations which drive a corresponding cycle of economic growth is certainly attractive to anyone that is searching for quantitative regularities in technological change. But whether or not cyclic waves exist must be decided by an appeal to data.

FREQUENCY OF INNOVATION

A count of technological innovations as a function of time was used by Mensch to support his theory of the connection between technology and the economy. These are shown in graphical form in Figure 1-1 of Chapter 1. Mensch's own graph of these data extends back to 1740 and used the number of innovations in each decade instead of a running average. The earlier data were omitted here because there were so few data points before 1800. The curve seems to lead to definite conclusions. Not only do innovations cluster, they also occur in regular cycles and, in keeping with the belief that the pace of innovation is accelerating, successive peaks are ever higher. Furthermore, the peaks in innovation frequency are right in the middle of major world depressions. This supports Mensch's contention that basic innovations start a long-range boom, that economic growth is driven by the adoption and exploitation of new technologies, and that when the potential of these technologies is exhausted and their markets saturated, growth stops and depression sets in. This contention is in agreement with the Schumpeterian thesis, even though Schumpeter's model is more detailed and extensive than this and even though Schumpeter's and Mensch's models differ in some respects. I will return to Schumpeter's thesis after a consideration of empirical data on innovations.

Before accepting Mensch's data as proof of the Schumpeter–Mensch thesis, it should be examined in more detail. Actually, a detailed criticism of Mensch's use of data on innovations has been given by Freeman, Clark, and Soete (1982), with the conclusion that the data are not appropriate for testing Mensch's conclusions. Three important points are that the innovations are not a statistically representative set, their dates of introduction are ambiguous, and they are of widely differing types. On this last point, Mensch's list includes both far-reaching major innovations, such as the diesel locomotive, radio, television, and the jet engine, as well as improvements on major innovations, such as the hydraulic clutch and power steering. He even includes the ballpoint pen and the zipper, which are just not in the same class as the great major innovations. Considering that Mensch is trying to connect large-scale economic trends to important technological changes, this is a serious criticism. But despite valid criticisms of his use of data, Mensch has laid bare profound relationships between the economy and technology. He has publicized, modified and, to a degree, confirmed the Schumpeterian view.

If Mensch's data are insufficient, can we not use some other data set? Another list has been compiled by van Duijn (1983), who was also dissatisfied with Mensch's data and therefore drew up a list of 160 major innovations for the nineteenth and twentieth centuries. While not quite as heterogeneous as that of Mensch, his list does include the fountain pen, the safety match, and the Bunsen burner along with the telephone, the motorcar, and the steam turbine. Nevertheless it is a list of important innovations specifically worked up to determine the degree of bunching of innovations. Van Duijn's data are shown in Figure 3-1, which shows the nine-year running sum of innovations as a function of time. The clustering is obvious. Some decades are periods of much more innovative activity than others.

I myself have constructed a list of innovations in connection with the preparation of this book. These are given in Appendix 2. In compiling this

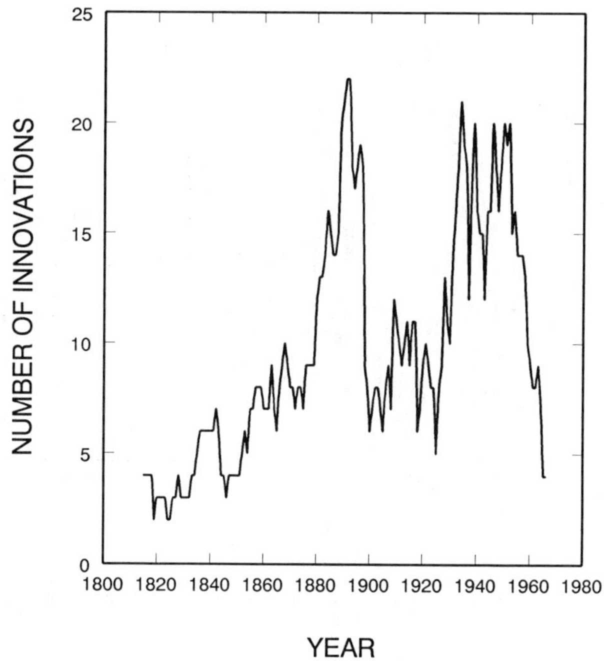

Figure 3-1 Nine-year running sum of major innovations from data of van Duijn.

list I had no program in mind other than identifying what I thought were the important innovations since the Industrial Revolution. The results are shown in Figure 3-2, in which the nine-year running sum of innovations is plotted as a function of time.

While they differ in detail, Figures 3-1 and 3-2 both show pronounced clustering and, where they overlap, both show similar periods of high innovation. Both curves show that the periods from 1930 to 1950 and from 1870 to 1890 exhibited higher than average innovation rates. Figure 3-1 shows a peak centered on 1840, while the corresponding peak in Figure 3-2 is centered on 1830. In addition, Figure 3-2 shows peaks centered on 1750, 1770, and 1790.

Both Figures 3-1 and 3-2 represent heterogeneous data sets in that all types of innovations from all kinds of industries and products are included. It is of interest to examine more restricted sets such as those pertaining to a single industry or a single technology. Figure 3-3 shows the five-year running sum of innovations in the farm equipment industry listed by Sahal (1981, pp. 332–333) from a Michigan Agricultural Experiment Station compilation by W. A. Cromarty in 1959. The plot shows pronounced clustering, even though it is superimposed on a rising trend.

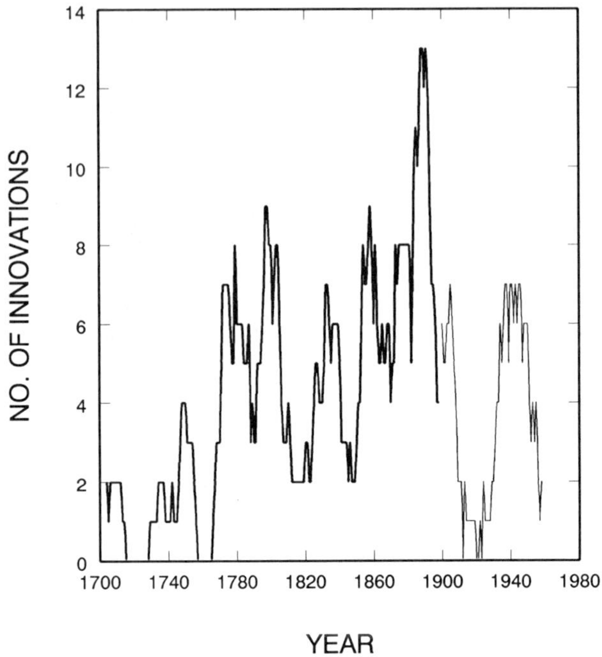

Figure 3-2 Nine-year running sum of innovations from data of Girifalco.

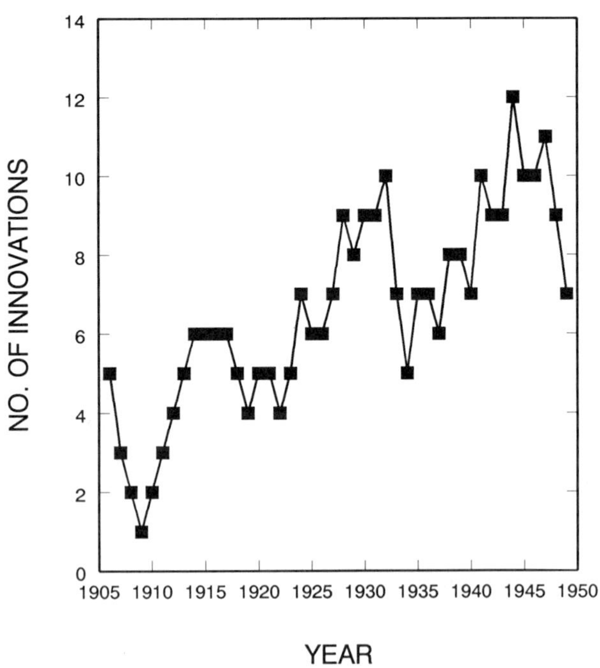

Figure 3-3 Innovations in the farm equipment industry: five-year running sum.

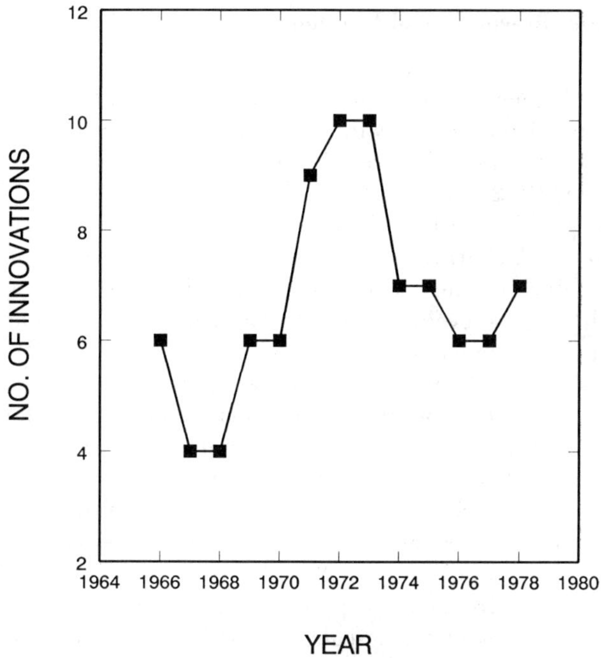

Figure 3-4 Major semiconductor product innovations: five-year running sum.

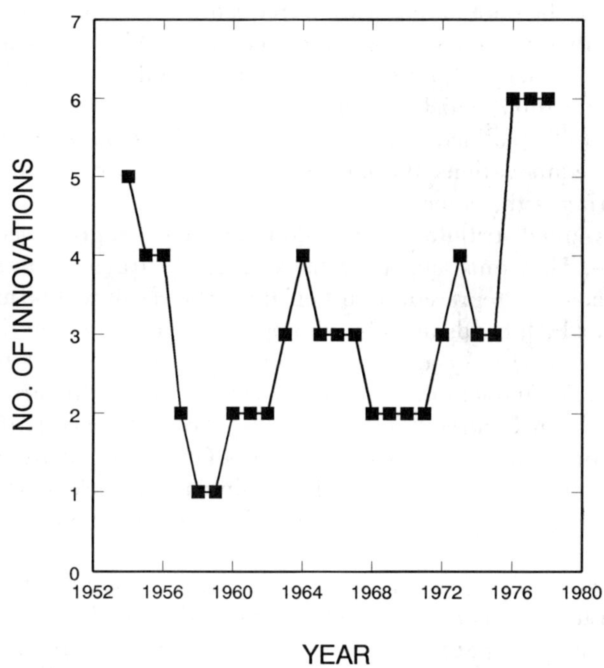

Figure 3-5 Major semiconductor process innovations: five-year running sum.

Data on the frequency of innovations in the semiconductor industries are shown in Figures 3-4 and 3-5, which display the five-year running sum of product and process innovations respectively using data taken from Freeman, Clark, and Soete (1982, pp. 116, 117). Finally, Figure 3-6 shows the number of innovations per decade for television technology (Walter Prusak, 1984, term paper for L. A. Girifalco). This is not strictly a plot of innovations, since commercialization did not start until the mid-1930s. It is a plot of the important technological events in the development of television and includes both inventions and innovations.

The results shown in Figures 3-1 to 3-6 are general. There is a clustering of innovations no matter how the innovations are classified. Major innovations, improvement innovations, product, process, industry, or specific technology innovations all show clustering. Innovations do not flow in a smooth steady stream, but in a jerky, fluctuating manner. Furthermore, the clustering occurs at all time scales, whether measured in centuries as in Figures 3-1, and 3-2, in decades as in Figures 3-3 and 3-6, or in years as in Figures 3-4 and 3-5. Innovation frequencies appear to have a random, fractal structure of irregular fluctuations at every scale of classification or time scale.

THE ORIGIN OF CLUSTERING

Examination of the data on innovation frequency immediately provokes the thought that they represent a random process in which innovations occur purely by chance. We will put aside the difficulty in determining the specific date of an innovation, which is great enough to cause substantial disagreement among well-qualified scholars. This difficulty is inherent in the very attempt to date innovations. But no matter what dates are chosen, the nature of the clustering is the same.

Random events distributed in time do not, of course, give rise to an even flow of events. They must cluster. This is easily illustrated by tossing a coin and letting each toss represent a unit of time. The result will be an irregular time series in which heads or tails will appear in clusters of varying sizes. If heads is taken to be the occurrence of an innovation and tails taken to be the absence of an innovation, then series will be obtained that are completely similar to those in Figures 3-1 to 3-6. The only difference is the possible rising trend that can be discerned in some of the plots of the number of innovations as a function of time. This is similar to stock market analysis in which a random series of price fluctuations is superimposed on a long-range rising trend.

Does this mean that innovations are uncaused, spontaneous events? Of course not. A historical analysis can be performed for each innovation which traces its origins, the path it takes, and the influences on its development. Far from an innovation being uncaused, it normally has a myriad of "causes." That is, a great many factors influence the entire path of innovation from inventions to precursors to commercialization, as was emphasized in Chapter

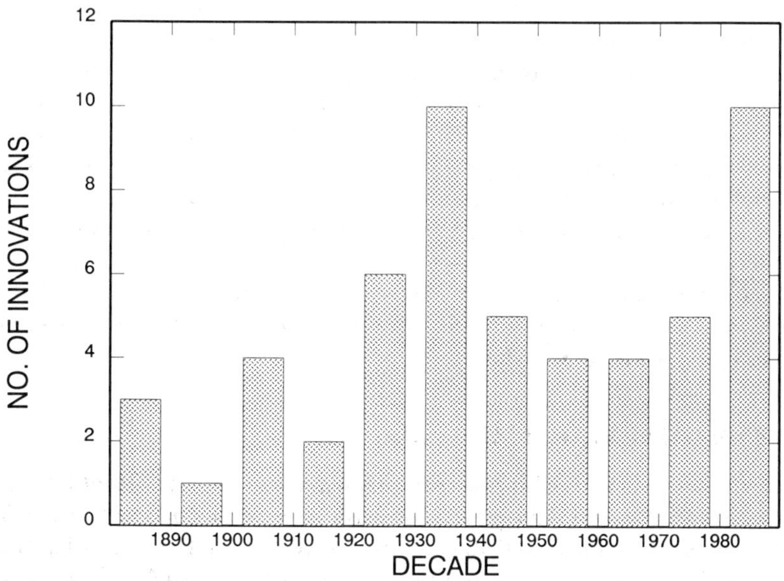

Figure 3-6 Number of television innovations per decade.

1. Furthermore, various innovations are subject to a different mix of influences with differing intensities. Some innovators are independent inventors while others belong to large organizations; some depend on other innovations; some are delayed by budgeting problems; some may require long testing periods because of government regulations. Even a brief reading of the history of technology exposes the great variety of factors that determine not only the date of an innovation but whether or not it will appear at all. This is analogous to the flipping of a coin, which is certainly not an acausal event. Whether or not the coin will come up heads or tails is determined by the precise action of the fingers in making the toss; the angle at which the coin leaves the hand, the force of the flip, the attitude of the coin resting on the fingers, the points on the coin where the force is exerted, the air temperature and pressure, the presence or absence of wind, and so on. Each toss is determined according to the laws of physics, but each toss is unpredictable because it depends on a great variety of factors that can differ from toss to toss. If every aspect of the conditions preceding the completion of a toss could be accurately known, then the outcome would be known. But this would not change the irregular nature of the results of a series of tosses since the variation in these conditions from toss to toss would still exist. In the same way, a detailed historical case study of an innovation would reveal the social and economic forces that led to it, but a series of innovations would still yield an irregular random structure for a plot of innovation frequency as a function of time.

The origin of clustering is therefore simple. It arises as a statistical consequence of the multitude of influences operating on the process of innovation. It must be remembered, however, that the natural clustering of a sequence of random events is reinforced by the fact that innovations tend to initiate other innovations. For example, the development of lasers stimulated the development of optical fibers for communications; the deployment of satellites spurred worldwide microwave transmission; and advances in photolithography made possible the microchips which led to the microcomputer.

A mathematical statistical analysis of the innovation process based on these ideas has been given by Sahal (1981, Chapter 3) and was shown to be in agreement with the data.

INVENTIONS AND TIME LAGS

Innovations exist for which no previous invention can be identified. Minor, and even substantial, changes in manufacturing processes, for example, can take place without any earlier event that can be called an invention. But the great majority of important innovations depend on an earlier invention. This can be true even of process innovations in which the new process itself is an invention if it represents a change in technology that must be tested, modified, and can be patented. Innovations are normally preceded by inventions. The frequency of inventions and their relation to innovations are therefore of great interest.

A comprehensive list of dates of invention can be compiled by taking the data of Mensch as corrected by Freeman, Clark, and Soete (1982, Chapter 3) and adding lists that they put together from the work of Jewkes, Sawers, and Stillerman (1969). From this was constructed a nine-year running sum of inventions which are plotted as a function of time in Figure 3-7. This plot shows the same type of clustering that is exhibited by innovations. The reasons are the same: the appearance of an invention is influenced by a very large number of factors that vary from one invention to the next.

In the list used to construct Figure 3-7, the dates of innovation as well as those of invention are available. It is therefore possible to compute the time lag between invention and innovation and to examine its change with time. This was done for each of the innovations in the list, and the time lag between invention and innovation was plotted in Figure 3-8. The time lags show a great variability as would be expected from two irregular series. But Figure 3-8 suggests that the maximum and minimum time lag, as well as the average time lag, has been decreasing for the past two centuries. To confirm this, the average time lag was computed for each decade from 1790 to 1970 and plotted in Figure 3-9. Even though some decades had many more points in them than others, and therefore the decade averages suffer from the uncertainties of the statistics of small numbers, the downward trend is obvious. At the beginning of the nineteenth century, the average time between an invention and its commercialization was over eighty years. At the midpoint

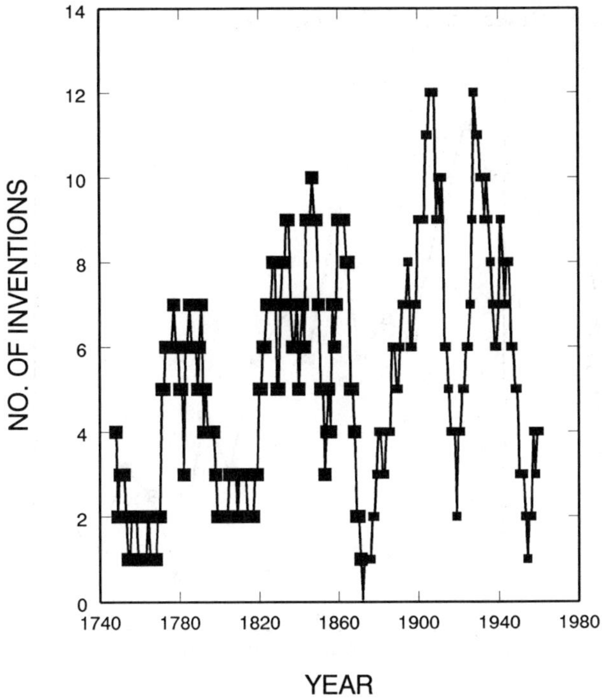

Figure 3-7 Number of major inventions: nine-year running sum.

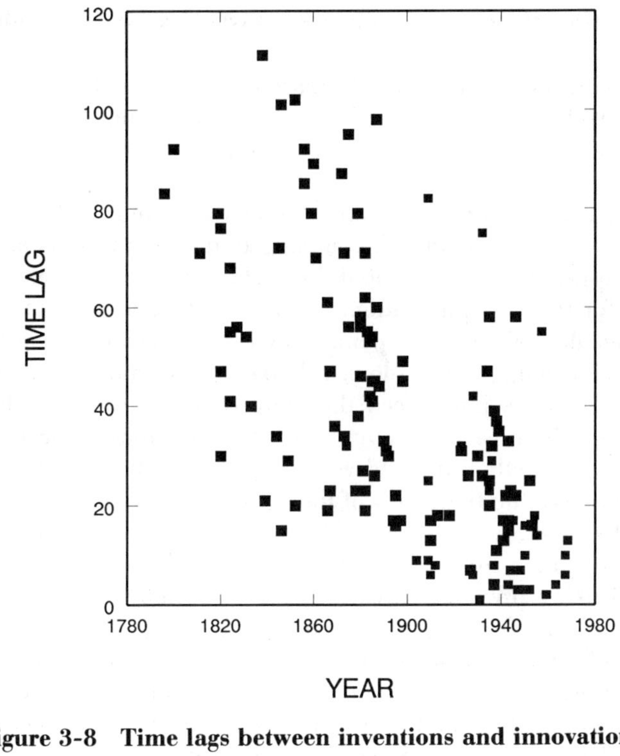

Figure 3-8 Time lags between inventions and innovations.

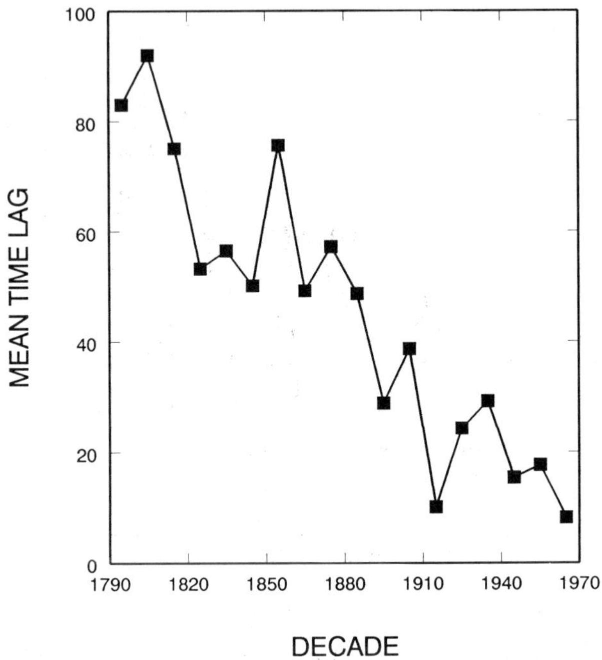

Figure 3-9 Mean time lags between inventions and innovations.

of the twentieth century this had decreased by a factor of four to five years. The pace of technological change is certainly increasing. Not only are there more innovations in recent years, but the time it takes for an invention to make it to commercialization is much less.

An innovation is a new combination of elements that includes materials, earlier devices, processes, and ideas arranged in such a fashion as to provide a new or improved technical capability. These elements might be created specifically for the new innovation, but many of them were already in existence. Even new elements depend heavily on previously available techniques and knowledge. Technological knowledge is cumulative, so that the more innovations are introduced the more elements exist for the next innovation. Also, the means of communicating scientific and technical information has continually improved with time. It is to be expected, therefore, that as time passes the frequency of innovation would show a general, rising trend. Furthermore, as experience of bringing technology into practice accumulates, the process of reaching into the general pool of technical knowledge and of using previous inventions for new innovations becomes more efficient. Thus, just from the general idea of learning, the time lag between invention and innovation would be expected to get smaller as time passes. But beyond this, there is a critical difference between the innovation process of the twentieth century and that at the early stages of the Industrial Revolution. In the eighteenth century, the impact of contemporary science on technical innovation was indirect. Innovation was carried out almost independently of science. This started to change in the later part of the nineteenth century, and the twentieth century saw the systematic exploitation of science for technical ends. After the successes of the applications of science in World War I and World War II, organized research and development became widely adopted by government and industry as a tool to accelerate the progress of technology. It was therefore inevitable that innovations would follow inventions more rapidly.

INNOVATION AND BUSINESS CYCLES

The data presented so far do not give unqualified support to the idea that technological innovations drive long cycles of economic activity. It is true that innovations cluster, but whether or not they occur in regular cycles depends on the data set chosen and only Mensch's data exhibits an unequivocal cyclic regularity. However, there is no doubt that there is an intimate relationship between technological change and the economy. Traditionally, economic productivity was ascribed to inputs of capital and labor and many economists were surprised when, in the 1950s, studies showed that the traditional inputs could only account for a minor fraction of economic growth. The rest was called the "residual" and then ascribed to technological change. It is astonishing that it took so long for the central role of technological change in economic growth to achieve widespread recognition. In a very real sense, all of economic growth is the result of innovation. Just imagine carrying on life without any tools whatsoever; no lathes, drills, or computers; no railroads, airplanes, or automobiles; no printing presses, radio, or television; no tractors, looms, or sewing machines; not even any hammers, spades, or hoes. The difference between that life and reality is the result of technological change.

Of course, not all innovations that contribute to economic growth are technological. Organizational and social innovations are also important. Industrial production and distribution of goods; the generation and distribution of electricity; transportation and communication systems; banking systems and securities markets; all these require social and institutional structures that were innovations as vital as those of technology. And there is a reciprocal relationship between technology and the social environment. The particular kind of social, institutional, and political conditions has a strong effect on the nature and timing of innovations, while technological advance calls forth changes in social conditions and organizational structures. While all this is true, technological change is at the core of economic growth because it removes physical limitations and produces capabilities that did not previously exist. Without this expansion of technical possibilities, social or organizational innovation would have to take place within static limits and would have no way of inducing continual economic growth.

Technological change is the ultimate creator of wealth, and even the traditional economic measures of capital and labor arise from technology. Capital exists because of a surplus of production over consumption, which is the direct result of technology, without which no surplus is possible. Also, the effectiveness of labor is a direct result of technology, without which it would be a simple and constant muscular input. The contributions of capital and labor to economic growth must be understood as a process in which the surplus fruits of earlier technology are contributing to the present economy.

It must be emphasized that it is a *change* in technology that induces economic growth and creates wealth. If there were no inventions, no innovations, and no technological diffusion, there would be no possibility of economic growth. Even if diffusion continued but innovations were to suddenly stop, the economy would become static. It could continue to grow for a time as it absorbed and made increasing use of the existing stock of technology, but this growth would slow down and stop as technological diffusion approached completion.

The Long-Wave Theory

With this in mind, let us return to the Schumpeter–Mensch theory of the long wave and describe it in a little more detail. The starting point is a number of important innovations that appear at about the same time. The reason that they cluster in a particular time period is that economic conditions are right. The traditional industries are not very profitable, people are not buying the traditional products, and the economy is in depression. Further investment in the traditional industries is obviously a bad idea, and both capital and labor are available for new ventures. The vanishing profits in the traditional industries, along with the general despair of a depression, overcome the resistance to change that is characteristic of entrenched enterprises with the result that investments are made in new technologies. The new technologies stimulate consumer interest, raise employment, provide good returns on capital, and drive a long-range economic boom. But sooner or later, the growth associated with the new technologies must end. After their initial successes, the new technologies induce a "swarm" of improvement innovations which extend their capabilities and make them ever more attractive in the market place. At the same time, imitators follow the daring innovators that introduced the innovations because of the profits to be made. The innovators have already broken the ground, thereby making it easier for others to follow. Both the improvement innovations and the imitators hasten the diffusion of the new technologies. But the performance of a technology has a natural limit set by the laws of nature, and the time comes when further improvements are not cost effective or cannot be made at all. Also, markets are finite, and as products are refined to the point that they become static, markets become saturated. The technology then stops being a contributor to economic growth and addresses only replacement markets. By this time the "new" technologies have become mature technologies in which industry has an overcapacity because all competitors added extra capacity during the growth phase in attempts to capture market shares. Now money is lost instead of made and unemployment rises. This condition persists for a time because of the enormous investment in the mature technologies. Technical know-how, marketing skills, customer base, physical plant, and personal careers are all tied up with the mature technologies. Only after considerable pain in the resulting depression are people ready to try something new. This gives rise to a new set of innovations, and the cycle starts all over again.

The appearance of major innovations does not exhibit a periodic cyclic character even though innovations tend to be more frequent in times of major depressions, but this does not invalidate the long-wave theory. The cycle of economic growth does not require a cycle of innovations; it only requires that some major new technologies come along that lead the economy. For this purpose, the timing and rate of the diffusion of the technologies is much more important than the dates at which they are first introduced. There is no doubt that some major technologies grew in tandem with long waves of economic growth. Textile machinery, iron, and steam were the driving technologies from the depression of the latter part of the eighteenth century to the next depression of the 1820s. Railroads and steel drove the wave of depression to recovery to prosperity to depression from the 1830s to the 1880s. Electric power, chemicals, and the automobile were the technologies for the 1880s to 1930s cycle, and electronics, polymers, and the jet engine were important technologies for the wave that started in the 1930s. For each wave, a number of leading-edge major technologies grew and diffused together and approached maturity together.

A fundamental difficulty is that there are so few long waves and the uncertainties in both the economic and technological sets of data are large. The data alone cannot, therefore, be used to decide unequivocally whether or not long-wave cycles exist and whether or not economic and technological waves are linked. The observed correlations might be statistical artifacts with no fundamental significance. But many of the elements of the technology–economy connection outlined above make eminent sense and are compatible with what we know about technological change. In fact, the value of the Schumpeter–Mensch description does not depend on whether or not the world economy rises and falls at precisely regular periods in synchronism with periodic cycles of innovation. Underneath this macro scale there are less global economic and technological activities to which a picture similar to that described above certainly applies. At the national and regional level, at the level of specific industries and even specific firms, economic success is determined by the life cycles of technologies. It cannot be otherwise in a high technology economy. As briefly described in Chapter 1, technologies are born, grow, mature and die in an organic cycle of slow start, rapid growth, slower growth, and final stasis. Any economic unit that depends on a particular technology or group of technologies must therefore go through the process of growth and stagnation. There are multiple examples of this, from the economy of entire nations that could not easily change from an old to a new technological base, to the fortunes of mature industries and firms that could not rapidly adapt to new technologies. Recent examples include the tire industry that could not switch to radials, consumer electronics in the United States, and steel throughout the world. In the 1970s even the steel companies would not invest in steel.

The wave model that seems to make sense is therefore as follows: at every level there are waves of innovation in the sense of technology life cycles. These technology waves drive economic waves for the economic units that are tied to them. There is a degree of correlation among these waves because both technologies and subeconomies are mutually reinforcing. Technical progress in one area often makes possible technical progress in another, and economic success in one industry can spur progress in another. In a highly interacting techno–economic system it cannot be otherwise. The degree of synchronization varies throughout the different levels of the economy and through time. But there is enough correlation to induce waves of economic growth in individual industries, in nations, in regions, and throughout the world.

Whether or not regular cycles of growth and depression are a permanent feature of the economy is another story. A complete acceptance of the long-wave theory would imply that a depression must occur about every half century. Accordingly, a depression must have been scheduled for the 1970s or 1980s. The downturn of the early and mid-seventies was nowhere as severe as that in the 1930s, and by comparison does not deserve the name of depression. As of 1990 there are no signs of a major depression of this type. Of course, even the most committed long-wave theorist would not insist that the timing of the wave be accurate to fractions of a decade. But it still remains to be seen whether or not a depression will occur that is in keeping with the long-wave theory. Even if we accept the theory for the past, that does not mean that it will apply in the future. World War II marked a real change in both economics and technological progress. Governments were much more active in trying to smooth out economic ups and downs after the war and took more complete control of such factors as interest rates and money supply. At the same time, the process of technological change became very different. The systematic application of R&D with heavy governmental participation became really widespread only after the War. Whether or not these changed conditions have broken the old cycle of boom and bust is still an open question.

Of course, the relation between innovation and the economy is not restricted to long waves; shorter time scales are also important. Mansfield (1968) has investigated the connection between innovation and concurrent economic conditions in the iron and steel, petroleum refining, and coal industries. His fundamental assumption was that current economic conditions were of primary importance in the decision of a firm to innovate. As a measure of the business conditions, Mansfield took the percent of capacity actually being utilized by the industry. He then searched for a statistical correlation between percent capacity being utilized and the frequency of innovation over the time period from 1919 to 1958 by assuming a quadratic relation between innovation frequency and percent utilization. Process and product innovations were treated separately. While the results of such a calculation are necessarily rough, they did show that the maximum rate of process innovations took place when the utilization of capacity was about two-thirds for the steel industry and about three-quarters for the petroleum and coal industries. If the industries were operating at either full or at half capacity, the rate of innovation was only about one-third the maximum rate. Mansfield's rationale for this is straightforward. When the utilization of capacity is low, firms are not doing a lot of business and the risks of innovation are high. When the utilization of capacity is very high, business is very good, facilities are not available for investigating changes in manufacturing, and firms have no reason to innovate since they are selling everything they can make. At some intermediate level, the firm is making enough money to take some risk, would like to use more of its capacity, and is willing to innovate in the hope of increasing market share.

There was no correlation between percent of utilized capacity and the innovation rate for product innovations. However, Mansfield stated that the statistical test was weak for product innovations so that no conclusions could be drawn from this result.

It seems clear that the economy and technological change are closely related at all levels and for all time scales.

THE ORIGINS OF INNOVATIONS

In searching for the origins of technological innovations, we are looking for something more than the fact that innovations are the products of human minds. An understanding of how innovations arise would presume a knowledge of the characteristics of the people who invent and innovate, of the process of the birth of an idea to its concrete implementation and commercialization, and of the effects of social conditions, institutional forms, and economic factors on that process. The objectives here will be more modest; we will try merely to outline the role of individuals versus that of organizations, the differences between the innovation processes in our time relative to that in earlier eras, and the relationship of large corporations to technical innovation. In general, we will replace the spectrum of the importance of innovations by dividing them into three types: major innovations, lesser innovations, and improvement innovations. The major innovations are of the type listed in Appendix 2. They represent no technologies or dramatic extensions of technology giving rise to new products or processes or extensive advances in existing products or processes. With hindsight, they are recognized as having large economic effects and being large advances over existing practice. Lesser innovations represent a genuinely new advance but are not very far-reaching and normally do not require the extensive changes in manufacturing and marketing associated with a new technology. The improvement innovations are the continual advances of performance made in a major innovation.

In the analysis of the origins of innovations, it is the major advances that are normally considered. The classic work in this field is that of Jewkes, Sawers, and Stillerman (1969). They examined a large number of innovations, including the diesel–electric locomotive, the gyrocompass, insulin, the jet engine, polyethylene, radio, radar, and xerography. Many of the conclusions given here are taken from their work.

A reading of the history of technology in the eighteenth and nineteenth centuries reveals that most of the great technical achievements of that time originated with individual inventors and entrepreneurs. That is, invention and innovation arose from the thoughts and efforts of people acting on their own, in control of their own time and of the resources needed for their work, rather than from the purposeful, planned activity of large organizations. Often, these individuals created organizations of their own which evolved into large-scale enterprises but, in general, their inventions and innovations were not the result of corporate objectives and activities in already existing large firms. Textile machinery, steel making, the steam engine, the steamship, the railroad, the internal combustion engine, electric power and light, the telephone, and vulcanized rubber were all the creations of men who worked independently and were not directed by the structure of large organizations. This is in keeping with the idea of a past "heroic" golden age of invention, in which lone inventors struggled and succeeded against all odds to bring technological gifts to an unappreciative world. The heroic age is said to have ended after the great spurt of American invention in the latter part of the nineteenth century. According to this idea, invention and innovation are fundamentally different in the twentieth century because they are now dominated by research and development in the large corporations.

Jewkes et al. have documented the origin and development of a large number of important inventions in the twentieth century and find limited support for the idea that invention has been taken over by large firms. At least half of the innovations they studied had their origins in individuals acting outside of large institutions in the same sense as the mythical heroic inventor. These people included those who worked on the gyrocompass, Bakelite, cellophane, radio, penicillin, streptomycin, catalytic cracking, xerography, air-conditioning, television, and the jet engine. Even when the invention has its origin in an industrial laboratory, it is often the work of individualistic people working on a low budget with little corporate control over their activities. Carothers working on nylon at du Pont is a case in point. Although there is no doubt that the lone inventor has decreased in importance in this century, it is not true that the process of invention and innovation has been completely taken over by the large corporation. Hamberg (1966) extended Jewkes' work to include twenty-seven more innovations and found only 26 percent of major innovations in the twentieth century originated in the laboratories of large firms. The rest came from individual workers, small companies, and people in university or government laboratories. In a study of the aluminum industry, Peck (1962) found that only one of seven major innovations came from a primary aluminum producer and three came from independent inventors. Enos (1962) found that all seven major inventions that are the basis for petroleum refining were the work of independent inventors.

The twentieth century certainly saw important changes in the process of innovation. The widespread growth of the industrial research laboratory started at the turn of the century. But perhaps a more significant development of this century is the increased involvement of government in the innovation process. It was World War II that marked the great change in this regard, not the turn of the century. The success of the atomic bomb project was only the most dramatic of the wartime technical achievements. Others included the electronic computer, radar, numerically controlled machines, and the development of sulfa drugs. The organization of science and technology for the war effort was so successful that the federal government has been heavily involved in the conduct and support of research and development ever since.

Another way that modern innovation is thought to differ from that of the past is in the part played by science. In modern times, invention and innovation are said to rest on solid scientific foundations and to flow directly from scientific research, whereas in the past invention was an empirical matter owing little or nothing to science. The truth is rather more complex than this. Many of the great inventors of the past were well versed in contemporary science and, when not scientists themselves, knew the important scientists of their time. James Watt (1776) was a university instrument maker who knew Joseph Priestley and the other distinguished British scientists. His knowledge of science, particularly the properties of gases and of the atmosphere, certainly helped him in his work on the steam engine. Charles Parsons (1884), developer of the turbine, was well educated in science, having studied mathematics at Cambridge. Parsons stated that he was led to the steam turbine by hydrodynamic calculations of the flow of fluids under small pressure differences. Edmund Cartwright (1787), inventor of the power loom, was an Oxford Fellow; Eli Whitney (1793), inventor of the cotton gin, had a Yale education and was in close contact with the intellectual elite of New England; Wheatstone (1837), who made great contributions to the telegraph and the dynamo, was a Professor at King's College; even Edison, who is often taken as a prototype of the lone inventor working independently of science, had many scientific contacts and organized a research laboratory which employed scientists to work on his inventions.

While the impact of science was certainly much more indirect, the scientific content of innovation in the eighteenth and nineteenth centuries does not seem to be much less than that of modern times. It was true then and it is true now that great technological advance depends on the existence of a pool of scientific knowledge upon which inventors can draw. On the other hand, it is no more true now than it was then that innovation starts with basic research which leads to applied research which in turn gives rise to inventions. This straight line route is too simple and practically never realized. Basic research and the replenishment of the stock of scientific knowledge is essential, but its connection to innovation is complex, and there is seldom an obvious route between specific scientific discovery and an innovation. Having said all this, it is nevertheless true that the relationship of science to innovation is different now then it was before the twentieth century. Not only is it tighter and more pervasive, but a great deal of innovational activity is directly based on scientific results to a much greater extent than ever before. The compact disc, based on lasers; all microelectronics, based on solid state semiconductors; fiber optics, based on materials science and quantum optics; these are just some examples of modern innovations that arose directly from scientific results.

Although the rise of large-scale research and development did not displace the individual inventor for major innovations, it certainly has almost completely taken over the improvement innovations that follow the introduction of major products or processes. Organized industrial research is also very successful at introducing lesser innovations that, while not merely improvement innovations, are not in the same class as the major innovations discussed above. They differ from the major innovations in that they generally take a shorter time to work out and their economic benefits are apparent and readily realized. Examples of these lesser innovations would be a new stain remover, ultrasonic testing of metals, high speed bottle fillers, continuous annealing, electrostatic loudspeakers, microwave ovens, disc brakes, flexible contact lenses, and polyvinyl chloride siding. Industry is very good at such innovation. A study by Gellman Research Associates for the National Science Foundation ("Indicators of International Trends in Technological Innovation"; NSF-C889; 1976) found that the independent inventor was not a major factor in innovations introduced between 1953 and 1973. West Germany exhibited the largest fraction of inventions that were ascribed to individual inventors (34 percent), while Japan and the United Kingdom exhibited the least (7 percent and 2 percent respectively). In France and the United States only about 16 percent of the innovations had their origins in individual inventors. The bulk of the innovations studied by Gellman Research Associates were of the type that I have labeled lesser innovations, and the results strongly support the idea that most of them originated in industry.

The conclusions from historical and case study analyses can be summarized as follows: the inventions that lead to innovation originate from a variety of places. For major innovations, the individual inventor is still an important source, although this is somewhat less so now than in previous centuries. The bulk of lesser and improvement innovations originate in industry, but for major innovations individuals, universities, and government laboratories are more important as sources of invention than large companies. In general, the pool of scientific knowledge is important for innovation, even though there is not always a direct line of basic research to applied research to invention. Inventors and innovators usually have a good scientific or engineering background and have access to current scientific information. This seems to have been true in earlier times as well as in the present, although at present the involvement of science in innovation is more important than ever.

THE EFFECT OF CORPORATE SIZE

The growth of large, multinational corporations and their obvious role in the marketing and use of new technology in the twentieth century has led to claims that innovation depends on the existence of large firms. The argument for this point of view is based on the fact that technical innovation is often expensive and that only large companies have the resources to develop new technologies. Also, the results of research and development are often varied and uncertain, and only large corporations with their interest in a variety of products and processes can take full advantage of innovation. The argument for the opposing view, that large firms are not conducive to innovation, is based on the well-known propensity for established institutions to adopt a conservative posture and not to take any actions that threaten to change the status quo. Before deciding between these two viewpoints, the issue must be made more precise by stating it as follows: do very large firms innovate to a larger or a lesser degree than would be indicated by their share of the industry? To answer this, a measure of firm size such as sales or capacity or assets must be chosen. Then if it is found that large firms introduce a greater percentage of innovations than their share of the industry, it can be concluded that large firms indeed are better at innovating than small firms.

The data on the origins of invention and innovation strongly suggest that the initial inventions that lead to major technological change are not usually the result of work in large corporations, but that development often is the work of large firms and that large firms are more active in originating modest and improvement innovations. This does not mean that smaller firms produce no improvement innovations. The Gellman Research Associates study prepared for the NSF shows otherwise. Of the 319 innovations studied, they found that 23.5 percent of them were introduced by firms with 100 or fewer employees and only 34.5 percent were introduced with firms having more

Table 3-1 Percent Share of Industry and of Innovation for the Four Largest Firms

Industry	Time Period	Percent of Industry	Percent of Innovations	Ratio
Steel	1919–1938	62	30	0.48
Steel	1939–1958	63	51	0.81
Petroleum	1919–1938	33	54	1.64
Petroleum	1939–1958	39	43	1.10
Coal	1919–1938	11	18	1.64
Coal	1939–1958	13	27	2.08
Ethical drugs	1935–1949	50	37	0.74
Ethical drugs	1950–1962	33	27	0.82
Chemicals	1930–1950	67	58	0.87
Chemicals	1951–1971	57	41	0.72

than 10,000 employees. This, however, does not answer the question as to whether or not large firms are responsible for a disproportionate share of technical innovation because Gellman Research Associates did not report the share of the market as a function of firm size.

The most careful studies of the relation between firm size and innovation were done by Mansfield and his coworkers. They investigated this question in detail for the iron and steel, petroleum refining, and coal industries (1968), the ethical drug industry (1971), and the chemical industry (1977). Some of their results are shown in Table 3-1, which compares the fraction of innovations with the industry share for the four largest firms in each industry. The last column in the table gives the ratio of the percent of innovations to the percent of the industry. For the steel, petroleum, and coal industries, capacity or output was taken as a measure of firm size. For the ethical drug and chemical industries, firm size was measured by dollar sizes and assets respectively.

While the innovations studied were not trivial, many of them were not major technological advances. They were chosen from the responses of industry representatives and experts as to the important innovations in their industries, and vary in their degree of radical change. In the petroleum industry, for example, new cracking methods and hydrogenation were listed along with advances in solvent extraction and dewaxing, although these represent two different classes of technical advance. In general, most of the innovations were important but not of the type that revolutionizes industries.

The results in Table 3-1 show that the relation between firm size and innovation is different in different industries. For the petroleum refining and coal industries, the four largest firms introduced more than their fair share of innovations. The four largest coal companies, for example, introduced more than twice the number of innovations after World War II than would be expected if innovations were proportional to size. For the steel, ethical drugs, and chemical industries, however, the percentage of innovations arising from the four largest firms was significantly less than would be expected from their share of the industry. Before the war, for example, the four largest steel companies had less than half of their fair share of innovations, and this increased to only 81 percent after the war.

As shown in the table, Mansfield et al. separated the data into two different time periods, the time of demarcation corresponding to the decade of World War II. Table 3-1 shows no consistent difference between the time periods studied. However, an analysis of the contribution of large and small firms to innovation before 1939 and after 1939 shows that small firms became less important as innovators in the later time period (Mansfield 1968).

FUNCTIONS OF R&D

In modern society, technological advance arises from massive investments in research and development. *Industrial R&D* is that complex of activities aimed at acquiring new knowledge, improving materials and processes, developing new engineering designs and functions, and creating new products, all with a view to some ultimate application. The socioeconomic function of research and development is to provide the flow of technical progress on which a technological society is based. Of course, not all innovation is the result of what is normally thought of as R&D. A great many improvements are made by workers, craftsmen, operators, and others on materials and processes in the course of their ordinary work. Learning by doing and improving as you go along are characteristic features of modern industry. But even though the changes from this kind of activity are important over the long run, the innovations involved are generally very modest. It is organized R&D that provides the major basis for most innovation.

Research and development activities are usually divided into different categories depending on how close they are to a final application. *Basic research* is the pursuit of knowledge for its own sake, without any practical application in mind; *applied research* is aimed at a practical application, while *development* is the generation of new products or processes. This is essentially the classification adopted by the National Science Foundation. Obviously, this classification is a bit fuzzy; the difference between basic and applied research, for example, is often only in its intent and not the way in which it is done. Nevertheless this description is useful and provides a model with which to describe the process by which our stock of knowledge ultimately produces new and useful products.

In its simplest form, the model states that basic research provides the pool of fundamental knowledge without which no technological progress is possible. Applied research draws on the results of basic research to study possible products or processes and provides the basis for development, which uses the results of applied research, converting them to useful ends. That is, the model defines a spectrum of sequential activities. Unfortunately, this model is too simple. Technological innovation often occurs even when the basic science is not fully understood; basic research must sometimes be done at the applied research, or even the development stage, when it is belatedly found that some necessary knowledge is missing; applied research is sometimes done in particular fields with only the vaguest idea of specific applications. Innovation is usually not a clear-cut process that lends itself to simple descriptions. But it is certainly true that the stock of basic knowledge is a critically important resource for innovation, even though its connection with applications is often complex, long-range, and hard to unravel in specific cases. It is also true that applied research generally differs from basic research in that it is not driven by the content of the research itself, even though it may be discovering new basic knowledge. The prime directive of applied research is toward application. If a choice is to be made between following up an interesting lead or staying within the applied objectives, applied research chooses the latter.

A useful distinction is that between generic and targeted research. Targeted research is that which is aimed towards a particular product or process, while generic research is the acquisition of knowledge for a particular field of interest which has the potential for generating a large group of products or processes. Both types can be either basic or applied, although generic research is more likely to be basic while targeted research is usually applied.

For a corporation, the fundamental purpose of industrial research and development is to maintain and improve the position of the firm by contributing to growth and profitability in both the short term and the long term, either directly or indirectly. In knowledge-based industries, R&D should be intimately linked to corporate strategy and planning in a reciprocal interaction. R&D should be guided by strategic objectives, and strategy should be formulated with a full appreciation of the dynamics of technological innovation. While the first of these conditions is well recognized, the critically important role of R&D and innovation in strategic planning is often not given sufficient consideration.

The overall function of industrial R&D is easily identified, but this identification is not enough and amounts to little more than a truism. Research and development is a complex set of activities and performs a variety of functions, all supposedly aimed at fulfilling the general corporate imperatives of growth and profitability. These functions have been given a variety of specific descriptive labels such as: to improve existing products and processes; to stay competitive in current lines of business; to develop new products or processes; to provide a window on external science and technology; to maintain personnel for problem solving and consulting purposes; to exploit discoveries made elsewhere; to protect patent positions; to anticipate external technical challenges to existing products and markets; to present a high-tech image; to advance the pool of knowledge.

The ways these activities are described and classified are important because, to a considerable extent, they reflect and determine how R&D is viewed, the expectations from R&D, how it is organized, and even the kind and amount of R&D that is performed.

For strategic and policy purposes research and development should be described in a way that relates it to the flow of technological innovation, particularly as embodied in the evolution of specific technologies and the growth of technology-based enterprises. The usual division into basic research, applied research, and development is not adequate for this, and a different, complementary description is needed. To this end, the multiple functions of R&D are conveniently described in terms of the classification of technology as incremental, defensive, aggressive, or radical. This classification represents overlapping peaks in the spectrum of research and development just as it does in technological innovation. Even at the risk of oversimplification, however, the classification is useful because it provides a vocabulary through which the connection between R&D and corporate or social purposes can be clarified.

Most research and development is of the incremental kind, in which existing products or processes are continually improved. Over a period of time, the change in the product or process might be quite substantial, but each step along the way is relatively small. The functions of this kind of research are clear and easily justified. They include: keeping customers happy, decreasing costs, adding features, improving quality, and meeting or beating the competition. In short, the purpose of incremental R&D is to maintain a competitive position in an existing technology.

Defensive R&D has a similar purpose, but its origins are different. Incremental research is the result of the exploitation of the inherent technical possibilities of a particular technology. Defensive R&D, however, is a response to conditions that are outside the stream of the normal course of the growth of the technology. These conditions may arise from regulatory requirements or new technological challenges and may result in R&D that would not otherwise be done.

Aggressive R&D seeks to advance technology in larger steps, and its driving force is the result of a different approach than that of either incremental or defensive research. Rather than responding to immediate market conditions or other external forces, aggressive R&D takes advantage of technological possibilities and tries to implement them. Incremental R&D is often identified with the "market pull" approach to innovation, while aggressive R&D is often of the "technology push" variety.

The purpose of radical R&D is to supersede or go beyond the capabilities of existing technology. Rather than playing in the arena of gradual improvement, it seeks to create new products or processes based on scientific or engineering principles not previously in common use. Its function is to provide the base for entirely new products and even entirely new industries. Radical R&D can be thought of as an extreme form of aggressive R&D. We will reserve the term "aggressive," however, for research which stays within a given scientific and technological framework, even though it seeks significant advances.

The utility of this classification of the functions of R&D is that it can be directly related to strategic and planning issues. There are times when it is desirable that a firm concentrate on incremental R&D and other times when aggressive or radical research is necessary. The R&D portfolio, R&D costs and benefits, the measure of success of R&D, and even its organizational structure are heavily dependent on which of the multiple functions of R&D are most important at a given time.

DEFINITIONS OF SUCCESS IN R&D

There are three important measures of the degree of success of corporate research and development: technical, commercial, and economic (Mansfield, 1977). For a specific project, technical success is achieved if the technical goals of the project are realized. Synthesis of a new compound, achieving a certain signal to noise ratio, and developing a corrosion resistant coating may all be successful results of R&D projects. But many projects that are technically successful go no further than the laboratory. Commercial success is not achieved until the technical success results in a product that is brought to the market or a process that is put into use. Again this is not the whole story because the profitability of new products and processes is highly variable. Some are only marginally profitable, some even lose money, while others remain highly profitable for many years. Economic success of projects that are both technically and commercially successful is measured by the rates of return on the R&D expenditure from the point of inception to the point of its introduction to the market.

For an individual project, the application is simple. Technical and commercial success is achieved when the project's R&D objectives are met and when it results in something new in the market. Economic success is achieved when the rate of return on the new product or process is equal to or greater than the rate of return that could be obtained from other investments. Definition of the overall success of a group of projects or of a laboratory is slightly more complex. What is of interest here is the fraction of the R&D activity that is successful. Accordingly, we define the probability of success as the ratio of successful projects to the total number of projects. To normalize projects of different size, these probabilities can be weighted according to their costs. Note that what we are defining are conditional probabilities. The probability of economic success is a fraction of projects that are technically and commercially successful. The probability of commercial success is a fraction of technically successful projects. The term "conditional probability" arises from the fact that in computing the probability of commercial success, only those projects that satisfied the condition of being technically successful are counted.

These ideas have a simple quantitative form. Out of a total of N projects, let N_t be the number that are technically successful, N_c the number that are commercially successful, and N_e the number that are economically successful. Then the probabilities of technical, commercial, and economic success, P_t, P_c, and P_e respectively, are defined by

$$P_t = N_t/N; \quad P_c = N_c/N_t; \quad P_e = N_e/N_c \tag{3-1}$$

The cost-weighted probabilities are defined in a similar manner as

$$P_t^* = C_t/C; \quad P_c^* = C_c/C_t; \quad P_e^* = C_e/C_c \tag{3-2}$$

where C is the total cost of all the R&D projects, and C_t, C_c, and C_e are the costs of technically completed, commercialized, and economically successful projects respectively.

While these definitions are simple, they must be used and interpreted with some care. Defining project success requires that a selection of projects be made, and this selection should be done with a view to the kind of information being sought. Aggressive and radical research and development projects are inherently riskier than those that focus on incremental or defensive objectives. Thus, lumping all projects together may be appropriate to get an overall picture of the returns on R&D, but more information is obtained by treating different kinds of R&D separately. Furthermore, a recognition that some classes of R&D are inherently more risky than others is needed by management if it is to properly assess the effectiveness of its R&D programs.

These three measures are straightforward and specific and, as was done by Mansfield (1977; see Section N), can be applied to an econometric analysis of R&D projects. It would be a mistake to think, however, that they provide a complete picture of the definition of success in industrial research and development. In any large, knowledge-based corporation there are many research projects that are economic, commercial, or even technical failures. But if the experience gained from these projects contributes to the expertise and competence of the corporation in a technology that provides the base for a major part of its business, then the R&D may be a strategic success. Similarly, a project may be successful technically, commercially, and economically, but if it commits corporate resources to a technology that is on the way out rather than to one that is growing, the corporation's competitive position may ultimately be severely damaged. The R&D is then a strategic failure in spite of its success by other measures. It is important, therefore, to consider the degree of strategic success of corporate R&D, which may be defined as the degree to which R&D satisfies and contributes to appropriate strategic objectives. This is often difficult to quantify and often requires informed qualitative judgment, but an understanding of the strategic success of R&D is critical. Often, of course, strategic failure of R&D arises from an inadequate understanding or specification of strategic objectives by management outside the R&D structure. This is why the coupling of R&D to strategic planning is so important.

TECHNOLOGICAL CYCLES AND R&D

The strategic setting in which research and development is done varies strongly from one firm to another. However, the dynamics of innovation itself has a strong effect on the probability of success of different kinds of research and on the type of R&D that should be pursued at any given time. A brief recapitulation of the process of technological change will serve to put the description of industrial research and development in the context of the evolution of technical progress.

Technological growth is organic. It consists of a set of interacting and overlapping cycles of different kinds that, taken together, define the rate and direction of technological progress. The most important of these are the cycle of R&D on specific products or processes, the product life cycle, the cycle of progress of specific technologies, the cycle of technological diffusion, the cycle of technological substitution, and the growth cycle of clusters of technologies. Each of these cycles can often be described as an S-shaped curve in which the growth rate is initially slow but then increases rapidly until it reaches a maximum, after which it decreases to very low values. Progress in R&D, measures of technical performance, market share of new products or new technologies, the evolution of major technologies or clusters of related technologies, and the overall technical progress of a technological society can all be described in terms of a model of organic growth as a function of time, in which there is an initial slow growth followed by a period of rapid advance which then slows down and eventually stops.

There are three important features of these phenomena that have wide applicability. The first is that the time scales of technological change are rather long even though they vary considerably. It took less than eight years for detergents to go from 10 percent to 90 percent of market share, but synthetic rubber did not reach 90 percent replacement of natural rubber until sixty years after it reached the 10 percent level. For a large number of technological changes, the takeover times are between twenty and fifty years. The cycle of growth for major technologies seems to be about fifty years. The time lag between invention and innovation can also be substantial. Although the average lag is currently less than a decade, it can be as high as half a century. This poses a particularly vexing problem for the management of technological industries in which the needs of the short and the long term must be reconciled. Both rewards and punishments in our business and political system depend on short-term perspectives. Promotions and elections for individuals take place every year to every six years; annual reports, budgets, and raises are on one-year cycles, while performance reviews are even more frequent; the quarterly report has a notorious effect on assessments of corporate performance. Stock market prices are reported every day, and there is no doubt that many corporate decisions are made with an eye on the price of the company's stock. With the time constants of technological change being measured in decades, there is a serious mismatch between the time scales of technological change and those of short-term success or failure.

The second feature is that growth curves for new technologies are strongly nonlinear. It is natural to think that progress will always be slow when an innovation first appears and, during rapid growth periods, it is easy to believe that growth will always continue. This natural human tendency is reenforced by the long time it takes for a technology to go through its cycle. It is therefore easy to forget the third important point, which is that there is a limit to how much a technology can spread. It is obvious that when a technology has saturated its market or reached its maximum potential it cannot grow any further, but there is often a tendency to ignore this fact.

For research and development, it must always be kept in mind that the productivity of R&D is related to the degree of maturity of the technology being studied. The intrinsic nature of the R&D process permits it to be described in terms of a cycle of progress versus effort expended. The first thing to note in this regard is that for any new line of development progress is initially slow simply because little is known. But as more information is acquired, advances are made at an ever-increasing rate and exponential progress seems assured. As time goes by, however, the inherent limits on which the work is based come into play and productivity of the R&D goes down. For any given technical approach, these limits always exist. They arise from natural laws and engineering constraints such as thermodynamic efficiency, strength of materials, optical resolution, power densities, and so on. This means that when performance is plotted against cumulative effort expended to develop and improve a given product or process, we should expect an S-shaped curve as shown in Figure 3-10. The derivative of this curve is a measure of the R&D productivity and is shown in Figure 3-11. This thesis was developed by R. N. Foster (1979) to explain the declining research productivity in the chemical industry during the seventies. From his extensive experience, Foster concluded that not only the chemical industry, but other industries as well, were continuing work on mature technologies for which the point of maximum productivity was past. The closer one approaches the limits of the technical possibilities, the smaller the advance for a given degree of effort. This is a natural consequence of the R&D process and cannot be mitigated by such devices as reorganization or changes in personnel.

Figure 3-12 shows how much easier it is to make advances in a new technology relative to one which is mature. This curve displays the ratio of the productivity of two technologies, one of which starts out at only 4.7 percent along the way to its ultimate performance potential when the other has already achieved 90 percent of its potential. It was assumed that both technologies have a logistic performance curve with the same growth constant. The productivity is defined as the derivative of the logistic curve. The ratio of the productivity of the new to the old technology is plotted against effort, which can be measured in terms of man-hours or dollars. (For computational convenience the abscissa is the rate constant times effort plus

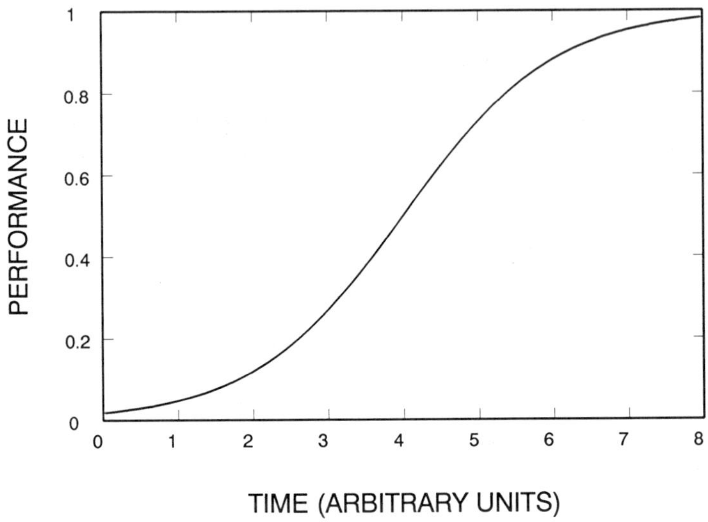

Figure 3-10 Ideal technological growth curve.

four.) The advantage of the new technology is small at first, but it grows rapidly. When the new technology is 50 percent complete, it has a fifty-fold edge over the old. This means that fifty times as much money must be spent on the old technology for it to achieve a given percentage improvement as would be needed for the new. It is no wonder that the introduction of new

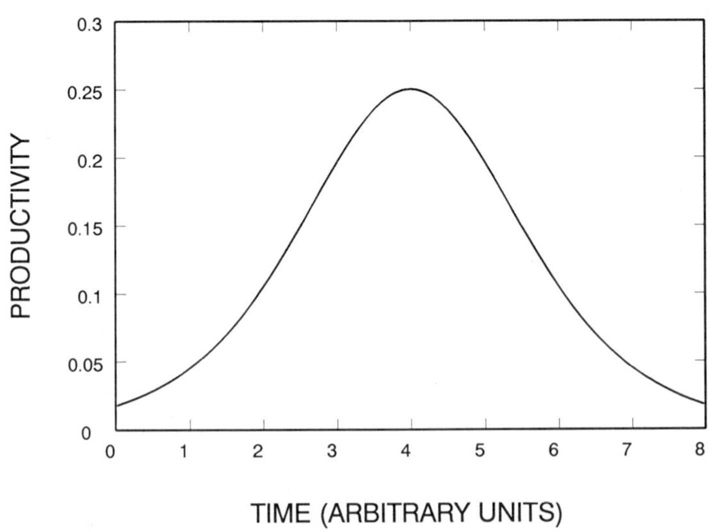

Figure 3-11 Ideal technological productivity curve.

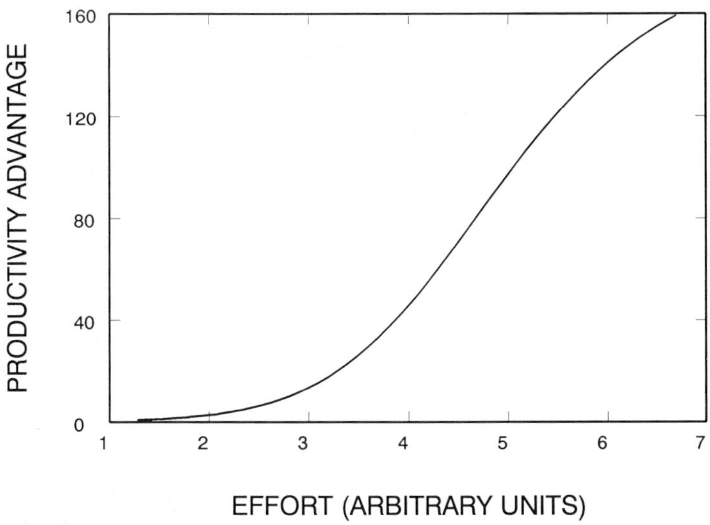

Figure 3-12 Productivity advantage of a new technology.

technologies has such drastic consequences for those that cannot adopt them. Of course, as the new technology approaches its limits, the curve would start to bend over and the advantage of the new technology would disappear as it becomes mature.

Technical progress curves such as Figure 3-10 are normally plotted against time, even though cumulative effort would be more informative, because time is a more readily available parameter. In general, the effort expended in developing a technology rises rapidly with time as more money and people are allocated to promising projects. Plotting technical performance against effort expended would give a more elongated S-shaped character and a slower rise to the maximum than plotting against time.

THE TECHNOLOGY-DRIVEN CORPORATE CYCLE

It is useful to construct a model that describes the history of a corporation whose origin and growth depends on technological innovation. Such a model will necessarily be abstract and will ignore many of the features of doing business in the real world. Very few companies can be precisely described by the model; nevertheless, it will display the primary issues that must be met in relating corporations to technological change. In particular, it will shed light on the kind of R&D that is appropriate at different stages in the history of a knowledge-based firm.

The model is based on our knowledge of technological change and is embodied in the following assumptions:

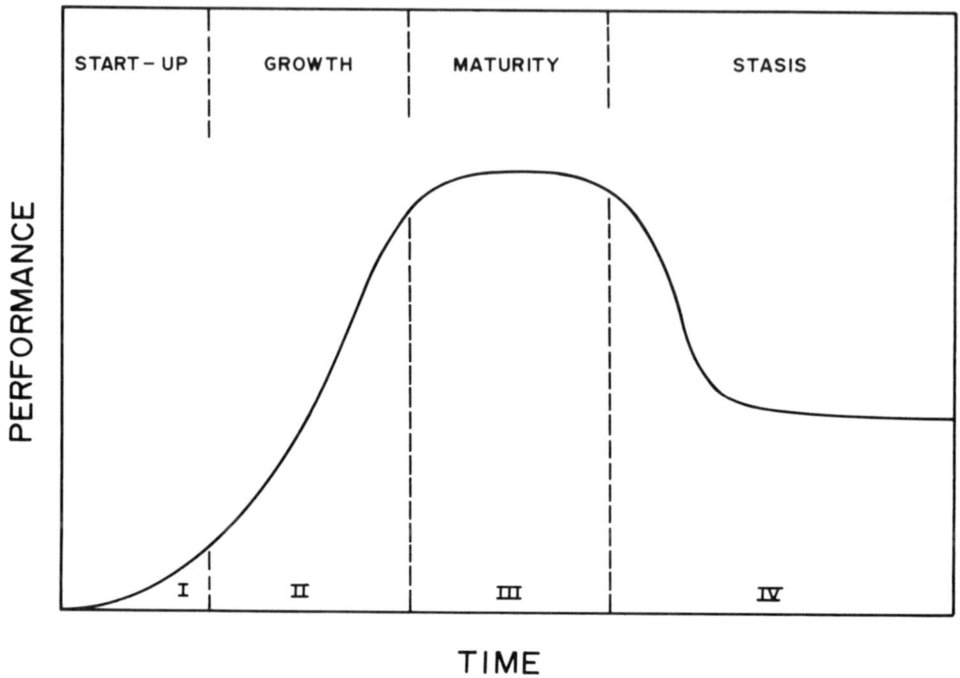

Figure 3-13 Model corporate growth curve.

1. The corporation is founded to commercialize a new invention which is based on a new, relatively undeveloped technology.
2. The new invention leads to several new, related products that are capable of great improvement.
3. The corporation does not diversify and stays with its technology throughout its history.
4. The corporation has normal, free market competition but no new technologies are introduced, either within or outside the firm, that would render its product line obsolete.
5. The corporation is adequately capitalized and top management always acts to maximize profits.

If a measure of growth (i.e., total sales) is plotted against time, our model corporation exhibits a history represented by Figure 3-13.

It is convenient to divide this corporate history into four phases: start-up, growth, maturity, and stasis. The characteristics of each of these phases are displayed in Table 3-2. During the start-up phase, the corporation introduces a product based on a major technological innovation. The product undergoes a number of major changes and improvements which are stimulated by users'

Table 3-2 Characteristics of the Technology-Driven Corporate Cycle

	Phase I	Phase II	Phase III	Phase IV
Type of Innovation	Major product changes	Incremental product change; major process change	Incremental process change; cosmetic design change	None
Stimulus of Innovation	Users' needs; users' feedback	Internal technical capability; cost and quality improvements	Cost and quality improvement	
Competitive Emphasis	New product performance	Some product variation; cost and quality improvements; market share	Cost reduction; standard products	Defensive; protectionism
Market	Market niches	Growth in mass market	Market becoming saturated	Replacement market
Organization and Management	Small; flexible	Expanding; highly organized	Rigid	Rigid

needs, and the interaction between the firm and the users is very strong. Since the new technology is initially more expensive than the old, the firm must find market niches in which the new or improved performance of the new product is of overriding importance. Under these circumstances, it is natural that there is considerable feedback from customers and that the corporation responds to this feedback. During Phase I the corporation is small, entrepreneurial, flexible, and R&D oriented.

During Phase II the technology is improved, several related products are developed, and market penetration increases. Innovation in the products becomes more incremental and, as time goes by, product improvements become smaller. This innovation is directed towards decreasing costs and improving quality of the existing product line. As sales increase, the corporation expands to reap the benefits of scale, so process innovations become more important than product innovations. Innovation results from the internal R&D capability, which expands in size and sophistication. While market share and total profits increase during Phase II, profit margins may go down as the firm tries to increase market share by lowering prices. During this growth phase the corporation becomes highly organized with each unit having well-defined responsibilities and functions.

After a time the market starts to saturate and growth levels off. The firm enters Phase III, which is one of maturity. Innovation consists of small changes in processing and of only cosmetic design changes in products. The organization is rigid and the corporation focuses on cost reduction in trying to sell its standard products.

Ultimately, the market becomes fully saturated and the corporation can sell only replacements. Sales decline, there is no innovation, the technology has reached its potential, and the corporation is in a condition of stasis.

This history is a natural consequence of the dynamics of technological change. Abernathy and Utterback (1978) examined the relation between innovation and corporate operations and found that these relations were what one would expect from the cycle described above. Thus, while not many firms follow this cycle in detail, it provides a rational framework to analyze the role of R&D in corporate strategy.

TECHNOLOGICAL DISCONTINUITIES

The model corporate cycle shown in Figure 3–13 represents the flow of events that are dictated by the market and internal technological factors. But often, a corporation is not left to itself to develop in this way. Rather, a technological threat appears in which some other corporation introduces a competitive product based on a new technology. General Motors introduced diesel–electric locomotives in 1934, thereby undermining the position of traditional steam locomotive manufacturers. Lockheed lost its market share when the turboprop Electra could not compete with the superior technology of Boeing and Douglas. Japanese entry into the American consumer elec-

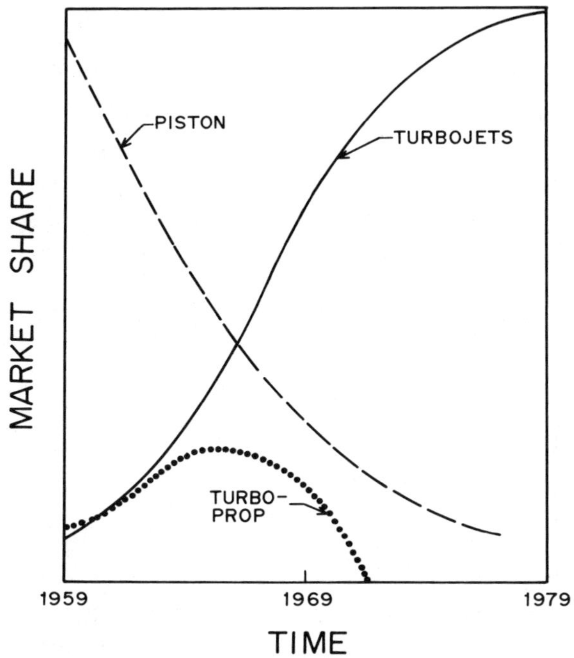

Figure 3-14 Market shares of piston, turboprop, and jet engine aircraft.

tronics market was based on using the transistor to develop pocket-sized radios while American manufacturers stayed with large, heavy "portables." These, and many other examples (Cooper and Schendel 1976), show that sooner or later a company can expect to meet a technological discontinuity. That is, it will be faced with a new technology superior to its own and it will have to jump onto that new technology rapidly and efficiently or lose its business. The evidence suggests that many firms do not have the ability to do this. In a study of twenty-two separate firms presented with major technological threats, Cooper and Schendel found that only two of them successfully entered the new technology as independent firms.

When a new technology wins, an old one loses and the S-curve is the result of the new growing at the expense of the old. An example of this is shown in Figure 3-14, which is a schematic representation of the market shares of piston engines, turboprops, and turbojets in the aircraft industry in the decade of the sixties (Blackman 1974). Curve II shows an S-curve for the takeover of the market from piston-driven planes (curve I), while curve III is the market share for turboprops. Before 1950 it was clear that the piston-driven aircraft had reached a high state of development and that the next generation of planes would not rely on the piston engine. Douglas and Boeing met this challenge by developing the DC-8 and the 707 respectively, both of which were turbojet-driven, while Lockheed entered the mar-

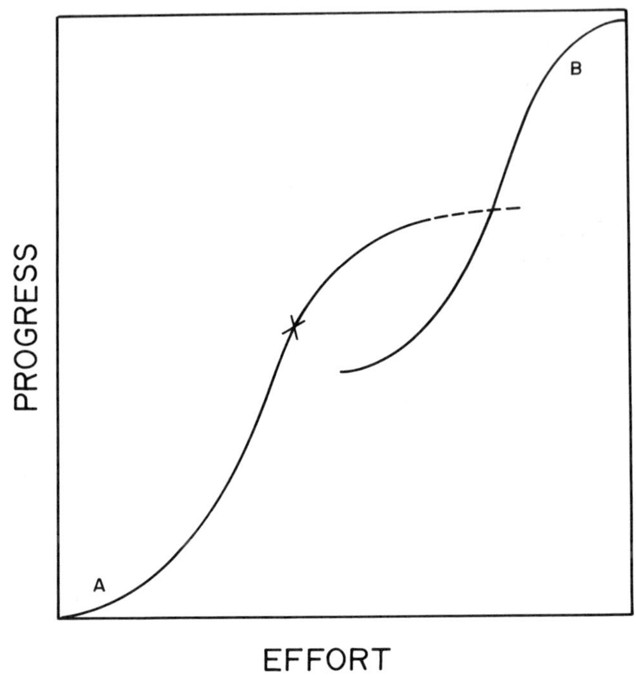

Figure 3-15 The technological discontinuity.

ket with the Electra, which was a turboprop. The turbojets had significant advantages over the turboprop; they had more power, greater fuel economies, higher capacity, longer range, and higher maximum speeds. The turboprop retained too many of the disadvantages of the reciprocating engine-driven planes. The results were inevitable. Boeing and Douglas dominated the market while Lockheed got into serious trouble. Lockheed's choice of technology was much less aggressive than either Boeing's or Douglas'. It was more incremental in nature in that by retaining the propeller and much of the physical configuration of earlier aircraft, it did not make a successful transition to a new and superior technology. This sort of thing happens often.

Lockheed ran into a technological discontinuity. Such a discontinuity is illustrated in Figure 3-15. A company moves along its technology curve A up to the point X where it meets competition from a better technology. If it stays with the old technology it obviously will be in serious trouble. To continue its success, it must cross the technological discontinuity from X to Y. This is very hard to do. The company has been moving happily along its performance curve, which it was able to do because of enormous investments in time and money with a concomitant organization of people having the proper expertise, a full knowledge of the technology, the appropriate marketing strategy, and a command of all of the relevant business factors. Suddenly it is being challenged to change all that and to do it fast. It needs to write off investment, acquire new kinds of people, reevaluate its markets, and learn a new way of doing business. If it fails, someone else takes over the leadership position. Such discontinuities occur not only as a result of competition but also because a technology becomes mature. A mature industry is a troubled industry because the demand for its products levels off and falls. Unless it can cross a discontinuity to a new product line, it stagnates.

THE R&D PORTFOLIO

Choosing the right R&D programs to pursue at the right times is certainly one of the most difficult tasks facing those who are responsible for guiding the course of technological innovation. What areas of basic research are important? Which embryonic technologies should be developed? What are the bottlenecks to further progress and which are the most critical? Of all possible innovations, which would be the most useful or the most profitable? How do the R&D programs fit in with institutional capabilities? How should funds be allocated among basic research, applied research, and development? What kind of research (radical, aggressive, incremental, or defensive) should be emphasized? The answers to questions such as these have a great effect on the rate and direction of technological change, the social impact of technology, competitive positions, marketability of products and services, and the health of corporations.

Industrial R&D programs can be put together in a variety of ways. In general, senior management defines corporate objectives, the general fields of business, the overall technical goals, and the amount of money to be spent on R&D. Within the framework of senior management guidelines, R&D management defines areas of investigation and the general allocation of funds among different kinds of programs. Specific R&D projects are formulated by the R&D scientists and engineers. These projects are coordinated and rated for priorities by a planning staff and reviewed by upper management for approval or rejection. The actual process is sometimes more complex than this since it can also involve interaction with production and marketing personnel, the needs of various corporate divisions, targets of opportunity, "product champions," the relative strength of personalities, and so on.

At any given time, most R&D is an incremental outgrowth of what has gone before. The inner dynamics of innovation injects a degree of determinism and momentum into the R&D process such that many future projects are a natural result of past work. Also, there is a good deal of imitation in the sense that different laboratories in the same industry work on the same or closely related problems. The reason for this is that the technological possibilities in a given field are the same for everyone. Also, by pursuing similar goals firms reduce the risk of competition, even though they may reduce the chances of unusually large advances.

There is no formula for constructing, or even defining, an optimal R&D

program in detail. But the flow of technological change and its effects on the corporation can provide a setting that clarifies the strategic success or failure of R&D. To start with, let's reconsider the idealized corporate cycle described above. The cycle starts with the initial results of aggressive or radical R&D. The R&D remains aggressive until the firm enters Phase II when it becomes more and more incremental. A real firm would like to avoid all the negative effects of maturity and stasis that result when the technology and the market are fully developed. This cannot be done as long as the firm stays with the old technology. New products or services are needed, and these can only come from new lines of technological innovation. The right time to start looking for new developments is during the growth phase of the corporate cycle. The natural inclination during this phase is to improve the products incrementally and to reap as much market share and profit as possible. The great satisfactions of success inhibit the long-range view and not much attention is likely to be paid to aggressive research for the sake of speculative new kinds of business. But technological extrapolations are not linear, and aggressive or radical research is needed at precisely the time of rapid corporate growth so that a new growth curve can be started as the old one inevitably dies. If a firm can bring itself to forego some immediate profits and to recognize that all cycles must end sooner or later, it can increase the chances of continued growth by the aggressive development of new technologies just at the time that it seems that the new is not needed. Also, the habit of thinking of innovation in these terms reduces the chances of being caught by a technological discontinuity. With the recognition of the dynamics of technological change in the firm's strategic thinking, and with the presence of people that are at the forefront of new work, the firm may even be fortunate enough to create a technological discontinuity for others instead of being the victim of one.

Even in our idealized firm, with only one line of products based on one technology, it is clear that several different kinds of research are needed if the sad results of Phases III and IV are to be avoided. During the growth phase both incremental and aggressive (or radical) R&D should be done. As the technology and the firm climb higher on the growth curve and both the technology and the markets start to approach their limits, the amount of R&D performed on new technology should increase at the expense of the old. For a diversified firm, each line of business needs to be treated separately according to its degree of technological maturity.

These ideas have strong strategic implications. Petrov (1982) has tried to capture some of them by constructing a technology portfolio model in which the technologies underlying product lines are identified and two parameters are defined. The parameters are relative technological position (RTP) and technological attractiveness (TA). The RTP measures the strength of the firm in a technology relative to its competition. (Petrov proposes the cumulative sum of R&D and capital investments as a suitable measure of RTP.) The TA measures the potential return for investing in the technology. This leads

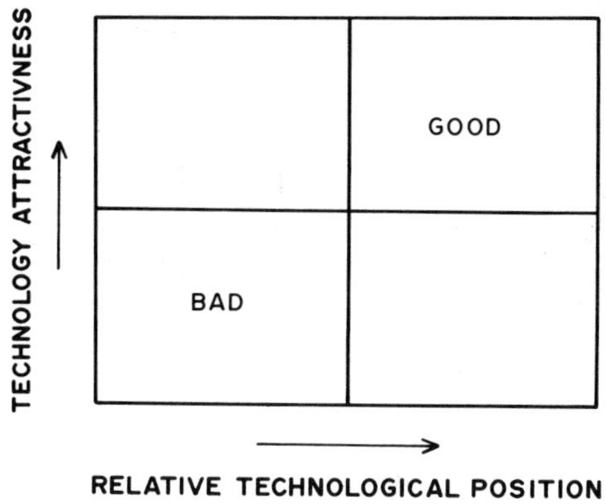

Figure 3-16 Technology portfolio of Petrov.

to a two-dimensional matrix as shown in Figure 3-16. The implication is that if one of the technologies is in the upper right-hand quadrant of the matrix, the firm is in a strong position which should be exploited. When both RTP and TA are low, the technology should be dropped. If RTP is high while TA is low, it may not be worthwhile to continue in that technology. When RTP is low but TA is high, the firm must decide if it wishes to continue in that technology and must invest accordingly.

From the viewpoint of technological dynamics this approach is inadequate. The two basic strategic factors in a technology are its degree of technological maturity and the extent of its radicalness. Newly emerging technologies have very different strategic implications from those that are mature, and the corporate life cycle is closely related to the technological life cycle. Also, the impact of a technology depends strongly on how radical it is. Incremental improvements in existing products have less strategic importance than innovations that can lead to entirely new products. Furthermore, the degree of technological maturity and the rate of innovation are often linked, since there is more opportunity for extensive change when a technology is new. A matrix representation that would connect technological and business considerations would be multidimensional, but a simple two-dimensional representation is useful for discussion. This is shown in Figure 3-17, in which technological maturity is plotted against a market index. If the purpose is to describe the current situation, then the market index would just be the market share. If the purpose is to explore future possibilities, then the market index could be potential market size or growth rate. For strategic analysis both kinds of indices should be used.

The matrix is divided into six regions to illustrate how the evolution of a

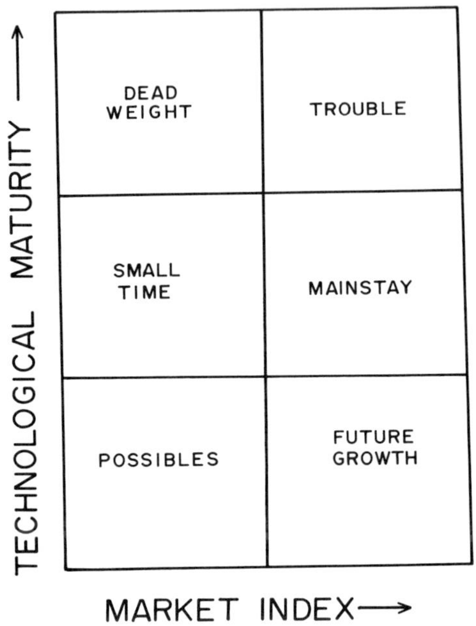

Figure 3-17 The dynamic technology portfolio.

technology bears on strategic issues. A technology that is mature, with corresponding products that have a large market share or are a large part of the firm's business, poses serious problems. This part of the matrix is labeled "Trouble." The market can become saturated or a new technology can appear that challenges the old, thereby confronting the firm with a major discontinuity. If the firm has not prepared to replace the old technology, it can be subject to severe trauma.

If the technology is mature and the market index is small, the corresponding business should probably be divested. While not exposing the firm to cataclysmic dangers, it doesn't hold much future promise. This region is labeled "Dead Weight." A technology that is in the middle of its evolution and has a high market index represents the large, growing part of the firm ("Mainstay"). If the market index is small, the business will never be worth much ("Small Time").

The regions that are of great interest for the future are those that are technologically young. This is where new products and entire new businesses come from. For these, the market indices that estimate future size and growth require special attention. The region "Possibles" refers to those new technologies in which the firm has a low market share or its estimate of future markets is pessimistic. In either case, close monitoring is required because estimates of future markets for new technologies are notoriously inaccurate and often much too low.

The exciting possibilities are those for new technologies where the market index is high. These are the most likely sources of the future welfare and growth of the firm.

An ideal R&D portfolio is one in which incremental R&D is being carried out on "Mainstay" technologies which are carrying the major business of the firm, "Trouble" is being phased out because the technology is near its full potential, and aggressive (or even radical) research is being done on "Future Growth" technologies. This is a dynamic process in which technologies continually move up the scale of technological maturity while being replaced by new technology at the bottom of the scale. A good part of the reason for the recent transitional chaos in American industry is that this technological flow has not been nurtured and planned for. By the 1970s a number of important technologies were quite mature. These included steel, rubber, jet aircraft, household appliances, and automobiles. Unfortunately, many firms had been acting as if the growth in these technologies would continue indefinitely. But of course this could not be so and the resulting distress was unavoidable.

CHARACTERISTICS OF INDUSTRIAL R&D

The bulk of research and development in the United States is done in private industry. In 1986, industrial R&D expenditures amounted to $85.6 billion dollars, which was 73 percent of the national total. Of this expenditure, 36 percent of the funds were from federal sources. The distribution of R&D is far from uniform in American industry. Six industries account for nearly three-quarters of the total industrial research and development effort. These are aerospace, electrical and electronics, instruments, machinery, computers, and chemicals (including allied products and petroleum). Roughly two-thirds of the industrial expenditure is for development, about one-quarter for applied research, and about one-tenth is spent on basic research. (NSF Science Indicators, 1982; NSF Science and Engineering Indicators, 1987).

The total amount of money spent on R&D in 1982 was about $79 billion, which was about 2.6 percent of the Gross National Product. In 1986, the total expenditure had grown to $116.8 billion, or 2.72 percent of the GNP. But this includes expenditures on defense and space R&D. Civilian expenditures were 1.76 percent of the GNP in 1982 and 1.85 percent of GNP in 1986. This was up from 1.23 percent in 1962 when the total R&D expenditure was 2.72 of GNP. In terms of the Gross National Product, there was a general downward trend in the ratio of R&D to GNP to a low of 2.23 percent in 1977 and 1978 followed by a steady rise. This was caused largely by fluctuations in R&D for defense and space. The civilian R&D expenditures as a percentage of GNP rose from 1.23 percent to 1.49 percent for the five-year period from 1962 to 1967, stayed roughly constant until 1978, and then started to rise to its current level.

The remarkable concentration of R&D in a small number of industries arises from several factors. In the first place, these industries are the direct outgrowth of developments that started during the Industrial Revolution and were reinforced by the technological innovations centered in the decades of the 1890s and the 1930s. These innovations included the internal combustion and diesel engines, electric lighting and power, electric communications, organic synthesis and drugs, petroleum refining, plastics and polymers, jet engines, and advanced electronics. One reason for R&D concentration is that it reflects the natural momentum induced by the particular innovations of the past. Another reason is that it reflects the kinds of current innovation that society is willing to pay for, such as increased health care, greater mobility, and ease of communication. Also, some industries lend themselves to innovation more easily than others. Housing, for example, does not enjoy the high rates of innovation exhibited by electronics. The industry is dispersed among a large number of small enterprises, zoning and construction regulations are very conservative and vary widely from place to place, and the housing stock turns over slowly.

But the results of innovation are not as concentrated as its origins. The innovative, knowledge-based industries permeate the economy and are responsible for introducing innovation into industries that do not do much R&D. Most innovation in electric utilities, for example, originates in the R&D work of electrical suppliers. In construction, new materials, new heating and cooling systems, and construction machinery arise from innovations introduced outside the construction industry.

Industrial R&D is concentrated in yet another way; most of it is done in very large firms. Small firms normally have limited product lines and limited sales so they simply cannot afford very much R&D. Large firms, with their large markets and diversified product lines, not only can afford R&D, but also are in a better position to reap its rewards. These considerations have led to the proposition that the larger the firms, the greater the rate of innovation, and that giant corporations are necessary for progress in a technological society (Schumpeter, 1939; Galbraith, 1952). This proposition was based on the ideas that the cost of modern innovation is so great that only the largest firms could afford it, that R&D must be done on a large scale so that failures can be outweighed by successes, and that a firm must have a market share large enough to make innovation worthwhile. The analyses of the innovation process by Scherer (1965), Jewkes, Sawers, and Stillerman (1969), Mansfield (1968, 1971), and others do not bear out this thesis. The evidence strongly suggests that, in general, the largest firms do not do as much R&D or as much innovation per dollar of sales as smaller firms. Of course, if a firm is very small it does little R&D per sales dollar. The most R&D intensive firms are those that are in the second tier; they are large, but not the largest in the industry. Also, small firms are responsible for more innovation than the R&D expenditures would indicate. In fact, many small firms are formed precisely for the commercial exploitation of a new technology. The initial stages of innovation (basic or applied research and product prototypes) are much less expensive than the later stages, when large sums of money must be spent on all the details of refinement, development, and gearing up for production. Individual inventors and small technological firms are still important and provide a significant portion of the innovations that are ultimately developed by large firms.

The nature of industrial R&D expenditures suggests that most technical progress in industry is incremental and that industrial research leading to major innovations is rare. This is confirmed by studies of the origins of technological innovation. Of sixty-one major innovations studied by Jewkes and Stillerman (1969) plus 21 analyzed by Hamberg (1966) introduced from the beginning of this century to 1955, only nineteen (26 percent) originated in the laboratories of large companies. The importance of independent people in the innovation process has already been mentioned. In the aluminum industry, only one of seven major innovations originated with a major producer, and that was undertaken in cooperation with an aircraft manufacturer (Peck 1962). During 1940 to 1955, thirteen major innovations were introduced in the steel industry. None came from the American steel industry and seven were initiated by independent inventors (Hamberg, 1966). In an examination of the origins of twenty-five major innovations of the Du Pont company between 1920 and 1950, Mueller (1962) found that fifteen originated outside the firm. Given the functions and objectives of private industry, none of this is very surprising, but it does raise the question as to whether the distribution of R&D expenditures is optimal.

The conservative, incremental nature of most industrial R&D is confirmed by analysis at the project level. A number of studies have been performed of the characteristics of R&D projects in industrial laboratories (Mansfield et al. 1968, 1977), (Baker, Siegman, and Larson 1971) which show that most projects result in relatively small advances. Furthermore, management aims for and expects small advances that have a high chance of being successful. In one large laboratory, the expectation of technical success for seventy projects in 1963-1964 was 80 percent (Mansfield 1968). Upon completion, it was found that only half the projects were successful in terms of meeting their objectives on time. However, closer analysis revealed that for many of the "failed" projects, the reason for failure was not technical but the result of a change in objectives or transfer of people to other projects. Correcting for this gave the result that 75 percent of the projects were technically successful, which is in remarkable agreement with the management's estimates before the projects were begun. Since larger advances are more risky, these results are in accord with the bulk of the projects having modest technical goals. Mansfield (1968) found that, indeed, projects that were described as more incremental by research managers had a higher success ratio than those that were regarded as trying for larger advances in the state of the art.

As would be expected, the time horizon for such projects is rather short. Many projects were expected to be completed in a year or two, and the bulk of them were planned to be done in four years or less with an expectation that they would be implemented in a matter of months after completion. None of this means that no long-range work leading to totally new products, or even industries, is ever done in industrial laboratories. Nylon and the transistor are prime examples of products arising from basic research in industrial laboratories that led to entire new industries. But such work is relatively rare. Most R&D is done with much more modest goals in mind.

On the average, the rates of commercial and economic success are also high, although there is considerable variation among firms. In his study of R&D projects in sixteen firms for which the rate of technical success was greater than one-half, Mansfield found that the rate of commercial success varied between 12 percent and 100 percent with an average of 57 percent, while the rate of economic success varied from 59 percent to 100 percent with an average of 74 percent. R&D is clearly a profitable activity. In a study of seventeen innovations introduced by firms between the mid fifties and the early seventies, Mansfield and coworkers found that the average pretax rate of return on the total cost of introducing an innovation was 34 percent. This compared quite favorably with the 15 percent considered to be a good rate of return on investment at the time. The fluctuations around the average were high: rates of return ranged from 0 percent to 214 percent for individual innovations. These results were for the firms introducing the innovations. A successful innovation, however, often saves money for others as well as the innovator. A major motivation for innovation is to offer products to customers at lower prices and to improve processes so as to decrease costs. Also, industrial customers often have clear ideas on the type of changes they would like their suppliers to implement so that they can reduce their own costs. It would be surprising if an innovation did not yield a benefit to customers. Indeed, it was found that the seventeen innovations were as beneficial to customers as to innovators. The total rate of return averaged 78 percent when all beneficiaries were included in the calculation.

These analyses refer only to innovations that were a success, in the sense that they were commercially implemented. They did not address the question of the impact of failures on the returns from innovation. To get some information about the return on the total costs of R&D, Mansfield examined the results of R&D for a major industrial laboratory from 1960 to 1972, using all corporate costs that could be ascribed to investment in innovations emanating from that laboratory. He found that the private rate of return (benefits only to the innovator) varied greatly from year to year, ranging from 2 percent to 31 percent, but the average for the thirteen-year period was 19 percent. The uncertainties of innovation can make it risky in the short term, but it is clearly quite profitable in the long term.

The R&D projects and innovations being discussed here were neither trivial advances nor major breakthroughs, but were the "normal" kind of advances one expects from the bulk of industrial R&D that we have labeled as incremental.

DETERMINANTS OF SUCCESSFUL R&D

Research and development is a complex activity whose success or failure depends on a large number of interacting factors. Sorting these out for any specific laboratory or for a particular segment of R&D would require a detailed case study, which would include company policy and strategy, status of the technologies being pursued, decision making methods, nature of the personnel, facilities available, competitive position of the firm, R&D budgets, the mix of different kinds of R&D, laboratory organization, relation to marketing and production, management expectations, and effectiveness of communications. Nevertheless some general comments can be made about some of the more important factors that have an impact on the degree of success that is realized from industrial R&D. Three different levels of organization need to be considered in analyzing the success or failure of R&D. These are the corporate, laboratory, and project levels.

Corporate Level

Overall R&D strategy and its relation to corporate objectives is determined at the corporate level. The nature of the firm's business and its underlying technology, the size of the R&D budget, organization of the R&D function, and expectations of how R&D contributes to corporate objectives are the responsibility of the central management of the firm. At this level, success is measured by the extent that R&D contributes to corporate strategic objectives and to the overall profitability of the firm, both in the short term and the long term. If the strategic position is sound, the stage is set for successful R&D. If the strategy is flawed, R&D can be a failure for the firm even though the laboratory and its projects are successful. It is at the corporate level that the most careful attention must be given to the R&D portfolio to ensure that new technologies are developed as the old ones mature. Simple, short-term profitability calculations are insufficient to measure the corporate success of R&D, and different expectations of risk are required for different kinds of R&D. High rates of success can be expected for incremental research during the growth phase of a technology, but as the technology matures lower returns are realized for a given R&D expenditure. For new fields the uncertainties in the performance of the technologies, the nature and size of markets, and the costs of development and production combine to make the risks significantly greater than for incremental R&D. Yet, new fields must be pursued because the long-term risks of being left behind with dwindling markets and vanishing profitability are much greater. The important point is that the same criteria of commercial and economic success that are appropriate for incremental R&D cannot be applied to radical and aggressive R&D. The strategic objectives of radical and aggressive R&D can be listed as follows:

1. To explore new technologies for possible lines of future business
2. To keep up with the competition and even gain a lead in emerging technologies
3. To avoid being confronted with a technological discontinuity or even to create one's own technological discontinuity

In short, radical and aggressive research is meant to provide for the long-term viability of the firm. This is very different from the purpose of incremental R&D which is to contribute to the profitability of the ongoing, major parts of the firm's business. In the long term, of course, the success of radical or aggressive R&D can be judged by the usual financial criteria. But it often takes many years for this kind of work to come to fruition, and in the meantime qualitative judgment is often the only way to decide if the R&D is meeting the corporate objectives.

Aggressive and radical research is often given insufficient emphasis, particularly when firms are on a growth curve for their existing products. A contributing factor to the economic turmoil of the 1970s and 1980s was that a number of technologies had become mature and were saturating their markets. The firms depending on these technologies had not done the aggressive R&D that would have prepared them for this maturation. But a new wave of technologies based on computers, robotics, telecommunications, nonlinear optics, and molecular biology is creating new leading economic sectors. The current transition period is chaotic and uncertain, but there is a general recognition that the transition is taking place. Aggressive and even radical R&D is now widely recognized as important. Large numbers of large and small firms are engaged in extensive R&D on the new technologies. Many of these efforts will fail, but those that succeed will generate large returns.

To summarize, the important determinants of the degree of R&D success at the corporate level are the ways in which R&D is integrated into corporate strategy, the acceptance of different risk levels for different kinds of research, and the adequacy of funding on a continuous basis to support the R&D objectives.

Laboratory Level

Laboratory management has the responsibility for pursuing the corporate R&D objectives. At this level, determinants of success include the particular lines of inquiry selected for the R&D, the talents of the people hired, nature of laboratory services, support personnel and equipment, and above all, the effective communication of corporate goals to the R&D scientists and engineers.

This last point is critical. The best people doing basic and applied research were educated in an environment which prized independence of thought and individual creativity. The mark of success was favorable peer judgment based on contributions to knowledge without any serious consideration of practical applications. The American system of graduate education in the sciences is remarkably successful in producing individuals who can think for themselves and are self-motivated to do good work. But the culture is academic in every sense of the word. The task of laboratory management is to nurture the individual creativity and initiative while guiding it towards the achievement of corporate R&D objectives. The classification of different kinds of R&D is again useful here. The more aggressive or radical the R&D, the greater the need for self-determination on the part of the investigator, not only in how the work is done but also in the selection of projects and even fields of study. The more incremental the R&D, the greater the extent to which it will be driven by the natural momentum of the technology.

Engineers, particularly those working in development, have a different education and different attitudes from those of pure science towards their work that makes it much easier to incorporate them into industrial R&D environments. But even here some care must be exercised. The best engineering schools have become quite scientific and much of their work would be classified as applied or even basic research in industry. This produces engineers with a much more extensive and rigorous foundation in fundamentals than ever before. They have close ties to the sciences and can attack a wider array of problems in a greater variety of specialities. The value of a broader approach to their own careers and to industry is obvious, but this has been achieved at the expense of practical industrial content in the curriculum, which must be taken into account by R&D management.

Research, by its very nature, depends on the individual imagination, drive, and zest of the investigator. It is a personal affair and cannot be programmed or managed in a highly structured and predictable manner, as might be the case with production or marketing. The greater the degree of control, the more likely that research will produce pedestrian results. Laboratory management's task is a delicate one in which the need to conform to corporate goals must be reconciled with the freedom of inquiry and the ability to pursue the unexpected that is necessary for high quality, productive research.

The Project Level

Success at the project level obviously depends on the choice of projects and on the abilities of the investigators. It is at this point that management must demonstrate an understanding of the research process and the characteristics of research investigators. A paper by Parmenter (1964) is illuminating in this regard. Groups of research workers, group heads (the first level of management), and laboratory directors (the next higher level of management) at RCA Laboratories were asked to discuss the criteria for research project selection and to rank their importance. For the researcher, sixteen criteria were identified as being important for project selection. The three at the top of the list were: enthusiasm of the researchers, scientific significance, and the enhancement of understanding. Note that none of these is directly related to corporate objectives. In fact, none of the first ten most important criteria for project selection listed by researchers contained an explicit statement of direct benefit to the firm. Profit potential was listed as eleventh most important and the interests of management were fifteenth. But probability of technical success was first on the list prepared by group heads whereas it was sixth for researchers. For laboratory directors, profit potential was first. This is not unexpected to anyone that is in the R&D community, but it does stress the point that motivations differ at different rungs of the management ladder.

It would be a mistake to think that all of this means that management should try to force researchers to choose projects using criteria more closely related to commercialization and profitability potential, particularly for long-range, aggressive, or basic research. In the first place, research is uncertain enough so that it is very hard to tell what will be profitable in the long run. But more importantly, the researcher needs to be immersed in the enterprise of discovery in a way that draws out the best work he or she can do. This means that the researcher must be led by the rhythm of the work itself, its possibilities, its state of the art, and its unknowns. Only by being at the edge of the unknown and dealing with it as a personal confrontation will the researcher have a chance of finding something of value. Management's job, in fact, is to encourage the culture of exploration while guiding the work along the avenues leading to corporate goals, using a looser hand for basic and aggressive research than for incremental R&D.

In his studies of the success rates of incremental R&D projects, Mansfield was able to draw further conclusions that are of interest for the management and conduct of R&D. Some of these conclusions are intuitively obvious, but others are not. An important conclusion was that the impact of marketing and production on probability of successful innovation was substantial. Estimates made by executives of eighteen firms indicate that if marketing and production personnel had properly understood the potential of the innovations, the overall probability of their success (technical, commercial, economic) would have increased from 32 percent to 51 percent, on the average. R&D, marketing, and production executives all agreed on this point. In fact, the estimates of marketing and production executives were somewhat higher than those of R&D executives.

It was also found that early and careful review of R&D projects enhanced the chances of success at all three stages: technical completion, commercialization, and economic return. Also, projects that start with real market needs ("demand pull") are more likely to be technically and commercially successful; but given technical and commercial success, they are not necessarily more profitable. Projects that emanate from R&D personnel ("technology push") are likely to have a lower chance of technical completion, but those that are completed and commercialized have a significantly higher chance of being economically successful. That is, R&D personnel start riskier projects, but the payoff is higher. Coupling R&D with marketing considerations is certainly important and, as would be expected, its major effect is to improve the chances of commercialization. However, it seems to have no effect on the chances of economic success.

A rather surprising result, at least at first sight, is the effect of quantitative R&D project selection techniques. Such techniques improve the chances that a project will reach technical completion and the chances that it will be commercialized, but the chances of economic success are decreased. The clear implication is that quantitative planning (which is based on such factors as pay-back periods and return on investment) pushes R&D toward safer projects, and these do not have a large market impact.

There are, of course, many factors that affect the success of R&D that are not neatly classified into the three major management levels. Chief among these are organization and communications. The problem is that different structures are needed for different kinds of R&D. For basic research or aggressive R&D, the researchers need to have considerable independence from the ongoing business of the firm. Their job is to create the knowledge base for the future business of the firm, and too close a relationship to the current business can limit their sphere of thought, involve them in incremental R&D and generally deflect them from their main mission. I am familiar with one major, knowledge-based firm, in which most R&D was centralized in one location with only the later stages of development being done in operating divisions. There was no conscious separation of R&D into radical, aggressive, or incremental components, although it was classified as being either basic or applied. The R&D staff was strongly encouraged to maintain close ties with the operating divisions. In fact, the project selection system provided for projects requested by divisions and for which the divisions would pay. This did couple the R&D to the operational and marketing requirements of the firm, but it had a negative effect on long-range, aggressive research. The division projects were often of short duration with tight schedules so that in any given year an investigator was likely to have responsibility for two, three, or even more projects with differing and onerous reporting requirements. Furthermore, the work for the divisions became incorporated into the laboratory budget so that it became necessary to acquire a certain number of division projects just to support the budget, and the laboratory was in the position of actively trying to sell its services to the divisions. Such a situation is certainly not conducive to the pursuit of the most imaginative kind of work of which research people are capable.

The organization of R&D must satisfy requirements that often conflict, and no ideal structure exists. Basic and aggressive long-range research needs to be organized according to discipline or field of technology with little pressure from the day-to-day conduct of the business of the firm, while incremental R&D must be sensitive to existing and developing products and be closely attuned to the current production processes and marketing efforts. This would seem to dictate a structure in which long-range work is done centrally and funded by the corporate center while shorter term R&D is supported by operating divisions and conducted in a more production- and market-oriented environment. This structure is actually adopted by a number of firms, but it does not solve all the problems of R&D organization. There is still a need, for example, for the long-range work to be brought to a stage where it can produce products sooner or later; communications among long-range research, short-term R&D, production, and marketing are therefore critical.

In general, the organization of R&D needs to optimize the following desiderata:

For basic, long-range, radical, or aggressive research:

1. Freedom of project selection within broad areas compatible with corporate goals
2. Freedom from responsibility for the current technical needs of the firm
3. Organization of R&D groups along disciplinary lines or by fields of technology
4. Measures of success based on the discovery of new knowledge, new phenomena, and peer review rather than the immediate development of new products or processes
5. Gentle and personal guidance by management into areas that have promise for future applications

For incremental, short-term R&D:

1. Close coupling with production and marketing
2. Measures of success based on the value added to existing products and processes and the current technologies by R&D
3. Organization of R&D groups by lines of business or by products and processes
4. Freedom to pursue R&D projects as the investigator deems best within the constraints of the firm's current business
5. Careful monitoring of schedules and budgets to insure cost effectiveness of the R&D

These two sets of desirable conditions represent two ends of a spectrum. Many R&D activities fall somewhere in between the aggressive and incremental so that the extent to which each of these conditions is established requires judgment and compromise.

The above conditions apply to the working groups in R&D and are designed to encourage the best work that they can do, but they are not sufficient. The organization of R&D must also facilitate intrafirm technology transfer. That is, the results of R&D must actually lead to commercial applications; the technology developed in the laboratory must be transferred to production and marketing. Particular attention must be paid to this in the case of basic and long-range research since the conditions for doing the best research are not compatible with the conditions for easiest technology transfer. It is too easy for new knowledge that can ultimately have important practical applications to go unrecognized. Also, when incremental or development work is organized along product lines or lines of business it runs the risk of losing contact with the more basic or aggressive research that can enhance its effectiveness. All products and lines of business are based on generic technologies that are in a constant state of change, and it is necessary that even those doing incremental work be aware of the state of the art in these technologies. The R&D organization should bring managers of different kinds of R&D together on a regular basis, along with those responsible for production and marketing, specifically to explore the question of technology transfer.

The degree of success depends heavily on the extent to which the interrelations of the different kinds of R&D are understood throughout the firm and on the measures taken for continual cross-communication and mutual education among those involved in all aspects of the R&D process.

4
Technological Diffusion and Substitution

TECHNOLOGICAL DIFFUSION

A new technological innovation takes time to spread to its potential users. It took twenty years for the basic oxygen steelmaking process to get to the point where it accounted for 90 percent of steel production in the United States. The earlier open hearth process did not reach the 90 percent point until forty-three years after its first introduction and synthetic rubber took nearly sixty years. While many technologies take decades to achieve widespread use, there are others that penetrate the market with astonishing rapidity. In 1950, less than 10 percent of the electrically wired households in the United States had television; by 1960, this number had increased to over 90 percent. The time for synthetic detergents to go from a substitution level (for soap in clothes for laundering) of 10 percent to 90 percent was less than nine years. A major task of technological dynamics is to describe the rate of the spread of new technologies and to determine the reasons for the rates that are observed.

As always, some precision of nomenclature is important. A variety of terms have been used to describe the spread of technology. These include "rate of adoption," "rate of imitation," "rate of diffusion," and "rate of substitution." These terms are often used interchangeably but are sometimes used to describe somewhat different concepts. For our purposes, we will define two different terms, diffusion and substitution, which are sufficient to present the subject in a quantitative framework.

The concept of technological diffusion has two simple elements: the extent of use and time. That is, the extent of diffusion of a technology at any given time is defined by the degree to which it is in use at that time, so that the process of diffusion is the gradual growth of the use of a technology as time passes. While this is a simple definition, it contains a number of complexities that must be dealt with in specific cases. First, there is the familiar problem of defining a technology. In studying the spread of electronic amplifiers, for example, we might simply count their number as a function of time, thereby defining an overall technology of amplifying electrical signals. But solid state amplifiers are different from their vacuum tube predecessors and each can be taken to represent different technologies. Again, audio and radio frequency amplifiers are different and serve different purposes; audio amplifiers for home use are different from those for public address systems and so on. Furthermore, even with subcategories technical improvements take place which make a later product different from its earlier counterpart. The only way out of this difficulty is to recognize that it is just a matter of definition. A technology is defined by a cluster of similar functions based on a cluster of similar scientific and engineering principles. Since both functions and principles can be complex and multileveled, technologies can be defined in a variety of ways. There is nothing wrong with this as long as the definition of a technology makes sense technically and as long as any conclusions from a study of technological diffusion are compatible with that definition.

The extent of use of a technology can be defined in a variety of ways even for a given technology. Thus, the rate of the spreading of numerically controlled machine tools can be measured by counting the number of companies that have adopted them or the number of tools that have been purchased or the number that are in use or the fraction of production output in which they are used or in the dollar value of the output. Each of these measures is different and gives different information. Part of the definition of the extent of diffusion is the specification of the population in which diffusion is taking place. If, for example, we wish to follow the diffusion of continuous casting of steel, the relevant population is the set of all companies that cast steel. On the other hand, for a study of the diffusion of home appliances, the population would be the number of electrically wired households. In many cases, the actual population of users is not directly available and in other cases some other measure is more useful. Tonnage steel production, for example, is a good measure of the diffusion of steel use. Everyone uses steel so that the number of people using steel is not a useful measure. Diffusion therefore must be defined as the growth of steel use in this case. Tonnage per capita is therefore a better measure of the diffusion of steel use. We can still retain the nomenclature of "population," even though in this case it is rather abstract, by recognizing that a technology can diffuse by the increase of extent of use by each member of the population as well as by an increase in the number of members adopting the technology.

A description of the diffusion of a technology therefore should start with the following:

1. A definition of the technology
2. A specification of the population or its proxy within which the technology diffuses
3. A choice of the parameter which measures the extent of diffusion

Very often, the measures that can be used for diffusion analysis are limited by the nature of the data that are available. To measure the diffusion of microcomputer technology, for example, it would be desirable to know the number of people who use micros at least once a week. There are no data on this, but the number of micros that are sold each year can be obtained and used as a measure of diffusion of the technology. Convenience and availability of data often dictate the measures of diffusion that are actually used. In these cases, we cannot start with the first two of the above specifications and then choose a diffusion parameter. Rather, the two specifications become questions that must be asked relative to the diffusion parameters that are available. Given the available parameter, what is the technology it describes and what is the population it is diffusing into?

Once these definitions have been established, technological diffusion can be determined from analysis of the time series for the diffusion parameter that measures the extent of use of the technology. As indicated above, setting up these definitions is not always a trivial problem, and the definitions affect the kinds of conclusions that can be drawn as well as defining the framework of the analysis. A major difficulty of studying technological change is that of obtaining data; the diffusion parameters that can be used are often limited by the kinds of data available.

After a population and a diffusion parameter have been defined there are still two major issues that can be met. The first of these is that the population may not be constant in time. The number of firms, the number of electrically wired households, and the number of consumers all change with time. Indeed, we can expect that most populations in which technological diffusion takes place change as time passes. The time series for the diffusion must then be supplemented with a time series for the size of the population. This can be used to define a new diffusion parameter, which is the ratio of the population that has adopted the technology at a given time to the total population of potential adopters at that time, to give a time series for the fractional penetration of the technology. The curves for the penetration of television into American households shown in Figure 1-5 display measures of this type. The number of households increases with time so that the fractional penetration measures the diffusion in a growing population.

The second issue is that the time series for the diffusion parameter can show large irregularities so that it is difficult to interpret the temporal trends. Sometimes the use of the fractional penetration smooths out these irregularities but sometimes it does not. As an example, consider the diffusion of the open hearth process for making steel, which started before 1880 and was essentially complete by 1935. Two possible choices for the population of diffusion are the number of furnaces and the quantity of steel produced. The corresponding diffusion parameters are the number of open hearth furnaces and the amount of steel produced by the open hearth process. Both production and investment in new equipment are strongly affected by economic conditions, competition, and wars so that strong irregularities exist for both of these diffusion parameters. The fractional penetrations, however, behave differently. Dividing the number of open hearth furnaces by the total number of furnaces gives a fractional penetration parameter for furnaces and this still gives an irregular time series. The reason for this is that the number of furnaces is not large, the construction of a furnace represents a substantial investment, and the decision to build one is strongly influenced by idiosyncratic factors. But the ratio of open hearth production to total production gives a remarkably regular time series that is easily subjected to mathematical analysis. This is, therefore, a convenient parameter for the study of the position of open hearth technology with respect to the systemic trends of steel production, while the furnace penetration data are more suitable for studying the relation of the adoption of the technology to the changing conditions of war, peace, competition, and economics. In general, time series for diffusion or substitution are more regular for parameters that refer to large numbers of units.

The diffusion of a technology is always preceded by the diffusion of information about that technology. In ancient times, information spread slowly and the rate of adoption of a technology was often controlled by the rate of diffusion of information. Modern travel and communications have enormously increased the rate of information transfer, but even today information on new technologies is not available to everyone instantaneously. While advertising reaches large numbers of people very rapidly, it often takes more information about a new product than can be presented or absorbed in an advertisement before a decision to buy can be made. (This, of course, depends on the cost and complexity of the new product. The simpler and cheaper the product, the easier it is to make a decision.) Also, many innovations require a rather detailed knowledge before they can be judged by the potential users. This is the reason for the existence of marketing departments, technical sales representatives, and the entire apparatus by which firms make the capabilities of their wares known. Nevertheless, it is true that information is much more widely available than ever before and that its rate of diffusion is much less of a barrier to the adoption of new technology than in the past. We will return to this point when we present the factors that limit or accelerate the diffusion of technology.

Diffusion focuses on a particular technology and measures its spread independently of any other technology. However, a new technology often fulfills the same function in a better way than one already in use. The new technology then grows at the expense of the old. This is the phenomenon of technological substitution. The open hearth process, for example, grew at the expense of the older Bessemer process, and the diffusion of open hearth technology was also the substitution of open hearth for the Bessemer technology. Similarly, transistors substituted for vacuum tubes, electric lighting substituted for gaslight, and detergents substituted for soap. The relevant parameter is the ratio of use of the new technology to the total use of both the new and old technologies. This ratio often shows a gratifyingly regular progression with time so that substitution analysis is an attractive way of tracking technological change. Proponents of substitution analysis have gone so far as to say that all technological changes can be reduced to substitutions since basic human wants are few and every new technology can be interpreted as a new way of satisfying an old need. There is some merit in this view, but carrying it to extremes can obscure important factors. Motor transport, for example, can be thought of as a substitute for the horse since it was just a new transportation technology. But this ignores the fact that motor transport is so superior to the horse in speed, comfort, capacity, range, and cost that it is a qualitatively different technology that has totally transformed the nature of society. Also, long after horse transport had all but vanished, motor transport continued to diffuse in the sense that it was used for new purposes such as long distance food hauling in refrigerated vehicles. Also, motor transport led to a national highway system and became a major economic sector in a way that was not possible for the horse. We therefore adopt the view that substitution analysis is most useful when the new and the old technology are relatively close in purpose, function, and effects.

The discussion to this point is fairly general and provides a starting point for the description of the diffusion of innovation. But there is no single descriptive framework that can be applied to all cases of technological diffusion. The reason for this is that there are large differences between different kinds of innovations and the different populations within which they spread. It is useful to classify technological diffusion into various categories that display these differences, even though such a classification cannot be unique or exhaustive. We will therefore distinguish among technological diffusion of the following kinds: diffusion in business and industry; diffusion of industrial process innovations; diffusion of consumer products; diffusion of services; diffusion in government or the military; diffusion of "low-tech" innovations; diffusion of "high-tech" innovations; diffusion of major, radical innovations; diffusion of incremental innovations. A little reflection reveals that each of these has different characteristics. It is worthwhile to comment on some of these differences.

Diffusion in business and industry is more likely to depend on purely economic factors than diffusion in government or among consumers. Also, populations in which technology diffusion takes place are usually rather small for innovations that address industrial processes or equipment. Innovations that are industry-specific affect a relatively small number of firms, often as few as several dozen or less. There are, however, a large number of consumers, even when broken down into market segments. Data for diffusion of consumer products are therefore usually more satisfactory from a statistical point of view than data for industrial innovation. This is particularly evident in some process innovations which can be profitably used only by a small number of larger firms so that the population in which diffusion takes place is limited. Of course, not all innovations in business and industry have this character. The spread of microcomputers in the business world has more the character of the diffusion of a consumer product even though the diffusion of large mainframe computers does not.

The spread of technology in the government or the military is ultimately driven by government policy on national needs so that economic factors, while important, are secondary. Furthermore, technological advance is often the result of explicitly stated government specifications. While industrial users of a technology often play an important role in inducing change and innovation among suppliers, the government has a much more direct involvement in the specification of the technologies it uses.

An important distinction is that between "low-tech" and "high-tech" innovation. The common usage of the term "high-tech" implies advanced technology based on recent and sophisticated science and engineering. It is commonly applied to such fields as computers, telecommunications, and robotics. From the point of view of innovation, however, we need a more careful definition. A low-tech innovation will be taken to be the one that is essentially fully formed when it is first put into use. This means that there will certainly be a period of some changes and improvements and of learning how to use the technology most efficiently, but this period is not long; it is a relatively small fraction of the life cycle of the technology. A high-tech innovation, on the other hand, is more complex and capable of extensive improvement and change over a large part of its life cycle. It requires attention from highly trained technical people for its installation, use, and development; its full potential is not realized until long after its first introduction. Clearly, high-tech innovations will diffuse in a different way and at different rates than low-tech innovations. The terms high and low technology define two ends of a spectrum and many innovations are somewhere between, but the distinction is still a useful one, particularly in times of major technological transitions such as we are currently in.

Radical innovations such as the automobile, the telephone, and the computer encompass the technological changes that dramatically change society. They generally are based on major scientific and engineering advances and they create entire new industries as they evolve over many decades. Their diffusion takes place as they evolve, and they give rise to ancillary or subsidiary industries. Incremental innovations, on the other hand, are those that contribute to the major technology; they have shorter life cycles and often take place only within a subsection of the industry of the major technology. Usually, process innovations are of the incremental kind. Of course, this is not always true. Catalytic cracking of petroleum, for example, while not a radical innovation in the sense of the transistor, is far from a merely incremental advance.

Enough has been said to show that technological diffusion is a set of quite heterogeneous phenomena. The particular nature of the technology and the circumstances in which it exists and grows are basic determinants of the causes, mechanisms, and rates of its diffusion. However there are some obvious generalities that can be made. First, for a technology to diffuse, there must be some advocates of the technology who act as the primary agents of change. Second, there must be some adopters who accept and implement the technology. Third, there must be reasons for both the advocacy and the acceptance. Fourth, there must be a mechanism of communication between the advocates and the adopters. Fifth, there is a quantity and quality of information that must be attained before a decision to adopt can be made. Sixth, for a given level of information, there is a probability of adoption that may be different for different adopters. Seventh, technological diffusion takes place in a changing technical, economic, political, and social environment. The nature of this environment has a profound effect on the rates of diffusion and even on the types of technologies that will diffuse. While this is only a bare framework of the adoption process it illustrates that technological diffusion is a complex technical, economic, and social phenomenon.

Our interest is in describing technological diffusion quantitatively and in elucidating its causes and mechanisms. It is desirable to have a mathematical description of diffusion that displays whatever underlying constancy of causes and circumstances exists. Real-time series for diffusion often have a regularity that makes such a description possible. But just as often they show fluctuations that reflect changes in causes or the environment in which diffusion takes place. We will start by examining the analytic framework for diffusion and substitution that seeks to represent some of the regularities of technological change.

DIFFUSION ANALYSIS

The first empirical question to be asked about data on the diffusion of technology is whether or not time series for specific instances of diffusion display any regularities. The answer is a qualified yes. Analysis of a great many cases supports the general thesis that technological diffusion is a result of both a smooth evolution in time and erratic factors. That is, it is often possible to construct a smooth curve to represent the time series, but the data points can sometimes deviate substantially from the curve. It is important to emphasize that these deviations have a different character from those of experimental error encountered in laboratory measurements. Of course, there are problems of measurement and of acquiring accurate data in describing technological diffusion. But even when these are minimized, large deviations from a smooth curve can still exist that cannot be removed by getting better data. Furthermore, these can often be identified with specific changes in the business, political, or social environment or with the specific differing characteristics of the adopters. This identification is an important part of the effort to understand the origins of technological change. That is, there are real departures from regularity that have nothing to do with methodological errors and that must be explained. The size of these deviations varies greatly depending on the specific case being studied and on the nature of the parameters used to describe the extent of diffusion. For some cases the deviations are so large that hardly any general conclusions can be drawn, while for others the data points fall on a very smooth curve. For example, the data for residential consumption of electricity per customer in the United States fall on quite a smooth curve with minimal fluctuations. On the other hand, as pointed out earlier, the tonnage of steel produced by the open hearth method is very erratic from year to year. About all that can be said in the latter case is that over the decades there has been a substantial increase in production. From year to year, however, production fluctuates widely in response to political and economic conditions. It is in terms of these short-term factors that the tonnage production of steel must be understood. The data on consumption of electricity represents the continual and inexorable electrification of American society, which is not strongly affected by short-term changes in economic or social conditions, while steel production responds very quickly to such changes. The critical role of the nature of the diffusion measure is illustrated by the fact that a smooth curve is obtained when the fraction of steel produced by the open hearth process is plotted against time. Thus, while steel production is controlled by short-term considerations, the importance of open hearth technology relative to other methods of making steel exhibits a long-term regularity.

Given all this, it is still true that a great many instances of technological diffusion can be represented by smooth, continuous curves and described by growth equations of the type described in Appendix 1. The most common of these are S-shaped curves that start out slowly, accelerate, and then slow down as they approach a constant level, the slope of the curve approaching zero at both short and long times, with a maximum rate somewhere between.

A rationale for the ubiquity of the S-curve can be developed as follows: Consider an innovation for which the total number of potential adopters is N_T and let $N(t)$ be the number that have adopted at time t. At any given time, let the probability that there will be an adoption during the time interval from t to $t + dt$ be $P(t)dt$. Then the total number of adoptions during dt is given by the probability of an adoption multiplied by the number of remaining potential adopters. If we call this number dN, then

$$dN = P(t)\,(N_T - N)dt \tag{4-1}$$

or

$$\frac{dN}{dt} = P(t)\,(N_T - N) \tag{4-2}$$

which is the differential equation for the diffusion rate.

Note that as N approaches N_T, the rate goes to zero so that the upper limit of the S-curve defines the ceiling of possible diffusion.

The specific form of the adoption equation depends on the form of $P(t)$. How can the probability of adoption be expected to vary with time? In most cases, we would expect $P(t)$ to be small in the early stages of adoption and to increase as more and more of the population adopts the innovation. Only a few venturesome souls will take up an untried product or process; costs for an innovation generally decrease with time; innovations usually improve considerably as time passes; the natural resistance to change is overcome slowly; it takes time for information about the innovation to be absorbed; its increasing utility and spreading acceptance induce more people to adopt it; as the innovation becomes more widely adopted it becomes familiar and the resistance to change weakens. All these factors support the idea that the probability of an innovation being adopted is small at first but increases with time. When $P(t)$ is an increasing function, the rate in Equation (4-2) is the product of an increasing and a decreasing function. The number of adoptions per unit time is then similar to a bell-shaped curve and, when integrated, gives an S-shape for $N(t)$.

Another interesting rationale for the S-curve, based on the effectiveness of information transfer, has been given by Casetti (1969), who assumes that adoption results primarily from messages about the innovation from those that have already adopted it, that potential adopters have different degrees of resistance to adoption, and that resistance is overcome by a sufficient number of messages about the innovation. Then the probability per unit time that a potential user will adopt is proportional to the number of messages, which is proportional to the number that have already adopted. That is,

$$P(t) = k(N)N \tag{4-3}$$

where $k(N)$ is a proportionality factor that measures the effectiveness of the message in inducing adoption. Casetti assumes that $k(N)$ decreases with increasing number of adopters because those with higher resistance to change will adopt later and require more messages to be induced to accept change. Now Equation (4-2) becomes

$$\frac{dN}{dt} = k(N)N(N_T - N) \tag{4-4}$$

For constant N_T this always leads to an S-curve.

Note that for $k(N) = k$, a constant, Equation (4-4) reduces to the logistic:

$$\frac{dN}{dt} = kN(N_T - N) \tag{4-5}$$

In Casetti's scheme, this means that the messages are equally effective for all potential adopters and that the message rate is constant.

It is easy to imagine situations that do not conform to Casetti's hypothesis. The main channels of communication, for example, might be media advertising, a fixed number of salesman, trade journals, and so forth. The results would then be different. For a message rate and effectiveness that is independent of the number of adoptions, $P(t)$ is constant and Equation (4-2) becomes

$$\frac{dN}{dt} = k(N_T - N) \tag{4-6}$$

which is the equation for the decaying exponential rise. This might occur, for example, as the result of constant media advertising to a population of individuals whose susceptibility to adoption is a constant independent of the number that have already adopted.

The more common interpretation of the logistic curve is in terms of the theory of contagion. For a disease that spreads by contact, the rate of increase of its incidence is proportional to the frequency of binary contacts between infected and uninfected individuals. Whenever there is such a contact, there is a finite chance that the uninfected individual will contract the disease. This immediately yields Equation (4-5). In the same way, those that have not yet adopted an innovation are presumed to be induced to do so by those that already have. That is, when a nonadopter meets an adopter there is a probability that the nonadopter will take on the innovation because of the perceived benefits being enjoyed by the adopter. This interpretation is clearly compatible with Casetti's approach.

These various interpretations of the logistic equation undoubtedly contain elements of truth in specific cases of technological diffusion. But no universally applicable interpretation is valid. It must be remembered that the logistic is a straightforward result of the truncation of a series representation of the general growth function as derived in Appendix 1. In fact, the earliest applications of the logistic curves to technological diffusion were empirical treatments. Griliches (1957), in his classic analysis of the diffusion of hybrid corn, regarded the logistic as just a convenient representation of the data, while Mansfield (1961), in his pioneering work on the diffusion of twelve industrial innovations in four industries, derived the logistic equation from a series expansion independent of any mechanism just as in Appendix 1. The nature of the process of innovation diffusion can be understood only from detailed information ancillary to the data itself although the data can suggest lines of inquiry.

Now let us examine the logistic and some of its generalizations. For this purpose we rewrite Equation (4-5) in the reduced form

$$\frac{dy}{dt} = \beta y(1 - y) \tag{4-7}$$

where

$$y = N/N_T \text{ and } \beta = kN_T \tag{4-8}$$

The solution of Equation (4-7) is

$$y = \{1 + \exp[\beta(t_{1/2} - t)]\}^{-1} \tag{4-9}$$

where $t_{1/2}$ is the time at which $y = 1/2$.

This is often written as

$$ln[y/(1 - y)] = \beta(t - tt_{1/2}) \tag{4-10}$$

since then β and $t_{1/2}$ can be computed from a rectilinear fit of a plot of $ln[y/(1 - y)]$ vs. t.

These equations have been discussed in Appendix 1, but there are several points worth emphasizing here. The first is that while y becomes very small with decreasing t, it never actually becomes zero for any finite time, since $y \rightarrow 0$ as $t \rightarrow -\infty$. This means that the logistic can be applied to technological diffusion only after at least one adoption has taken place. However, this is more of a theoretical than a practical issue. The second point is that while y approaches unity for large times, it never actually gets there since $y \rightarrow 1$ as $t \rightarrow \infty$. This is in accord with many observations on technological diffusion, which seldom goes to 100 percent. Not all electronic equipment is completely based on solid state circuitry, for example; there are still vacuum tubes in use. There is even some horse transport in use after over eighty years of growth of motor transport. The final point is that $t_{1/2}$ marks the inflection point in the logistic curve of Equation (4-9) and the maximum of the logistic rate of Equation (4-7); the logistic has a symmetry about its half time. This limits its flexibility in describing technological diffusion.

The logistic gives a reasonable fit for a number of diffusion phenomena, but in a number of cases there are systematic deviations between the data points and the best fitting logistic that show that some other equation might give a better fit. This has given rise to several generalizations of the logistic.

The simplest generalization of the logistic is to recognize that not all members of a population are potential adopters. Then, instead of Equations (4-7) and (4-9) we have

$$\frac{dy}{dt} = \beta y(Y - y) \tag{4-11}$$

$$y = Y\{1 + \exp[\beta(t_{1/2} - t)]\}^{-1} \tag{4-12}$$

where Y is the maximum fraction of the population that will ultimately adopt the innovation. This is actually the form used by Griliches (1957) in his study of the diffusion of hybrid corn, since he found that in many cases not all available acreage was ultimately planted in hybrid corn. In principle, the introduction of an upper limit less than 100 percent changes nothing since it only means that the nature of the adopting population was incorrectly defined initially. This is easily corrected by dividing Equations (4-11) and (4-12) through by Y to recover the original logistic. However, it is not always possible to determine what the upper limit will be. Then Equations (4-11) and (4-12) must be used, treating Y as a disposable parameter to be determined from the data. Thus while the generalization of the logistic to the case of an upper limit less than 100 percent is theoretically trivial, it can be quite important in practice. Martino (1972) has investigated the errors induced in estimates of the half time $t_{1/2}$ and rate constant β by errors in estimates of the upper limit. He finds that if the upper limit is underestimated, $t_{1/2}$ and β are overestimated and vice-versa.

A generalization of the logistic has been given by Sharif and Kabir (1976) that includes several equations as special cases. This generalization is:

$$ln[y/(Y - y)] + \sigma/(Y - y) = C + kt \tag{4-13}$$

where σ is a parameter that can take any value from zero to unity, depending on the set of data to be represented. When $Y = 1$, and $\sigma = 0$, Equation (4-13) reduces to the ordinary logistic; when $0 < Y < 1$ and $\sigma = 0$, it reduces to the logistic with an upper limit less than 100 percent; when $\sigma = 1$, it reduces to Floyd's equation (1968). The motivation for developing (4-13) was that the logistic ($\sigma = 0$) overestimated projections of trends from early data, while Floyd's equation ($\sigma = 1$) underestimated them. A judicious choice of σ between zero and unity can improve this situation.

The rate equation for the generalized logistic is obtained by differentiating Equation (4-13) to get

$$\frac{dy}{dt} = ky(1 - y)[(1 - y)/(1 - y + \sigma y)] \tag{4-14}$$

where we have let $Y = 1$ since this can be done without any loss of generality. When $\sigma = 0$, Equation (4-14) becomes the differential equation for the ordinary logistic. Note that for small y the effect of σ is small since the term in the square brackets is then near unity. As y becomes larger, however, this term becomes less than unity and the rate slows down relative to that for the ordinary logistic. For this reason, σ is called the delay coefficient.

The generalized logistic has a point of inflection and a maximum rate, but these do not occur at the time when $y = 1/2$, as is the case for the ordinary logistic. To determine the point of maximum rate, differentiate Equation (4-14), set the result equal to zero, and solve for y^*, the value of y when dy/dt is a maximum and when y goes through its inflection point when plotted against time. Doing this, we get a quadratic equation for y^*.

$$2(1 - \sigma)y^{*2} - 3y^* + 1 = 0 \tag{4-15}$$

whose solution is

$$y^* = [3/4(1 - \sigma)]\{1 - [1 - 8(1 - \sigma)/9]^{1/2}\} \tag{4-16}$$

We choose the negative root in (4-16) because it gives the correct inflection point $y^* = 1/2$ when $\sigma = 0$.

As σ varies from zero to unity, y^* decreases. For $\sigma = 1, y^* = 1/3$, as can be seen directly from (4-15) or by taking the limit of (4-16). The generalized logistic therefore allows for a variety of S-curves, but it is still a limited class with inflection points between 1/3 and 1/2. Several of these curves, for different values of σ, are shown in Figure 4-1. They show that the delay

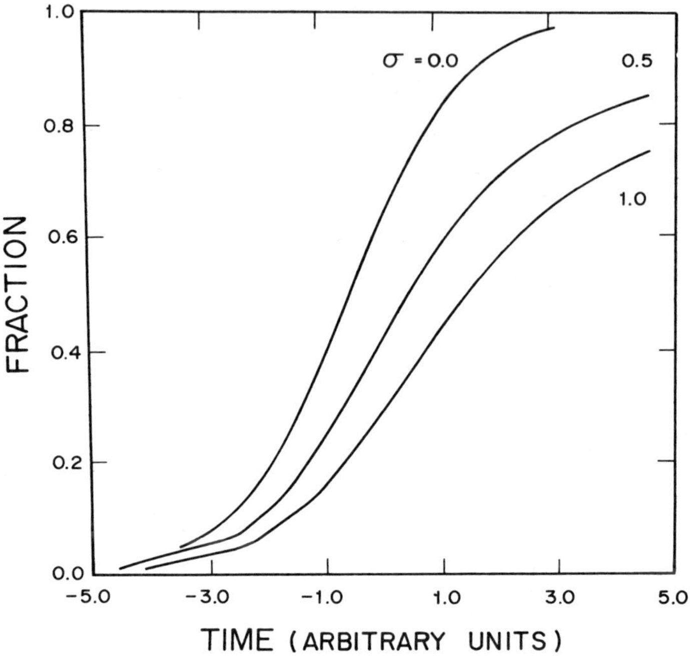

Figure 4-1 The generalized logistic with different delay coefficients.

coefficient, σ, slows down the diffusion rate relative to the logistic, especially for times beyond the inflection point. That is, the probability of adoption is not strictly proportional to the number that have already adopted but is lower than this at later times. This is in accord with Casetti's hypothesis that later adopters have a greater resistance to adoption.

An alternate approach to the mathematical representation of S-curves is based directly on probability distribution functions. This was developed in an important paper by Pemberton (1936), who studied the diffusion of three social innovations: postage stamps among countries, legal limitations on taxation by local governments among states, and compulsory education among states. For all of these, Pemberton found the data to display an S-shape when plotted against time, with relatively small deviations from a smooth curve. A variety of S-shaped growth equations were already available. The logistic had been extensively used in biology to represent the growth of individuals and of populations (Pearl 1925) and the Gompertz equation had been used by economists (Kuznets 1930). Even the cumulative normal curve adopted by Pemberton had been used to describe social progress (Lehfeldt 1916). Pemberton's choice of functional form was not made on the basis of goodness of fit. He pointed out that any of the available S-curves would fit the data reasonably well. Rather, his choice was made on the basis of his view of the factors controlling adoption. His reasoning went as follows:

When a new innovation is available, a number of factors enter into the decision to adopt. Depending on the nature of the innovation, these factors can include income, social standards, degree of competition, age of capital stock, profit level, interest rates, market prices, political conditions, currency exchange rates, local laws, Federal regulations, transportation costs, existing product mix, and so on. These factors are of different intensity for different members of the adopting population so that at any time there will be only a fraction that are prepared to adopt. Furthermore, the chance that all factors are favorable for adoption at very early times is low. Conversely, the chance that a very long time has to pass before the factors are favorable is also low. In fact, we can expect some average time to exist for which the fraction of the population subject to factors sufficiently favorable to adopt is a maximum. Also, since the factors are many and vary incrementally among the members of the adopting population, it is reasonable to assume a normal curve for the fraction adopting at a time t. That is, the number of adoptions between t and $t + dt$ is assumed to be

$$dN = N_T(1/\sigma\sqrt{2\pi})\exp[-(t - \tilde{t})^2/2\sigma^2]dt \tag{4-17}$$

so the rate of fractional adoption is

$$\frac{dy}{dt} = dN/N_T dt = (1/\sigma\sqrt{2\pi})\exp[-(t - \tilde{t})^2/2\sigma^2] \tag{4-18}$$

where $t^{\sim}$ is the average time of adoption and σ is the standard deviation. (Do not confuse this with the delay coefficient in Equation (4-13). I have used the same symbol for the delay coefficient and the standard deviation so as to conform to common usage.)

Equation (4-18) is the familiar bell-shaped curve of probability theory. Its integration yields the fractional diffusion of technology as a function of time;

$$y(t) = (1/\sigma\sqrt{2\pi})\int^{t} \exp[-(t' - \tilde{t})^2/2\sigma^2\, dt']. \tag{4-19}$$

This is the cumulative normal curve. It is really not very different from the ordinary logistic since it is a symmetric S-curve with $\tilde{t}$ being the time at which $y = 1/2$ and at which the rate of diffusion is a maximum.

SUBSTITUTION ANALYSIS

Substitution analysis is concerned with tracking the replacement of one technology by another when both the old and the new technologies are designed to perform the same function. Examples are the replacement of wood floors by plastic tile, the replacement of vacuum tubes by transistors, the replacement of soap by detergent, and the replacement of sailing ships by power-driven ships. The relevant parameter is the fraction of the use captured by the new technology. Thus if N_A is the amount of the market captured by the new technology and N_B is the amount retained by the old technology, then the substitution measure is defined by

$$f = N_A/(N_A + N_B) = N_A/N_T \tag{4-20}$$

where N_T is the total amount.

N_A and N_B can be measured in terms of product units, dollar volumes, production quantities, or any other convenient measure of the extent of use of the technology.

Diffusion and substitution are related, but different, concepts. *Diffusion* is concerned with the rate at which a technology spreads and approaches its saturation limit. *Substitution* is concerned with the rate at which one technology replaces another. There is one special case when diffusion and substitution are identical, namely when only two technologies exist for a given market or function and when the market or function is fully saturated by the sum of the two technologies throughout time. In this case f of Equation (4-20) is the same as the diffusion measure defined previously.

The relation between diffusion and substitution can be obtained by differentiating Equation (4-20) with respect to time to get

$$d\ln N_A/dt = d\ln f/dt + d\ln N_T/dt \tag{4-21}$$

Clearly, if N_T is constant then the differential equations for the logarithms of the diffusion and substitution parameters have the same form. In general, however, N_T is not constant and the rate of diffusion depends on the rate of change of the sum of the shares of the two technologies as well as on the rate of substitution.

In many cases, not only is N_T variable, but it changes erratically in time because both N_A and N_B fluctuate irregularly from year to year. Often, the fraction f is a smooth function of time in spite of the fact that the N_A and N_B may not be. This is one of the reasons that substitution analysis is so useful.

Substitution analysis gives a mathematical description of the interaction of two technologies and shows not only how the new technology grows but also how the old one declines. Since f is the fraction of new technology, $q = (1 - f)$ is the fraction of old technology and

$$dq/dt = -df/dt \tag{4-22}$$

Diffusion analysis, on the other hand, is concerned only with the rate of adoption of the innovation and pays little attention to the divestment or displacement of an old technology.

By definition, f varies from zero to unity. Empirical studies show that, when plotted against time, f often exhibits an S-shape. In fact, the S-curve seems to be more firmly established for substitution than for diffusion. The same equations described above for the representation of technological diffusion are used to describe substitution. Indeed, some of them (such as the generalized logistic with delay coefficient) were first used for substitution

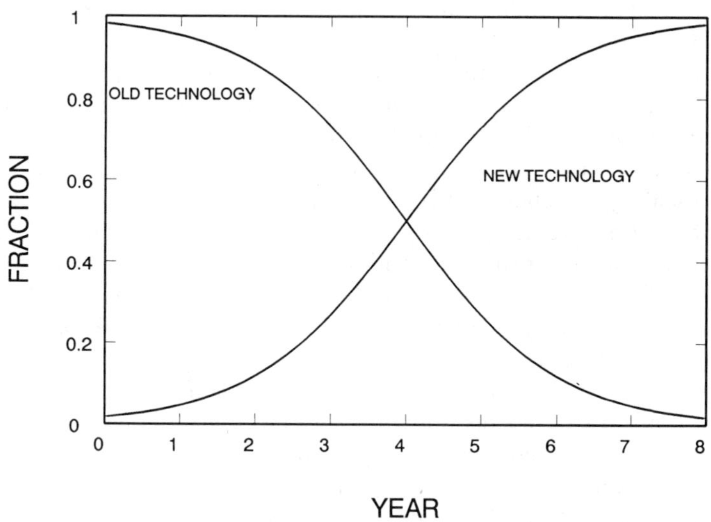

Figure 4-2 Fraction of new and old technologies during a substitution.

analysis. Also, the qualitative rationales for explaining the S-shape of diffusion curves carry over essentially unchanged as rationales for the S-shape of substitution curves. As a rule, substitutions are more accurately represented by the S-curve equations than are diffusion data.

The logistic is widely used to represent substitution. If we adopt it here, then the rate of increase of the new technology is given by

$$df/dt = \alpha f(1 - f) \tag{4-23}$$

and the rate of decrease of the old technology is

$$dq/dt = -\alpha q(1 - q) \tag{4-24}$$

These equations have the solutions

$$f = \{1 + \exp[\alpha(t_{1/2} - t)]\}^{-1} \tag{4-25}$$

$$q = 1 - f \tag{4-26}$$

A plot of the fraction of the market captured by the new technology and the fraction retained by the old technology is shown in Figure 4-2, as computed from these equations. The use of substitution in displaying how an old technology decays as another grows is evident.

MULTILEVEL SUBSTITUTION

The analysis given above assumes substitution is binary in that there are only two technologies competing with each other, a well-established technology being displaced by a newer and better one. This binary model is very useful and yields valuable results. In general, however, it is an oversimplification. If often happens that a series of technologies for a given objective are introduced at different times, and two, three, or more technologies may be competing with each other at any given time. That is, substitution may be a multilevel phenomenon in which an initial technology may be substituted for by a second while the second is being displaced by a third and so on. Mathematical approaches to this more complex situation have been developed by Sharif and Kabir (1976), who studied the overlapping competitive substitution of piston, turboprop, and turbojet aircraft, and by Girifalco and Simozar (1985) in their analysis of the competition among the Bessemer, open hearth, basic oxygen, and electric furnace processes of making steel. The method given here is that of Girifalco and Simozar.

There are two alternative approaches that can be used for multilevel substitution. Both are based on reducing the multiple interaction among the technologies to binary competition between technologies or groups of technologies. In this way, the methods developed for simple binary substitution can be easily extended to the more complex case of multilevel substitutions. The mathematical details of both approaches are given in Section I of Appendix 1.

The first approach assumes that a superior technology displaces all inferior ones as a group. Thus, if we order the technologies in a series of increasing superiority and choose one member of this sequence, then the binary model is applied to a binary competition between that member and all earlier members in the sequence treated as a group. This leads to a series of binary equations that describe multilevel substitution. The second approach assumes that all technologies are in pairwise competition with each other so that the binary model applies to each pair and multilevel substitution is the set of pairwise substitutions.

The two approaches not only lead to different equations, they also imply different conditions of decision making. In the first approach, the decision to adopt a new technology is driven primarily by the technology itself, and it is not important which old technology the adopter currently holds. Then the superior technology displaces all inferior technologies without discriminating among them. In the second approach, adopters of a new technology behave differently depending on their current technology, which affects the rate at which they adopt the new. The decision to adopt is really a set of binary decisions, and the adopter does not regard all previous technologies as equivalent when considering a new technology. The new technology therefore competes with each of the other technologies separately while they in turn compete with each other. The second approach also implies that adopters will make their changes at different rates depending on which technology they are currently using.

It is entirely possible that for a given set of substitutions, some will take place according to the first approach while others will take place according to the second. Which is the best approach for particular substitutions must be determined empirically from the data and from the nature of the technologies.

An interesting point to note is that the sequence of substitution is generally taken to be temporally determined. That is, the later technologies are assumed to displace earlier ones. But this is not necessarily always true. A technology can displace some earlier ones, thereby gaining market share initially, while it is in turn being displaced by another which could have started earlier. It is not necessary that the sequence of displacement be the same as the temporal sequence of innovation. An important case in point is that of steelmaking, in which the electric furnace is displacing the basic oxygen method in spite of the fact that the electric furnace process is much older than the basic oxygen process.

Let's examine a simple illustration in which there are three technologies labeled 1, 2, and 3, such that technology 2 displaces technology 1 while technology 3 displaces 1 and 2 as a group. Then the fraction of technology 2 that is replacing technology 1 is given by

$$f_2 = n_2/(n_1 + n_2) \tag{4-27}$$

and the fraction of technology 3 replacing both technologies 1 and 2 is

$$f_3 = n_3/(n_1 + n_2 + n_3) \tag{4-28}$$

and because the fractions must add up to unity, we have

$$f_1 = 1 - f_2 - f_3 \tag{4-29}$$

where n_1, n_2, and n_3 are the market shares of the three technologies with

$$n_1 + n_2 + n_3 = 1 \tag{4-30}$$

These are just special cases of Equations (A-71) and (A-72) in Appendix 1.

For a tri-level substitution, Equations (A-74) in Appendix 1 give the n's as

$$\begin{aligned} n_3 &= f_3 \\ n_2 &= f_2\,(1 - f_1) \\ n_1 &= 1 - n_2 - n_3 \end{aligned} \tag{4-31}$$

Figure 4-3 shows an example of the shares of three competing technologies as described by Equations (4-31), when the f's are computed by logistic functions with different half-times. As expected, the first technology starts at a high level but decreases as the second technology grows at its expense. As time passes, the third technology becomes important, thereby decreasing

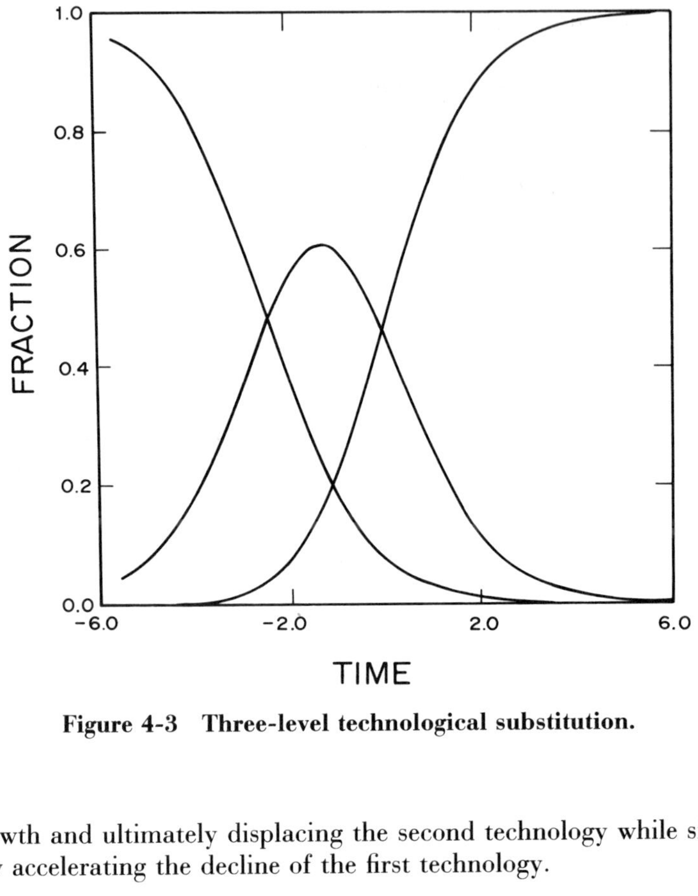

Figure 4-3 Three-level technological substitution.

the growth and ultimately displacing the second technology while simultaneously accelerating the decline of the first technology.

THE DIESEL LOCOMOTIVE: AN EXAMPLE OF BINARY SUBSTITUTION

For over a century the steam locomotive was the dominant railway motive power. From George Stephenson's "Rocket," built in England in 1829, to the great engines of the twentieth century, the steam locomotive grew in size, power, and efficiency. The enormous expansion of railway systems that united entire continents was accomplished with the steam locomotive. Yet, as for all technologies, the steam locomotive had inherent limitations. It was noisy and dirty, it did not have a high energy efficiency, and the open fire and high steam pressure made it unsafe. Furthermore, maintenance was frequent and costly.

The first alternative to steam was electric traction in the form of the Siemens locomotive in 1879. However, the ancillary systems of power generation, substations, and overhead wires or third rails effectively limited electric traction to urban systems. The steam locomotive retained its dominant position until the advent of the diesel–electric.

In 1912 the diesel-Klose-Sulzer Company built a 1,000-HP locomotive. The first commercially successful diesel in the United States was built by the American Locomotive Company in 1925. Its first unit was a switching locomotive which was in continuous service from 1925 to 1957. But it was not until the 1930s that the diesel was developed to the point that ensured its success. A major technical problem was that of transferring power from the diesel engine to the wheels. The diesel cannot quickly adapt to rapidly varying loads, so that direct power transmission is very limiting and mechanical; hydraulic and electric transmission systems were used to try to solve this problem. This was accomplished by General Motors, which in 1934 built a diesel–eletric that was smaller, lighter, and more powerful than any to that date. In the diesel–electric the diesel engine is linked to an electric generator which drives electric motors to transmit power to the wheels. The success of this design was the starting point of the major spread of diesel locomotive technology.

The advantages of the diesel over steam are many. The diesel is quieter and cleaner, it does not require any feed water, and its energy efficiency is much higher. The diesel has a smoother ride and smoother acceleration, which results in greater comfort for passengers and less wear on the track system. Acceleration is more rapid, it can run at higher speeds, and standby costs are less since the engine can be turned off even for short idle periods as there is no need to maintain a head of steam. A great economic advantage is that the diesel locomotive has very low maintenance requirements and can make trips of thousands of miles without servicing. With these advantages, the widespread diffusion of the diesel-electric locomotive was assured.

A quantitative investigation of the spread of the diesel was performed by Mansfield (1961). As part of a study of twelve innovations in four industries, he determined the dates at which each of twenty-four major railroads first adopted the diesel locomotive and plotted the cumulative number of first adopters as a function of time. This, of course, is a very incomplete measure of diffusion, but Mansfield was primarily interested in the conditions that led a firm to first adopt a new technology. After a first adoption, it takes time for a new technology to spread through a company, so Mansfield also studied the rate of intrafirm diffusion by plotting the fraction of all locomotives that were diesels for each of thirty railway firms (1968). This was really a study of the substitution of steam locomotives by diesel-electrics. The spread of a technology within firms still does not give direct information on the spread throughout the country. I therefore supplemented Mansfield's work with an analysis of the total number of diesels in use and the fraction of locomotives that are diesel as a function of time, using readily available data from the "Historical Statistics of the United States; U.S. Department of Commerce."

Mansfield's data in his Table 9.1 (1968) are plotted in Figure 4-4, along

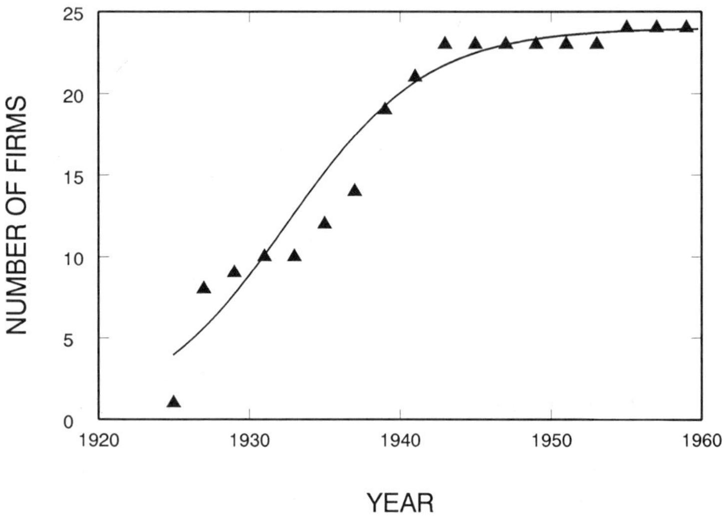

Figure 4-4 First adoptions of the diesel locomotive.

with a logistic curve obtained by fitting the data to Equation (4-10) using the standard linear regression least squares method. The curve is given by $ln(N/24 - N) = -417.011 + 0.21579t$, which corresponds to a half-time of 1932 and a 90 percent takeover time of twenty years. (This result is slightly different from that reported by Mansfield since it uses the data in his table, which is slightly different than that in his earlier work.) It therefore took some time for firms to decide to use the new technology. The first firm adopted in 1925, but it was not until 1955 that all twenty-four had done so. After the first few firms had adopted a diesel in the 1920s, the number of adoptions leveled off until the diesel–electric was introduced in 1934 by General Motors. The number of adoptions then increased greatly. Note that while the logistic represents the data fairly well overall, it underestimates the number of first adoptions in the earlier years and overestimates them in the middle years.

Using his data base on the rate of first adoption of twelve innovations in four industries, Mansfield searched for the factors that determined the decision to innovate by assuming a linear relation between these factors and the rate constants in the logistic, and computing a correlation coefficient to test for statistical significance. These factors included the degree of profitability of the innovation, the size of the investment needed to innovate, the durability of the equipment being replaced, the growth rate of the firm at the time of adoption of the innovation, the phase of the business cycle (up or down), and the year in which the innovation was adopted. This last factor was meant to represent such effects as increased effectiveness of communications or increased sophistication of business analysis methods over time. Only the first two factors showed a statistically significant correlation with the rate constants. The rate was higher for higher profitability and lower for higher investment cost. The fact that a correlation could not be found for the other factors does not necessarily mean that they are not important, since the data base was small.

In his later study, Mansfield analyzed the rate at which thirty different railroad companies acquired diesel locomotives after the date of their first adoption. This was an intrafirm substitution analysis of diesel for steam, since he plotted the fraction of all locomotives that were diesel for each firm, fitting the results to a logistic function. Most of the data fit the logistic fairly well, although there were half a dozen or so for which the fit was poor. The results showed that the 90 percent takeover times ranged from 3.3 to 17.6 years with an average of 11.3 years. The half-times, however, showed less variation, the bulk of them being between 1940 and 1944 with an average of 1942. This suggests that competitive pressures were such that no company could afford to remain behind for very long.

Just as in the case for first adoption, Mansfield looked for correlations between the logistic rate constants and factors that might affect that rate. The factors he examined were firm size, profitability of replacing steam by diesel, firm liquidity, and degree of risk. Profitability, liquidity, and risk all had the expected correlations (positive, positive, and negative), but the correlation between rate and size was not statistically significant. Mansfield had expected a negative correlation on the grounds that small firms could not afford to maintain systems for two different types of traction power.

Now let us look at the national statistics for the total number of diesel locomotives in service as a function of time. These are shown in Figure 4-5 along with a logistic curve fitted to the data whose equation is $ln[(N/30{,}500 - N)] = -745.237 + 0.32811t$.

This corresponds to a half-time of 1950 and a 90 percent takeover time of 11.5 years. The logistic fits well, its correlation coefficient being 0.9968, but there is a clear deviation at the latest time. In fact, the number of locomotives decreases from its maximum value, a fact that cannot be represented by the logistic. To what degree this is the result of competition of other forms of transport or increasing efficiency of the diesel would require further analysis. Actually, the total number of locomotives was decreasing since 1925, when there were nearly 68,000 in service, to 1960 when the number was only about 31,000. In the early days, most of this decrease was the result of improvements in the capabilities of steam locomotives and in the efficiency of their use. Later, the increased power of diesels, as well as competition from trucks, decreased the total number of locomotives even further.

The substitution of diesel for steam is readily computed as the fraction of all locomotives that were diesel. This is shown in Figure 4-6, which gives the increasing percentage of diesels and the decreasing percentage of steam locomotives as a function of time. The curves are obtained from a logistic fitted to the diesel data, the curve for steam being just the difference between

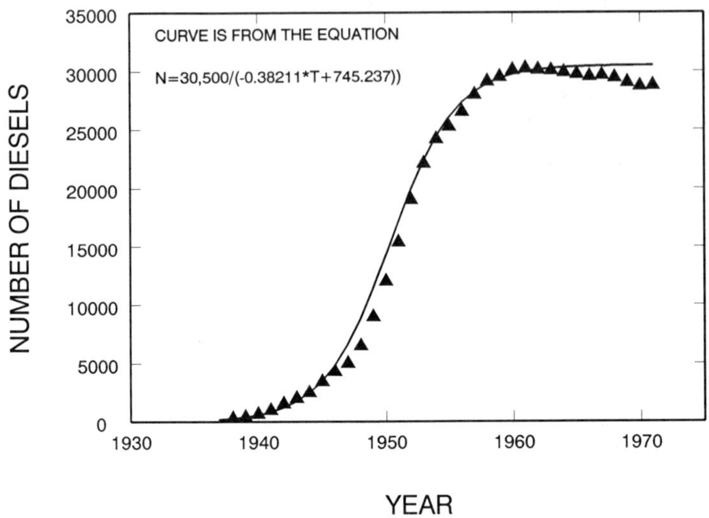

Figure 4-5 The number of diesel locomotives in service in the United States.

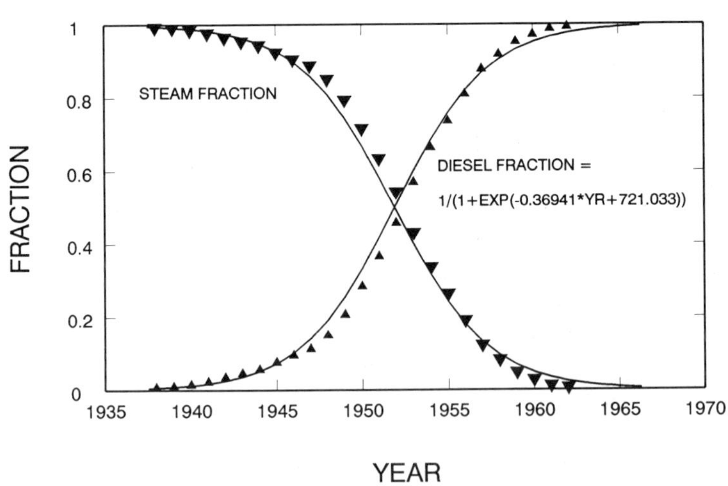

Figure 4-6 Fraction of diesel and of steam locomotives in service in the United States.

the diesel percentage and 100 percent. The equation for the diesel logistic is $ln[f/(1 - f)] = -743.378 + 0.38099t$, corresponding to a half-time of 1951 and a 90 percent takeover time of 11.5 years. As expected, the substitution curve is much smoother than the diffusion curve. Also, the substitution curve is represented by the logistic very well with some small deviations just before the saturation level. The curves for the national diffusion and substitution of diesels display the strong effect of General Motors' diesel–electric. In spite of the fact that diesel locomotives had been built as early as 1912, it was not until the late thirties, with the advent of the improved transmission technology, that any extensive diffusion took place.

It is useful to summarize and compare the time constants for the various plots of the spread of diesel locomotives, and we do this in Table 4-1.

This table shows that first adoption was slower than the rate of increased use of diesels once they were adopted. This is expected, since first adoption is the introduction of a new technology into the firm, after which it becomes more familiar and resistance to its growth is less. The table also shows that national substitution was delayed relative to intrafirm substitution, the half-times being 1951 and 1942 respectively. This confirms Mansfield's speculation that smaller firms would adopt more rapidly.

While this analysis of the spread of the diesel locomotive was brief, it does illustrate some important points. First, it demonstrates the typical growth curve for a technology which starts out slowly, gathers ever increasing speed, and then slows down as it approaches a natural limit. Second, it shows that there are dynamic measures of technological change that can indeed be approximated by smooth curves. The national substitution of diesel for steam locomotives is a good example of this. Third, it shows that the data can be approximated by simple functional forms, the logistic being sufficient to describe the spread of diesels to a good approximation. Fourth, it shows that idiosyncratic as well as systemic factors determine technological change. The curve for the number of first adoptions can be approximated by a logistic, but there are significant deviations from it. These deviations are not the result of "experimental error" in the data, but represent points at which actual firms made decisions. The fact that the points do not all fall on a smooth curve shows that factors specific to individual firms exist, while the fact that a smooth curve gives a reasonable approximation is the result of

Table 4-1 Time Constants for the Spread of Diesel Locomotives

Process	t (.9)	$t_{1/2}$
First adoption	20.4	1932
Intrafirm substitution (average)	11.3	1942
National diffusion	11.5	1955
National substitution	11.5	1951

more general systemic factors. In this case, the systemic factors include the inherent superiority of the diesel locomotive, the strong competition between firms, the human resistance to change, the large investment already made in steam, and so on. Finally, we must point out that this is far from a complete picture of the change from steam to diesels in railways. Not only are there other measures that could be analyzed, such as the growth of technological performance, but there is an entire substructure of human and social history, an array of subtechnologies, and a wealth of individual and institutional achievements that attends the growth of any technology but is necessarily left out of a purely quantitative analysis. The fact that aspects of technological change can be described quantitatively, however, is remarkable and useful.

THE DIFFUSION OF RADIO BROADCASTING

By World War I, all the technological elements needed for radio broadcasting were in place. Guglielmo Marconi had already succeeded in making long distance wireless telegraphy a reality using a spark gap transmitter and a coherer-decoherer for a detector by the turn of the century. Through the work of Sir Oliver Lodge, Marconi, and John S. Stone, adjustable tuning was available by 1900. Fleming invented his diode in 1904 and De Forest had the triode by 1906. In the same year Reginald A. Fessenden demonstrated the use of a "continuous" wave that could be modulated to carry sound signals. This was a great improvement over the discontinuous output of spark apparatus. The principle of heterodyning was known quite early and became practical with the development of the vacuum tube oscillator in 1912. Edwin H. Armstrong applied for a patent for his feeedback circuit in 1914 and for the superheterodyne circuit in 1919. Commercial radio telephony was technically possible since at least 1910 but, as is often the case, actual implementation was much later. It was not until the early 1920s that commercial radio broadcasting started to grow.

Of course, wireless radio telegraphy was well on its way before World War I. Maritime wireless in particular was well established. Marconi had developed this market as an obvious niche for which there was no competition. The value of wireless telegraphy was well appreciated by navies as well as commercial shipping. An additional boost to maritime telegraphy was given by the *Republic*, which sank in 1909, and the *Titanic*, when it sank in 1912. The *Carpathia* responded to the SOS signal from the Titanic and rescued 700 survivors. But the *Carpathia* was 58 miles away from the *Titanic* when it heard the SOS and took over three hours to reach the *Titanic*. It was later learned that a freighter was only about 25 miles away at the time but had no wireless, and that another ship was less than 20 miles away but her wireless operator had gone to bed. A number of maritime nations then passed laws requiring ships to carry wireless.

The other major prewar radio activity was that of amateurs who bought and built transmitters and receivers and operated them as a hobby. These amateurs were a significant factor in the development and growth of radio. Of course, they created some chaos as well as being enthusiastic promoters of the new technology. Their radio traffic often interfered with official transmissions, and the Navy in particular was distressed at its inability to get through the confused clamor of large numbers of amateur transmissions. The extent of amateur radio activity was such that when the first licensing law was passed by Congress in 1912 nearly a thousand transmitters were licensed. The number increased rapidly, and by 1917 there was 8,562 amateur transmitters licensed in the United States (Barnouw 1966; pp. 32–33).

Until 1920 most radio, and practically all commercial wireless, consisted of point to point transmission. Indeed, the pulse-modulated spark equipment pioneered by Marconi was suitable only for point to point communication. Pulse-modulated sparks were useful for Morse code communication but not for the transmission of voice or music. The Marconi systems really were wireless replacements for telegraph and were primarily used to communicate between points that were not connected by wire.

But the idea of broadcasting, in which a powerful transmitter would send music, news, and entertainment to a large number of receivers over a large area, arose early in the history of radio. It was generally recognized that this could not be done with pulse spark transmission and that a continuous electromagnetic wave had to be generated that could be modulated to carry the voice or music information. Several schemes were tried, but it was first the alternator of Fessenden and then the vacuum tube oscillator that provided the necessary continuous wave. A number of experimental broadcasting stations were on the air before the war. Fessenden broadcast a program which included a violin solo (played by himself) on Christmas Eve of 1906. In 1908 De Forest broadcast recorded music from the Eiffel Tower which was heard 500 miles away, and in January 1910 he broadcast the voice of Enrico Caruso from the New York Metropolitan Opera House. Other early broadcasters included Thomas E. Clark, C. D. Herrold, and H. J. Power.

All these efforts, however, were in the nature of tests or demonstrations. Broadcasting as a growing commercial enterprise had to await two important and interrelated developments: government regulation and the takeover of radio by big business. The dividing line was World War I. Before the war, radio technology was developed by independent inventors and commercialized by individual entrepreneurs who were often also inventors. There was little governmental control over radio, amateurs were a major source of transmissions, and there was a chaotic mix of interfering frequencies on the air. After the war, most technical development was carried on in the research departments of large companies, and large firms dominated commercial radio. Transmission was regulated by government, which assigned frequencies and set standards. This transformation of the commercial and governmental environment determined the way in which broadcasting grew.

A critical event was the formation of the Radio Corporation of America, which was inspired by government in response to perceived national interests. With the advent of war, the Navy took over all radio stations and found them so useful that it attempted to acquire control over all radio after the war. American Marconi had emerged as the largest and most powerful firm in radio even before the war. But this was a British controlled firm and there was serious concern that Britain would dominate radio just as it dominated cable, therefore expanding its influence while the United States would be a poor second in communications. Finding it politically impossible to acquire a government monopoly over radio, the Navy sought to Americanize it and make it responsive to government. To this end, Commander S. C. Hooper and Rear Admiral W. H. G. Bullard cooperated with Owen D. Young of General Electric to create RCA, which was incorporated on October 17, 1919. American Marconi was asked to transfer its assets to RCA, with individual stockholders getting RCA stock in exchange and General Electric purchasing the shares held by British Marconi. The Marconi companies could not refuse. They were in desperate need of Alexanderson alternators for continuous wave transmission and these were General Electric devices. Also, the Navy still held all of American Marconi's land stations and strongly implied that they would not be returned if the offer were to be refused. It was clearly better to hold stock in a company that was favored by the government than in a company that was the subject of governmental hostility. The formation of RCA was also intended to solve the problem of patent conflicts. Both General Electric and American Telephone and Telegraph held important patents and had a strong interest in radio, General Electric through equipment manufacture, and AT&T because of radio telephony. By mid-1920 a patent cross licensing agreement had been worked out among RCA, GE, and AT&T. The group was soon joined by Westinghouse which had allied itself with International Radio and Telegraph, which held the Fessenden patents and acquired control of the Armstrong patents. With this essential technology, Westinghouse could not be left out. Later, United Fruit was invited to join because it held patents on crystal detectors and loop antennas. At that point, the patent pool contained about two thousand patents. This was meant to assure a quasi-monopolistic development of radio by large-scale industry.

A number of attempts at broadcasting were made before the war and some of these were successful. Harold J. Power, for example, started a station at Tuft's College in 1915 that was backed by J. P. Morgan, and De Forest was broadcasting from New York in 1916. But these initiatives, many of them by amateurs, were interrupted by the war. The United States Navy had taken over all transmitting facilities on April 7, 1916, and did not relinquish control until 1919. The growth of broadcasting was then dominated by large corporations.

The first successful commercial broadcasting station was KDKA in Pittsburgh by Westinghouse. It was inspired by the operation of 8XK, an amateur station run by Frank Conrad, a Westinghouse engineer. The station was licensed by the Department of Commerce on October 27, 1920, and broadcast the returns of the Harding–Cox election on November 2. The growth of radio broadcasting had begun, but not as smoothly as one would expect from the existence of a strong governmental interest and a patent pool controlled by large corporations. The problem was that the importance of broadcasting relative to point to point transmissions was not widely realized. Broadcasting not only led to disagreements among patent pool members, but also exposed the totally inadequate Federal role. The only legislation in existence was a licensing requirement dating from 1912. But there was no provision in the law for refusing a license; licensing was merely a form of registering stations. The result was that there were so many stations on the same or nearly the same frequencies that clear reception was difficult and often impossible. No Federal regulatory power existed until the Radio Act of 1927, which established the Federal Radio Commission. In 1934, the Communications Act was passed which subsumed the Radio Act of 1927. Interstate and foreign wire communications were included under its jurisdiction as well as local and national broadcasting. The number of people on the Commission was raised from five to seven and it was renamed the Federal Communications Commission. By then, some order had been imposed on the chaos by assigning frequencies and regional responsibilities. Without the Radio Acts of 1927 and 1934 the widespread diffusion of broadcasting would not have been possible.

The spread of radio broadcasting consisted of two distinct but interacting and sometimes conflicting parts: the diffusion of amplitude-modulated broadcasting and the diffusion of frequency modulation. Early broadcasting was by modulation of the amplitude of a continuous carrier wave of high frequency generated either by the Fessenden–Alexanderson alternator or, later, by the vacuum tube oscillator. Amplitude modulation was fairly straightforward and capable of supporting long distance transmission with equipment of reasonable cost. However, it had the defect of being prone to static noise. Despite the best efforts of radio engineers during the 1920s this defect could not be overcome. It was, in fact, inherent in the very nature of AM that it would be noisy. Any source of electromagnetic radiation of either radio or audio frequencies could alter the amplitudes of the broadcast wave so that everything from x-ray equipment to static and atmospheric electricity could and did introduce static into the signal. Increasing the power of the transmitter helped, but if pushed too far this led to interference among stations. Filter circuits were not very effective because much of the noise had frequency components similar to those of the broadcast signal. The solution was found by Edwin F. Armstrong, who started to examine the static problem in 1923 and had a working FM system by 1933. Since FM works by imposing the frequency variations of an audio signal on the radio frequency of the carrier wave, it was not subject to noise arising from the distortion of amplitudes. Armstrong was not the only one to realize that a radio wave could be modulated by frequency variations, but he was the only one to believe it was feasible. Both RCA and AT&T were doing research on frequency modulation. But the conventional wisdom dictated that the band of frequencies be as narrow as possible so as to enhance sensitivity and decrease the interference of reception between stations. This logic was correct for AM and was carried over to FM. The result was that narrow band FM was also prone to static. Armstrong's insight was to use a wide band. All the audio information could then fit in the band and noise was eliminated.

In addition to having much less noise than AM, FM had other important advantages. There was virtually no interference between two FM stations operating at different frequency bands, and filters could now be effective because they only had to cut out unwanted frequencies, not unwanted amplitudes. Also, the dynamic range of the audio signal could be much larger than in AM. Finally, FM could be easily multiplexed to carry two signals at a time, thereby making stereo broadcasts possible. With all of these technical advantages it would appear that the immediate adoption of frequency modulation and its substitution for AM was assured. Such was not the case. FM broadcasting was not immediately embraced by the radio industry for two nontechnical reasons. The first was that FM broadcasting required changing the entire system of transmission and reception. There was in existence a large investment in AM equipment which would have been rendered valueless if FM were introduced. If FM, or any other method of eliminating static, could be achieved by simply adding a component to receivers or transmitters, it would have been welcomed. In fact, what David Sarnoff of RCA had called for was a "black box" to do just that. Armstrong's invention was not a black box, it was an entire broadcasting system. This raised a financial barrier to its adoption.

At first, RCA supported Armstrong's work on FM, and he worked with RCA engineers from 1933 to 1935 at the RCA labs in the Empire State Building. But in April of 1935, RCA told Armstrong to remove his equipment and stopped supporting his work. Two weeks later, RCA announced a major research program to develop television. This was the second reason for not adopting FM. Sarnoff believed that the next wave of technical progress would be in television and wanted all of RCA's resources put into its development. It would have been much easier to develop FM broadcasting, and FM actually was in place and widely used before television became a commercial success.

Frequency modulation went through experiences that could easily be interpreted as outright hostility to the new technique. The FCC would not authorize commercial FM stations until 1940, although there were twenty-two experimental stations in existence by then. FM seemed to get a boost in 1940 when the FCC took the 42–50 MHZ band from television and assigned it to FM broadcasting. But in 1943 the FCC ruled that AM stations that also had FM stations could duplicate the programming on both kinds of stations. Of course, these stations would not put on programs for which FM has the great advantage of high fidelity. This had two effects that were highly detrimental to FM. The first was that the incentive for the consumer to purchase an FM set was lowered. Even worse was the fact that AM station owners essentially gave away advertising on their FM facilities, thereby undercutting the independent FM stations.

The position of FM was further damaged in 1945 when the FCC moved FM from the 42–50 MHZ slot and assigned it the frequency range of 88 to 108 MHZ. This was done to accommodate television broadcasting. About half a million receivers and fifty-three transmitters were immediately made obsolete. Later in the same year the FCC imposed the single market plan on FM which limited antenna heights and maximum power output, thereby limiting the range of FM broadcasts. This again decreased advertising revenues, since advertisers preferred the longer range of AM broadcasts. The effect on independent FM stations was predictable. As a group, they did not report a profit until 1976.

Armstrong initiated a long series of law suits against RCA and other companies that had infringed his patents without paying him royalties. This consumed a great deal of his time and money. On January 31, 1954, he commited suicide by jumping out of a 13th story window. Shortly thereafter, RCA settled out of court, but it was the only company to do so.

Before his death Armstrong had developed FM multiplexing, which made stereo broadcasting possible and gave new energy to the FM industry. Furthermore, in 1972 the FCC reversed its 1943 ruling and now ruled that AM stations could not duplicate programming on their FM facilities. FM had arrived, and it captured half of all listeners by 1979.

The Federal Communications Commission was dominated by people who were committed to the existing technology; namely amplitude modulation, and who believed that the next important advances would be in television. Erickson (1973) believes that there was, in fact, a kind of "loose conspiracy" among industrial and regulatory personnel to stifle the competition of frequency modulation. The effect on the diffusion of radio broadcasting is consistent with this thesis.

Figure 4-7 shows the number of licensed commercial broadcasting stations as a function of time. The rapid growth of AM broadcasting immediately after 1920 slowed during the Great Depression but resumed after World War II. FM broadcasting started later and did not start to grow until after the war. Even though FM broadcasting had been delayed well after it was ready, its performance was so superior to that of AM that the number of stations rapidly climbed to match and even exceed the number of AM stations. But the impact of the shift in frequency assignments in 1945 is clearly evident in the curve. The number of FM stations decreased rapidly and did not start to rise again until 1960. By then there were about 3000 AM stations in existence and although the rate of growth of FM was high, it was not until the 1980s that the numbers of FM and AM stations were comparable.

We can only speculate on the course of the diffusion radio broadcasting would have taken in a more favorable industrial and regulatory climate. It is possible that FM would have substituted for AM and that AM broadcasting would now be a minor enterprise. Certainly the difference in technical per-

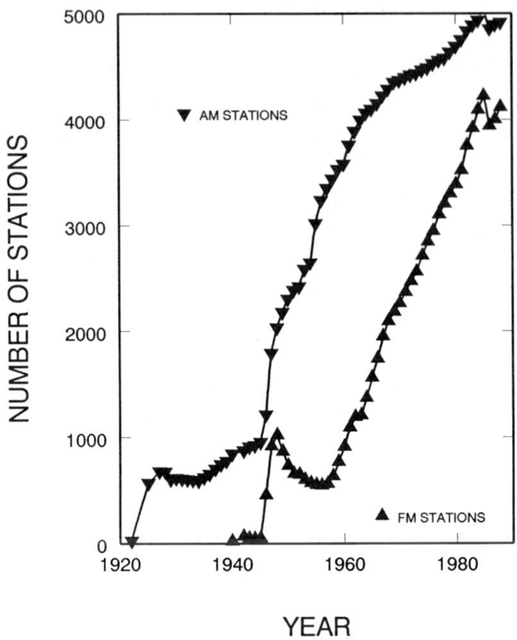

Figure 4-7 Number of commercial radio stations in the United States.

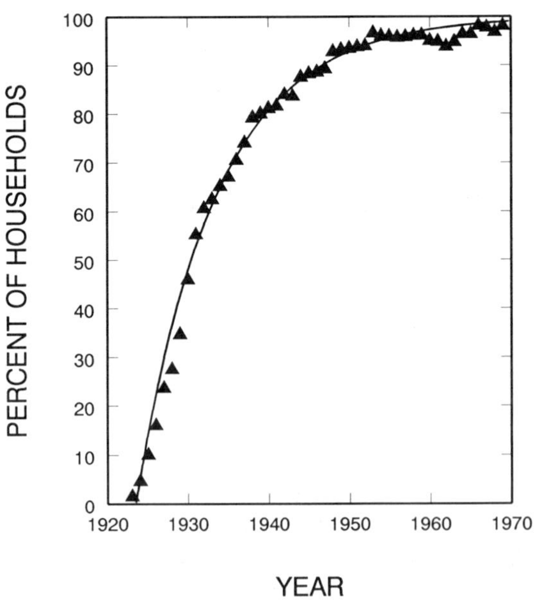

Figure 4-8 Percent of households with radios in the United States: curve computed from $P = 1 - \exp(-0.09938{*}\mathrm{YR} + 1923.4)$.

formance of the two modes should have led to that result. At any rate, it is certain that the growth of broadcasting was profoundly influenced by massive idiosyncratic intervention in the form of government policy. The systemic growth would surely have favored FM. It should be noted that the Depression and the war also constituted idiosyncratic factors that delayed the spread of broadcasting.

Idiosyncratic factors are not at all evident in the growth of consumer acceptance of broadcasting. Figure 4-8 displays the growth in the percentage of American households with radios. The growth was quite regular and quite fast. There was no incubation period, and the data are well described by a decaying exponential rise with an initial growth rate of nearly 10 percent. The public did not need any more time to become familiar with radio. Wireless telegraphy had existed since the turn of the century and radio had captured the public imagination. Radio clubs and radio amateurs were everywhere, broadcasting experiments were highly publicized, and large numbers of people were caught up in the new, miraculous technology. Broadcasting found a ready market.

STEELMAKING: AN EXAMPLE OF MULTILEVEL SUBSTITUTION

Before mathematical models can be applied to technological substitution, a decision must be made as to which technologies are in actual competition. This sometimes requires some care, especially for multilevel substitution. Steelmaking technologies provide a good illustration of this.

The steel industry has been the subject of extensive study. The recent book by Gold et al. (1983) provides a comprehensive and scholarly economic and technological analysis of the industry in this century. It discusses the evolution of steelmaking processes in detail. Adams and Dirlam (1966) have performed a study of the adoption of the basic oxygen process in the United States with emphasis on economic factors, while Maddala and Knight (1967) compared the growth of this process in thirty-seven countries. Girifalco and Simozar (1985) performed a substitution analysis of the major steelmaking processes, and it is that work which is discussed here.

The four processes considered here, Bessemer, open hearth, basic oxygen, and electric arc furnace, all produce steel and are therefore part of a general steelmaking technology. However, the electric process is rather different than the others. It was introduced before either the open hearth or the basic oxygen processes contributed significantly to steel production; its initial importance has been in the production of high quality and alloy steels. It first grew at the expense of the crucible process, which made products whose quality could not be matched by either Bessemer or open hearth steels. Its flexibility in both process and end product is greater than that of the other processes; it does not need molten iron as an input; it can be adapted to a large range of production volumes and is cost effective for small volumes; its costs are largely driven by the costs of electricity and scrap. These are factors that define a rather different market than that for other processes, which are not well suited for smaller runs or special alloy steels. Also, as costs decreased and flexibility increased, electric steel was able to make inroads into some of the traditional submarkets of the other production methods. In fact, the growth of electric arc steelmaking in the last few decades is largely the result of direct successful competition for such markets. The other processes, however, were similar in the types of products and markets they were addressing in that they were large scale, they produced steel for the high volume applications, and were generally associated with highly integrated iron- and steelmaking enterprises. It makes sense, therefore, to ask whether the basis of competition for electric steel can be considered to be the same as for the other processes.

Here, we make the judgment that electric steel indeed competes differently from the others. We assume that Bessemer, open hearth, and basic oxygen steels compete with each other, addressing similar markets, and that adoption decisions for these technologies were driven by the economic and technical factors inherent in them. For convenience, we call these methods chemical combustion processes. They successively displaced each other as the technical advantages of the later processes became evident. For electric steel, however, we assume that changing product mix and changing industry structure (more small mills) are important. We regard the slow but continual growth of electric steel production as the result of a competition between the electric process and the chemical combustion processes as a group.

For the mathematical analysis, therefore, the competition between the electric process and the others is a binary substitution, whereas the competition among the others is a three-level process in which open hearth substitutes for Bessemer and basic oxygen simultaneously substitutes for open hearth and Bessemer. This means that we use the partial fraction model discussed in Multilevel Substitution (page 145) for the substitution sequence electric → basic oxygen → open hearth → Bessemer. Note that electric steel is not in the same order for substitution as for its temporal introduction. Empirically, this was done because doing the mathematics in the same order as the date of introduction gave results that did not fit the data as well. In fact, using the temporal order led to curves that gave a decreasing market share to electric steel in recent years, contrary to the data, while assuming that electric steel competes with all the others as a group correctly reproduced the rising trend for electric steel. This supports our idea that electric steel is in binary competition with the chemical combustion processes as a whole. In fact, it was the failure of the mathematical analysis based on the temporal sequence to describe the data that led us to examine the special characteristics of the electric arc method and conclude that it competed with the combustion methods as a group.

Steelmaking provides a good test for any multilevel substitution model. The tonnage production of steel, while showing a large general rise for the past two centuries, varies erratically on shorter time scales. Wars, international competition, recessions, and booms have a profound effect on steel demand. Also, the product mix has been changing continually, the important applications have been changing, and even the nature of the iron ore has not remained constant. Yet technological substitution is quite regular when expressed in terms of fractional share of production by different processes.

The mathematical procedure was straightforward. Data on the tonnage production of steel by the four processes were used to compute the partial fractions defined by Equation (A-71) in Appendix 1. At this point, a choice for the functional form of the binary substitutions defined by the partial fractions had to be made. Both the logistic and the Gompertz equations (See Appendix 1) were tested, and the best results were obtained by using Gompertz equations for the substitution of basic oxygen for open hearth and for the substitution of open hearth for Bessemer, while using a logistic equation for the substitution of electric steel for those made by the chemical combustion processes.

The Gompertz and the logistic were fitted to the data using a least squares procedure directly on the equations rather than on their linearized forms to give more accurate results. The values of the partial fractions computed from the equations were then used in Equations (A-74) to compute the market share for each process, which was then compared to the empirical data. The results are shown in Figure 4-9, which includes the empirical data points and the computed curves. The rate constants and 90 percent takeover times for each substitution as defined by the partial fractions are shown in Table 4-2.

The first point to note is that the simple model used here gives a remarkably good representation of the data for a complex series of substitutions. There is, of course, fine structure in the data that is not well reproduced by the equations. From 1880 to 1890 the curves underestimate the fraction of open hearth production and overestimate the production by the Bessemer process, while from 1895 to 1925 the reverse is true. From 1925 to 1955 the curve overestimates the fraction of production by the open hearth process, and the turndown in basic oxygen steel production after 1975 is sharper than that given by the curve. However, these deviations are not large, considering that the analysis covers more than a century and that idiosyncratic factors had to be important during that time.

The data for the growth of the basic oxygen process merit some comment. It is often stated that the reason for the deterioration of the competitive position of the United States relative to other countries was the result of a combination of high labor costs and the industry's delay in adopting the new technology. Adams and Dirlam (1966) in particular have sharply criticized American steelmakers for their failure to adopt the basic oxygen process more quickly. However, the data show that the adoption of the basic oxygen process in the United States was quite rapid. The rate was much faster than that for any of the other steelmaking technologies, and by 1970

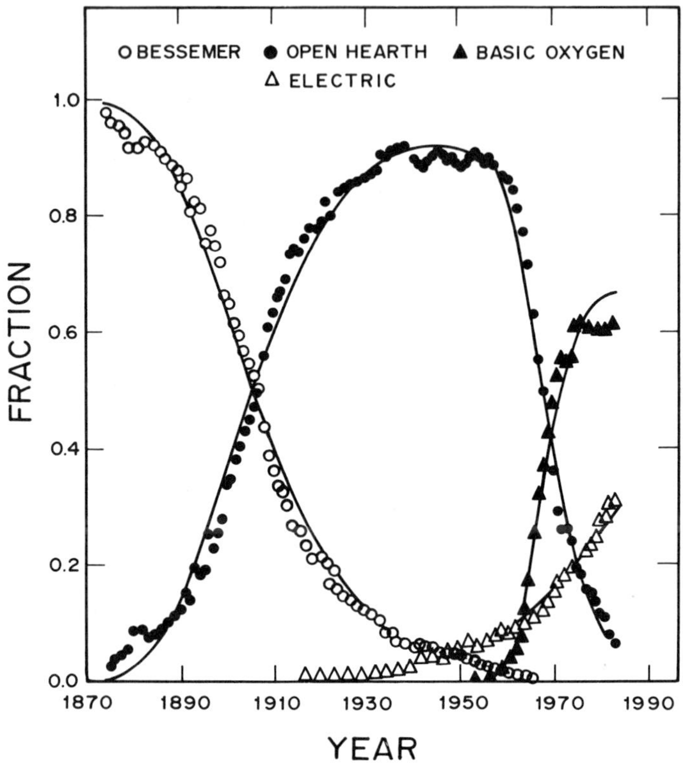

Figure 4-9 Multilevel substitution of steelmaking technologies.

basic oxygen accounted for half of the production by the chemical combustion methods. The 90 percent takeover time was only nineteen years for basic oxygen while it was forty-three years for the open hearth. This seems contrary to the idea that lack of adoption of the basic oxygen process was a major factor in the steel industry's problems. Maddala and Knight (1967) found that the United States was a leader among twenty-one countries that had adopted basic oxygen by 1964. For the period of 1954–1964, basic oxygen production as a percent of total increase in production was greater for the

Table 4-2 Multilevel Substitution Parameters for Steelmaking

Substitution Process	Equation	Rate Constant	t (90%) Years
OH → B	Gompertz	0.0176	43
BO → OH + B	Gompertz	0.172	19
EL → BO + OH + B	Logistic	0.0649	68

United States than for any other country. However, this rapid adoption was essentially for *new* capacity. A great many of the old furnaces were retained.

The market share of electric steel has shown a slow but steady growth for over a century. If the data are projected by the equations given here, they indicate that electric steel will have half the market by the year 2000 and will be dominant (90 percent or more) shortly after 2030. Such a projection must be viewed with care. The future growth of electric steel production depends on the availability of scrap and on cheap electricity. Nevertheless, the data suggest that electric steel will continue to increase its market share.

RESULTS OF DIFFUSION AND SUBSTITUTION STUDIES

Many studies of the diffusion and substitution of technologies have been performed, and references to a number of these are given in the bibliography. In this section we review the results of a few of these that are of particular interest. We choose Mansfield's (1961) early work on the diffusion of industrial innovations because this work is a classic in the field that set a standard for later studies; Fisher and Pry's substitution analysis (1971) because it defined the substitution approach for a wide audience, and Blackman's results (1974) because they display interesting facts on the spread of consumer durables in the American home.

The results of Mansfield included in the analysis of the spread of the diesel locomotive given above were part of a larger study which analyzed the rate of adoption of twelve innovations by firms in four industries. For each of these innovations, Mansfield acquired the date of its first adoption for the major firms in its industry and computed the fraction of firms that had adopted as a function of time. He then fit the data to a logistic equation and determined the rate constants for the equations, along with correlation coefficients as a measure of goodness of fit. It should be noted that Mansfield arrived at the logistic equation by the series expansion method given in the Appendix, and that this was the first presentation of this derivation in the literature of technological change. Mansfield's interest is in economics, so he tried to relate the rate of adoption to economic factors by searching for linear correlations between the rate constants and economic factors that might influence the adoption decision. The results of this for the diesel were described above.

Table 4-3 lists the innovations, the number of firms for each innovation in Mansfield's study, the half-time and 90 percent takeover time computed from the logistic function, and the correlation coefficient for the fit.

Table 4-3 shows that the correlation coefficients are high enough so that it can be concluded that the logistic curve is a good fit. However, examination of Mansfield's plots for the empirical data points show that in many cases the data points exhibit considerable scatter about the smooth curve. Thus, while the logistic is statistically verified, actual data points can deviate from it significantly. This again illustrates the difference between idiosyncratic

Table 4-3 Data From Mansfield's Study of Adoptions

Innovation	No. of Firms	Half-time	t (90%) Years	Corr. Coeff.
Railroads:				
Diesel locomotive	25	1933	22	0.89
Centralized traffic control	24	1938	23	0.94
Car retarders	25	1936	31	0.90
Iron and steel:				
By-product coke oven	12	1909	26	0.98
Continuous wide strip mill	12	1931	13	0.95
Continuous annealing	9	1950	26	0.93
Bituminous coal:				
Shuttle car	15	1942	14	0.95
Trackless mobile loader	15	1941	14	0.97
Continuous mining machine	17	1951	9	0.98
Brewing:				
Tin container	22	1935	2	0.96
High speed bottle filler	16	1957	12	0.97
Pallet loader	19	1953	8	0.97

and systemic factors. The small number of data points for many of the innovations should be noted; presumably a larger data set would result in less scatter and higher correlation coefficients.

The 90 percent takeover times range from 2 to 31 years. There is a clear difference among the four industries with railroads and iron and steel having longer takeover times than bituminous coal and brewing. Mansfield noted the differences among industries when he computed the linear relation between the rate constants and firm size and profitability of the innovation. He found that when the rate constants are referred to constant size and profitability, they were about the same except for the brewing industry, for which the rate was considerably higher. As stated above for the diesel, Mansfield investigated the effects of profitability of the innovation, firm size, durability of equipment being replaced, growth rate of the firm, the business cycle, and the time at which the innovation was introduced. He found that increasing profitability and decreasing firm size both increased the rate of first adoption. For the other factors, the results were not statistically significant although there was some tendency for the rate to increase with the firm's growth rate, during expanding phases of the business cycle, and for later dates of introduction of the innovation.

The work of Fisher and Pry was of a different kind. While Mansfield focused on the factors affecting first adoptions, Fisher and Pry tried to construct a general model of technological change based on substitution. They assumed that in many cases technological change consisted of the

Table 4-4 Fisher-Pry Results for 17 Substitutions

Substitution	Half-time	*t* (90%)
Synthetic/natural rubber	1956	58
Synthetic/natural fibers	1969	58
Plastic/natural leather	1969	57
Margarine/butter	1957	56
Electric/open hearth specialty steels	1947	47
Water/oil house paint	1967	43
Open hearth/Bessemer steel	1907	42
Sulfate/tree-tapped turpentine	1959	42
Ti/Pb-Zn paint pigments	1949	26
Plastic/hardwood home floors	1966	25
Plastic/other pleasure boat hulls	1966	20
Organic/inorganic pesticides	1946	19
Synthetic/natural tire fibers	1948	17.5
Plastics/metal in cars	1981	16
BOF/open hearth steel	1968	10.5
Detergent soap (U.S.)	1951	8.75
Detergent soap (Japan)	1962	8.25

replacement of an old technology by a new one to satisfy essentially the same need and that the substitution could be described by the logistic equation. Apparently, they were unaware of the earlier work that had used the logistic, such as that of Griliches (1957) and Mansfield (1961). Their paper had a major impact since it explicitly stated the assumptions and rationale for the substitution process, carefully described the need to use proper units, and gave a successful mathematical description of seventeen innovations. The substitutions they studied along with the half-times and 90 percent takeover times computed from fitted logistic curves are shown in Table 4-4.

The takeover times vary from 8 to 58 years, and there seems to be no correlation with either the nature of the technology or the time at which substitution started. As pointed out by Fisher and Pry, the data suggest that our sense of increasing change as time goes by may be the result of experiencing more instances of change rather than of increases in the rapidity of individual changes.

A basic assumption of this method is that once a substitution has reached a few percent it will go to completion. However, applying this idea requires that the saturation level be known. Fisher and Pry recognized this and tried to account for it in some cases. For example, they recognized that detergents would not be used for personal hygiene, so they omitted face and bath soap from their calculation. But in some cases their choice of saturation level is doubtful. For example they assumed that the substitution of plastic for metal in cars and the substitution of margarine for butter would be complete. Both

Table 4-5 Diffusion of Appliances in American Households

Appliance	Half-time	*t* (90%)	90% Date
Radios	1938	37	1955
Washers	1942	35	1965
Electric refrigerators	1944	18	1954
Vacuum cleaners	1951	33	1968
Black and white television	1953	17	1960
Color television	1971	10	1976
Air conditioners	1972	21	1984
Electric freezers	1979	61	2009
Dishwashers	1980	34	1996
Garbarge disposers	1980	36	1997

these expectations seem unrealistic. In the case of the substitution of electric furnace for open hearth steels, they assumed that the saturation level was defined by specialty steel production since this was the product of the electric furnace. But, as we have seen, the electric furnace came to be used for more general products as well, and a correct substitution analysis should include all steel production. This factor, as well as the fact that the substitution of steel technologies is a multilevel process, makes the results of Fisher and Pry for steel technology substitutions different from those obtained by Girifalco and Simozar. In spite of these caveats, Fisher and Pry's work is important in that it developed the methodology for substitution analysis and determined the directions of later work.

The next set of data we will examine is from Blackman (1974), who determined the logistic functions that fit data on the growth of household appliances. The measure he used was the fraction of electrically wired households that had acquired the appliance as a function of time. His results are given in Table 4-5. In addition to the half-time and 90 percent takeover time, the table also shows the date at which takeover is 90 percent complete as computed from the logistic equation.

Blackman published his paper in 1974, and a number of the appliances listed above had not penetrated into households very completely at that time. If more recent data were to be included in the analysis, some of the numbers in Table 4-5 would undoubtedly change. Also, he assumed that all electrically wired households would eventually acquire each of the appliances. Nevertheless some interesting conclusions can be drawn from the data. The first is that labor-saving home appliances have become widespread only in rather recent times, so the technological contribution of liberating housewives from excess drudgery to the changing position of women is of recent origin. It is also interesting to note that the spread of television was the most rapid, only refrigerators coming close to the short takeover times of either black and white or color television. Blackman's takeover time of radios was more than twice that of black and white television and over three and a half times that of color television. But before any important conclusions are drawn from this, the limitations of Blackman's analysis must be noted. In the first place, Blackman used a limited number of points; in the second place, he fit all data to a logistic curve. Figure 4-8 shows that an analysis using extensive data leads to the decaying exponential rise, not the logistic, as the curve that describes the diffusion of radios. The more extensive data show that the 50 percent point was reached at about 1930, and that it took about thirteen years for the number of households with radios to go from 10 to 90 percent. Also, the data for the diffusion of television in Chapter 1 show that the 90 percent takeover times for black and white and color television were also thirteen years. The implication from Blackman's work, that television diffused much more rapidly than television, is incorrect. This error arose from applying the logistic to a limited data set.

These reviews of past work show that quite a bit of information can be obtained from quantitative analyses of technological diffusion and substitution. It also shows that the work must be done and interpreted with some care and with a proper appreciation of factors that are external to the mathematical methods themselves.

NUCLEAR POWER: A CASE OF INTERRUPTED SUBSTITUTION

Once a new technology starts to replace an old one, the substitution usually goes to completion. If the new technology is good enough to capture 5–10 percent of the market in its initial stages, it can be expected to ultimately replace the old technology almost completely. The advantages in cost and function and the greater final potential of the new technology assure its continued success. Substitution is a systemic process whose long-term evolution is controlled by the socioeconomic competitive response to lower costs and technical superiority. As a result, substitution is among the most regular phenomena associated with technological change.

But this regularity is maintained only if the long-range systemic factors continue to operate. Regularity requires that fundamental conditions that affect substitution either remain constant or change slowly enough that adjustments can be made in a smooth continuous way. An important case in point is the substitution of nuclear energy for fossil fuels to produce electricity. The generation of electricity by nuclear power has aroused controversy for many years. Proponents of nuclear power claim it costs less, is cleaner and safer than fossil fuels, and contributes to national energy independence. Opponents claim that nuclear power is not safe and cannot be made safe, that it costs more, and that it poses unsolvable problems of waste disposal. Whatever the merits of these arguments, the substitution of nuclear for fossil fuels continued at a regular pace right up to 1979. As shown in Figure 4–10, the fraction of electricity generated by nuclear energy in the United States was accurately described by a logistic from 1960 to 1979.

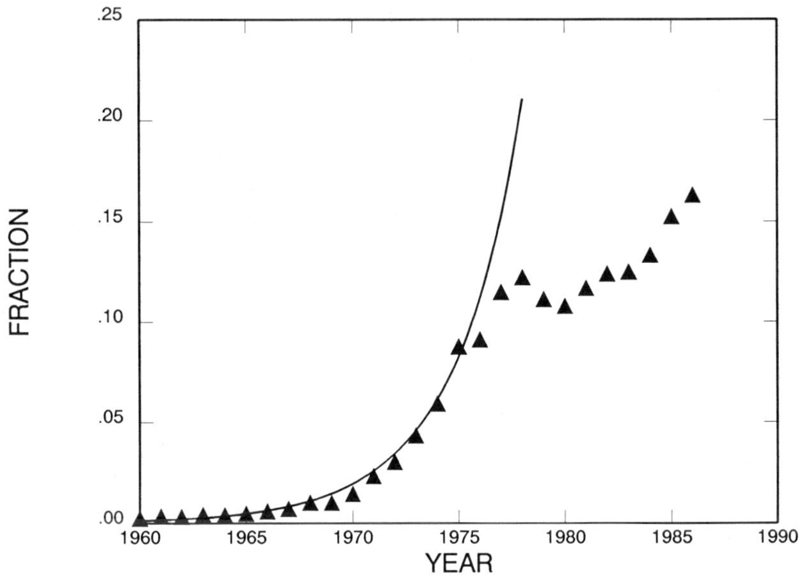

Figure 4-10 Fraction of electricity generated by nuclear energy in the United States: curve computed from $F = 1/[1 + \exp(-0.27519*\mathrm{YR} + 546.065)]$.

Furthermore, the rate of substitution was fairly rapid, indicating that until 1979 the controversial aspects of atomic energy did not have a large effect. But on March 28, the accident at Three Mile Island changed all that. In spite of the fact that there were no fatalities, only a small amount of radiation was released, and there is no real evidence of nuclear-enhanced medical problems from the accident, the substitution of nuclear for fossil fuels came to a stop. The combination of increased political influence of nuclear opponents, exposure of poor management procedures, increased government oversight and regulations, and increasing costs put a halt to further expansion of nuclear power.

Nuclear power is a case of interrupted technological substitution. The important point here is that substitution is regular only as long as systemic factors operate and are not negated by idiosyncratic events. There is an inevitability of the growth of a superior technology only as long as special events do not intervene. These may arise from popular perceptions and the political process, as in the case of nuclear power, or from other causes ranging from wars to weather.

DIFFUSION AND THE CLUSTERING OF TECHNOLOGIES

The initial event of the commercial introduction of a technology is certainly important, since it marks the beginning of the use and spread of a technology.

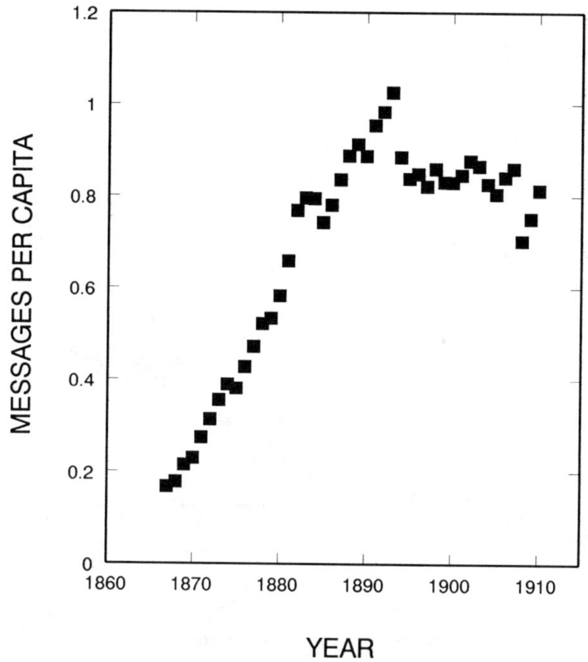

Figure 4-11 Annual telegraph messages per capita for Western Union.

An analysis of the dates of technological innovations shows that they tend to appear in clusters so that they depend and reinforce each other. But a new technology has an important impact only after it has spread among a large number of users. As a technology becomes more widely available, not only does it become cheaper and attain a higher level of performance, it also has an increasing interaction with other technologies. It is the simultaneous growth of the use of groups of technologies that leads to widespread advances of both an economic and technical nature. It is therefore pertinent to ask to what degree there is a clustering of the diffusion of technologies, in addition to the clustering of innovations.

For many cases, a simple measure of the clustering of technological diffusion can be obtained by taking advantage of the fact that technologies often diffuse at high rates soon after their introduction and slow down at later times. This is obvious for those technologies that diffuse according to the simple, ideal S-curve, but is also true for diffusion curves that are not so well behaved. Another type of curve is that of Figure 4-11, which shows the per capita number of telegraph messages per year. There was a rapid, fairly constant growth rate for about two decades, after which the growth stopped and even reversed. Technological diffusion also often displays a growth curve similar to that in Figure 4-12, which shows the number of incandescent lamps produced per year in the United States per capita as a

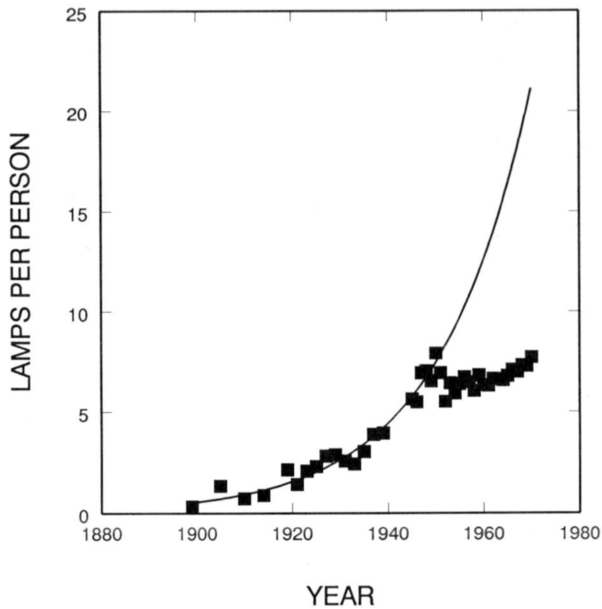

Figure 4-12 Incandescent lamps per capita produced in the United States.

function of time. The points follow a well-defined exponential curve with a 5.2 percent growth rate for five decades, after which saturation sets in abruptly.

It is usually possible to find a period during which the growth of the technology is greater than at any other times. It is the coincidence or overlapping of these time periods that we take as measures of the clustering of the diffusion of technologies. A list of such time periods is given in Table 4-6.

This table shows that there indeed is extensive coincidence and overlap of the diffusion of major technologies, the most obvious cluster being farm tractors, electric service, automobiles and rayon-acetate fibers. Additional data would surely reinforce this conclusion. For example, the growth in the use of steam power was parallel to that of railroads.

It must be pointed out that there are some cases for which an analysis of this kind cannot be made. For example, the growth in production of sulfuric acid per capita in the United States was nearly steady for over seventy years, except for fluctuations downward as a result of the 1930s depression and upward because of World War II. In such case, a period in which growth is more rapid than in other areas either cannot be found or is so long as to lose meaning for a clustering analysis. But the number of technologies that exhibit periods of more rapid diffusion is large, and a consideration of the clustering of the spread of technologies is meaningful.

Table 4-6 Periods of Highest Growth Rates for the Diffusion of Technologies in the United States

Technology	Period of Maximum Growth Rate	Midpoint Year
Railroads	1830–1870	1845
Telegraph	1870–1890	1880
Telephone	1875–1890	1882
Steel	1890–1910	1900
Tractors	1910–1920	1915
Electricity	1905–1925	1915
Autos	1900–1930	1915
Rayon and acetate	1910–1930	1920
Radio	1920–1930	1925
Incandescent lamps	1900–1950	1925
Aluminum	1940–1960	1950

SOME COMMENTS ON CAUSAL FACTORS

The literature on technological diffusion and substitution is large, and the bulk of it not only describes, but also tries to determine the reasons for, the particular rates at which technologies spread. Sometimes the causes are studied by statistical analysis between them and rate constants, as in the Mansfield work. As pointed out by Gold (1981), the results of such studies have not been very satisfactory. Either the data base is too small or the only factors that show statistical significance are obvious ones such as profitability. (It will generally be the case that adopted industrial innovations will be profitable.) More often, causal factors are sought by historical and case study methods or from a knowledge of the characteristics of the innovations and the industries that adopt them.

The number of factors that have been identified as affecting the rate of diffusion or substitution is large. For enlightening discussions of these, see Fliegel and Klein (1966), Nabseth and Ray (1974), Davies (1979), and Gold (1981). Economic factors have received particular attention and we will begin our comments with these.

The profitability to a corporation of adopting an innovation is difficult to calculate. A new process can result in widespread changes in a firm, including changes in product mix, inventory methods, depreciation schedules, tax status, marketing methods, personnel, labor costs and productivity, transportation, nature of materials and supplies, and even organizational structure. Furthermore, costs change with time as learning takes place and will be different for corporations that adopt at different times. The task of sorting this out so that the costs and benefits attributable to the innovation can be identified, and then comparing this to the pre-existing conditions, is enormous, and it is doubtful if the true profitability of any one innovation can ever be determined with any accuracy. Also, there is no way of knowing just how well an innovation will turn out before it is adopted. It can only be the expected profitability that plays a role in the decision to adopt. But, as Gold pointed out, the complexity and inherent uncertainty of the exercise deters many managers from attempting the calculation to a level of detail that would identify the actual profitability to be expected from an innovation. Often, projections of costs and revenue are relatively simple or are merely based on the expected change in output relative to the new inputs. Clearly, relating the rate of adoption to ex post facto profitability is of no value in trying to ascertain causes of adoption or nonadoption.

An innovation is not a static entity. As it is introduced and used it undergoes continual improvement, with production costs and operating costs both decreasing with time. As the new technology diffuses and becomes better understood, both technological and economic risks decrease. But the technology being replaced is not standing still either. The greatest advances in gas lighting took place after the introduction of the electric light, and the smallest, most efficient, and cheapest vacuum tubes were produced after the transistor was on the market. Potential adopters therefore face dynamic conditions in which decisions must be made as to whether to adopt now or later when entry is cheaper and there is a greater fund of experience to draw upon. This decision will depend on conditions that are specific to each adopter.

The structure of the industry and the posture of the corporation will have important effects on the probability of adoption of a new technology. In highly concentrated industries, or when the cost of entry into the business is high, there is likely to be less competitive pressure and management can afford to take more time in introducing innovations. In a highly competitive environment, however, rapidity of adoption is often essential. Late adopters may find that they must adopt just to survive and then must incur the costs of innovation without the benefits reaped by earlier adopters. The corporation itself has a planning horizon for capital expenditures and must consider the availability of funds as well as managerial and labor resources, organizational problems, and the possibility of achieving its ends by nontechnical versus technical means. A major characteristic of a mature corporation is that it has a particular style which is closely intertwined with the old technology. The physical plant, the training and interests of personnel, the ability to market, reputations and prestige, union contracts, and the positions of management are all tied to the old technology. If the change to a new technology is too radical, the resistance to that change can be considerable. On the other hand, if a corporation and its markets are growing and if management has an entrepreneurial bent, innovations will be adopted more quickly.

These are only a few factors that affect the rate of diffusion of a new technology. Others include the vintage of existing equipment, economies of scale, the cost of lagging behind competitors, legal and licensing restrictions, governmental testing requirements, Federal and local regulations, interest rates, local tax and zoning laws, foreign competition, antipollution requirements, and the prices of common stocks. The list can include any of the wide number of changing conditions that industry faces on a daily basis.

The spread of new consumer goods depends on an equally complex set of factors, including the distribution of disposable income, social acceptance, intensity of advertising, and comparative costs.

What are we to make of all this? It is clear that to unravel the causal factors for the spread of a particular innovation requires a set of detailed case studies of those that have adopted it. Furthermore, doing this successfully for one innovation does not mean that general principles have been found that are necessarily valid for all innovations, although some broad similarities can be expected. In fact, we are back to the position at the beginning of this chapter. The causes of technological change are many and their intensity varies considerably for different adopters. At any given time, there are a great many forces both for and against adoption. The rate of adoption depends on the imbalance between those factors that favor adoption and those that deter it. For an unsuccessful innovation, this balance is negative. For a successful innovation, the balance is slightly positive in the early stage, becomes greater as the innovation spreads and proves itself, and finally vanishes as its potential is exhausted. The S-curve is the natural result. The fact that idiosyncratic scatter is imposed on this curve, especially when the adopting population is small, is the direct result of different factors acting differently on different adopters.

5
The Dynamics of Energy Usage

THE ENERGY TRANSITION

In the complex of changes that define the Industrial Transition, increasing energy use occupies a central position. An industrial society consumes massive amounts of fuels to transform massive amounts of materials into useful products, to transport large numbers of people and huge quantities of goods, and to maintain comfortable temperatures in living and work spaces. Extensive use of energy is a necessary condition for the wealth, variety, and mobility of an industrial society.

The importance of energy to human development is dramatically illustrated by a large-scale view of the per capital energy use as a function of time. This is shown in Figure 5-1, which traces the use of energy at different stages of the evolution of human societies. Primitive man used only the 2,000 calories or so per day taken in as food. With the mastery of fire, this more than doubled in hunting societies, and with the advent of primitive agriculture it doubled again. Gradual improvements in agriculture were accompanied by another doubling so that by 1400, advanced agricultural societies were using about 26,000 calories per capita. These increases were gradual and took place over millenia. But with the coming of the Industrial Revolution energy use increased very rapidly until industrial societies were using ten times as much energy per capita as had been used in the most advanced agricultural societies. Energy consumption continued to increase after the Industrial Revolution was well under way. Figure 5-2 shows this for the United States in which the per capita energy consumption was about 70,000 calories per day per person until the rapid growth starting at the turn of the century, after which it increased by a factor of three over a period of three-quarters of a century.

But even this underestimates the energy intensity of advanced societies. Since the Industrial Revolution we have learned to use energy with ever-increasing efficiency. Energy efficiencies have increased by at least a factor of twenty in the past 200 years so that the effective per capita command of energy has been increased by at least a factor of 200 by industrialization.

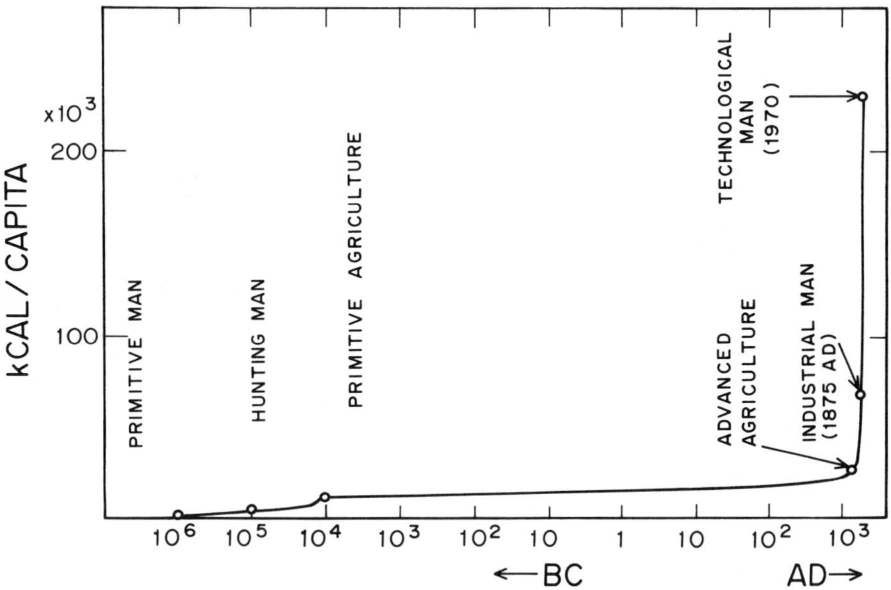

Figure 5-1 Long-term energy usage versus time.

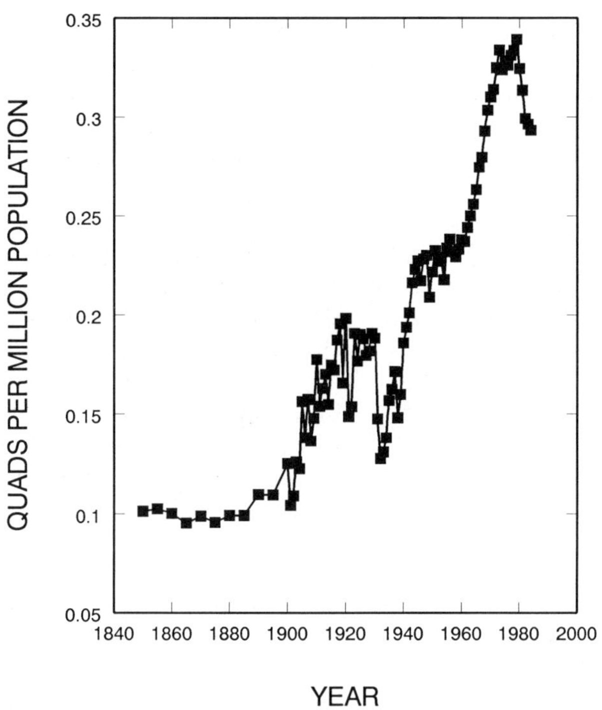

Figure 5-2 Annual energy consumption in the U.S. per capita.

The most remarkable aspect of Figure 5-1 is its rapid rise in modern times. From a long-range point of view, the growth in energy use was negligible for nearly the entire existence of the human race. But very recently, over a very small period of time, the growth can only be described as explosive. Of course, the increases before the modern era were important. The improvements in agriculture, the growth of cities, and the development of civilization were all based on the increasing availability of energy. The foundations of the modern world were laid in these times and the preindustrial achievements were immense. But the changes were slow, being measured in centuries and millenia. The rapidity of the rise starting in the eighteenth century is without precedent; its rate and magnitude has totally and irreversibly altered the conditions of human life.

This change would not have been possible without the utilization of new, rich energy sources. Agricultural society depended primarily on wood and animal power, with some help from wind and water. Industrialization was accompanied by a shift to coal, and later, by the use of fuel hydrocarbons. These provided high energy density fuels in large amounts. Wood, human and animal muscle, wind and water simply could not supply the huge quantities of cheap energy needed in convenient, concentrated, and transportable form. Mineral fuels could and did.

The shift to new energy sources in the United States is shown in Figure 5-3, which shows the fraction of energy consumed from mineral fuels (coal, oil, and gas) relative to total fuel usage (coal, oil, gas, and wood). Other sources, such as animal power, solar, wind, geothermal, and nuclear, are not included in this curve since the primary substitution of mineral fuels was for wood and these other sources made only a small contribution to total energy usage.

It took over six decades for mineral fuel use to go from 10 percent to 90 percent of the total. This coincided with the rapid industrialization of the country. The use of wood has not completely vanished; even today wood supplies about 4–5 percent of our fuel energy. However, this small amount arises from special circumstances such as the use of pulp and chips in lumber and paper mills and would hardly be missed if no longer available. Our dependence on mineral fuels has become all but complete.

The replacement of wood by mineral fuels was slow at first but then increased rapidly until it approached saturation and then leveled off. In fact, the substitution followed a classic logistic pattern, represented by the curve in Figure 5-3. This curve was obtained by fitting the logistic to the data up to 1926, by which time the substitution was over 90 percent complete. The equation of the curve, in logarithmic form, is: $\ln [f/(0.96 - f)] = -127.228 + 0.067604t$, where f is the fraction of energy from mineral fuel and t is the date. The half-time was 1882, and the time for replacement to go from 10 percent to 90 percent was sixty-five years. Note that as a result of the fact some wood is still being used, the limiting value in the substitution equation is 0.96.

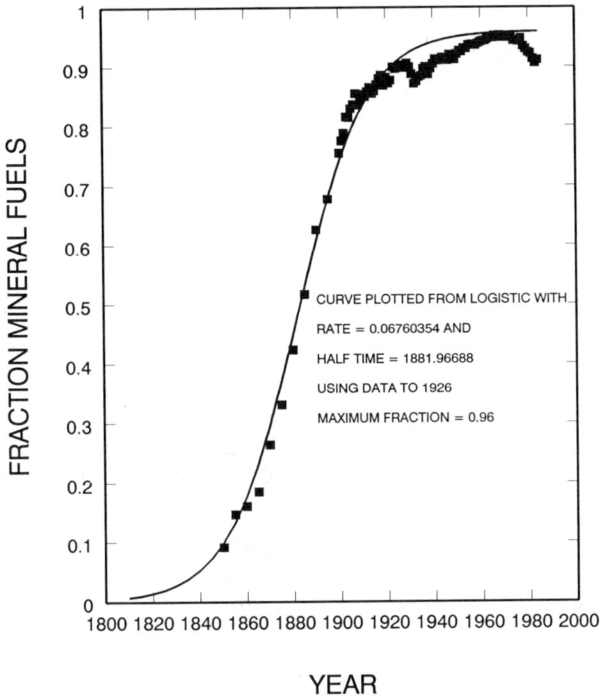

Figure 5-3 Fraction of fuel use from mineral sources in the United States.

A more detailed picture of the energy transition is shown in Figure 5-4, which shows the fraction of energy use from the major energy sources in the United States. (Minor sources such as geothermal, solar, and wind energy are neglected since they collectively amount to less than 1 or 2 percent of the total.)

Figure 5-4 shows that the early substitution was primarily coal for wood. The fraction of coal use was a maximum in the first decade of the twentieth century, at which time coal supplied over three-quarters of our energy. By this time, the use of oil and gas was starting to grow until now the fluid hydrocarbons account for about two-thirds of United States energy use. Coal is not a trivial contributor, however, making up about a quarter of the total. Note that the use of coal increased at the expense of oil and gas in the 1970s and 1980s in response to the actions of the OPEC oil cartel. Hydropower is used to generate electricity, and this has been constant at about the 5 percent level for many years. Nuclear power, again for electricity generation, started to contribute in 1960 and is still only a small fraction (about 5 percent) of our energy use.

The substitution of one fuel for another is driven by both technical and economic factors. The advantages of coal over preindustrial energy sources

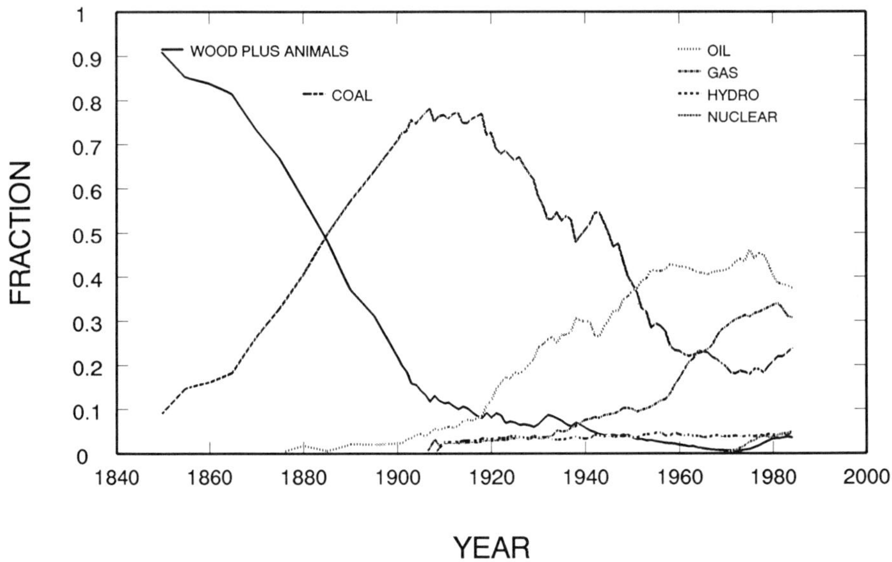

Figure 5-4 Fraction of energy consumption by source.

were overwhelming, and it was available at a reasonable cost. The increasing use of coal was an inevitable concomitant of industrialization. It was both a cause and a result of the growth of industry, taking part in a feedback loop in which its availability made possible the growth of energy-intensive industry and the growing industrial activity increased the demand for coal. Wood, water, wind, and muscle were just not enough to fuel the industrial transition because the supply was too small and too localized. Coal was abundant, has a high energy density, and is transportable. The supply of wood is limited and, even in well-forested regions, is a dispersed resource that must be collected, dried, and transported. Its dispersion, water content, and lower energy density (BTUs per unit volume) as well as its relative scarcity put it at a distinct disadvantage relative to coal.

The rise in the use of oil is primarily due to its use in internal combustion engines, to the convenience of liquid fuels, and to the fact that petroleum can be processed to yield a variety of fuel products suitable for a variety of uses. Before the advent of the internal combustion engine, petroleum was chiefly used to produce kerosene for illumination and heating. Gas, and later electricity, became serious competitors for these applications, and the petroleum industry might have died had it not been for its use in engines. But the advantage of liquid fuel relative to coal is considerable. It can be accurately metered, easily stored and transported, is clean, and leaves no ash, a fact which is essential for internal combustion engines. Petroleum therefore became the basis of transportation. Ultimately, it even replaced coal in rail transport when the diesel-electric replaced the steam engine. From a functional point of view, coal today is the least desirable fuel, being used primarily in large-scale operations such as utilities.

Natural gas is the cleanest of all mineral fuels. Not only is it ash free, but it does not produce the large amounts of sulfur and nitrogen compounds that are liberated in the burning of petroleum products. Because of its low energy density, it would have to be highly compressed for use in mobile internal combustion engines and is therefore not well suited for automotive use. But it is an excellent fuel for industrial processes and for space heating.

In spite of their functional superiority, the fluid hydrocarbons did not replace coal immediately. An important conclusion from Figure 5-4 is that major energy systems take time to develop to the point where they can make large contributions to our energy needs. It took sixty years for coal to replace wood and sixty years for oil and gas to take over from coal. The production, distribution, and use of energy resources involve massive and highly complex technological, economic, and social systems. It is naive to think that any new energy source can be exploited to an extent that it can make a major contribution to our national energy needs in time scales less than decades.

The transition from low to high energy use is the hallmark of industrialization. But not all peoples participated equally in this transition, and in fact the rates of industrialization for different countries have been very uneven. The annual per capita energy consumption for the world since 1920 is shown in Figure 5-5. It approximately resembles the energy growth curve for the United States (Figure 5-2) except that the energy consumption per capita for the world is much smaller than that for the United States. In fact, the per capita world energy use today is less than that of the United States a century ago. This is a dramatic illustration of the fact that a good portion of the world has not gone very far on the path to industrialization.

A more detailed picture of world energy consumption is given in Figure 5-6. This is an integral distribution plot. That is, the y-axis gives the number of people who annually consume more energy than represented on the x-axis. The energy units are in kilograms of coal equivalent, one kilogram of coal being equivalent to about 25,000 BTUs. Distribution curves are given for 1960 and for 1979. The first point to note is that there has been a significant increase in per capita energy consumption from 1960 to 1979, the average going from 1,425 to 2,271 kilograms. In spite of the fact that the industrialized countries made a major contribution to this increase, it also represents progress for the lesser developed countries. This is demonstrated by the fact that the entire distribution curve shifted to the right from 1960 to 1979. Energy growth has more than kept pace with population growth; detailed examination of the data on which the distribution curves are based shows that in virtually every country in the world per capita energy consumption increased from 1960 to 1979. In 1960 only about 500 million people consumed more than 3,000 kgms coal equivalent per year. This was a little more than 16 percent of the world population. By 1979, the number of people using more than 3,000 kgms was 1.2 billion, which was over a

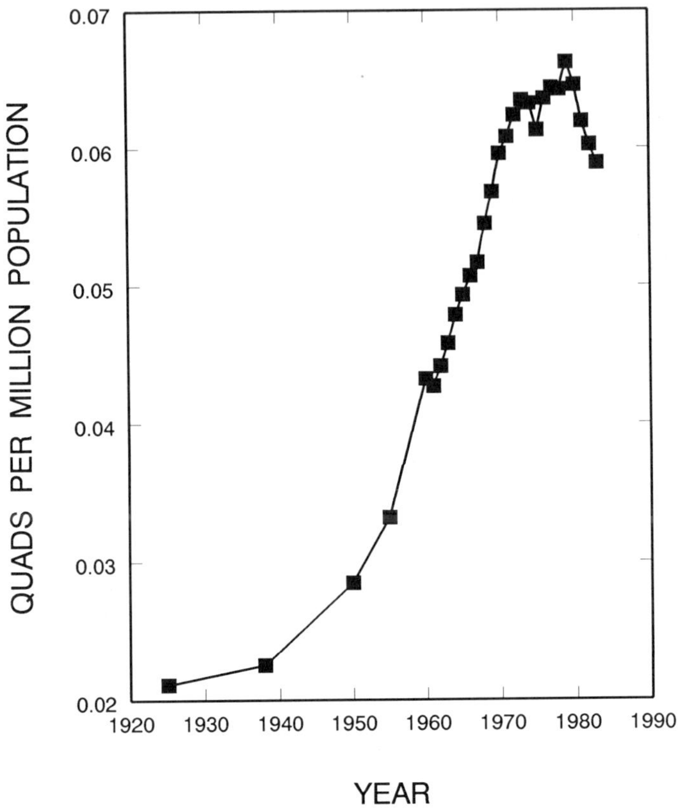

Figure 5-5 World energy consumption per capita.

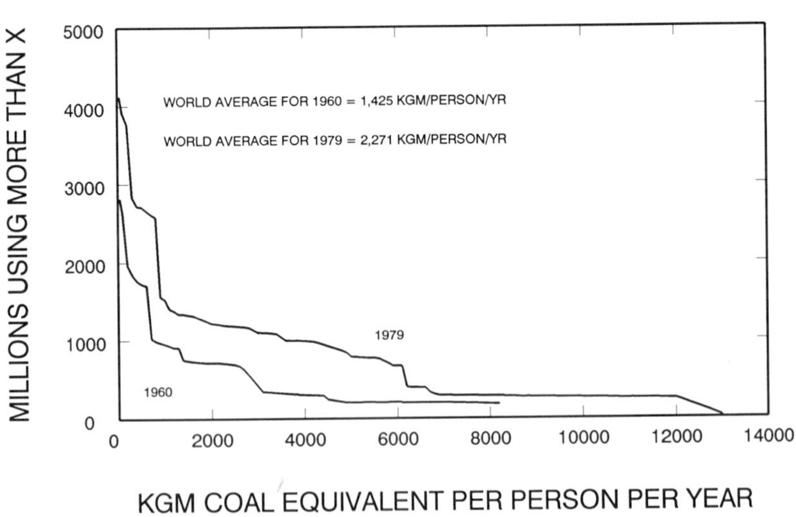

Figure 5-6 Integral distribution of per capita world energy consumption.

quarter of the world population. Nevertheless the disparities of energy use between rich and poor are stark. As late as 1979, three-quarters of the world population had an energy usage less than that of the highly industralized Western societies in the second half of the nineteenth century. In fact, almost one-third of the world population in 1979 had an energy consumption less than 1500 kgm of coal equivalent per capita. As noted above, the energy use for preindustrial agricultural societies was just about 1500 kgms/year/person (26,000 calories/year/person). Clearly, about one-third of the world is industralized, one-third is on the path to industrialization, and one-third hasn't yet started.

A few technical points must be mentioned concerning Figure 5-6. The first is that the per capita energy data on which the figure is based included only "commercial" energy; that is, the data reflect only fuels that are bought and sold. The energy consumption of the poorer countries is therefore underestimated since the figures do not include fuel that is gathered and used by individuals outside the market mechanism, such as wood and dung. The next point is that the distribution curves in Figure 5-6 were built up from the average per capita consumption for each country in the world. Since there is a distribution of energy consumption among individuals in each country, the curves are not precise measures of the distribution of energy use among all individuals of the world. The precise definition is that the curves give the number of people living in societies in which the average energy consumption per capita is greater than a given amount. Neither of these points has any serious effect on the conclusions that were drawn from the data.

ENERGY SOURCES

The mineral sources of energy that are currently being exploited are primarily petroleum, natural gas, coal and uranium. Relatively small amounts of energy are obtained from other sources that include water power, wood, solar energy, wind, and geothermal formations. Altogether these contribute a small fraction to the energy consumption of a modern society. Mineral energy sources constitute the foundation of modern technological society, and a summary of their properties is presented here.

Liquid crude oil varies greatly in color, viscosity, and density. Some is similar to gasoline, but most range from viscous green-red to black liquids. The green-red has a density of 0.97 gm/cc; the black is lighter with a density of 0.76 gm/cc.

Crude oil is a mixture of hydrocarbons. Half the weight of average crude consists of light compounds containing from three to fourteen carbon (C) atoms per molecule.

Small molecules (up to four C atoms) constitute the gases; molecules with five to ten c atoms constitute gasoline; ten to fifty C atoms per molecule make up the light fuels. Heavy molecules, containing 50 to 100 C atoms, make up heavy fuels, waxes, and asphalts. Saturated, unsaturated, ring, and chain molecules are all present. Widespread commercial use of petroleum started in about 1860, and the important products were illuminating oil (kerosene) and lubricants. But with the growth of automobile use the demand for the lighter gasoline fractions increased rapidly. This led to the development of cracking methods by which molecules with long hydrocarbon chains were broken down into smaller ones, thereby permitting the use of much larger portions of the crude oil.

Traces of oxygen, sulfur, and nitrogen are also present in crude oil. The sulfur and nitrogen are contaminants which are responsible for some of our more noxious forms of pollution, and a great deal of effort is expended in trying to remove them and nullify their effects.

Natural gas is normally found in association with petroleum or in areas where petroleum has been found. Natural gas is usually about two-thirds methane. It also contains a small amount of hydrocarbons with five or more c atoms per molecule (about 5 percent) and traces of air and carbon dioxide. The rest is ethane, propane, and butane. This is the natural product and is called wet gas. It is processed to remove the liquifiable portions to give dry gas, which is primarily methane with some ethane and propane and only small amounts of higher hydrocarbons.

The hydrocarbons extracted during the production of natural gas are called natural gas liquids (NGL) which consist of hydrocarbons with five or more C atoms per molecule.

Liquified petroleum gas (LPG) is a mixture of propane and butane. It is liquified by storing it under pressure.

Natural gas is a desirable fuel for many applications. It is clean, having very little of the sulfur and nitrogen contaminants that plague the use of petroleum and coal. Also, since methane has a lower carbon to hydrogen ratio than hydrocarbons with larger molecules, the combustion product has a proportionately lower carbon dioxide content. Natural gas therefore makes less of a contribution to the global greenhouse effect per BTU of heat generated.

Coal comes in a variety of types that are classified by carbon content and volatile matter as illustrated in Table 5-1, which gives data for coals from specific regions. Note that while these values are typical, coal from other regions may vary. For example, the subbituminous sample was from Washington; a sample from Colorado had a heat content of only 6,690 BTU/lb and a carbon content of 30.2 percent. These data are from Penner and Icerman (1981, p. 26).

The chemical compositions of coal vary widely and include trace amounts of many elements in addition to carbon and hydrogen. These range from sodium to arsenic. Coals from the Eastern United States usually contain appreciable amounts of sulfur, which poses serious pollution problems, while Western coals are relatively sulfur-free.

The hydrocarbons in coal differ from those in petroleum in that they have

Table 5-1 Properties of Various Coals

	Fixed Carbon %	Volatiles %	Water %	Ash %	Heat Content BTU/lb
Anthracite	81.8	4.8	4.4	9.0	13,130
Bituminous					
L.V.	65.8	19.6	2.3	12.3	13,200
M.V.	63.6	23.4	3.1	9.9	13,530
H.V.	46.5	43.8	5.9	3.8	13,150
Sub-bituminous	41.0	34.2	13.9	10.9	10,330
Lignite	30.2	27.8	36.8	4.7	8,580

a higher average molecular weight and contain primarily aromatic (ring structures) rather than aliphatic (chain structures) compounds. The fact that it is a solid and leaves a solid ash makes coal a dirtier and less convenient fuel to use than the fluid hydrocarbons.

The only naturally occurring isotope fissionable by slow neutrons is U-235. This constitutes 0.71 percent of natural uranium, the predominant isotope being U-238. After capturing a thermal neutron, an atom of U-235 fissions and produces two to three more neutrons. These in turn are captured by other nuclei of U-235, liberating more neutrons so that a chain reaction can be sustained.

A U-235 reactor must use highly enriched fuel since so little of the fissionable isotope is contained in natural uranium. However, the non-fissionable U-238 can be converted to plutonium. Plutonium can be made from U-238 by placing it in a U-235 reactor, thereby converting it to Pu-239, which is fissionable. Similarly, thorium-232 can be converted to the fissionable U-233. This is the basis for breeder reactors which enable all natural uranium and thorium to be used as nuclear energy resources. One gram of U-235 releases about 8.2×10^{10} joules of heat, which is equivalent to about 13.4 barrels of crude oil or about 2.7 metric tons of coal. Figures for Pu-239 or U-233 are similar.

ENERGY IN THE ECONOMY

The role of energy in economic life is so vital and pervasive that an economy cannot function at all without the use of energy resources. A detailed analysis of how energy is used to serve economic ends would be enormously complex and amount to a complete description of the entire economy. But the critical economic factor that affects all aspects of energy use is its cost. If energy is sufficiently cheap, then fuel is available to drive the growth of the modern economy. If it is too expensive, then the economy must stagnate.

Energy is big business. The value of the production of fossil fuels in the

Table 5-2 U.S. Energy Costs and Expenditures in 1985 (Billions of Dollars)

Fuel	Primary Cost	Fuel Expenditure
Coal	16.4	29.7
Gas	46.0	72.9
Oil	124.1	223.6
Total fuel cost	186.5	326.2
Electricity from fossil fuels		154.5
Electricity from nuclear power, wood, water, etc.		2.9
Expenditure for fuel by electric utilities		42.5
Total energy expenditure		441.1

United States in 1987 was $97.3 billion. In addition to this, $46.5 billion worth of fossil fuels were imported in that year. If the $7.6 billion in exports are subtracted from these figures, we get $136.2 billion as the cost of primary fuels for the United States in 1987. This does not include the costs of other, relatively minor energy sources such as uranium, wood, and water power.

Of course this is not the total cost of the use of energy in the economy. It is just the cost of the fuels at the initial stage of production; at the wellhead or mine. To be used, the fuels must be processed into appropriate fuel products and distributed to the points of consumption. When this is taken into account, we find that the expenditures for fuels and electricity in the United States was $441 billion in 1985.

Some important facts emerge from a closer look at these costs and expenditures, which are provided in Table 5-2. The first point to note is the special status of oil. We spend more on oil, both at the primary and consumption stages, than for the other fuels combined.

In 1985, $42.5 billion was spent on fuels by electric utilities, which was converted into $154.5 billion worth of electricity. In computing the total expenditure of energy consumption, the cost of fuel to utilities was subtracted from the expenditure for electricity so as not to count its effect twice. After this subtraction, electricity expenditures constitute about one-quarter of the total.

The ratio of expenditure on fossil fuel products to primary cost was 1:75, whereas the ratio of total energy expenditure to primary cost was 2:37. The difference between these two ratios is the result of the cost of converting heat to electricity. Electricity therefore seems to be an expensive form of energy relative to direct use of fuels. A true estimate of the relative costs of various energy forms, however, must include an analysis of the efficiency with which they can be used for their designated purposes. Electric motors and electric lamps, for example, perform functions that are not performed as well using other energy forms.

For economic growth, it is not the absolute cost of energy that is important but its cost relative to other costs and relative to total economic activity. A convenient measure of this is energy costs relative to the Gross National

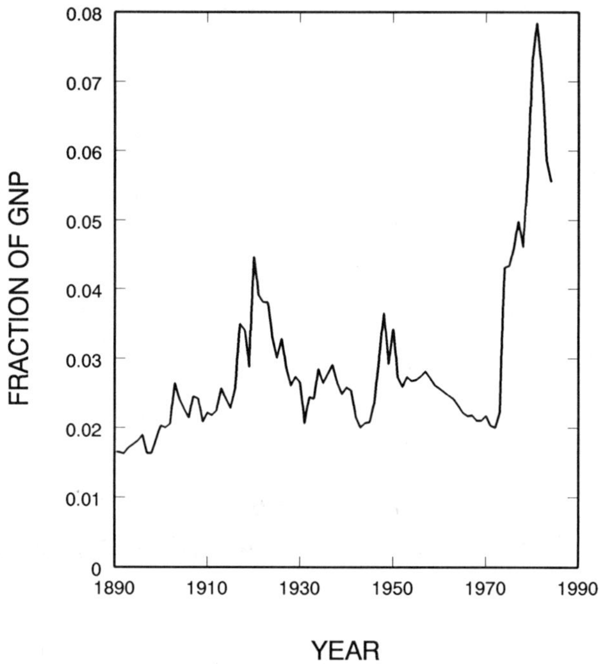

Figure 5-7 United States fossil fuel costs as fraction of GNP.

Product. This is shown for the United States in Figure 5-7, in which the fraction of the GNP spent for primary fuel costs is plotted against time. For most of our history, fossil fuels have indeed been cheap, fluctuating between 2 and 4 percent of the GNP. With the Arab oil embargo and the ability of OPEC to set oil prices in the 1970s, the cost of energy rose rapidly and reached nearly 8 percent of GNP in 1981. The drop after 1981 was the result of a combination of increased efficiencies of energy use, conservation measures, and increased exploitation of non-Mideast resources.

Historical total energy expenditures follow the same ratio as that for 1985. That is, total expenditures are roughly twice the primary fuel costs. This is shown in Figure 5-8, in which the fraction of GNP for total energy expenditures as a function of time are seen to be about double the fraction for primary fuel costs in recent years.

The low cost of fossil fuels for over a century was an important factor in the rapid industrialization and economic growth of the United States. The increase in these costs after the OPEC embargo certainly contributed to the economic dislocations of the 1970s and the question arises as to whether or not the economic growth rates that we have become accustomed to in the past can be maintained when primary energy costs are much above 5 percent. The answer to this is not yet clear. Such increases in energy costs certainly absorb large sums that could have contributed to economic growth, but it

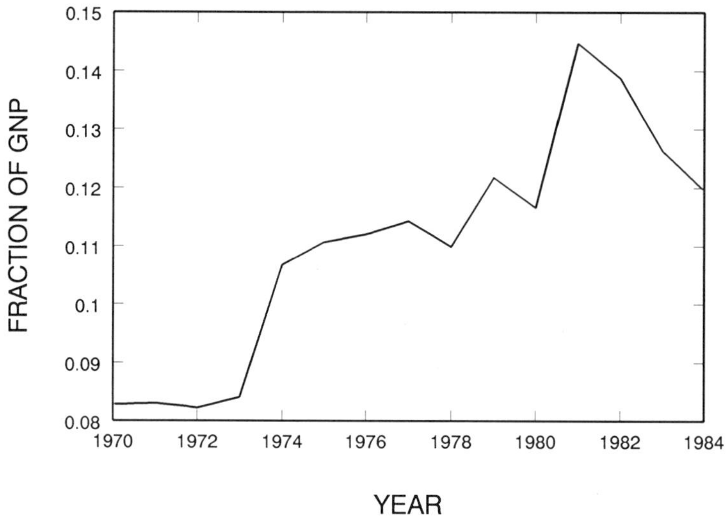

Figure 5-8 Expenditures on fuel and electricity in the United States as a fraction of GNP.

should be noted that relative fuel costs have been decreasing for the past five years. Also, the size of the GNP must have an effect on this question. The GNP is much larger now than in the past and increases in fuel costs must have a smaller effect than when the GNP was smaller. Furthermore, many of the new technologies which are making major contributions to economic growth are much less energy-intensive than those on which the economy of most of this century was based. These technologies include information processing, telecommunications, microelectronics, robotics, genetics, and biomedical engineering, all of which require much less energy per unit of output than steelmaking, chemicals, plastics, jet aircraft, and textiles, which are typical of the highly energy-intensive industries on which previous economic growth depended. In the long run there is no energy shortage, but energy supply has become such an important political issue that short-run projections can only be speculative.

The importance of energy in a modern economy is so striking that some analysts have claimed that there is a direct causal connection between energy use and GNP. That is, it is sometimes claimed that a certain number of BTUs of energy must be consumed to produce a dollar of Gross National Product. From a statistical standpoint, this question can be investigated in two ways; by horizontal or longitudinal analysis. The horizontal analysis consists of comparing the GNP and energy consumption of a number of different countries in a given year, while the longitudinal analysis compares the GNP–energy relation for a given country over a period of years. Let us first consider a longitudinal analysis for the United States by simply plotting the GNP

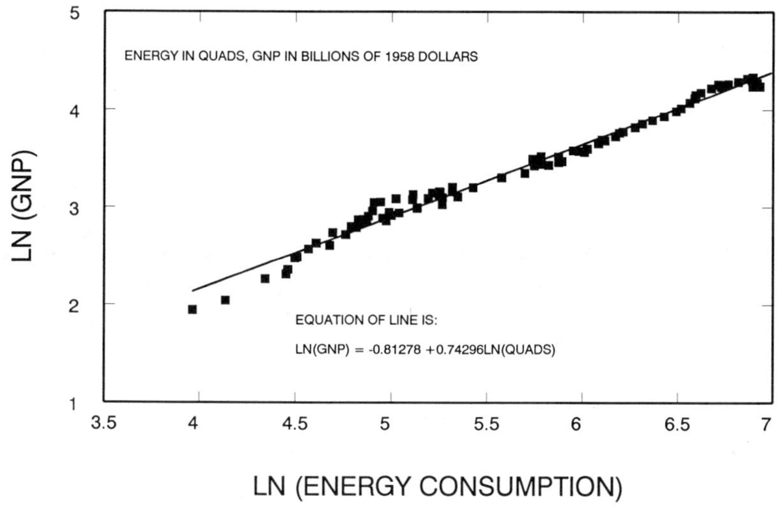

Figure 5-9 Relation between GNP and energy consumption in the United States.

versus energy consumption for a period of years. This is done on a log–log scale in Figure 5-9 for data covering the years 1890 to 1984. The correlation appears to be excellent, and this is often cited as evidence of a direct causal relation between economic growth and energy useage.

To aid in interpretation of these data, we write down the simple equations of growth as applied to GNP and energy consumption. These are:

$$ln(G/G_o) = g(t-t_o) \tag{5-1}$$

$$ln(E/E_o) = e(t-t_o) \tag{5-2}$$

Where G and G_o, E and E_o, are the GNP and energy consumption at times t and t_o respectively, and g and e are average growth rates defined by

$$g = 1/(t-t_o) * \int (dlnG/dt')dt' \tag{5-3}$$

$$e = 1/(t-t_o) * \int (dlnE/dt')dt' \tag{5-4}$$

where the integrals are taken over the time period$(t-t_o)$.

By defining these time averages of the rates, the equations for growth are put into a form similar to that for exponential growth regardless of the forms of the time dependence of the GNP and energy consumption. However, the "rate constants" in Equations (5-1) and (5-2) are constant only if the growth laws for the G's and E's are indeed exponential. In all other cases, g and e depend on the time period over which the averages were taken and therefore depend on the time.

Elimination of the time between Equations (5-1) and (5-2) gives the GNP-energy relation as

$$ln(G/G_o) = A\ ln(E/E_o) \tag{5-5}$$

where A is the ratio of the growth rates.

$$A = g/e \tag{5-6}$$

Equation (5-5) provides a rationale for the log–log, plot, but a straight line would be expected only if A were constant. However, the fact that Figure 5-9 approximates a straight line is not sufficient to let us conclude that the ratio of growth rates for GNP and energy use is precisely constant. The plot is not sensitive enough for this because the growth rates in Equations (5-1) and (5-2) are time averages. For each succeeding year, the effect of the previous years on the averages increases. For later years, therefore, any changes in growth rates are masked by the heavy contribution of earlier years to the averages. We would expect this to result in a curve that approaches a log–log plot at later years no matter what the relation between GNP and energy use. In fact, it is apparent from Figure 5-9 that the data fit the curve better at later years than at earlier ones. Nevertheless, the data extend over a long enough range to establish that the growth rate ratio is *approximately* constant, that is, $dlnG/dlnE \approx$ constant.

A direct comparison of GNP to energy use can be made by calculating the ratio of energy use to GNP as a function of time. This is shown in Figure 5-10. Clearly, this ratio is not constant. Until 1920 this ratio increased, but after 1920 it has shown a strong downward trend right to the present. An interpretation of this graph is that we have learned to use energy more efficiently so that less of it is needed for a given level of economic activity.

It may seem as if Figure 5-9 and 5-10 are not compatible, but in fact they are. A linear log–log plot is equivalent to the equation $E = KG^{-A}$; so the ratio of energy use to GNP is $E/G = KG^{-1-A}$.

This means that if the GNP per capita increases with time, as is certainly the case, then the E/G ratio decreases, as observed after 1920. Note that no conclusions can be drawn from the earlier data because the log–log plot of Figure 5-9 is poorest before 1920.

Now let us turn to a horizontal analysis in which the relationship between energy use and the level of economic activity is examined for a variety of different countries at a particular time. The mathematical framework for interpreting the data is similar to that for longitudinal analysis except that it is applied to many countries at one particular year instead of to one country at many years.

Consider a group of countries, each being labeled by an index j. The Gross National Product G_j and energy consumption E_j can always be described as functions of time by the equations

$$ln(G_j/G_{jo}) = g_jt \tag{5-7}$$

$$ln(E_j/E_{jo}) = e_jt \tag{5-8}$$

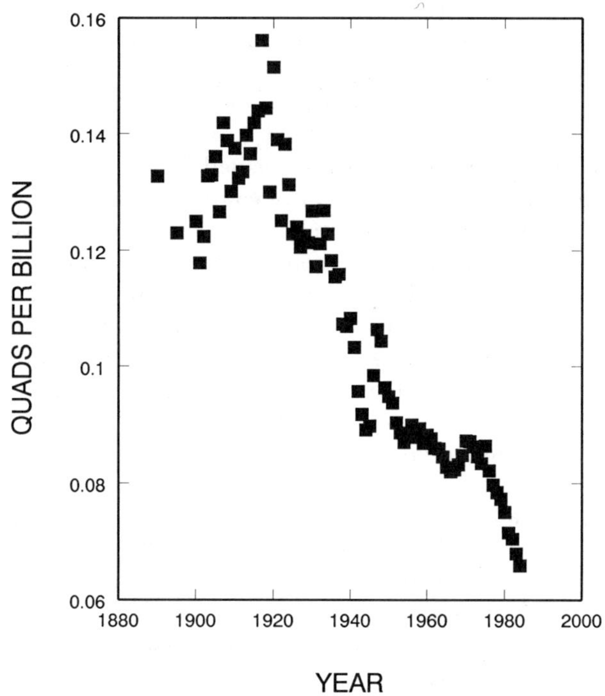

Figure 5-10 Energy–GNP ratio for the United States.

where G_{jo} and E_{jo} are the GNP and energy consumption for the jth country at time zero, and g_j and e_j are averages of the growth rates over time t for the growth of GNP and energy consumption respectively and are defined by integrals just like those of Equations (5–3) and (5–4), except that the subscript j is used to identify different countries.

The horizontal relation between GNP and energy consumption is obtained by eliminating the time between Equations (5–1) and (5–2) to get

$$ln(G_j/G_{jo}) = A_j \; ln(E_j/E_{jo}) \tag{5-9}$$

where

$$A_j = g_j/e_j \tag{5-10}$$

Just as in the longitudinal analysis, the rate "constants" are, in general, functions of time.

From a purely mathematical viewpoint there is nothing in these equations that requires that there is or is not a correlation between GNP and energy consumption. If the ratio of the growth rates A_j are the same for all countries then a log–log plot for a set of countries at a given time would yield a straight line according to Equation (5-9). This would mean that in the year for which the plot is made, an incremental increase in energy consumption would be accompanied by a proportionate increase in level of economic activity for all countries. Furthermore, if linear log–log plots are obtained for different years, this would mean that the ratio of the growth rates is independent of time, and this would support the idea that there is a fundamental connection between energy use and economic activity.

To make a horizontal plot, data are needed for energy consumption and for GNP per capita for a large number of countries during a given year. The energy consumed by different countries is a straightforward measure, since fuels have the same energy content no matter where they are used. But comparing the GNPs of different countries is an entirely different matter. The traditional method of such a comparison makes use of official currency exchange rates so that the GNPs can be reduced to comparable units in a single currency. But the exchange rates are often determined by political factors and currency market speculations. Their relation to "real" costs is not at all clear. All horizontal analyses based on exchange rates are therefore suspect.

The most thorough and reliable international comparison of purchasing power is that of Kravis, Heston, and Summers (1982), as updated by Summers and Heston (1984, 1988), in which they determined national incomes and products for thirty-four countries by a detailed comparison of prices in local currencies for an extensive list of products, commodities, and services. The extent of the errors possible in using exchange rates is indicated by the fact that these authors found that incomes and growth rates could be underestimated by a factor of two to four for low-income countries. In our horizontal analysis of the relation between energy use and national economic levels we therefore use the results of Summers and Heston. They computed Gross Domestic Product but did not determine Gross National Product. The GNP is the total market value of all goods and services, while the GDP excludes the net income from abroad and measures the economic activity within the national boundaries. For large countries the difference between GNP and GDP is small, but for small countries the difference can be substantial. Since the GDP is what is available, it will be used for the horizontal analysis. Plots for the logarithms of GDP and energy consumption are shown in Figures 5-11 and 5-12 for the years 1960 and 1979 respectively. The energy consumption data were taken from the 1981 World Development Report of the World Bank.

Figures 5-11 and 5-12 show that there certainly is a log–log correlation between GDP and energy consumption. The correlation coefficient is high for both years, but the scatter of the points is considerable. The number of countries in these plots is large, and they include poor, rich, underdeveloped, and industrialized nations. They also include countries with centrally planned as well as free market economies. The correlation does not seem to depend on national characteristics of wealth or market system.

The slopes of the two plots are not identical but they are not too different. Their ratio is 1.3. Given the scatter in the data, it is hard to conclude whether

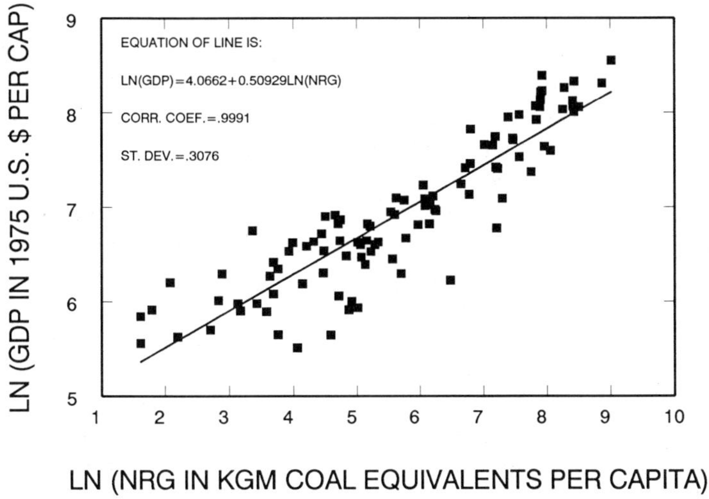

Figure 5-11 Relation between energy consumption and GDP for 102 countries in 1960.

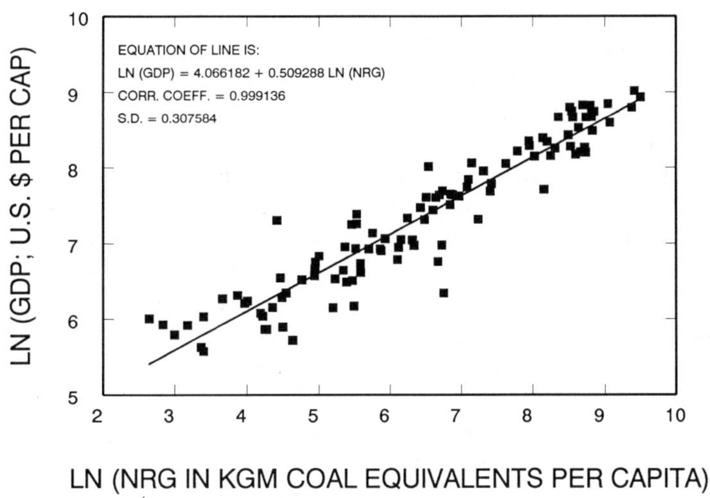

Figure 5-12 Relation between energy consumption and GDP for 106 countries in 1979.

or not this difference is significant. A log–log plot of all the data for both years (not shown) exhibits considerably more scatter and is not as good a fit as the plots for the years taken separately. This suggests that the difference in slopes for the 1960 and 1979 plots is real.

The most reasonable conclusion from these curves is that the level of economic activity and the consumption of energy indeed do go together, but only to a first approximation. There is a systemic correlation on which is superimposed idiosyncratic fluctuations among countries. The origins of these fluctuations are many. The distribution and efficiency of energy use varies widely among countries. Climatic conditions dictate that the per capita use of energy for space heating will vary among countries; advanced industrial nations are more efficient in energy use and use a greater fraction of energy in industry than underdeveloped countries; travel requirements, energy needs of agriculture, and fraction of energy-intensive industries differ widely from one country to another. The idiosyncratic national characteristics inevitably lead to the observed scatter, even given that there is an underlying systemic relationship.

This does not at all imply that there is a simple causal relationship between energy use and the national product in the sense that all that is needed to raise the level of economic activity is to use more energy. The national product is the result of a complex mix of many factors, ranging from market structures, financial mechanisms, government policies, and institutional organization to educational levels, availability of resources, industrial technologies, and the rates of innovation. But through it all, energy must be used to convert raw materials to finished goods, for transportation, and for heat, light, and power. The systemic forces that drive economic growth must also result in increased energy use.

The United States is often accused by energy analysts of being excessively profligate in the use of energy relative to other nations. If the question of the excessive use of energy by some countries has any significance, then it must mean that those countries whose points lie below the fitted line in Figures 5-9 and 5-10 are wasteful while those above the line are efficient. This interpretation is not necessarily valid because of the aforementioned differences among national circumstances. But even so, the points for the United States lie close to the line in both the 1960 and 1979 plots. The per capita consumption of energy in the United States is quite appropriate given the level of economic activity. This is an illustration of how inadequate data can yield misleading conclusions. All previous plots of energy use versus GDP or GNP were based on comparing a smaller number of nations and by using exchange rates to obtain consistent monetary units. But the work of Kravis, Heston, and Summers shows that these are not realistic measures by which to compare international incomes or gross national products. Furthermore, energy–GNP plots are often plotted linearly and usually using a small number of countries. Such analysis easily leads to the wrong conclusions.

THE EFFICIENCY OF ENERGY UTILIZATION

Energy is always utilized in some kind of a conversion process in which one form of energy is transformed into another. At some point in the process a useful result is effected that is valued by human beings. For space heating, the chemical energy of fossil fuels is converted into heat energy that maintains comfortable temperatures. For transportation, the chemical energy is converted into heat energy and then into mechanical energy of motion. For electric light and power, the chemical energy of fossil fuels or the nuclear energy of fissionable materials is converted first to heat, then to rotary motion, and then to electricity. The electricity is finally converted to mechanical, light, or heat energy depending on the application. In all these processes the amount of energy captured for the useful application is always less than the energy content of the original energy source.

The efficiency of energy utilization is the critical technical factor that is important for all fuels and energy sources. Efficiency determines whether or not the use of a particular energy source for a given application is economically feasible; an increase in efficiency is equivalent to an increase in supply and a decrease in price. Increased efficiency has the effect of using less fuel for the same result so that it is a major means of conserving energy and mitigating pollution as well as saving money. Advances in the technology of energy use have resulted in a continual improvement in efficiency which has made energy-intensive societies possible.

There are several different definitions of efficiency that are useful for different purposes. These definitions are easily understood by starting with the description of how fuels are actually used. Whether it is coal, gas, oil, wood, or uranium, the first step in using a fuel is to burn it to produce heat. If combustion is complete so that the final products are not combustible, the efficiency of the conversion of the chemical (or nuclear) energy to heat energy is unity. In practice, combustion is never perfect, and incomplete combustion results in efficiencies less than unity. The efficiency for combustion is defined as

$$\varepsilon(\text{comb.}) = (\text{ht. energy out})/(\text{chem. energy in})$$

or, in shorter symbols,

$$\varepsilon_{co} = H/E \tag{5-11}$$

This is a First Law efficiency, so named because the First Law of Thermodynamics requires that energy be conserved. Perfect combustion would therefore yield an amount of heat equivalent to the chemical energy content of the fuel.

The heat generated by the burning fuel can be used in several ways. First, consider its direct use to heat living spaces or in industrial processes. The heat is carried in a working fluid such as air or steam which transmits the heat energy to the living space or to the vessels in which the industrial processing takes place. In either case, the task is to maintain a temperature T in a given space. We will focus on space heating, but the general principles apply also to industrial process heat. If the space were perfectly insulated, then the only energy required would be an initial amount to bring the space to the desired temperature. Since perfect insulation means that no heat could escape, the temperature would be maintained indefinitely without further heating. But there is no perfect insulation, and even if there were, no one could live in such a space. At the very least, some energy must be expended to provide a certain level of humidity and air flow for ventilation and comfort. In addition, incomplete insulation permits heat to be lost through conduction. A continual injection of heat into the space is therefore needed to balance these losses and maintain a given temperature. A practical measure of space heating efficiency can be constructed by defining a standard to which real situations are referred. Let's do this by defining a standard space (house or room) in which the rate of heat loss is just enough to provide a comfortable level of humidity and ventilation. Call this heat rate R_s. In a real house, the ventilation rate and humidity may be more than what is just needed for comfort, and there is also conduction through walls and ceilings. The actual heating rate is therefore $R = C + V$, where C is the rate at which heat is conducted to the outside and V is the actual heat loss sustained from air flow and humidification. The efficiency of space heating can therefore be defined as

$$\varepsilon_p = R_s/(C + V) \tag{5-12}$$

This is a practical definition of efficiency that deals with the realities of existing space heating. To illustrate its use, consider the result of Ross and Williams (1978) that, for a typical house, 60 percent of the heat loss is through conduction and 40 percent is lost through ventilation and humidification. Also, the ventilation rate could be reduced to 20 percent of the value for the typical house with no loss of comfort. For a typical house, therefore, $R_s = 0.2V$ and $V/C = 2/3$; from which, $V = 5R_s$ and $C = 7.5R_s$. Putting these results in Equation (5-12) gives $\varepsilon_p = 1/(5 + 7.5) = 0.08$.

The practical efficiency for heating a typical house is only 8 percent, and this ignores the incomplete combustion in typical furnaces. For a furnace with 70 percent combustion efficiency, this reduces to 5.6 percent. If perfect combustion could be achieved, if perfect insulation could be installed, and if only enough ventilation and humidity were provided for adequate comfort, only 8 percent of our current consumption of fuel for residential heating would be needed. Of course, perfect insulation is unattainable, but Ross and Williams quote an American Physical Society study which concluded that the conductive losses could be reduced to one-third of the then current values by the use of good insulation and well-designed windows. This conclusion can give us a rough measure of the potential for achieving practical increases in space heating efficiency. The heat loss rate of such a house would be $(1/3)C + 0.2V = (1/3)C + 0.2(2/3)C$.

Dividing this by $(C + V) = C + 2\,C/3$ gives the fraction of heat needed for the improved house relative to the typical house. The result is 0.28. This shows that improved insulation and reduction of unnecessary air flow can reduce the heat needed for space heating to 28 percent of its current value. Since the data all refer to 1979, this is not true nationwide because many houses have been improved since then and new construction is better insulated than it used to be. Nevertheless, the calculation shows that great savings can be made in conventional space heating.

The efficiencies defined above are First Law efficiencies in that they deal only with the energy balance arising from the Law of the Conservation of Energy for a given device or system. They do not reflect the total inefficiencies in our use of energy because they do not refer to the most thermodynamically efficient way of transforming heat into work. Thermodynamics shows us that there are certain ideal processes (called reversible processes) in ideal heat engines that maximize the amount of work that can be obtained from a given amount of heat and minimize the amount of work required to transfer a given amount of heat. The ideal heat engine operates by transferring heat from a high to a low temperature using a fluid to do work in the process. The heat transfer is done reversibly; that is, in such a way as to maintain the system in equilibrium at every step of the process. No heat engine can do more work per unit heat transfer than this ideal, reversible heat engine. Thermodynamics tells us that even for ideal engines only a fraction of the heat energy can be converted to work. That fraction is the ratio of the work performed to the heat transferred and defines the Second Law efficiency, $\varepsilon_{s.l.}$ as

$$\varepsilon_{s.l.} = A_{max}/H = 1 - T_o/T \tag{5-13}$$

where H is the amount of heat transferred from the higher temperature T to the lower temperature T_o thereby performing an amount of work A_{max}. No real engine can do more work than this and $\varepsilon_{s.l.}$ is the maximum efficiency permitted by the laws of Thermodynamics.

Now consider the task of heating a room or a house by transferring heat to it from the outside. To do this requires work, since heat must be transferred from a colder to a hotter environment. This work is done by a heat pump which is just a refrigerator run backwards. Again, thermodynamics tells us that there is an ideal heat pump which requires less work than any other device to transfer a given amount of heat from a lower to a higher temperature. This minimum amount of work, A_{min}, is related to the heat transferred, H, by the Second Law as

$$A_{min} = H(1 - T_o/T) \tag{5-14}$$

To make these ideas concrete, consider a house that is to be kept at a temperature of 75 degrees Fahrenheit when the outside temperature is 30 degrees F. Converting these to degrees Kelvin gives

$$A_{min}/H = (1 - T_o/T) = (1 - 272.05/297.05) = 0.08416$$

This means that only 8.42 percent of an equivalent amount of work must be done to transfer a given amount of heat. In other words, 11.9 times as much heat can be transferred using an ideal heat pump as would need to be generated in the house by perfect combustion. Of course, this assumes an ideal reversible heat pump, but real heat pumps achieve at least a threefold improvement over direct combustion. This means that all the efficiencies computed above for space heating, which were referred to the existing system of burning fuel to provide heat, need to be divided by at least three when referred to practical heat pumps and by 11.9 when referred to the best theoretically possible performance. For the practical efficiency of 5.6 percent computed above, these figures show that relative to theoretically achievable performance, current efficiencies are only 1.9 percent if heat pumps currently available are used and only about 0.5 percent if ideal heat pumps could be used. The potential for improved efficiency for space heating is clearly very large.

Now let's turn our attention to the efficiency with which heat is used to do work.

The First Law efficiency for the process of converting heat into mechanical work in a heat engine is the ratio of the work performed to the heat put in.

$$\varepsilon_w = W/H \tag{5-15}$$

where W is the work obtained from a heat engine from an amount of heat H. The "fuel efficiency" is less than this because combustion is incomplete, and is obtained by multiplying Equation (5-15) by the combustion efficiency.

The Second Law of Thermodynamics places an upper limit on the efficiency of the conversion of heat into work as given by Equation (5-13).

The maximum efficiency of Equation (5-13) can be increased only by increasing the difference between the hot and cold temperatures of the heat engine. The role of high temperature materials is therefore critically important, and there is a constant search for materials that can withstand ever higher temperatures.

Of course, the maximum efficiency is never reached. Conductive and frictional heat losses, for example, decrease efficiency below the theoretical Second Law value and the amount of work, A_{act}, that is actually realized from a heat engine is less than the possible maximum, A_{max}. This can be described by defining an efficiency ε_f by

$$\varepsilon_f = A_{act}/A_{max} \tag{5-16}$$

The overall fuel efficiency is therefore given by

$$\varepsilon = \varepsilon_{co}\, \varepsilon_f\, (1 - T_o/T) \tag{5-17}$$

Table 5-3 Efficiencies in the Steps of Converting Fuel to Work in an Average Automobile: 1978

Step	Efficiency (%)	Cumulative Efficiency (%)
Oil production	96	96
Refining	87	84
Gasoline transport	97	81
Thermal to mechanical	27	22
Transmission to wheels	55	12
Rolling efficiency	60	7

This displays the components of the efficiency of converting the chemical energy of fuel into mechanical work and shows that savings can be effected by improving the combustion process, decreasing frictional losses in the engine or the loss of heat to the environment, increasing working temperatures, or decreasing exhaust temperatures.

Note that some care must be taken in applying these concepts of efficiency. In some applications, such as the generation of electricity, we are interested in the efficiency of converting heat to rotary motion which is then converted to electricity. The losses in the calculation of efficiency then include only the frictional effects in the rotors and air friction. In auto transport, however, the losses include not only internal frictional effects in the engine and aerodynamic drag, but also rolling friction and losses in transmission of mechanical power from the engine to the wheels. Each application requires an analysis of the entire train of events from the burning of the fuel to the desired mechanical result.

An important instance of the conversion of the chemical energy of fuel to mechanical energy is embodied in the automobile. Fowler (1978) has given a table of efficiencies for the various steps in converting fuel energy to automobile motion. His data are reproduced in Table 5-3.

The figures in Table 5-3 are characteristic of an automobile that gets about 14 miles per gallon. New automobiles are capable of twice this, but there are still so many old vehicles on the road that the national average in the late eighties was still less than 18 miles per gallon.

Note that this table includes a calculation of the energy required to produce the crude oil, refine it to produce gasoline, and transport it to the point of sale. These are not normally included in calculations of automobile efficiencies, but they must be considered if we wish to account for the entire energy content of the oil taken from the ground. These are highly efficient operations but their effect is not trivial. If they were to be ignored, the efficiency of converting gasoline to automobile motion is 9 percent. That is, if gasoline could be made and delivered to automobiles without any energy expenditure, we would increase the overall efficiency from 7 to 9 percent and need almost 30 percent less oil than we actually use for automobiles.

Table 5-4 Possible Efficiency Improvements for Automobiles

Action	Efficiency Improvement (%)
Higher engine temperatures	100
Aerodynamic design	5
Radial tires	10
Improved transmission	10–15
450 lb weight reduction	18

These figures are for an average auto in the late seventies. The efficiency can be improved in a variety of ways, most of which have been implemented for new automobiles in the mid-eighties. These are shown in Table 5-4.

The overall effect of all the improvements shown in Table 5-4 would be to improve the efficiency of using gasoline from 9 percent to 22 percent. The largest effects are from decreasing automobile weight and increasing engine operating temperatures. Most of the weight reductions have already been made, but not very much progress has yet been made in increasing operating temperatures by large amounts.

The fist step in generating electricity is the burning of fuel in a heat engine to convert heat to mechanical work, which is then converted to electrical energy. The description of efficiency is the same as above except that another factor is introduced to measure the efficiency of converting mechanical to electrical energy. The electricity is then transmitted to the point of use, thereby incurring transmission losses, after which it is converted to heat (for electric heating, drying, or cooking), to mechanical work through electric motors, or to light. Each of these involves a series of steps whose efficiency is less than unity and contributes to the energy losses all of which are, in principle, similar to what has already been presented. The Second Law efficiencies appear whenever there is a conversion of heat to work or when the efficiency of heating space is computed. Otherwise, efficiencies at each step are measured by taking the ratio of energy out to energy in. The overall efficiency of the use of electric energy is the product of the efficiencies at each step from combustion to final use:

$$\varepsilon(\mathrm{el}) = \varepsilon_{\mathrm{co}}\varepsilon_{\mathrm{f}}\,(1 - T_o/T)\varepsilon_{\mathrm{me}}\varepsilon_{\mathrm{tr}}\,\varepsilon_{\mathrm{ap}} \tag{5-18}$$

where $\varepsilon_{\mathrm{co}}$ and ε_{f} are the combustion and "frictional" efficiencies defined above, T and T_o are the combustion and exhaust temperatures of the generator, $\varepsilon_{\mathrm{me}}$ is the efficiency of the conversion of mechanical to electrical energy, $\varepsilon_{\mathrm{tr}}$ is the efficiency of the transmission of electricity from the generating plant to the point of end use, and $\varepsilon_{\mathrm{ap}}$ is the efficiency of the final application of electrical energy.

It is instructive to track the efficiency of electricity use for three applications: large motors, which are important for industrial mechanical power; small motors, such as those used in home appliances; and electric lighting

Table 5-5 Efficiency of the Use of Electricity for Various Devices

Step	Large Motor	Small Motor	Incandescent Light	Fluorescent Light
Chemical to thermal	0.9	0.9	0.9	0.9
Thermal to mechanical	0.4	0.4	0.4	0.4
Mechanical to electrical	0.99	0.99	0.99	0.99
Transmission	0.92	0.92	0.92	0.92
Electrical to mechanical	0.92	0.65	N.A.	N.A.
Electrical to radiant	N.A.	N.A.	0.05	0.2
Net efficiency	0.3	0.2	0.02	0.07

by fluorescent or incandescent lamps. The efficiencies for these at the various stages of conversion to electricity and then to end use are shown in Table 5-5.

The product of the efficiencies for the first four steps in Table 5-5 is the same for every application and is the efficiency of generating electricity from fuel and bringing it to the user. This is about 33 percent. (Note that the energy cost of producing the fuel and transporting it to the generating plant is not included.) The figure of 0.9 for the combustion efficiency and 0.4 for the conversion of heat to work are average figures for modern, efficient plants. The conversion of mechanical to electrical energy by rotating machinery is highly efficient, but there are significant transmission losses. The figure of 0.92 transmission efficiency (8 percent loss) is an average national figure for the mid-eighties.

Larger motors are significantly more efficient than small ones. There are, of course, further losses in the use of the mechanical energy of the motors, and these are also greater for small motors.

The overall efficiency of electric lighting is quite low, most of the losses occurring in the conversion of electric to radiant energy.

The examples given above show that the efficiency of energy use is low and that there is the potential for large improvements. But energy efficiency in the past was even lower, and the growing use of energy has been characterized by a continual increase in efficiency.

The longest record we have of increasing energy conversion efficiency is for steam engines. Starting with the Savery engine in 1698 to the present steam turbines used in steam–electric generating plants, the efficiency of converting fuel to mechanical work via steam has increased eightyfold. This progress is shown in Figure 5-13, which displays data on the best efficiency of steam engines as a function of time taken from Martino (1983) and Singer et al. (1958). The Savery engine had an efficiency of only 0.5 percent. This was more than doubled by the Newcomen engine which, by 1712, had an efficiency of 1.3 percent. The Watt engine doubled this efficiency again and by 1796 had an efficiency of 4.1 percent. Successive improvement involved

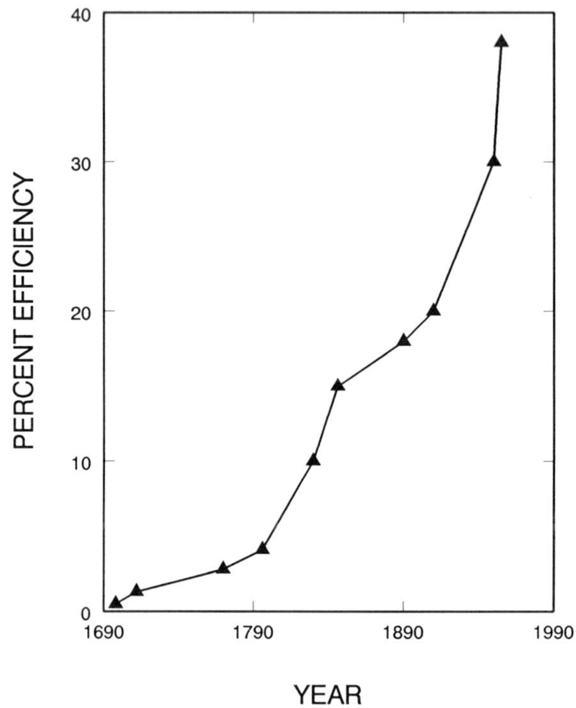

Figure 5-13 Steam engine efficiencies.

high pressure and multiple expansion cycles, and by 1906 the efficiency was up to 23 percent. Further progress depended on the development of the steam turbine which had an efficiency of 40 percent by 1955, a value that has not been significantly improved upon since. (Note that the data in this graph are not quite comparable. The early data were for overall fuel efficiency and included combustion efficiency, while the later data (steam turbines) did not. The data for the early years, however, are far from precise and the graph is sufficiently accurate to reflect the historic trends.)

Figure 5-13 shows the familiar growth curve of technological progress in which a slow start gives way to rapid improvement which ultimately saturates and can only be resumed by adopting another technology.

Another view of the increasing efficiency of energy use is shown in Figure 5-14, which gives the efficiency of electricity generation. This is an overall efficiency defined as the ratio of electricity generated to energy content of the fuels used in generating plants throughout the United States. It therefore includes the overall combustion efficiency. Again, there is a rapid rise with a saturation as technological limits are reached. The rise in the last few years is the result of using the more efficient gas turbines.

Figure 5-15 displays the energy costs of transporting freight in the United

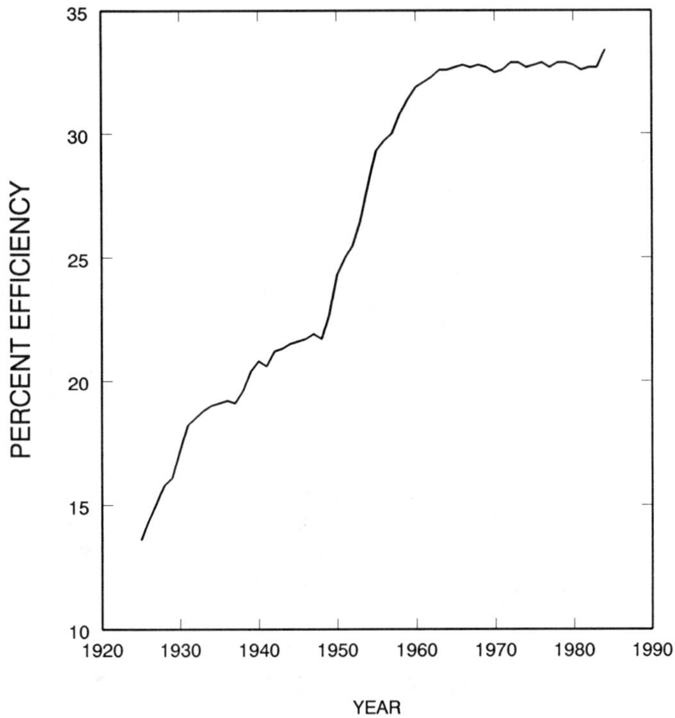

Figure 5-14 Thermal efficiency of electricity generation.

States in units of ton-miles of freight transport per thousand BTUs expended. This was relatively flat until about 1945 but then increased rapidly right to the present. Certainly the increased efficiency of the diesel–electric over the steam locomotive contributed to this, but it cannot account for the factor of fifteen improvement. Most of this was the result of advances in freight car design and in the management of car movements and scheduling. Many of these changes were each relatively minor but their cumulative effect was large.

One area in which we have not done well with regard to efficiency is that of passenger automobiles. The average miles per gallon for all United States cars is shown in Figure 5-16. From 1930 to 1975 there was a steady decrease in the average mileage. People were quite willing to pay for the delights of large, inefficient automobiles, particularly since gasoline was cheap. This turned around in the seventies as a direct result of the OPEC oil embargoes, price rises, and subsequent government pressure. Modern automobiles can be made with mileages up to 40 miles per gallon, and the upturn in the curve reflects the mileage of new automobiles, which is in the twenties. However, the fleet of high gas consumers is not yet completely replaced. Also, the stable oil prices of recent years have reduced the pressure to get

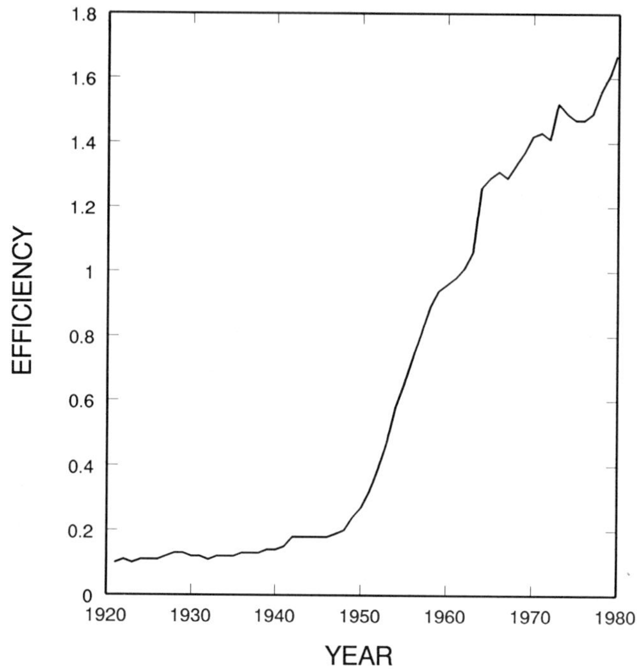

Figure 5-15 Rail freight transport efficiency in ton-miles per 1000 Btus.

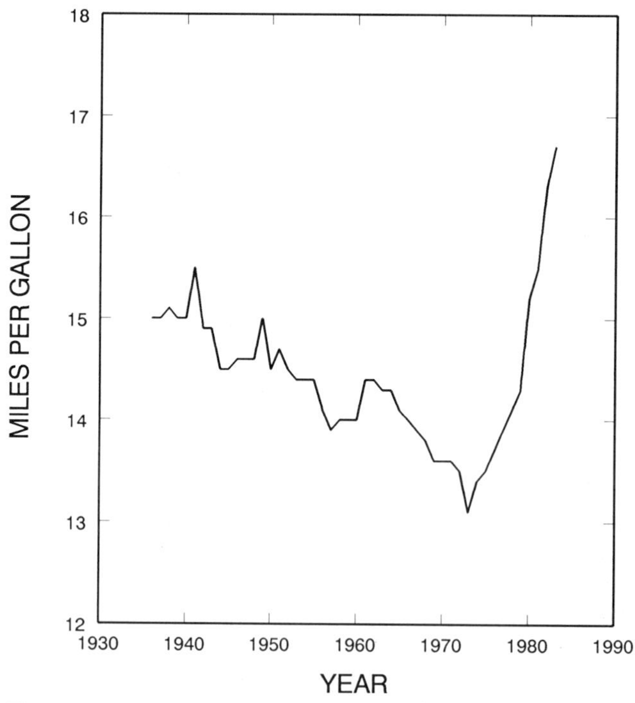

Figure 5-16 Miles per gallon of U.S. passenger vehicles.

more miles per gallon. Figure 5-16 does show that it is possible to increase auto efficiency rather rapidly. In five years the mileage of new cars doubled, while the overall efficiency of the average American automobile increased more than 25 percent. There is every reason to believe that these improvements can continue.

The historical examples given here all refer to First Law efficiencies in that they measure the amount of energy required to do a particular job. While the Second Law efficiencies are better for determining the potential savings in energy usage, the First Law efficiencies more directly reflect the historic trends of more efficient fuel usage.

Data on increasing energy efficiency can be multiplied indefinitely. There have been major improvements in the efficiency of incandescent lighting, and when this reached saturation because of technical limits, there were improvements in fluorescent lighting that doubled lighting efficiency beyond the best that could be done with incandescent lamps. On a longer time scale, the efficiency of artificial lights, from the candle in 1850 to the fluorescent lamp of 1940, increased by a factor of 550. In 1912, it took 1.22 tons of coke to make one ton of pig iron, while in 1960 it took only 0.75 tons (Gold et al., 1984). Similar improvements have been made in all aspects of energy use from metal processing and chemicals production to commercial space heating and air transport.

PATTERNS OF CONSUMPTION

The major sources of energy in use today are fossil fuels: coal, petroleum, and natural gas. In addition, some nuclear fuel is used to generate electricity and some wood is still used. These fuels are burned to produce heat which is used directly for space heating or industrial process heat, or converted to mechanical energy in heat engines. The mechanical energy is used directly, primarily in transport, or is converted to electricity. Electric energy is used by converting it to other forms in motors, electrolytic processes, lighting, and heating. The end result is that all the primary energy (the initial heat energy derived from the burning fuel) is ultimately converted to low-level heat.

In addition to the major fuel sources, some energy is supplied by unconventional sources such as direct sunlight, wind, tides, water, and geothermal sources. These are used primarily to generate electricity, although a small amount of sunlight is used for water and space heating. The most important of these sources is hydroelectric power, which is a significant contributor to electrical energy.

While primary energy is normally classified according to the fuel source, its consumption is classified according to the sector of the economy in which it is used. Of course, this can be done at a variety of levels of segmentation. The coarsest of these identifies energy consumption as taking place in the industrial, transportation, and residential–commercial sectors. Residential

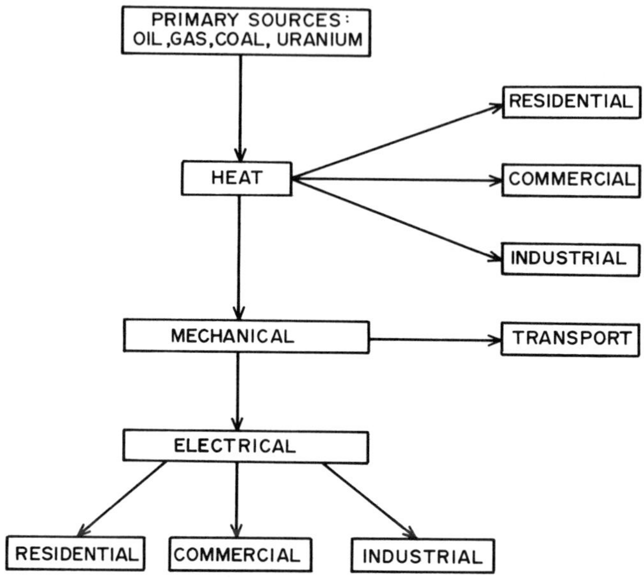

Figure 5-17 Flow of energy from source to end use.

and commercial uses are lumped together because they are very similar, namely to provide space heating and air conditioning along with services such as telephones and elevators. Sometimes, however, data are available that separate residential and commercial usage.

The flow of energy from primary sources to economic sectors is represented in Figure 5-17. Even in this highly simplified diagram it is evident that energy flow in a technological society is pervasive and complex. Each of the economic sectors shown is an aggregate of numerous, highly diverse activities that use energy in many ways. For example, residences range from modern single homes and condominiums with a full complement of energy-using devices to old farmhouses and inner-city apartments with inadequate insulation. Commercial buildings come in a wide variety of sizes and types, including small stores, theaters, large hotels, shopping malls, and office buildings; industrial uses encompass metal reduction, plastics manufacture, papermaking, and the entire array of modern industrial activity; transportation includes the movement of freight and people in planes, trains, and ships as well as automobiles and trucks. Going to just one more layer of classification would make the diagram too complex to read easily.

The classification of economic sectors into industrial, commercial, residential, and transport is convenient but is often too coarse. For example, it does not display as simple an idea as the energy used in agriculture. To answer such questions we must often go to other classification schemes. Nevertheless, the scheme shown in Figure 5-17 is a good starting point for examining the flow of energy use in society.

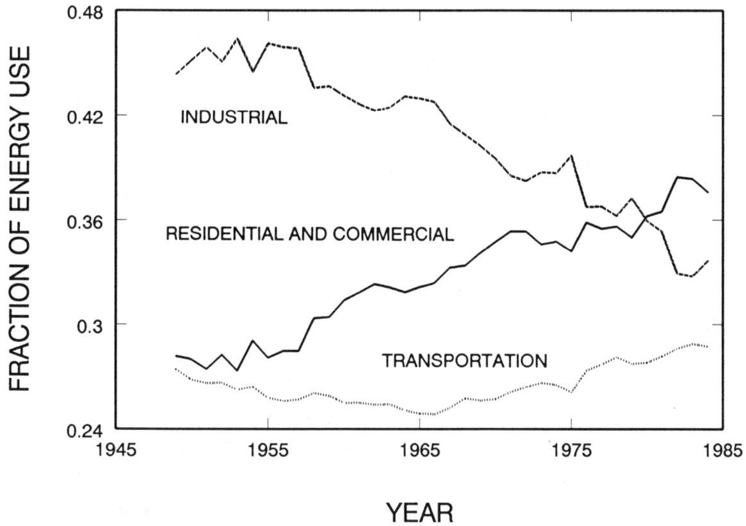

Figure 5-18 Fraction of energy use in economic sectors.

Figure 5-18 shows the fraction of energy use in the three major economic sectors for the United States. The curves are based on the total heat content of all the primary fuels that can be ascribed to energy use in each sector. If the data had been plotted in terms of total energy use for each sector, or even per capita energy use, all the curves would have shown a steady upward trend with time. Displaying the data as fractions shows the relative changes in intensity of energy consumption among the economic sectors. There has been a modest rise in the share of energy used in transport after the mid-1960s, following an even more modest decline. But there has been a large scale shift between the other two sectors. From the mid-fifties to the present there has been a decrease in the fraction of energy use in industry from about 45 percent to about one-third. Most of this decrease was accompanied by a corresponding increase in the fraction of energy used in the residential–commercial sector, which increased its share of total energy use from 28 percent to 38 percent. Two major factors contributed to this shift: increasing intensity of energy use in buildings and the increasing importance of services in the economy. The growth of television, air conditioning, and home appliances has greatly increased the household consumption of energy, and the shift of economic activity to services resulted in an increase in commercial buildings while their energy intensity also increased, not only because of air conditioning, telecommunications, and improved lighting, but also because of the increasing use of office machines such as copiers, electric typewriters, and computers.

About 92 percent of our energy consumption is in the form of fossil fuels, the remainder being primarily nuclear and water power with very small

Table 5-6 Percentage Consumption of Fossil Fuel Minerals by End Use for 1982

	Coal	Oil	Gas	Total
Residential and commercial	0.26	0.36	10.93	14.79
Industrial	3.36	11.78	9.73	24.87
Transport	0	27.01	0.91	24.87
Utilities	19.37	2.01	4.88	26.28
Non-fuel	0.06	5.33	0.75	6.14
Total	23.07	49.72	27.2	100

contributions from other sources such as wood and geothermal energy. The fossil fuels are therefore our current critical energy resources. The way their use is distributed among economic sectors is shown in Table 5-6, which is a matrix of type of fossil fuel versus type of use for the year 1982. Each entry in the table gives the percentage of type of fossil fuel used for each type of application. The percentages were calculated in terms of energy units (quads) so as to make the different fuels comparable. Consumption by electric utilities is separated out, and the non-fuel use of the fossil fuel minerals is also shown in Table 5-6. (The non-fuel uses are primarily chemical feedstocks with some lesser amounts for lubricants, waxes, and asphalt.) This is in contrast to Figure 5-18, in which non-fuel uses are not included and the industrial sector includes electric utilities. Table 5-6 gives a different view of energy use in that it presents the percentage of direct use of fossil fuels. Electricity, of course, is used in all sectors; but this has not been distributed to the other entries in the table.

The most important point shown in Table 5-6 is that the use of fossil fuels is not distributed equally among end applications. Coal is used primarily to generate electricity, transport depends primarily on oil, and natural gas is the major fuel for space heating. Electricity generation accounts for over 80 percent of coal consumption; over half of our oil use is for transportation, and nearly three-quarters of our natural gas use is for residential and commercial consumption.

Very little electrical energy is used in transport. Most of the energy used for transportation comes from liquid hydrocarbons. This fact presents us with one of our most critical energy issues. For space heating, electricity generation, and industrial process heat, one fuel can be substituted for another. Our transportation system, however, must have liquid hydrocarbons. In the short run, oil supplies are threatened by international conflict, particularly in the Middle East. In the longer run, oil will become scarce because it is a finite resource whose end is in sight.

Over 6 percent of fossil fuel use is for non-fuel applications, and most of this fuel is petroleum. This amounts to nearly 11 percent of our petroleum consumption. This fact is often overlooked but it is important because petroleum is a chemical feedstock for a wide variety of products, ranging from paints to plastics and textiles. However, future oil shortages would not be as serious for these non-fuel applications as for transportation because substitute chemical feedstocks could be developed from coal in time scales that are much shorter than those needed for massive changes in our transport systems.

Large-scale technological systems are the hallmark of highly developed societies. These systems include transport, communication, and energy systems to serve the three critical functions that constitute the underpinning of all technoeconomic activity. Such systems are pervasive and, to a considerable extent, define the modern, man-made environment of technology. They also have a powerful effect on the possibilities for change. Their size, the incredible investments they represent, the way they permeate all aspects of life, and the large numbers of people whose self-interest is bound up in them give them an enormous inertia. Any changes that imply large deviations from the existing conditions in these systems take a lot of time. The complex of equipment, people, and organizations devoted to the production, distribution, and use of electricity constitutes such a system.

The physical manifestation of the electrical power system consists of a number of central generating stations containing prime movers that turn generators, a transmission and distribution network, and end users. The prime movers are primarily heat engines, mostly steam turbines, although in recent years some gas turbines have been installed and some internal combustion engines are still in use. In addition to the heat engines, there are some generating plants that use water power, and a trivial amount of electricity is generated from so-called alternate sources such as wind, solar energy, and geothermal formations. The electricity generated at the central stations is fed over a transmission network to substations or local distribution points from which it is led to the end points where it is used. Along with this physical system, there is an organized management system that exercises centralized control.

Regional electric power facilities are model examples of industrial technological systems. They are large-scale and deal with massive amounts of materials and energy; they are centrally managed and controlled; they are complex and consist of a number of strongly interacting parts; a change in one of their parts has an effect on their other components. Most importantly, they have become essential to the well-being of the general population.

For many applications, electricity is the energy form of choice. It can provide heat, light, and mechanical work in a wide range of quantity. The flexibility and control that can be exercised over electricity makes it suitable for both small- and large-scale uses. The fractional horsepower motor made the electrical mechanization of the home possible, while large motors revolutionized the use of mechanical power in industry. Similarly, the divisibility of electric lighting made it possible to illuminate spaces of any size whatever in ways that were safer, cheaper, and cleaner than ever before. Electricity can do just about anything that any other form of energy can do and often do it more conveniently with less mess and at an acceptable cost. Only in transportation has electricity failed to be the most desirable energy form. The problem of creating mobile sources of electric power that are economic and have sufficient capacity has not yet been solved.

The present electric power systems are the result of an evolution that started over a century ago with the opening of the Pearl Street station in New York and the Holburn Viaduct station in London in 1882 to supply power for electric lighting using incandescent lamps.

It is significant that both stations were run by Edison companies, although the London company was merged with the Swan Electric Company, which made and sold incandescent lamps. These were invented by Joseph Swan, who developed the carbon filament lamp independently and at about the same time as Edison. Edison, more than anyone else, understood that to provide electric lighting an entire system for the generation, distribution, and use of electricity had to be developed. His technical work and economic considerations were always informed by the fact that all parts of the system interacted strongly and imposed requirements on one another. Other inventor-entrepreneurs focused on a single component, which they put together with components obtained from others when they commercialized their inventions. Edison's systems approach was an important reason for his success and put a structural stamp on electric power that remains to this day (Hughes 1983).

The Holburn Viaduct and Pearl Street stations completely transformed the use of electrical energy. Up until then the major application of electricity was in communications. In 1837 William F. Cooke and Charles Wheatstone installed the first telegraph line in London, and in 1838 they put in a telegraph on the Great Western Railway. Samuel Morse erected a telegraph line between Baltimore and Washington in 1843 and by 1866 there was a successfully functioning cross-Atlantic cable in operation. Electric lighting had been demonstrated and even sold commercially, especially using arc lamps. But the systems idea was necessary for the large-scale growth of electric power. Once central stations and integrated distribution systems were in place, the transition from using electricity for light to using it for mechanical power was easy and its growth was inevitable.

The growth was phenomenal. The Pearl Street station served a district of about one square mile, and in 1883 less than 9,000 lamps were in use. By 1899 25 million incandescent lamps per year were being produced in the United States; this became a billion per year by 1947. Both large and small practical electric motors that could be run from central station electricity were being developed at this time. A 15-horsepower motor was installed in a Boston elevator by Frank J. Sprague in 1888, and small motors were being manufactured by 1877 by S. S. Wheeler. In 1899 there were 484,000 horsepower of electric motors installed in American industry. This grew to 29 million horsepower by 1927 and 127 million by 1962.

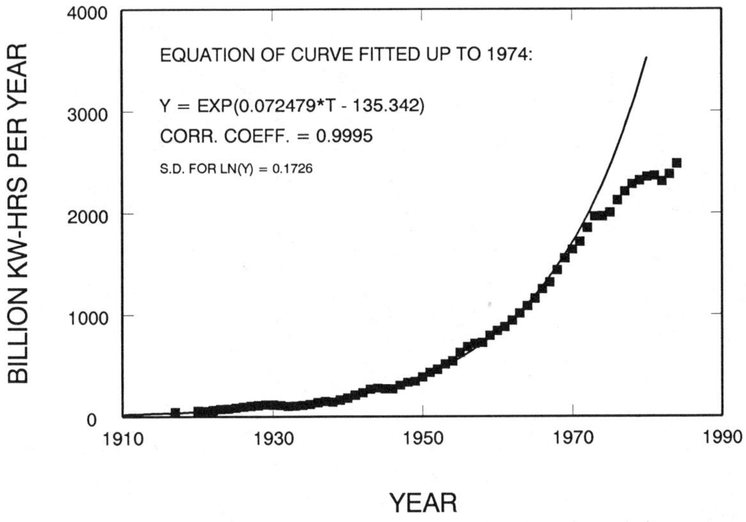

Figure 5-19 Electricity production in the United States.

The steady growth of the use of electricity is shown in Figure 5-19. There was a regular exponential increase of 7.2 percent per year right up to 1974. The Depression of the thirties and World War II had a negligible effect on this growth rate. After 1974 there is a considerable fall-off from the exponential curve which indicates that the use of electricity in the United States may be approaching saturation. Analysis of the per capita electricity production (not shown) does indeed show a distinct leveling off starting in the mid-seventies. This suggests that future growth of electricity will be more in step with population growth in the future. This is not a surprising result. As shown in Figure 5-20, practically all the households in the country were wired for electric service by the mid-fifties. The size of the electrical power distribution system relative to the number of dwellings was essentially at its maximum, and most households had a good complement of electrical appliances. The infrastructure is in place and the days of rapid growth are over.

Figure 5-20 bears a little more examination. It shows that the fraction of dwellings with electrical service grew exponentially at a rapid rate (9.7 percent per year) up to 1925, after which the growth was much slower. This is a common pattern in the diffusion of a new large-scale technology. A regular rapid growth is followed by a slower penetration of the new technology. This is a result of the fact that the users of a technology are not a homogeneous group but exist in segments that have inherently different adoption rates. For the diffusion of the home use of electric power, there are at least two market segments that adopt at different rates: farm dwellings and non-farm dwellings. Non-farm dwellings include urban and rural res-

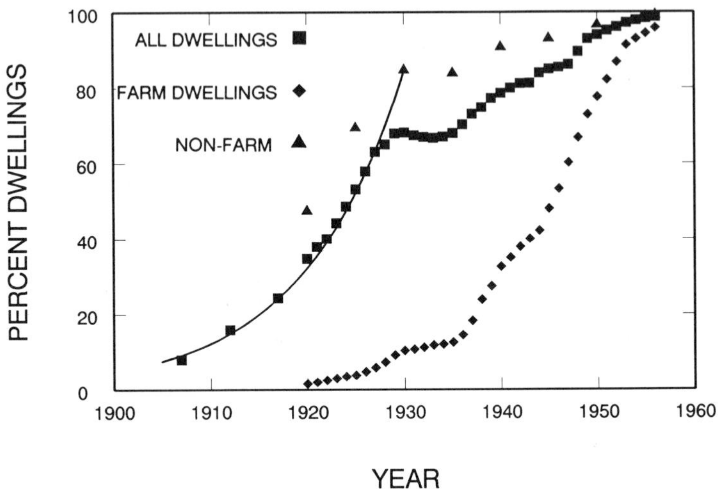

Figure 5-20 Percent of United States dwellings with electric service: curve computed from $P = \exp(0.96558*YR - 181.9165)$.

idences that are not farms. By 1925, 70 percent of the urban/non-farm rural dwellings had adopted electric service but only 4 percent of the farm dwellings were wired. The farm households constituted a quarter of the total in 1925, so the growth up to 1925 was practically all due to the electrification of urban and non-farm rural dwellings. After 1925, the rate of adoption of electricity was controlled primarily by the farm sector. Figure 5-20 displays the two market segments separately as well as the percent of total dwellings that were wired. Both the farm and non-farm curves are closer to the classical curves for technology diffusion than the curve for total dwellings, although there is a marked decrease in adoption rates during the Great Depression. The diffusion of electrical service in the farm dwellings was considerably slower than in the non-farm sector, the two curves combining to give a total diffusion curve that appears anomalous.

The fraction of national energy consumption that goes to produce electricity is shown in Figure 5-21. This fraction has been growing exponentially at a rate of 2.4 percent per year right up to the present and is projected to reach one-half before 1995. This trend cannot continue because transportation is a major energy use that cannot be replaced by electricity within our present energy system.

Within our present techno-economic system, oil is irreplaceable. Without it, modern transport stops. Raw materials could not be moved from their points of origin to mills and factories, goods could not be distributed from factories to wholesale and retail centers, foodstuffs could not be sent from farms to cities, and people could not travel the many miles they now do for employment and pleasure. A high technology society is a highly mobile

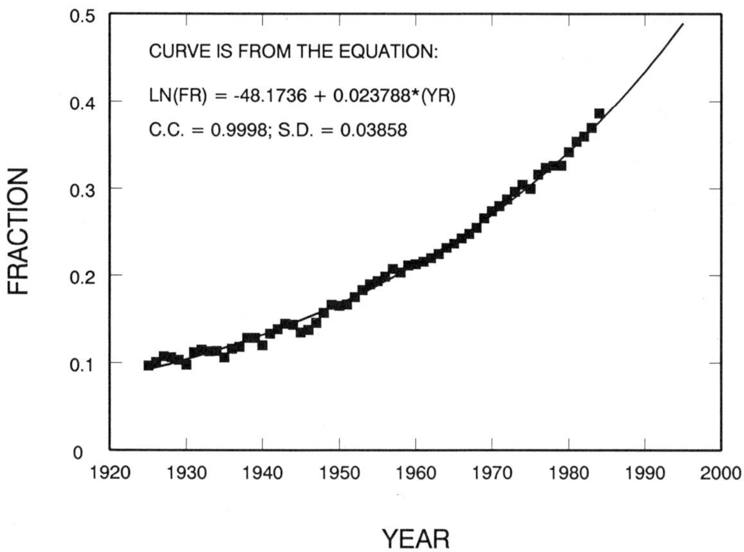

Figure 5-21 Fraction of primary energy consumption used for electricity generation in the United States.

society with a constant circulation of people and material. Without this mobility advanced technological societies cannot exist.

At the present time there is no replacement for liquid fuels derived from petroleum. Both coal and natural gas can be converted into liquid fuels, but the industrial systems to do this on a large enough scale do not exist. Furthermore, it would take years to create such systems. Electrically driven transport on a large scale is not at all practical as yet. Batteries are totally inadequate and must undergo extensive improvement before they can begin to approach the performance levels, economy, and convenience of the internal combustion engine. It is not only a matter of cost; batteries now available have a low energy density and low capacity and are technically incapable of satisfying our transport needs.

Alternative energy sources ranging from sunlight and wind to tides and biomass are often proposed as holding the key to our energy needs. But most of these alternate sources are suitable only for generating electricity. The others, such as biomass and oil-bearing plants, are so far from practical development that it is doubtful if they will ever help supply liquid fuels before petroleum supplies vanish.

Energy sources for space heating, for the production of electricity, and for many industrial processes are highly fungible. Coal, natural gas, or oil can be used interchangeably for many applications with relatively minor changes that can be made in a reasonable time. Wind, water, sunlight, waste, tides, and biomass, as well as conventional and nuclear fuels, can all be used to

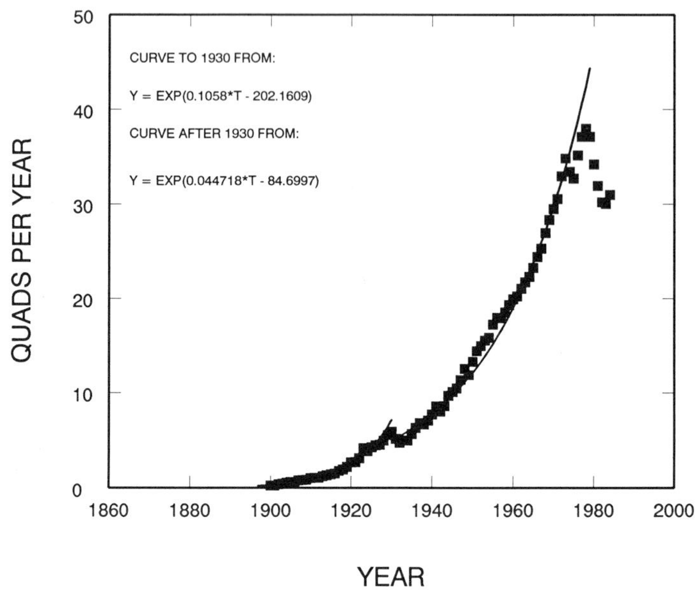

Figure 5-22 Total United States annual oil consumption.

generate electricity. But transportation is uniquely dependent on oil. Because of this, and because the supply is finite, the availability and use of petroleum forms the critical energy issue of our time. Here, we focus on the patterns of consumption.

The total consumption of petroleum in the United States is shown in Figure 5-22. There are two long time periods that show a remarkably regular growth pattern. From 1860 to 1930, the consumption data are well described by an exponential growth curve with an annual growth rate of nearly 10.6 percent. This stopped at about the beginning of the Great Depression and started out on another exponential curve with a growth rate of 4.5 percent per year from 1932 to 1978. After 1978 consumption started to decrease, and indeed dropped by over 16 percent from 1978 to 1984. This was the result primarily of conservation efforts and increased automobile mileage resulting from the OPEC oil embargoes. It is interesting to note that the major decrease in the rate of growth in consumption took place long before the OPEC oil embargoes. The growth rate was cut in half after the Depression. This was also true on a per capita basis. Data for the per capita consumption of petroleum (not shown) show that the per capita growth rate was 8.7 percent per year before the Depression and only 3.1 percent per year after the Depression up to 1978. The reason for this decreasing rate of growth of oil consumption can be traced to the changing growth rates of the use of motor vehicles. The total number of automobile registrations per capita as a function of time is shown in Figure 5–23. Before 1930, the growth rate

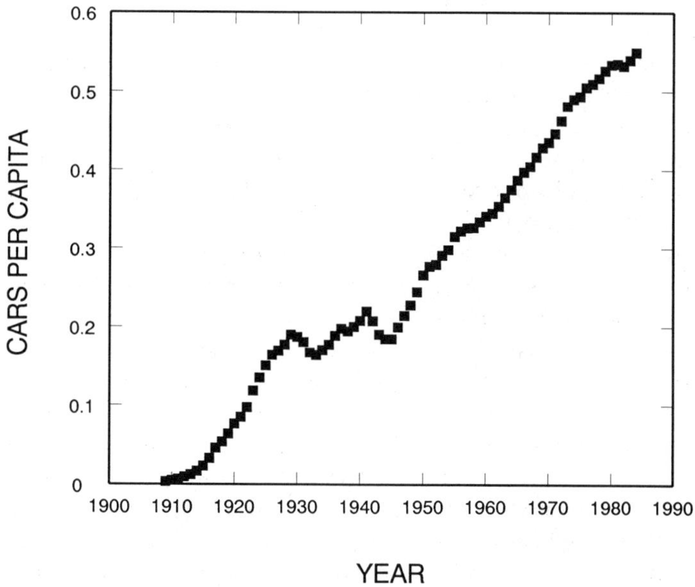

Figure 5-23 Auto registrations per capita in United States.

in number of registrations was over 20 percent per year. After 1930, it fluctuated around 4 percent. By 1925, the number of Btus burned in motor fuel was already one-quarter of the total BTU content of the petroleum being used, so the number of motor vehicles had a strong effect on the oil consumption. Figures 5-22 and 5-23 show how closely coupled oil is to transportation.

THE ENERGY INVENTORY

One of the most controversial issues concerning energy resources is that of their future availability. The importance of this question can hardly be overestimated since the very existence of a technological society depends on having a sufficient supply of energy. At more immediate levels, the degree of access to energy resources has strong effects on the power and economies of nations, the fortunes of corporations, and the well-being of consumers. Energy resource estimates are therefore of great interest in spite of their inherent uncertainties.

There are two broad classes of energy sources: nonrenewable and renewable resources. Nonrenewable energy resources are those that supply most of our energy. They include coal, natural gas, petroleum, shale oil, tar sand hydrocarbons, and nuclear fuels. Once they are used they cannot be replaced. Renewable energy resources include water power, wind, tides, sunlight, and biomass. They are renewable in the sense that they are replenished after they are used. The source that keeps renewing these resources is the energy flux from the sun.

Mineral fuels are finite, nonrenewable resources. It is important to know how much of these fuels exist and how much can be recovered for use. To approach these questions, a framework of definitions is needed that is in accord with the geologic, economic, and technological realities of resource extraction. This is given below and applies to all materials we take from the earth, not only mineral fuels.

Naturally occurring materials cannot be used until they are found and removed from where they happen to be. Furthermore, they will not be used unless the value of their use exceeds the cost of extracting and processing them. Most of the world's materials that we wish to use are hidden from view beneath the surface of the earth or the oceans, and require an economically sound technology for their extraction. There are thus two fundamental variables that must enter into the definition of resources: the degree of certainty of their existence and the cost of their extraction and processing.

A mineral deposit is defined as a region in which the concentration of the desired material is greater than its average concentration in the earth's crust. Deposits for a given mineral can occur in a wide variety of sizes and concentrations and in many locations. As a result of extensive explorations, the location and characteristics of many of these deposits are known. But there are many others whose existence is only suspected and yet others of which we know nothing. Failure to appreciate this simple fact has led to an astonishing amount of confusion and outright nonsense about the availability of resources for the future. A classification scheme that lent some precision of language to the issue of resource availability was not developed and widely adopted until the early 1970s.

In 1972, McKelvey based a set of definitions on the two important variables of certainty of existence and economic feasibility in an effort to organize and rationalize the nomenclature in use till that time. This was adopted, with some modification, by the U.S. Geological Survey (Brobst and Pratt 1973). The scheme divides resources into two broad categories: identified resources and undiscovered resources. Identified resources are specific deposits whose existence and location are known. A portion of these identified resources are known to have the characteristics (grade, depth, location, impurities, and so on) that permit them to be extracted economically. These are called *reserves* (also proven reserves). There are also deposits that have been identified but cannot be worked economically. The reason might be that the grade and conditions of the deposit do not allow extraction at a profit using current methods or that the characteristics of the deposit are not sufficiently known to determine its economic possibilities. These are called *conditional resources.*

We know full well that there must be resources that have not yet been found. Again these are classified into two types: hypothetical and speculative resources. *Hypothetical resources* are those that are not yet discovered but

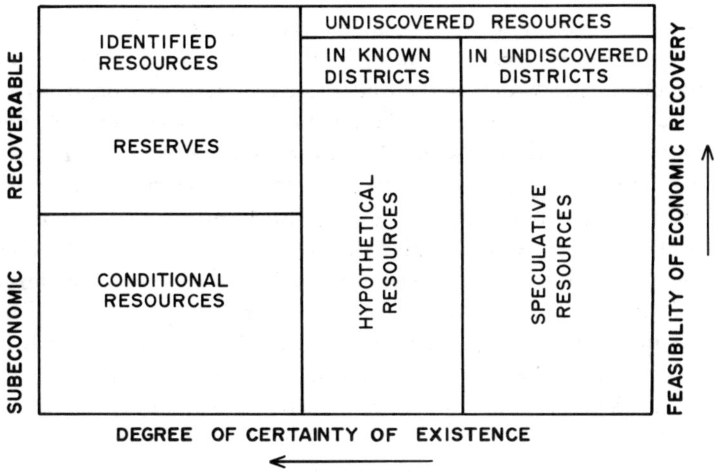

Figure 5-24 Classification of mineral resources.

can be expected to exist in known districts where other deposits have been found. *Speculative resources* are those that are not only undiscovered but are not in regions where discoveries have been made. They may be in regions with geological features similar to those in known districts or in regions whose potential has not yet been recognized. The speculative resources are those for which we have no real information.

The *total resource potential* is just the sum of these categories. It is obviously an uncertain number and subject to revision as new information is obtained.

This classification scheme is conveniently represented by a two-dimensional diagram in Figure 5-24, in which one axis represents economic viability and the other axis represents certainty of existence (McKelvey 1972; Brobst and Pratt 1973). The diagram is self-explanatory but it brings out some important points. The first of these is that estimates of resource availability must be made with great care in accord with the purpose of the estimate and with due regard for its inherent uncertainties. For purposes of short-range corporate planning, for example, the focus may be on reserves, while for long-range planning and public policy, conditional and even hypothetical resources should be considered. Attempts to forecast the long-range future availability of resources to determine whether or not they are sufficient to sustain a technological civilization must consider the total resource potential.

The most important point in connection with this classification scheme is that the resource picture is a dynamic one. As new discoveries are made, resources move from the undiscovered to the identified categories, and speculative resources can become hypothetical resources. Of even greater importance is the fact that, as technology advances, deposits that were uneconomic can be extracted at a profit, so that conditional resources become reserves. A dramatic modern example of this is the development of the pelletizing technique which permitted the extraction of iron from taconite at a profit. By 1975 iron ore in the Lake Superior district was nearly depleted, even though technological advance had allowed the use of lower-grade ores. Pelletization converted taconite from a worthless material to a valuable reserve of iron ore. For petroleum, secondary and tertiary recovery techniques have converted hitherto unavailable oil into proven reserves. Price changes have large effects on what resources are considered reserves. The large price increases following the OPEC oil embargo suddenly put shale oil into the position of being a reserve, but later price declines converted it back to a conditional resource. Prices and the balance of supply and demand play a major part in defining the resource categories. For many materials, a price increase of 25 percent to 50 percent would move large amounts of resources from the conditional to the reserve category. Technological improvement and economic factors are not independent. Increases in prices spur new technical efforts to recovery, and new technology often decreases prices.

The McKelvey diagram was meant to describe resources and reserves of nonrenewable minerals and is directly applicable to mineral fuels. But the same concepts are applicable to renewable energy sources. As an example, consider water power. Hydroelectric generating plants in place are the resource being used. All known locations at which hydroelectric plants do not yet exist but could be installed and economically operated with current technology are analogous to the proven reserves of mineral fuels. There are water sites that cannot generate electricity profitably with current technology but conceivably could do so if technology improves. There are also sites that are too far from centers of population and sites that have not yet been discovered but may be suitable for electricity generation sometime in the future. These are analogous to the noneconomical, hypothetical, and speculative mineral resources. The point is that renewable resources are subject to the same conditions of changing technology, economics, and even discovery as the mineral fuels.

Industrial firms generally have a straightforward objective when making resource estimates. They want to determine the future availability of the materials they need to stay in business. The producers are concerned about the amount of coal, oil, or gas that they will be able to take out of the ground and sell; the refiners want to know how much of what kind of fuel they can get at what price for how long. The industrial energy consumers want to make sure they will be able to get the fuels to produce their products. Such estimates deal with reserves and perhaps conditional resources, and with explorations to create reserves from conditional, hypothetical, and speculative resources. Even such a clearly defined task is fraught with difficulties. Aside from the geological issues and the limits of the information in hand, there are problems in estimating future technological developments, changing consumer preferences, substitution of one material for another, and political decisions. For practical purposes, proven reserves as defined under

Table 5-7 Estimates of World Fossil Fuel Resources (in Quads)

Fossil Fuel	Original Resource	Remaining Resource
Coal	250,000	250,000
Oil	10,000	7,000
Gas	10,000	7,000
Tar sands	300	300
Shale oil	>100,000	>100,000
Deep gas	10,000(?)	10,000(?)
Brine gas	>50,000(?)	>50,000(?)

current economic and technological conditions is the measure most often used by industry.

At the national level, nations are often interested in resources from the point of view of national objectives, ranging from balance of payments to the ability to make war. The relevant resource analyses then attach major importance to the national origin of the resources, the possibilities of the political or military interruption of supplies, and the definition of strategic materials.

Our interest here is to estimate the total amount of energy resources that can ultimately be recovered. This assumes that the technology will be available to extract these resources economically.

Estimates of the total amount of recoverable fossil fuels in the ground have been made by a variety of methods ranging from extrapolation of past discovery and production data to geochemical analysis. All these methods are subject to considerable uncertainty. Hubbert (1978), Penner and Icerman (1981, Ch. 2) and Rose (1986, Ch. 5) have collected and reviewed estimates of the ultimate possible world production of mineral fuels. Attaching a precise meaning or measure of accuracy to these numbers is difficult since they depend on so many unknown factors. In addition to the inherent difficulties of counting what cannot be seen, there is the complication of the McKelvey diagram: new technology can convert a useless deposit into a resource or extract material that was previously not recoverable.

The purpose in estimating the future supply of energy resources is to enable us to make some projections of how long they will last, the consequences of their use, and whether or not energy-intensive, high technology societies can continue to exist. These are large questions and could not be answered accurately even if accurate physical data on resources were available. Nevertheless, the questions are important enough that an attempt should be made. In fact, even though the basic data are highly uncertain, important conclusions can be derived from such analyses.

From the above cited reviews of the available resources, I have extracted what seem to be the most reasonable figures for the total amount of fossil fuel resources originally in the ground. These are given in Table 5-7 to one, or at most one and a half, significant figures, with the understanding that they could be in error by as much as a factor of two or even more for unconventional sources such as tar sands or geopressurized natural gas brines.

Accurate data on the amount of fossil fuels used to date are hard to come by. For coal, the original resource is so large that the amount used to date is within the error of the total resource estimate. The best estimate for cumulative consumption is for oil. More extensive and more accurate data for the consumption of oil exist than for natural gas or coal, and these data indicate that about 3,000 quads of oil have been consumed worldwide to date. For a time, natural gas was wasted by "flaring" as it emerged from the ground at petroleum wells. We have assumed that the amount burned to date is the same, in quads, as the amount for oil. Given the uncertainty of the original total resource figures, greater effort or precision is not justified.

The first three entries in Table 5-7 represent the fossil fuels now being produced with technologies that are in place. These entries are estimates of the amount of the conventional fuels that can ultimately be extracted with the expected evolution of these technologies. The other entries are for fuels that are much more difficult to get at and process. Major environmental as well as technical problems must be solved before they can be used on a scale comparable to that of the conventional fossil fuels. Nevertheless, the resources are there, and history shows that when the demand exists, the technology to extract natural resources develops. The figures for the energy in unconventional fossil fuels are uncertain, but they are more likely to be too low rather than too high.

Nuclear fuels exist in substantial amounts, but their useful energy content depends very much on the state of nuclear technology. In light water reactors, U-235 fissions to release fission products, neutrons, and energy. Naturally occurring uranium is primarily U-238, only 0.0711 percent being U-235. The U-238 is of no use for a "burner" reactor which depends on fission of U-235. In a breeder reactor, however, U-238 is used to generate a fissile material from the absorption of a neutron which converts U-238 to plutonium 239, which can then fission to generate heat. The breeder reactor uses the predominant uranium isotope and therefore extends the usefulness of uranium enormously. Also, thorium 232 cam be converted to U-233, which is fissile, and breeder reactors can be based on a thorium cycle.

Fusion cycles can be based on either deuterium or lithium. The deuterium cycle produces helium by direct fusion of two deuterium nuclei or by the formation of tritium, which then fuses with deuterium to form helium. In the lithium cycle, lithium absorbs a neutron to form helium and tritium. The tritium then combines with deuterium to form helium. Table 5-8 gives estimates of the world supply of energy from nuclear sources. It should not be necessary to stress that the technology to exploit fusion does not exist and its possible development is highly uncertain. The breeder reactor is still in its infancy, and even conventional light water "burner" reactors are not widely used in this country.

Renewable resources are often regarded as benign alternatives to the

Table 5-8 Estimates of Nuclear Fuel Resources

Fuel	Amount (quads)
Uranium (lwr)	1500–30,000
Uranium (breeder)	0.2–4 million
Thorium (breeder)	15–150 million
Deuterium	10 billion
Lithium	250,000

conventional dirty, dangerous energy sources on which we currently rely. Unfortunately, this is not entirely true; renewable energy sources have problems of their own. The extensive use of solar energy requires that large land areas be covered with solar cells or collectors. The use of solar energy is a materials-intensive enterprise. The production and deployment of these materials is not pollution free. The extensive use of biomass for energy places land in competition with agricultural and ecological uses. Wind power requires the deployment of large numbers of noisy wind turbines. Whenever energy is generated in large amounts undesirable effects of some kind result from the mere magnitude of the enterprise, no matter what the energy source.

The renewable energy sources are water power, biomass, solar energy, wind, tides, and geothermal energy. Note that we adopt the nomenclature in which solar energy is defined as direct sunlight which can be used to generate electricity by photovoltaics or used directly as a heat source, even though water power, biomass, and wind energy all originate from the flux of solar energy.

Water power, biomass (in the form of wood), and wind were the major non-muscle energy sources before the advent of the use of fossil fuels. In fact, fossil fuels came into use precisely because the renewable energy sources of wind, water, and wood were inadequate, inconvenient, and expensive. Renewable energy resources are generally dilute resources; they are spread out over large areas or volumes and must be collected and concentrated to be useful. This is certainly true for biomass, wind power, solar energy, and tidal energy. Those that are relatively concentrated, namely water and geothermal energy, are not widespread and exist at specific locations. Furthermore, they cannot be transported except as electricity.

Another characteristic of renewable energy resources is that they are suited primarily for generating electricity. Biomass is the only renewable resource that can function as a fuel and to act as a source of liquid fuels, although it must be subjected to extensive chemical processing to produce such products as alcohols.

In general, the technologies for large-scale use of renewable resources are not in place. Only in the case of water power is there a body of technology that is well known and immediately applicable.

In spite of their disadvantages, renewable resources are important. Once the fossil fuels are gone, they are the only alternative to nuclear fuels. As fossil fuels are depleted and become more expensive, or if society starts to recognize the great dangers of continued use of fossil fuels, renewable energy resources will be a desirable alternative for many applications. It is therefore of interest to estimate how much energy can be supplied by these renewables.

The calculation of the amount of renewable energy that can be used each year is both easier and more difficult than the estimates for nonrenewable resources. It is easier because in many cases upper limits are easy to compute. Unlike oil and gas, wind, water, and sunlight are not hidden from view. But the amount that can actually be used is hard to estimate because the technologies for that use are not in place and in most cases are in the earliest stages of development.

Water power is a renewable resource with which we have considerable experience. While the water-driven mills and pumps that once powered so much of industry are gone, hydroelectric power is still very much with us. Water power was, in fact, an early source of electricity generation; the first hydroelectric plant was installed at Niagara Falls in 1895, only thirteen years after the opening of Edison's Pearl Street station. Today, water power accounts for over 13 percent of the electricity generated in the United States. This is about half of the total hydroelectric generating capacity potential in the United States, which is about 161,000 megawatts (Hubbert 1978). For the whole world, the rate at which hydroelectric energy could be generated if all potential water sources were used is about 90 quads per year. Only about 10 percent of this is developed, most of the undeveloped hydroelectric capacity being in Africa, South America, and Southeast Asia.

Wind power enjoyed a renaissance for a time, particularly in the seventies after the OPEC price increases made the generation of electricity by wind nearly competitive with oil. Federal and state subsidies were sufficient to spark a great deal of interest in the installation of wind turbines, but this interest evaporated when oil prices decreased. Using the criterion that electricity generation by wind requires winds with average velocities of 14 to 16 miles per hour, it is possible to determine the sites that can be used and the wind energy that can be converted to electricity. The result of such a calculation is that for the entire world a maximum of 2000 quads per year can be generated by wind. This is an uncertain figure because accurate wind maps for the entire world are not available. For the United States, the maximum wind energy potential is about 35 quads per year (Koeppl, "Putnam's Power from the Wind," 1982). The United States regions where wind energy is feasible are along the East coast to Northern New Jersey, along large stretches of the West coast, some restricted areas in the Midwest, and all of Hawaii.

The possible contributions from tidal and geothermal energy are small. The amount of geothermal energy down to a depth of 10 kilometers has been estimated to be about 40,000 quads, but the amount that can be extracted is probably not more than 5 quads per year.[1]

Tidal power is an even smaller resource, amounting to about 2 quads per year (Penner and Icerman 1981; Hubbert 1978).

Direct use of solar energy is regarded by many to be the only sensible energy option. The advantages, disadvantages, economics, and technology have been discussed by many. (See Rose 1986, Chapter 8, for a particularly clear analysis.) Here, I am interested only in the maximum rate at which sunlight can be captured for energy use by either photovoltaic means or by direct heating. In either case, I will assume a conversion efficiency of 25 percent. This is rather high. The best photovoltaics have efficiencies that approach 25 percent, but these are under laboratory conditions where dust, dirt, and scratches do not lower the amount of light falling on the solar cell. Also, there are energy losses associated with other parts of the conversion system and with the distribution of the energy. But I will make this assumption, even though it is optimistic at present, since the overall efficiency of conventional energy systems is of this magnitude and technological progress can be expected to continually improve efficiencies. At any rate, the conclusions to be drawn about the maximum availability of solar energy are not sensitive to the precise values of efficiency. However, these become quite important when economic factors are considered.

On a cloudless day, the amount of energy falling on the earth when the sun is directly overhead is about 1000 watts per square meter, which is 333 w/m^2 when averaged over 24 hours. The world average for a solar day (500 minutes) when averaged over 24 hours is about 145 w/m^2. I will take 200 w/m^2 as an optimistic number for computing the amount of light falling on the earth that is available for conversion to electricity. This means that it would take 64.5 square miles of solar cell surface to generate one quad of electricity per year. For 25 percent efficiency, this increases to 258 square miles per quad. To generate the amount of electricity used in the United States in 1984 (7.8 quads) would then require 2,000 square miles of collection surface.

The collection surface needed to use solar energy to any significant extent is clearly enormous, but not so large as to rule out the use of solar energy on grounds of technical boundary conditions. Still, the areas are large. One hundred quads requires about 26,000 square miles while 1000 quads would require 260,000 square miles. This is nearly three times the area of Arizona.

---

[1]Geothermal energy is not, of course, a renewable resource in the sense that sunlight is. The maximum rate of use of 5 quads per year could probably be maintained for about 100 years. It is common, however, to include geothermal energy in the renewable category. Actually, there is some question as to whether or not water power should be called a renewable resource. Over a period of a century or more, silting of dams can become a problem, and the available power from hydroelectric facilities will decrease unless this is corrected. For this reason, these are often called alternative energy sources.

Surely the ultimate upper limit on the use of solar energy, even assuming an economical technology with 25 percent efficiency, must be less than twice this. The land requirement for collection surface alone would then be 0.5 percent of the world land area and 1.25 percent of the productive, habitable area. This does not include the area needed for ancillary equipment, buildings, and transmission, and neglects the energy requirements for producing the solar materials, operating the facilities, and distributing the energy. When all this is included an upper limit of 1,000 quads per year makes sense.

The important conclusion from this calculation is not that solar energy is impractical or the technology cannot be developed, but rather that the magnitude of the system needed to exploit it is immense and will take time to implement. The time scale for the extensive use of other energy technologies was half a century. Left to the usual evolutionary developments, deployment of solar energy will take even longer.

Intensive agricultural techniques can yield biomass with an energy content of 10^{11} BTU per square mile, so that it would take 10,000 square miles to produce one quad of biomass energy. The land requirements are much greater than those needed for solar energy. This was to be expected since the efficiency of energy conversion from sunlight to biomass is less than 1 percent. Furthermore, the land for biomass energy is in direct competition with land for other plant growth, much of it being for food crops. The ultimate amount of biomass energy that could be used is hard to estimate. One hundred quads per year would require a million square miles. Certainly an estimate of 50 quads per year is a generous estimate. At the low efficiency rates, it seems to make no sense to rely on biomass for any significant contribution to our energy needs. But biomass is the only renewable resource that has the potential to supply liquid fuels, and this may be its most important use in the long run.

Clearly, estimates of the rate at which renewable resources can be used are uncertain. The practical problems of such use are great and costs will almost surely be significantly greater than the current costs of using fossil fuels. The potentially largest renewable resources are solar and wind energy and, for both of these, estimates of the amounts of energy that can be used are highly uncertain. To express this uncertainty, the estimates in Table 5-9 include ranges in which the lower figures are thought to be feasible and the upper figures are thought to be absolute upper limits.

To put these numbers in perspective, recall that the energy production for the world in 1983 was about 280 quads. To replace this with renewable resources, solar and/or wind energy systems must be deployed. The lower estimates in Table 5-9 do not total much more than the present rate of energy use, and by the time the systems are in place the world energy use would exceed these estimates. The use of renewable resources for all future energy needs would require heroic feats of technological, economic, and political achievement. Even so, the replacement of liquid fuels by renewable resources is problematic. This cannot be done completely with biomass. The use of

Table 5-9 Estimates of World Renewable Energy Resources

Resource	Quads per Year
Hydroelectric	45–90
Biomass	20–50
Solar	100–1000
Wind	200–2000
Tides	1–3
Geothermal	2–10
Total	370–3150

electricity generated from sunlight or wind to synthesize fuels could conceivably be done, but the technology for this does not yet exist. An alternative is to cut back on the use of liquid fuels by using electric motors for transport, but the battery technology that would permit this is still in the future.

The renewable resources are primarily electricity generators, the most important of which are sunlight and wind. But both of these are intermittent sources and must be accompanied by energy storage systems, if they are to be practical, so that electricity can be supplied even when the sun is not shining or the wind is not blowing. Suitable storage systems are not yet available.

The conclusion from all this is that renewable resources can contribute significantly to energy supplies in the future but they are certainly no immediate solution to our energy problems.

The use of fossil fuels has been increasing steadily for over two centuries. Their production has been limited only by demand; as demand increased, it was possible to increase production so that an adequate supply was always available at a reasonable cost. For nonrenewable resources, this cannot go on forever and, indeed, the OPEC oil shocks of the 1970s were a strong reminder of this. Sooner or later, production of fossil fuels must decrease. It is important to have some idea of the changing rates of production as fuels become depleted.

PATTERNS OF PRODUCTION: THE HEWETT–HUBBERT MODEL

A model for the production cycle of fossil fuels has been developed by Hubbert (1978), based on the results of Hewett (1929) who studied the life cycle of mines and mining districts.

D. F. Hewett of the U.S. Geological Survey was fully aware of the transient nature of many mining regions and was familiar with the major problems confronting important districts in Europe. He was convinced that it was only a matter of time before these problems would grow to comparable importance in the United States. In 1926 he therefore visited twenty-eight mining districts throughout Europe to collect data and information that would clarify the future course of American metals production. The major points he examined for each mining district were: the number of mines being worked; the number of smelters or refining units in operation; the production from local ores; the amounts of ore exported, and amounts imported from and to the district. Hewett's general conclusion was that mines and mining districts went through an organic life cycle from youth to maturity to old age to death. In its early days, the number of mines grows and most of the production is exported. But as the smelting and refining capability grows, exports decrease and the production of metal increases. After a time, production tops out and starts to decrease with a corresponding increase in imports.

The production life cycle described by Hewett was adopted by Hubbert as a model for the rate of production of fossil fuels. The model simply states that the rate of production as a function of time is a curve that starts out slowly, grows with increasing rapidity until it reaches a maximum, and then decreases to zero. In its simplest form the curve is bell-shaped and can be represented by the derivative of the logistic equation. This model is quite reasonable. After a deposit is discovered and decisions are made to extract it, some time is required to begin and increase production. As facilities are installed, additional excavations are made, transportation needs are met, new customers are found, and production grows at an increasing rate. But as the mines extend deeper into the earth, as metal content decreases because of the geological gradients, and as the deposit is depleted, it becomes more difficult to maintain high production rates. Ultimately, the deposit becomes exhausted or the remaining ore cannot be extracted economically. Similar factors exist for oil and gas wells, in which the fluids become increasingly more difficult to extract as pressures drop, secondary extraction methods are required, and deeper wells must be sunk.

The production rate in the Hewett–Hubbert model is given by

$$dP/dt = kP(P_M - P) \tag{5-19}$$

where P_M is the amount of the resource that is ultimately recoverable, P is the amount that has been extracted up to time t, and k is a constant. The cumulative amount produced up to time t is

$$P = P_M/\{1 + \exp[K(T - t)]\} \tag{5-20}$$

where K is a rate constant given by

$$K = kP_M \tag{5-21}$$

and T is the time at which half of the recoverable resource has been produced.

Hubbert's most remarkable result using this model was the prediction, in 1956, that the production rate of petroleum in the United States would peak out sometime between 1966 and 1971. (The spread in the prediction corresponded to two alternate values of P_M, the total amount of recoverable oil in the United States. The half life of 1971 corresponded to a choice of 200

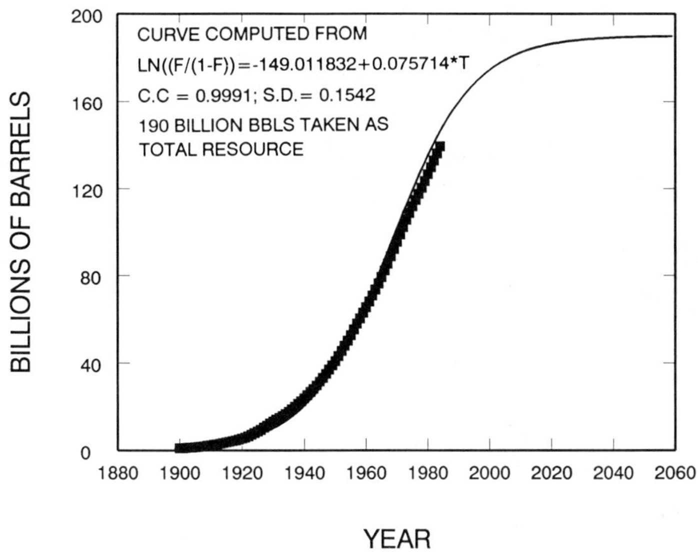

Figure 5-25 Cumulative oil production in the United States.

billion barrels of oil originally in place.) In fact, this prediction was fulfilled; production was a maximum in 1970 and has decreased since then. Using data up through 1984, I found a better fit by assuming that the total original resource was 190 billion barrels. The cumulative United States production of crude oil is shown in Figure 5-25, the curve being calculated from the logistic equation fitted to the data points. The annual production rate is just the difference in cumulative production for successive years. This is shown in Figure 5-26, both for the actual production data and for the model logistic curve. The half-life date (time of maximum production) obtained from the model calculation was 1968. It is interesting to note that the recalculation using more extensive production data did not alter Hubbert's original result in any significant way.

Hubbert used this model to compute the cycle of production of coal and natural gas as well as for oil. The results are shown in Table 5-10, which gives the time of maximum production of fossil fuels as computed from the model. (The result for U.S. oil is from my own calculation. All other results are from Hubbert 1978).

The predictions from the model are startling. Not only has the time of peak production of crude oil in the United States passed, but by 1997 ninety percent of our original endowment will be gone and production will be negligible. Essentially the same is true for natural gas. (Actual production of natural gas in the United States was a maximum in 1973.) For the world as a whole, production of crude oil will start to decline after the year 2000.

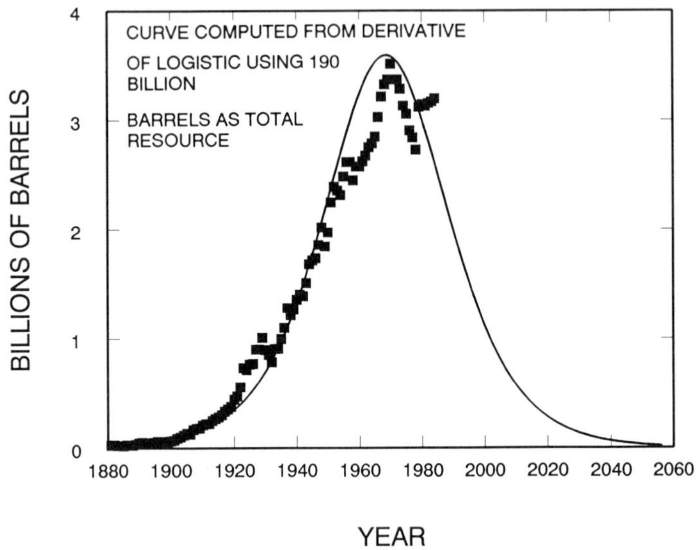

Figure 5-26 Annual production of oil in the United States.

Coal is our abundant fossil fuel. It can be produced in ever-increasing amounts for about two more centuries.

Given these results, it is important to understand the limitations and validity of the model. The essence of the model is the bell-shaped curve for annual production. The actual form of the equation used to describe the curve is of minor importance; the logistic is used because of its convenience. The basic question is: under what conditions is there is a life cycle of production which rises to a maximum and then decreases until it is negligible?

First, let us note that there is other evidence for the model. The cycle of production of silver from the Comstock lode and of iron ore from the Mesabi range both follow well-defined bell-shaped curves (Cook 1976). But these were modified by technological and economic conditions. In the case of the

Table 5-10 Half-Life in the Production Cycle of Fossil Fuels as Computed from the Hubbert–Hewett Model

Production Cycle	Year of Maximum Production
U.S. oil	1968
U.S. gas	1977
U.S. coal	2200
World oil	2000
World coal	2150

Comstock lode, the production consisted of three clearly distinct cycles. The second cycle started after the first was essentially complete and was the result of introducing a new technology that enabled lower grade ores to be processed. The third cycle was later yet and did not consist of mining per se but was the recovery of silver from tailings that had previously been rejected. In the case of the Mesabi range, the production rate of iron ore was interrupted by the Depression. Later, it was headed downward until pelletization techniques made it feasible to mine Taconite. A new production cycle then started. This illustrates a fundamental point: the model can be expected to hold for individual deposits or fields provided there are no rapid changes in technology of extraction or rapid shifts in economic conditions. But this is not the only requirement. The model also assumes that the use of the resource is limited primarily by supply, that the demand grows sufficiently to absorb the growth of production in the first half of the cycle, and that the decrease in the second half of the cycle is not because of lowered demand but because of greater difficulty of extraction.

With these provisos, the validity of the model for individual deposits is reasonable. The next question concerns its applicability to nations or to the world as a whole. For such large regions a number of different technologies are in place to extract a given resource, and each deposit or field has its own life cycle. But if we grant that the life cycle of each field has a bell shape, then their sum over a nation or over the world must also approximate a bell shape. Thus, we can expect the model to hold for large-scale regions providing it is applied to production by a given set of technologies that are changing only in an incremental way.

For the production of crude oil, the model calculations apply only to oil recovered by primary and some secondary recovery methods. These still leave a great deal of oil in the ground, and new cycles must be considered if tertiary methods become practical in a widespread way. The model also does not include new sources such as tar sands or shale oil.

Given its limitations, the Hubbert–Hewett model still illustrates an important point. The age of easily produced inexpensive oil will soon be over. While liquid fuels can be expected to be produced from coal, oil-bearing shales, or tar sands, these will be more expensive than in the past and will need a technological infrastructure that is not yet in place.

6

The Electronic Digital Computer

SIGNIFICANCE OF THE COMPUTER

For most people, arithmetic is not an exciting thing to do. Adding, subtracting, multiplying, and dividing have no intrinsic interest and are dull mental exercises. The only ability a digital computer has is to perform these basic arithmetic operations and to execute elementary logic by comparing numbers. And yet, computers have totally altered the nature of modern society. Mathematics, the sciences, education, war, manufacturing, transportation, insurance, publishing, finance, engineering design, agriculture, mining, entertainment; indeed every field of human activity has been, and continues to be, profoundly altered by computers. The changes attending the computerization of the world are not merely quantitative or incremental in the sense that we simply do what we did before more rapidly or more efficiently. The changes are also qualitative, in the sense that totally new capabilities are opened up and the way we think is being transformed. This is clear to everyone, and there is no point in providing a detailed exposition here, but two general developments must be mentioned that illustrate the power of computer technology in reshaping our world. The first is that instantaneous worldwide transmission of information and large-scale data processing capacity, coupled with cheap and rapid transport, is creating the first truly worldwide interdependent economy. The implications of this are just beginning to be appreciated, and the way political and social institutions respond to such widespread integration in a world of nation states will affect the welfare of all. The second general development is that computers give us the capacity to deal with complexity to a much greater degree than ever before. This capacity affects everything we do. Complex scientific problems that could not be addressed before are no longer intractable. Manufacturing, inventories, process control, and industrial optimization can be dealt with at an unprecedented level of complexity and sophistication. Computers make it possible to adopt the systems mode of thought, in which there are many related functions among interacting parts. We are no longer restricted to a linear, Newtonian approach in which there is a simple chain of causes and effects.

It is a remarkable thing that such exciting and far-reaching developments are based on machines that can only compare numbers and do arithmetic, and it is worthwhile to reflect on how this can be.

Certainly, arithmetic itself is an important function in a modern society. Part of the computer revolution is simply the capacity to do more arithmetic faster and thereby facilitate all the purely numerical components of human transactions associated with census taking, taxes, banking, retailing, accounting, insurance, reservations, and so on. But computers also do higher mathematics. They solve problems in differential equations, perform statistical analyses, and compute integrals. This is possible because all of higher mathematical analysis can be reduced to arithmetic. The evaluation of an integral or the solution of a complicated differential equation can be reduced to a series of simple arithmetic steps.

The power of arithmetic goes even beyond this. There is a close connection between arithmetic and logic, so that a machine that can do arithmetic and compare numbers can also deal with logical operations such as "if . . . then," "and," and "or." The physical embodiment of these capabilities resides in units that can exist only in one of two states ("on" or "off"; finite electrical signal or zero signal). The arithmetic is done in binary form by using a large number of such units that accumulate the results of arithmetic operations. The logical operations are performed by comparing the signals in several such units. For example, the logical operation "and" is embodied in a device (called a gate) that has an output ("on," or 1) only when it receives two inputs that represent "on" (or 1). If any of the two inputs is 0, then the output is 0 (or "off"). Each of the two inputs represents a statement when the input is 1, and the two are joined when both are present (the "and" operation). Analogous gates are constructed for the logical operations "or," "not and," "nor," and "exclusive or." An inverter is also used, such that the output of the gate is the opposite of the input. A computer can therefore be programmed to make choices among specified alternatives, and all the operations of logic can be performed.

The ability to do arithmetic and perform logical operations is coupled with the idea of symbolic coding to give computers their remarkable versatility. It is a simple idea. A letter can be represented by a number and the manipulation of letters therefore becomes an exercise in the manipulation of numbers. In fact, any symbol, be it a letter, word, punctuation mark, or dollar sign, can be represented by a number, and the rules for the manipulation of the symbols can be replaced by the manipulation of numbers. Computers can therefore handle language and graphics, as well as mathematics, by using numeric structures that correspond to the structures of the linguistic or graphic elements. By taking advantage of numeric structures that are isomorphic with other structures, the computer is no longer merely a processor of mathematics. It is now a machine that processes symbols and manipulates symbolic systems. Since all our explicit thought is based on symbols, it is hard to imagine an area of human thought that cannot be profoundly affected by computerization. At bottom, of course, the computer depends on its ability to process numbers at incredible speeds, but its function as a symbol processor is of critical importance since this is what makes the computer more than merely a high speed calculator.

Initially, computers were thought of primarily as mathematical calculators of great power and speed, even though the early developers of computers were fully aware of their logical structures and symbol processing potential. It was only after computers had proven their worth in scientific calculations and in business data processing that an appreciation of their use as symbolic machines became widespread. The fact that computers developed along digital rather than analogue lines facilitated the growth of symbol processing. In digital machines, the connection between arithmetic and logic is clear and straightforward; the connection was evident from the early days of the electronic precursors of the computer in which digital techniques were used for cipher analysis during World War II.

In an analogue computer, a number is represented by some continuous physical parameter, such as length or voltage, and the manipulation of the magnitude of this quantity corresponds to the manipulation of numbers. The slide rule is the prototype of the analogue computer; numbers are represented by lengths which are added or subtracted by sliding two lengths alongside each other. By appropriately marking the lengths with various scales, multiplication and division can be performed, and functions (such as sine and cosine) determined.

A digital computer represents numbers by the presence or absence of some physical parameter. The abacus is the prototype of the digital computer, since numbers are represented by the presence of beads. The important difference between a digital and an analogue computer is that in the digital system, a number is determined by counting discrete elements while the analogue representation is a measurement. In modern digital computers, the basic element is a bistable device; that is, a device that can exist only in one of two conditions, such as a magnetic field that can take on one of two directions or an electronic element that can be either conducting or nonconducting. Such devices are ideally suited to express numbers in the binary system and to expose, as well as use, the close relations between arithmetic and logic. The fact that computers developed along the digital path certainly facilitated their evolution into symbolic processors.

The origins of errors and the degree of reliability is different for analogue and digital systems. For digital systems, the accuracy is limited only by the number of bistable elements one is willing to use to represent a number. Each bistable element can represent the values 0 or 1; this is a bit of information, and the number of significant figures that can be carried depends on the number of bits used to define a number. Accuracy in an analogue computer, however, depends on the accuracy with which one can make a measurement of such a quantity as length, angle, or voltage. This is subject to the precision with which physical systems can be constructed, a limitation not present in digital computers. Also, bistable electronic elements can be switched between states at great speeds. This combination of speed, accuracy, and close connection to logical operations has made the digital computer an extremely powerful and versatile device.

The computer is the archetype of the new postindustrial technologies just as the steam engine was the archetype of the industrial technologies. The computer's function is to ease mental work and multiply human mental abilities, just as the function of the steam engine was to ease physical work and multiply muscle power.

THE PRECURSORS

The first fully operational electronic digital computer was completed at the Moore School of Electrical Engineering of the University of Pennsylvania in 1946. This was the ENIAC (Electronic Numerical Integrator and Computer). Like most other contemporary developments in machine computation, it was the result of wartime needs, which required numerical calculations on a scale and at a speed much greater than any existing capability could satisfy. The ENIAC was such an enormous advance over any previous work on automatic computation that it split the history of computing in two and began the modern era of computer technology.

Of course, mechanical computing of a sort did exist before World War II. There was, in fact, a distinguished tradition of the mechanization of arithmetic going back to the early seventeenth century with John Napier's logarithms (and his "bones" for doing multiplication) and William Oughtred's slide rule. In the early part of the twentieth century, desktop mechanical calculations were widely available. The most important of these were descendants of inventions by Dorr E. Felt (1884), William S. Burroughs (1888), and Stephen Baldwin (1911), which were derivatives of a design by Gottfried Wilhelm Leibnitz (1673). These were, at first, purely mechanical but later included electric motors for a power source. There were also available tabulating and accounting machines, along with punched card input and output equipment, that were electrically powered although the arithmetic was still done by mechanical means.

Although this equipment embodied ideas that were to be useful for later developments, none of them had the conceptual structure of the modern computer. This was remarkable in light of the fact that the computer architecture worked out by Charles Babbage in the 1830s was essentially that of the modern digital electronic computer. In spite of the fact that Babbage was one of England's most distinguished scientists and was well known not only in Britain but also in Europe, his work on computers had little long-lasting influence and had to be rediscovered.

A critical numerical problem of Babbage's time was the computation of mathematical tables. These tables were a necessary tool for scientists and engineers since their use was the only way to relieve at least some of the burden of the onerous arithmetic work imposed by an expanding science and technology. The large-scale publication of all kinds of tables, from addition to logarithms, had begun in the late 1700s and by the early 1800s they were a part of every scientist's library. The problem was that the tables contained many errors, in spite of the compilers' best efforts. Babbage's first contribution to mathematical aids was a table of logarithms of the natural numbers from 1 to 108,000, which was published in 1827. Before this, Babbage had been involved in the computation of actuarial tables. He was acutely aware of the large number of errors in existing tables and of the difficulties of achieving error-free results with human computation. He therefore set himself the task of devising a mechanical method of calculating and of printing the results. In his autobiography, he claimed that the idea of mechanical computing came to him as a student at Cambridge in 1812 or 1813. Babbage's first attempt at mechanical computing was the Difference Engine, which computed polynomials by taking advantage of the fact that if one takes successive differences of a polynomial for a set of equally spaced values of the argument, eventually a set of differences is obtained that is the same for all values of the argument. Note that from the highest (constant) difference, the next highest differences can be calculated for all successive values of the function if its initial value is given. It is therefore a simple process to work backward by adding differences in the proper order to obtain the value of the polynomial for all values of the argument, if its first $(n + 1)$ values are known, where n is the highest power of the polynomial. The number of successive differences that are needed to reach a constant set is equal to the highest power of the terms in the polynomial. The method of differences is more general than appears at first sight because any well-behaved function can be approximated by a power series. In fact, the method of differences was widely used by all table makers since it reduced the calculation of tables to problems in addition. Also, it permitted the calculations to be checked easily since the nth differences of the polynomial must all be constant. To check a calculation, it is only necessary to take the nth differences of the computed results and see if they are all the same.

By 1822, Babbage had completed a working model of a small machine which was capable of computing the values of quadratic functions up to six significant figures. With the aid of this model, and with the support of the Royal Society, Babbage received governmental funds to construct a full scale Difference Engine. It was designed to compute sixth-order polynomials and handle numbers of eighteen digits. The machine was never completed. The basic reasons were that the machining technology of the day was inadequate, and Babbage kept changing the design. Babbage actually contributed a great deal to tool making and machining in the process of trying to get the Difference Engine built and might have been able to complete it if only he had frozen the design at some point.

A fully working Difference Engine was finally built, but not by Babbage. George Scheutz, editor of a Stockholm technical journal, saw a popular article in 1835 written by D. Lardner which contained a general description of the Babbage machine. From this, Scheutz and his son Edvard undertook to build a Difference Engine. By 1854 they had a machine that computed fifth-order polynomials to fifteen significant figures.

In 1834 Babbage's work on the Difference Engine was interrupted by disagreements with his chief mechanic, Joseph Clement. During this hiatus, Babbage re-examined his ideas on mechanical computing and developed the concepts of his Analytical Engine, which were so close to those of the digital computer of more than a century later. His train of thought was spurred by the structure of the Difference Engine, which suggested that the addition registers were essentially all identical and could be thought of as storing numbers that could be sent from one register to another with arithmetic being done between any two of them. This idea developed into a structural concept of a machine consisting of a "store" which stored numbers, a "mill" which processed them, and a "control barrel" which specified certain basic instructions, These were the complete equivalent of a memory, a central processor, and microprogramming. The concepts also included input–output and programming via punched cards; in essence, Babbage anticipated the entire modern computer architecture. He adopted the idea of using punched cards from the Jacquard loom, which used them to control weaving patterns by permitting rods to pass through holes in the cards, thereby lifting the appropriate threads to make the pattern.

Babbage completely worked out the logic of the machine and actually constructed some small experimental parts but never built a full-scale Analytical Engine. After his death, the "mill," or central processor, was built through the efforts of his son Henry Babbage. It was completed in 1906 and worked as anticipated.

One of the few people to fully appreciate Babbage's work was Augusta Ada, the daughter of Lord Byron, the poet. She married William King in 1835 and became known as Ada, Countess of Lovelace. Ada had a consuming interest in mathematics and her mathematical abilities were formidable. She met Babbage in 1833 when she was seventeen years old at one of his famous parties and became fascinated with the idea of the Difference Engine and later with the Analytical Engine. She and Babbage developed a close relationship and she was the closest thing to a collaborator he ever had. They had an extensive correspondence concerning computation, and Ada undoubtedly made important contributions to the Analytical Engine, particularly to its logic and its programming.

Babbage was not the only one who tried to construct Difference Engines, although he was the first and had the most profound understanding of the theory involved. The greatest success of the nineteenth century was achieved by George B. Grant, who exhibited his Difference Engine at the Philadelphia Centennial Exhibition of 1876. It weighed over a ton, contained over 15,000 parts, and stood five feet high by eight feet long. After the exposition, it was given to the University of Pennsylvania. A copy was sold to the Provident Mutual Life Insurance Company of Philadelphia, where it was used to compute actuarial tables for twenty years.

Another form of the Difference Engine was devised in 1932 by L. J. Comrie in England, who had been using Hollerith-type punched card accounting and tabulating equipment for astronomical calculations. He had, in fact, been successful in performing Fourier Analysis on such equipment and had calculated the motion of the moon for the years from 1935 to 2000. By having his equipment modified so that the contents of one register could be transferred to another, he was able to construct a very large Difference Engine. IBM adopted the modification for its own equipment in the early forties.

The Difference Engine, the Analytical Engine, and the descendants of the Leibnitz calculator in the form of desktop calculators and comptometers were digital machines in the sense that they operated in discontinuous steps using geared wheels. But analogue computational aids also have a long history. These ranged from the astrolabe, which was used for astronomical calculations and was based on a stereographic projection of the celestial sphere onto a plane, to the planimeter, which was used to perform integrations by measuring the area under a curve. There is evidence that the astrolabe was known to the Greeks as early as 200–100 B.C. The planimeter was devised by Lord Kelvin and consisted of a rotating disc coupled to a pointer which was moved along the curve to be integrated. The disk in turn rotated a cylinder on which an engraved scale read off the value of the integral.

The Differential Analyzer of Vannevar Bush, which was built at the Massachusetts Institute of Technology in the 1930s, was essentially a large-scale application of the principle of the planimeter. In its original form, the planimeter could not be mechanized because the disc was moved by friction and any slippage would ruin the calculation. Bush solved this problem by inventing a torque amplifier based on the principle of the capstan. He had been trying to solve some differential equations for an electrical power network and decided it would be easier to build a mechanical integrator than to spend the time needed to integrate by hand. Although digital machines later became dominant, the Differential Analyzer played an important part in computer development even though it was an analogue machine. Its primary purpose was to solve differential equations, and in the decade of the thirties it was the only large-scale mechanism for doing this. Bush designed an improved version in 1935, a copy of which was built at the Moore School of the University of Pennsylvania, where the electronic digital computer was first developed. The later version of the Differential Analyzer weighed 100 tons. It included 2,000 vacuum tubes, several thousand relays, 150 electric motors, and 200 miles of wire, but it was still a mechanical analogue machine. The electrical components existed only to supply power and control for mechanical motions. The machine contained differential gears for addition and subtraction as well as integrators based on the principle of the planimeter. All told, at least five copies of the Bush machines were made before World War II. The Bush machines evolved along two lines that often merged: analogue fire control systems were developed for the automatic aiming of guns, and analogue machines were constructed that used electrical potentials instead of mechanical lengths and angles to represent mathematical quantities. For a number of applications, analogue machines were initially faster and cheaper than digital computers because they dealt with an entire mathematical entity, such as a curve, whereas in a digital machine, the curve had to be represented by a large number of discrete bits, each of which had to be processed. Special purpose analogue computers are still in use. The major disadvantage of analogue computers for mathematical applications is that the accuracy is limited by the cost of sophisticated circuits and the quality of components. This is especially damaging for sequential calculations in which errors can accumulate. Furthermore, the natural connection between discrete arithmetic and logic makes the digital computer a natural logical symbol processor.

Three other important developments in mechanized computing in the United States were started before World War II. These were the Bell Laboratory machines built by George R. Stibitz, the Harvard series of machines of Howard Aiken, and the large calculators built by IBM.

The Bell System had an urgent need for computing power because the design of electrical equipment was becoming more demanding. Desk top calculators were slow. Furthermore, electrical theory was heavily dependent on complex numbers, and it was hard to keep track of their two components using the calculators of the time. In late 1938 a project to build a relay-based "Complex Number Calculator" was well under way, and a design by Stibitz and S. B. Williams became a fully operational machine on January 8, 1940. This was a binary, digital machine that could add, subtract, multiply, and divide complex numbers, but it did not have a stored program capability. The physical embodiment of the binary system was a set of electromagnetic relays which represented 1 and 0 by being in a state of contact or noncontact, thereby either passing or not passing a current. The Stibitz–Williams machine was essentially an enormous calculator. It was in daily use until 1949, helping to design Bell Telephone equipment.

Stibitz built other machines during World War II. They were all relay machines and were motivated by the war and not by Bell's corporate needs or decisions. When the war started, Stibitz became associated with the National Defense Research Council and was involved with a project for automatic aiming of antiaircraft fire. Stibitz suggested the construction of a flight simulator run by paper tape which would have a large number of arbitrary target motions. To create the simulated flight paths on tape required large amounts of calculations, and Bell Labs was commissioned to construct a calculator for this data. It was called the Relay Interpolator and was completed in September 1943. After the war it was given to the U.S. Naval Research Lab where it was used until 1961.

A larger relay machine was completed in June of 1944 to compute shell trajectories, and a second of these was completed for the Naval Research Laboratory in March 1945.

The relay machines were successful enough that a contract was given to the Bell Labs to construct two copies of a larger machine, the Model V or "Twin Machine," so named because it could work in two modes; as a single machine on a large problem or as two machines on independent smaller problems. One machine was delivered to NACA, Langley, in 1946 and the other to the Ballistic Research Laboratory in Aberdeen in 1947.

After the war, Bell built one more relay computer which was installed at Murray Hill in November 1950. It was used for design calculations similar to those of the Complex Number Calculator and was a direct descendent of the Stibitz design, although by that time Stibitz had left Bell for the University of Vermont.

In the meantime, the first truly automatic computer was being built by Howard H. Aiken. It was based on IBM's mechanical technology in which all operations depended on the rotation of wheels. The degree of rotation, and therefore numerical values, was controlled by timed electrical signals that operated clutches coupling the wheels to a central rotating shaft.

Actually, Aiken led the construction of a series of machines, using technology ranging from almost all mechanical to electronic with ferrite core memories. He was not wedded to any particular technology and his engineering objectives stressed reliability rather than speed.

Aiken's interest in computing started when he was a graduate student at Harvard and had to deal with nonlinear differential equations whose numerical solution was beyond the available computing power. Unlike many other computer pioneers, Aiken knew Babbage's work, and this certainly influenced him in the design of control mechanisms. He also knew of the Thomas J. Watson Astronomical Computing Bureau at Columbia and of the punched card accounting equipment they used. His idea was to hook together a large number of the mechanical registers from such units, controlled by one central unit, to make a powerful calculator. In 1937 he convinced IBM to back the construction of a large-scale computing machine, with financial assistance from the Navy. IBM did not think there was a commercial market for this but thought the publicity was of value. Work started in 1939 at the IBM plant in Endicott, N.Y., and the machine became operational in January 1943. IBM engineers made substantial contributions, and B. M. Durfee, F. E. Hamilton, and C. D. Lake are co-inventors of the Mark I. Watson gave it to Harvard in the summer of 1943, where, under Aiken's direction, it was devoted to military uses for the Navy from May 1944 until after the war.

Aiken designed and built another machine for the Navy in 1945, which was finished in July 1947 and used by the Naval Proving Ground at Dahlgren, Va. This, the Mark II, was relay-based and much faster than the Mark I. Its successor, the Mark III, was finished at Harvard in September 1949. It used magnetic drum memory and was part electronic, part relay-based. Aiken's last machine was the Mark IV, which used magnetic ferrite core memories and was completed in 1952.

Aiken's Mark I was IBM's first venture into large-scale calculation, but in 1935 the company had already introduced its "multiplying punch," which was a relay-based machine that could multiply and record the results on punched cards. This evolved into the 604, introduced in 1948, which used vacuum tubes and could be programmed to a degree by the use of plugboard panels.

As a consequence of its work with Aiken, IBM designed its own version of a large-scale computer. It was relay-based and controlled by plug boards. Two were built in 1944–1945 for use at the Aberdeen Proving Grounds, and five more were made a year later, three of which went to the Naval Proving Ground at Dahlgren and two to the Watson Scientific Computing Laboratory at Columbia. The largest of the IBM precursors to modern computers was the Selective Sequence Electronic Calculator (SSEC), completed in 1948. It was a hybrid of relay and vacuum tube technology, although by this time the all electronic ENIAC was completed and widely known.

Electronic technology had already been applied to computation by John V. Atanasoff at Ames, Iowa. Atanasoff had a joint appointment in physics and mathematics at Iowa State College. His research interests led to a need for large-scale computing, and by 1937 he had conceived of the idea of applying vacuum tubes to computation. Working with his graduate student, Clifford Berry, Atanasoff constructed a number of computer components that were electronically based. He used capacitors mounted on rotating drums for memory, spot-charred punch cards for input–output, and vacuum tubes for the arithmetic and control units. A prototype of a machine to solve systems of linear equations was completed in 1942 and successfully did arithmetic, but the unit was never completed because Atanasoff went off to do war work. Atanasoff's work has given rise to heated controversies about priority claims as to who invented the electronic computer. The truth seems to be that there is more than enough credit to go around. Because he built various pieces which were not assembled into an integrated device, and because his device was quite limited in scope, it is hard to conclude that Atanasoff was the inventor of the electronic digital computer. Also, because of his memory and input–output systems, he could not take advantage of the inherent speed of the vacuum tubes. On the other hand, his achievement should not be underestimated. Without sponsorship and without funds, Atanasoff showed that electronic-based computation was possible.

There was, of course, other work going on in the United States which was trying to advance the state of the art of automatic computation. The most important of this was sponsored and conducted by the U.S. Navy, which had established a communications and code-breaking group in 1935. Statistical methods based on letter and word frequency distributions had been worked out that led to attempts to mechanize code breaking. The first Naval efforts in this regard used IBM punched card machines. The Navy contracted with Vannevar Bush to develop special high-speed digital machines for code breaking and also sponsored work at the National Cash Register Corporation's Dayton, Ohio, facility, where Joseph Desch and Robert Mumma had been working on the application of electronics to arithmetic computation since 1938. The Navy sponsorship at NCR was extensive, and engineers from other firms were sent to work at the Dayton installation. Other firms also undertook to build encryption and decryption machines, including IBM, Bell Telephone Laboratories, and Eastman Kodak. The importance of this work was twofold; it provided experience with the use of electronics for data manipulation, and some of the people associated with the work later helped to develop the American computer industry.

During this prewar and wartime activity, there was important work on computing going on in both Germany and England. That in Germany was particularly remarkable because it was carried out by essentially a single individual who implemented some of the most advanced concepts of computer science.

Konrad Zuse was a design engineer in the German aircraft industry with a degree in Civil Engineering, and it was the extensive calculations needed in the design of airframe structures that led him to automatic computing. By 1934 he knew that automatic computing needed only a control, a memory, and a calculator, and he believed he could embody his ideas in equipment. He quit his job at the Henschel Aircraft Company in 1936 to devote his time to the development of computers.

His first machine is called the Z1 and was controlled by punched 35-mm movie film. Its basic element was a mechanical gate constructed from sliding plates connected by rods. The purely mechanical construction created problems of signal transmission, so Zuse started on a relay-based machine before the Z1 was finished. This was the Z2. Helmut Schreyer, a graduate student working with Zuse, constructed a vacuum tube model of part of Zuse's machine. But in prewar Germany the parts for a machine containing large numbers of vacuum tubes were not available, so Zuse continued with the relay-based machine while still using a mechanical memory.

In 1939 the Z2 was nearly completed, but then Zuse was drafted. He was released from the Army a year later through the efforts of some friends and put back on aircraft design work. He completed the Z2 on weekends and gave a demonstration of it to the Duetsche Versuchsenstalt für Lufthart that convinced DVL that relay-based automatic computing was feasible. They agreed to finance a more ambitious project, but Zuse was again drafted and sent to the Eastern front. In a short time he was brought back again and continued work on the Z3, which became operational in December 1941. It was totally based on relay technology.

The Z3 was a prototype, and Zuse immediately started on the Z4 which had a longer word length and reverted to a mechanical memory. The Z4 was put into operation in April 1945 but was quickly moved from Berlin to Gottingen to Hinterstein, Bavaria, to escape air raids. It was hidden in a house in Hinterstein until 1948 when it was installed at the Federal Polytechnic Institute at Zurich. It was used there by Heinz Rutishauser, who contributed a great deal to the development of European computing, particularly in the fields of compilers and programming languages.

Immediately after the war, Zuse developed a sophisticated programming language for his machine called Plankalkul. The language allowed conditional statements, had variables and subscripts, and permitted the construction of compound statements. The only features of modern languages that were missing were references, pointers, and recursion. The language was not restricted to numerical procedures. Zuse showed this by treating Boolean expressions and checking chess moves with Plankalkul.

The work in England led to the COLOSSUS, an electronic machine that was the result of an effort to build automatic, high-speed processing equipment to help the British break German codes. This work was of sufficient importance to merit a separate treatment. Not only was it a great advance in computer technology, but it laid the foundation for the rapid development of computers in Britain after the war.

COLOSSUS

The COLOSSUS was a special purpose computer that used vacuum tube technology to perform cryptanalysis. It was a Boolean logic device rather than a number processor, although, because of the relation between arithmetic and logic, it could be used for arithmetic operations. This was once actually demonstrated, with the conclusion that while arithmetic was certainly possible on the COLOSSUS, it was not worthwhile because of the difficulty and complexity of the required program wiring.

The origin of COLOSSUS lay in the British response to the German military coding system. This response was based on a machine known as ENIGMA, which was derived from a design by an American named Hebern. The ENIGMA was a multiple substitution device which contained three wheels divided into twenty-six steps, one for each letter of the alphabet. The wheels were electrically connected and stepped in such a way that a letter typed in via a keyboard was permuted seven times; three times through the wheels, once upon reflection, and three more times through the wheels. The stepping ensured that the same substitution was almost never made twice for the same letter in a given message. The particular sequence of substitutions was determined by an initial setting. The number of combinations of the initial setting was so large that timely decoding could not possibly be done manually. In July of 1939, when it was clear that war was imminent, the Poles informed the British and French intelligence agencies that they had been successful in decoding ENIGMA messages with the aid of an electromechanical machine they called "bomba." The Polish Cipher Bureau had purchased a commercial version of the ENIGMA after they became aware of its use by the Germans in 1928, and by 1932 had made considerable progress in decoding it under the leadership of Marian Rejewski, a distinguished Polish mathematician. The British had established their cryptanalysis facilities at Bletchley Park outside of London and soon built their own versions of the "bombs." (These were so named because of the ticking sounds they made during operation.)

After the war began, the Germans used a more complex version of the ENIGMA and the "bombs" were simply not fast enough to provide timely decoding. Many of Britain's best mathematicians and logicians worked at Bletchley Park, including Alan Turing. They developed new deciphering techniques and provided the general requirements for a new generation of machine decoding aids. M. H. A Newman, a Cambridge mathematician, was now at Bletchley Park, and he got C. E. Wynn–Williams to design a machine that combined electronic and relay-based technology. The machine was completed in April of 1943. It was called the Heath Robinson, after a cartoonist who was the British version of Rube Goldberg. It was not a very reliable machine, and T. H. Flowers, from the telephone division of the Post Office, was called in as a consultant. Flowers was at Dollis Hill, the British equivalent of the Bell Labs, and was head of a group doing research on telephone switching. He had not been involved with the Heath Robinson project, although some Dollis Hill engineers had worked on it. Rather than an inadequate patching of the Heath Robinson, Flowers proposed an entirely new machine totally based on vacuum tubes. This ended up as the COLOSSUS and was a great success.

The first COLOSSUS was operational in December 1943. It had 1,500 vacuum tubes, which made it the most complex electronic device built to that time. The bistable tube circuits were capable of binary arithmetic, counting, and Boolean operations. The input was via paper tape which was read photoelectrically, and output was by electric typewriter. It was not really a general purpose computer but it certainly showed that automatic electronic computation was feasible. Several stages of improvement followed, and by the end of the war ten Mark V units were in operation at Bletchley Park. It was the Bletchley Park work that made the quality and extent of Britain's computer work after the war second only to that of the United States.

THE ENIAC

The pivotal role of the Moore School in the history of the electronic computer was the result of a confluence of several circumstances. These were: its prewar connection with computing and close relations with the Aberdeen Ballistic Research Laboratory; the pressing need of the Army for faster computation of firing tables; the existence at the Moore School of people who believed electronic computing was possible; and enlightened leadership at BRL that supported electronic computing even though many outstanding engineers believed it could not succeed.

The Moore School's connection with automated computing started with the construction of a Differential Analyzer which was sponsored by the Ballistic Research Laboratory. A condition of sponsorship was that the Army could use the machine in the event of a national emergency. The Army exercised that right in 1940.

Aberdeen had an intense interest in computation because it had to prepare firing tables for guns and, in fact, it had a Differential Analyzer of its own for this purpose. These were tables of trajectories computed from the theory of exterior ballistics. Since the trajectory was different for each angle of elevation, each size projectile, each charge, and each type of fuse, and since the variation of air pressure with altitude and temperature, and even the resistance of the soil to recoil, had to be taken into account, the amount of arithmetic needed for each gun was enormous. Since successful deployment of a new gun was dependent on a knowledge of trajectories, the need for fast, accurate arithmetic was urgent.

The Moore School was contributing to this need in two ways. The first was by solving for trajectories on the Differential Analyzer; the other was by training and using human computers. These were women recruited from graduates of the many colleges in the Philadelphia region. The Differential Analyzer worked well, particularly after it was modified by Cornelius Weygandt with the help of J. Presper Eckert so as to give it increased speed and accuracy. Weygandt was the Moore School faculty member responsible for running the Analyzer, and Eckert was a graduate student who played a major role in the construction of the ENIAC. It was clear to them that the capability of the Differential Analyzer had been pushed to a maximum and that greater speed and accuracy would require a different kind of machine. The analogue machine simply was not fast enough. A firing table for one gun would consist of several thousand trajectories and could take one month to compute on the Differential Analyzer if everything worked well. The human computers, of course, were even slower; while a multiplication of two ten-digit numbers on the D.A. could be done in half a second, a human being would take several minutes. There was no way that the computing facilities could keep up with the need.

The ENIAC got its start in March of 1943 when Herman Goldstine, the liason officer from BRL to the Moore School, talked with John Mauchly, a new member of the Moore School staff. Goldstine was a mathematician brought to Aberdeen by Oswald Veblen and assigned to the Moore School to organize and speed up the computing activity. Veblen was a mathematician (he was the nephew of the famed economist Thorstein Veblen) who had assembled an outstanding staff of scientists and mathematicians at Aberdeen. By April 9 the Moore School had presented a proposal to Aberdeen to build an electronic computer. The contract to build the Electronic Numerical Integrator and Calculator was officially signed on May 17, 1943.

The key Moore School people at this point were John Grist Brainerd, John Mauchly, and J. Presper Eckert. Brainerd was a Moore School faculty member who had the responsibility for research and became the administrative head of the ENIAC project. He was a major force in getting the Moore School to accept the project, which did so only on the condition that Brainerd accept full responsibility for its successful completion. His good relations with Goldstine, his administrative skills, and outstanding technical expertise were invaluable. The conceptual design and specific engineering were led by Mauchly and Eckert, with Mauchly concentrating more on overall design and Eckert more on specifications, standards, and circuits.

Mauchly had acquired an interest in automatic computing while a Professor of Physics at Ursinis College. He had been conducting research on the relation of the weather to solar activity. This work depended on extensive statistical analysis for which he had no adequate computational tools available. He was familiar with electronic counters of the sort used in cosmic ray and nuclear physics and investigated the use of electron tubes for computing. While at Ursinis he actually built some computing devices, including binary counters. These were based on neon tubes because they were cheap. They demonstrated that electronically-based computation was a distinct possibility. By December of 1940, Mauchly's work had progressed to the point where he fully believed he could have an electronic computer within a year. In that month, he also met Atanasoff, who had come to Philadelphia for a meeting of the American Association for the Advancement of Science. Atanasoff invited him to Ames. Mauchly accepted and went to Ames for a physics meeting in June 1941, where he saw the arithmetic machine that was being built by Atanasoff and Berry. That trip was a costly one, since it contributed to the loss of patent rights by Mauchly and Eckert and even contributed to a court denial that they were the inventors of the electronic computer.

Mauchly's connection with the Moore School began when he took a course in electronics there in the summer of 1941 and stayed on as a member of the teaching staff. He found a kindred spirit in Eckert, with whom he discussed the possibilities of constructing electronic computing machines.

The great barrier to electronic computing, in most people's minds, was the failure rate of vacuum tubes. An electronic computer would require large numbers of tubes, so that if they were to fail at their normal rate the computer would hardly ever be working. Practically all the time would be spent changing tubes. But Eckert was familiar with at least one device that used a large number of tubes. He was an expert on theater organs, which used hundreds of tubes and seemed to last forever. The secret was to run the tubes below rated power and never to turn them off. The failure rate decreased dramatically if the power at which they were used was significantly below their rating, and keeping them hot avoided the thermal stresses that caused filament breakage when they were repeatedly heated and cooled. Furthermore, the basic elements of the electronic circuits were to be bistable; that is, it was only necessary to determine if circuits were conducting or nonconducting. There was no need to maintain current or voltage at highly precise values. Eckert and Mauchly therefore thought electronic computing was quite feasible.

In August of 1942, Mauchly wrote a memorandum on his ideas for electronic computing which he distributed in the Moore School. In March of 1943, when Mauchly talked to Goldstine about high-speed computing, he found that his memo was lost. Apparently, it had not been reproduced and distributed as Mauchly thought. However, Mauchly got his secretary to reconstruct the memo from her notes. At Brainerd's suggestion, Mauchly and Eckert wrote out a more detailed proposal for an electronic computer, and it was on the basis of this that Aberdeen awarded a contract to the Moore School.

Although Mauchly and Eckert were the major creators of the ENIAC design, other gifted engineers made important contributions and, of course, the actual implementation of the design was a team effort. Arthur Burks, Kite Sharpless, and Arthur Shaw were senior engineers on the project, and the actual construction was supervised by Joseph Chedaker, who came from RCA and reported to Eckert. Burks was a logician–mathematician who had been in the Moore School electronic course with Mauchly in 1941 and, like Mauchly, was hired to the school's teaching staff immediately thereafter. Sharpless and Shaw were among a number of recent Moore School graduates hired to work on the new machine.

Many of the components of ENIAC were analogous to existing mechanical or electromechanical machines. While the basic operation of the machine was binary in the sense that it relied on bistable circuits, it was actually decimal in operation. For example, mechanical storage of a decimal digit in earlier calculators was accomplished by turning a wheel to one of ten possible positions. In the ENIAC, storage was analogous, in that there were ten bistable flip-flop circuits arranged in a ring. A digit was stored by having one of the ten circuits in a conducting state. The ENIAC had twenty-five computing units: twenty accumulators, three function tables, one high-speed multiplier, and one unit for division and extracting square roots. The accumulators stored numbers and did arithmetic by adding or subtracting (through the complement) to whatever was stored in it. The function tables added speed and flexibility to the machine by replacing calculations of functions (such as sines and cosines) by a look-up. A card reader and a card puncher obtained from IBM were used for input–output, and programming was derived from the IBM plugboard method. It was performed using switches and dials which wired in the required procedures. Stored programming was deliberately left out, as noted in a progress report to the Army dated December 31, 1943, because this would delay completion of the machine and because it was anticipated that many runs of a given program would be completed before it had to be changed. This shows that the Moore School engineers had conceived of the stored program at a very early date.

Within a year the design was frozen, and in May of 1944 two accumulators were successfully tested. By May of 1945, the ENIAC was essentially complete, only the square rooter remaining to be installed. The completed machine had 18,000 vacuum tubes, 70,000 resistors, 10,000 capacitors, and 6,000 switches. It took up 1,800 square feet, weighed thirty tons, and consumed 174 kilowatts of electricity. It was the most complex electronic device built to that time, and its components had to work to a reliability of one in a hundred trillion.

The first serious problem was solved on the ENIAC in the summer of 1945. Von Neumann, who by this time had become a consultant to the Moore School, had arranged for Stanley P. Frankel and Nicholos Metropolis to come from Los Alamos to determine whether a problem was soluble that was important for hydrogen bomb research. The problem concerned the fission bomb trigger for the H-bomb. The calculation was successfully performed, although the Moore School people did not know what the problem was about.

The public unveiling took place on February 14, 1946, and created a sensation. The first problems to be run on the ENIAC (except for the classified H-bomb work) included a calculation of the liquid drop model of the nucleus done by Frankel and Metropolis, boundary layers in compressible flow by Douglas Hartree from England, a shock wave calculation by Taube and Adele Goldstine, a calculation of thermodynamic properties of real gases by Goff, a University of Pennsylvania Engineering professor, a study of prime numbers by Lehner, and calculations in meteorological theory by von Neumann. The ENIAC was indeed a general purpose machine; practical high-speed electronic computing had arrived.

It is instructive to compare the Moore School's success with Babbage's failure. The first point to note is that the technology to build the ENIAC was available, and there was no technical reason why it could not have been built ten years earlier. The Moore School engineers were creative in everything from circuits to systems design but the basic electronic technology was already developed. Babbage, on the other hand, had to continually advance the state of the art of machining and mechanical technology. Indeed, his contribution in this area turned out to be of great importance even though he did not complete his computers. Another point is that the ENIAC team had no funding problems. The urgent need of the Army ensured an adequate funding level. But probably the most important point was that Mauchly and Eckert knew when to freeze the design and to leave improvements for a later machine. Furthermore, Eckert imposed a set of standards on the operating characteristics of all circuits that established a common set of design parameters throughout the machine. It was this disciplined approach to good engineering practice that contributed so much to the completion of ENIAC within a reasonable time. Babbage, on the other hand, was continually improving his machines, both in concept and in implementation.

The design of later machines did not closely follow that of the ENIAC. The experience on the ENIAC was essential for future development, but it was the computers built immediately afterward that determined the overall structure of digital computers to the present day. This structure was based on the logical requirements of automated computation, which were most cogently presented by Turing and von Neumann.

TURING AND VON NEUMANN

The versatility of the digital computer arises from its basic logical structure. The computer is the physical embodiment of an abstract conceptual system that relates logic, mathematics, and symbols. It requires that physical devices be created and related to each other in such a way that operations can be performed that correspond to arithmetic and logical processes. But this presumes that the abstract concepts of such operations and their relations exist. That is, to have a computer there must be a theoretical construct of a computing machine that includes structure, logic, and all the steps that enable computation.

Many individuals have contributed to the creation of the abstract theory on which automatic computation is based. In the 1830s Babbage already had a clear idea of the theoretical structure of computation and, although his efforts were focused on numerical computation, his ideas did include logic processing in that his Analytic Engine could perform conditional operations. But the total abstract structure on which modern computation is based was completed only in the twentieth century, and two individuals stand out as the major contributors. These were Alan Turing and John von Neumann. Certainly others were aware of the theoretical requirements of computing and realized that machines had to be designed to match these requirements. Konrad Zuse, for example, fully understood the connections between logic, mathematics, computation, and machine design. Mauchly and Eckert at the Moore School seem to have developed their theoretical base right along with the design of machines and were certainly aware of the possibilities of stored programming at an early date. In fact, it was not possible to build an automatic computer without an appreciation of the underlying logical and mathematical theory. Nevertheless, Turing and von Neumann are identified as the major creators of the theory of computers, not only because of the generality and importance of their ideas but also because of the force of their personalities, the respect they commanded for their overwhelming brilliance, and, especially in the case of von Neumann, the energy and clarity with which they presented their work.

Alan Turing studied mathematics at King's College, Cambridge, where he stayed on as a Fellow until 1936. He then went to the Institute for Advanced Study at Princeton where he stayed for a year. Church, Hardy, Einstein, Courant, and von Neumann were at the Institute at that time, and this must have been an exciting year for Turing, whose major interest was mathematical logic. The connection with von Neumann is of particular interest since it ensured that van Neumann became thoroughly familiar with Turing's ideas.

In 1936, at the age of 24, Turing published a paper that contained the concept of a general purpose computer that could not only process numbers but could also perform logical operations. His purpose in the paper was to show that Hilbert was wrong when he claimed that any mathematical problem could be solved by a fixed and definite process. Turing did indeed show that there were problems that could not be solved by purely mechanical means. The important point for computer development was that he showed that a universal automaton could be defined that could perform any calculation or any sequence of logical operations of any special purpose machine if it were given the appropriate instructions. These instructions constituted the program for the abstract Turing machine. The close relation to Babbage's work, which was known and admired by Turing, is evident.

The abstract Turing machine is a remarkably simple device. It consists of three parts: a tape marked off into squares, a control unit, and a read–write head that can write a symbol on one of the squares on the tape, or read a symbol that is already in the square and move one square over either to the right or left. The symbols in the square are called data. The control unit contains a finite set of "instructions." When the control unit is in a state to execute a particular instruction, the head reads the square under it and the combination of the instruction in the control unit and the data in the square determines the next instruction that the machine will use. Turing showed that this simple machine could compute anything for which a finite algorithm could be constructed. This is a powerful statement. It means that an automatic method can be devised to perform any mathematical or logical process that can be written down in a logical, well-ordered manner. Truly general purpose computers would not be possible if Turing's machine did not work.

Turing returned to Princeton in the fall of 1937, where he stayed until the summer of 1938. Von Neumann offered him a position as his assistant but Turing chose to return to King's College, and when war was declared he went to work at Bletchley park. The "bomb," which was the cryptanalytic machine built at Bletchley Park (about 1940) as a result of the Polish revelations, was evidently designed by Turing. He was the major figure in developing the logical and statistical techniques for breaking the German codes and played an important role in the design of the COLOSSUS.

There is evidence that the Bletchley Park group cooperated with the U.S. Navy's decoding group. Turing visited the States at least once during the war, and the Navy built relatively faithful copies of several British machines including the COLOSSUS. However, it appears that the Moore School team was completely unaware of the COLOSSUS and, conversely, the British became familiar with ENIAC only after the war.

In 1945 Turing joined the newly formed National Physical Laboratory, where he immediately started work on a computer design. By February of 1946 he had a complete design for a stored program computer which he called the Automatic Computing Engine. This was an advanced machine with mercury-delay line memory and used optimum coding. The National

Physical Laboratory had no facilities for electronic construction, and rather than create them it decided to contract the work to other agencies. This delayed construction which, along with disagreements among personnel at the NPL, led to Turing's leaving to go back to Cambridge. He returned to the NPL for a brief time in 1948 but quickly left to join the computer group at Manchester. A smaller version of Turing's design, called Pilot ACE, was built and became operational in 1950. Pilot ACE was conceived and built at the instance of Harry Huskey, who had worked on the ENIAC at the Moore School and was now on sabbatical leave at NPL. A full-scale version was completed in 1957. While this was still an advanced machine in concept, core memories were being used at about this time and the ACE rapidly became obsolete. Nevertheless, its design had an important influence on future computers.

John von Neumann became aware of the Moore School Computer project in June 1944. He was a consultant both to Aberdeen and to the Manhattan Project at the time and met Goldstine on the train platform at Aberdeen. Upon learning of the ENIAC, von Neumann arranged to visit the Moore School, which he did in early September 1944. He then had himself made a consultant to the project. He was intensely interested in computation through his work on applied mathematics in such areas as meteorology and hydrodynamics. He was particularly interested in implosions, which he was working on for Los Alamos for atomic bomb applications. His first visit to the Moore School was in early September 1944.

Although work on the ENIAC was far from complete, the design of a new and improved computer was already under way at the Moore School when von Neumann arrived. This was the EDVAC, the Electronic Discrete Variable Automatic Computer, and it was really for the design of this machine that von Neumann served as a consultant.

On June 30, 1945, a paper from von Neumann arrived at the Moore School. Its title was "First Draft of a Report on the EDVAC." It was an outstanding synthesis and extension of the thinking that had gone into the EDVAC with descriptions of functionality, logic, and architecture that have acted as a base for computer design ever since. But the paper caused a serious rift between von Neumann and Mauchly and Eckert. It contained hardly any reference to the Moore School work and to the fact that concepts such as that of stored programming were conceived by Moore School people. Furthermore, the Moore School work was classified, so Mauchly and Eckert could not publish to establish their priority. Clearly, the von Neumann document endangered future possible patent positions that Mauchly and Eckert might wish to establish.

Goldstine had the paper reproduced and sent copies not only to the Moore School staff but also to interested scientists throughout the United States and to Britain. This amounted to general publication of a landmark computer paper without due credit going to the Moore School.

Von Neumann's genius and force of personality led to his being identified as the leader of computer development, at least in the minds of the less knowledgeable. This was a development he apparently did not discourage. On January 11, 1946, the *New York Times* ran an article describing plans for a new electronic computer that was being proposed to the government by von Neumann and Vladimir Zworykin of RCA. Again there was no mention of the Moore School.

By March of 1946, von Neumann had put together a team to construct a computer at the Institute for Advanced Study in cooperation with Princeton and RCA. The project leader was Julian Bigelow, who had been working with Norbert Wiener at MIT. The team included Arthur Burks, who left the Moore School in March to join von Neumann. The machine was essentially done by January of 1951 and was officially dedicated on June 10, 1952.

The IAS machine embodied significant advances. It was a "parallel" machine in the sense that it processed all bits of a word simultaneously. Although in the "First Draft" von Neumann had strongly stated that computers must be serial, he had become convinced that the Selectron, a memory device developed by Jan Rajchman at RCA, could be used as the basis of a parallel machine. In the Selectron, storage was accomplished by having small steel rings embedded in mica that could be maintained at one of two voltages. At one voltage, an electron beam could pass through the ring; at another voltage, the beam was deflected. Forty Selectrons were planned on to process 40-bit words. A 4,000 word memory was planned.

In practice, the Selectron could not be made to work properly. The project was saved when the IAS team heard about the Williams tube from Hartree, who visited them from England in the Spring of 1948. This was a cathode ray tube whose memory function depended on the fact that when an electron beam strikes the phosphor, not only is light produced, but an electrostatic charge at the spot is generated that lasts about 0.2 seconds. It was developed by F. C. Williams and his assistant T. Kilburn at Manchester University after the war. The possible use of a cathode ray tube for a memory had been advanced in one of the Moore School lectures in the summer of 1946 by Eckert. Although Williams did not attend these lectures, he certainly knew people who had, and this could have initiated the train of thought that led to a successful electrostatic memory. Forty Williams tubes were used to build a parallel memory, but it had only 1,024 words capacity instead of the 4,000 in the original design. The machine was a great success and a number of copies were made by various institutions.

Both Turing and von Neumann not only formulated the basis of machine logical structure but also impressed their ideas on the actual construction of computers. In this latter regard von Neumann was more successful than Turing. He had carefully and extensively organized the theory of computer structure in the EDVAC report and in a 1946 paper with Arthur Burks and Herman Goldstine entitled "Preliminary Discussion of the Logical Design of an Electronic Computing Instrument." He was also a consultant for government agencies, IBM and other firms, and gave many lectures on computers. This, along with his widespread influence and greater degree of personal charm, and the fact that the United States had many more centers of computer development than Britain, made his influence greater than Turing's. The design of the modern digital computer is still called the von Neumann architecture.

DOWNFALL OF THE MOORE SCHOOL

In early 1946 the Moore School was the most important center for computer science and engineering in the world. It had successfully demonstrated the ENIAC on February 14 and had the most competent team of computer engineers in existence. The Moore School had all the ingredients to be the leading force in computers for decades to come. Instead, by the end of March the team disbanded and the Moore School gave up its lead to others. The Moore School never again became the dominant academic institution in computers.

The major reason for the Moore school decline was disagreement with respect to patent rights. This disagreement was exacerbated by differing views towards commercialization by the various people involved with ENIAC and by conflicts over the issues of credit for the inventions and developments related to electronic digital computing.

Three factions can be identified in the conflict that led to the breakup of the ENIAC team. The first was led by Eckert and Mauchly, who regarded themselves as the major inventors of the electronic computer and were interested in its commercial possibilities from an early date. The second consisted of von Neumann, Goldstine, and Burks. Von Neumann's orientation was completely scientific, in that he valued the computer for its possible contributions to science and not as something for commercial use. Although he had no objection to commercial activity per se, he strongly felt that academic science had priority and that it should be Eckert's and Mauchly's primary responsibility. Goldstine was a great admirer of von Neumann's and tended to follow his lead and to resolve doubts on questions of inventive priority in his favor. Burks eventually joined von Neumann and Goldstine at the Institute for Advanced Study to build the IAS machine, which was started while von Neumann was still a consultant for the EDVAC. The third group consisted of academics at the Moore School, who believed that commercialization by school staff was a violation of academic ethics and that academic pursuits should not include making money from work done at the school. Brainerd was the proponent of this point of view, and furthermore believed that the inventions involved were group efforts and could not be assigned to any one or two individuals.

Eckert and Mauchly were beleaguered from several sides. Von Neumann was rapidly acquiring the reputation for being the creator of computers and computer science. Being younger and less experienced, with no prior great accomplishments to their credit, and certainly without the enormous prestige of von Neumann, Eckert and Mauchly felt that recognition of their achievement was threatened. This feeling was reinforced by Brainerd, who thought their claims constituted an act of disloyalty to the Moore School.

In 1944 the Army was pressing the university to take out patents on the ENIAC because it feared that patent applications by Bell Telephone Laboratories on calculating circuits would endanger the Army's rights to the ENIAC technology. Eckert and Mauchly were already thinking of patent rights, and when the University did nothing about patents, they decided to file for themselves. On March 15, 1945, an agreement was completed and signed by the president of the university that granted license-free use to the government and commercial rights to the ENIAC technology to Eckert and Mauchly, who shared these rights equally. The stage for a three-way conflict was therefore already set by early 1945. This was exacerbated by several incidents. In February of 1945 Brainerd, as project supervisor, was asked for a report on electronic computers by the Office of Scientific Research and Development, which was the government's chief research and development agency. The request was not acted upon for nine months while Brainerd waited for clearance from the Army Ordnance Department. This came at the end of September and the report was due on November 30. Eckert and Mauchly did not learn of the request for this report until Brainerd had almost finished writing it, and they were most concerned that the report, without their authorship, would deny them the credit of their contributions. Reid Warren, a member of the Moore School who was now supervisor for the EDVAC, prevailed on Brainerd and arranged to have Eckert, Mauchly, and Goldstine added as coauthors. This attempt at peacemaking had only a temporary success. Von Neumann's "First Draft of a Report on the EDVAC" arrived on June 30 of that year, and it had a devastating effect on the perception of Eckert and Mauchly's accomplishment. Another conflict arose in October of that year, when Mauchly was the only major member of the ENIAC team not to be invited to an important conference on calculating machines to be held at MIT. Again, Warren intervened and Mauchly was invited, but Mauchly believed that the snub was directly caused by Brainerd.

Yet another conflict arose in January, 1946, when the *New York Times* article appeared. The article reported on a meeting between von Neumann, Vladimir Zworykin, and people from the Weather Bureau at which the possible application of electronic computing to meteorology was discussed. The article strongly implied that von Neumann and Zworykin were the pioneers and inventors of electronic computers. They were, of course, working on the Institute for Advanced Study machine by then, but did not see fit to mention the Moore School efforts.

It was becoming increasingly clear that the Moore School had to come to grips with the issues attending government sponsorship of research. To this end, Dean Pender appointed Irven Travis to be supervisor of research in

January 1946. Travis had been on the faculty but was called to the Navy in 1941 and returned in 1946. He felt that the Moore School would be more effective in attracting government funds if it had a clear patent policy, and he quickly moved to establish such a policy modeled along industrial lines. On March 15 each member of the Moore School staff was asked to sign a release form giving all patent rights to the University. Mauchly and Eckert refused to sign. On March 22 they were informed by a letter from Pender that they must give up their patent rights or resign from the Moore School, and that they must make the decision by 5:00 P.M. on that day. Mauchly and Eckert submitted their resignations effective March 30, 1946. Later, other Moore School people, particularly Shaw and Sheppard, left to join Eckert and Mauchly, and Burks left to join von Neumann. The Moore School's pre-eminence in computers was at an end. The Moore School administration demanded patent rights from Mauchly and Eckert in spite of the fact that they had a previous agreement with the president of the University. This demand was made possible by the unique relation of the Moore School to the university. The Moore School was a nearly autonomous organization whose relation to the university was circumscribed by the conditions of a trust which provided that the Moore School have its own Board and control over its own endowment.

One year after he had implemented the new patent policy, Travis left the Moore School to accept a position as a vice president of the Burroughs Corporation, where his responsibility was to establish an electronic computer division. In 1947 Kite Sharpless left to take part in the formation of Technitrol, a firm founded to make and sell electrical components. This completed the decline of the Moore School in the field of computers.

Although the result of all this was tragic for the Moore School, it had the effect of putting Eckert and Mauchly rapidly into the commercial world, with the result that computer technology was immediately transferred from academe to industry, thereby hastening the widespread practical application of computers.

Eckert and Mauchly rendered one more service to the diffusion of computer technology and to the Moore School in the summer of 1946, even though they were no longer on the Moore School staff. Carl Chambers had organized a course for that summer with the purpose of presenting the latest information on electronic computing. The lectures were given by many of the leading computer scientists and engineers including Eckert, Mauchly, Goldstine, and von Neumann. The lectures were attended by people from the major organizations in the United States and Britain that had interests in computing and were a potent force for the diffusion of computer technology.

POST-ENIAC MACHINES

The modern design of digital computers started with the computers developed immediately after the ENIAC, not with the ENIAC itself. The catalysts for these newer developments were the design of the EDVAC, the Moore School summer lectures of 1946, and the von Neumann papers on the logic and structure of computers.

The EDVAC was the first computer to be designed with a stored program capability. There has been some controversy as to the origins of the idea of stored programming, with the admirers of von Neumann attributing the concept to him. It is plausible that von Neumann would think of this. However, the historical evidence clearly points to the fact that Eckert and Mauchly had considered stored programming before von Neumann arrived on the scene and, in fact, had consciously left it out of the ENIAC in the interest of rapid completion. At the least, Eckert and Mauchly were independent originators of the stored programming concept. At any rate, the stored programming of the EDVAC, along with its structure of von Neumann architecture, made it the precursor of modern designs.

The EDVAC was not the first stored program computer to be built. The disarray caused by the breakup of the Moore School team interrupted the momentum of its work and the EDVAC was not completed until 1952. The lead in new computer construction passed to the British, not only because of the Moore School problems, but also because several of the British groups were oriented towards constructing usable machines for scientific work rather than just to advance the art of computing.

The three centers at which the most important British work was done were the National Physical Laboratory, Manchester University, and Cambridge University. The NPL work resulted in the computers known as Pilot ACE and ACE, as has already been noted. The Manchester work ended up in a machine built by Ferranti Limited in 1951 usually called the Mark I.

After the end of the war a group of outstanding computer scientists and engineers had gone to Manchester. These included people with experience on COLOSSUS, memory devices, and digital electronics, who were also familiar with the American work. M. H. A. Newman, F. Williams, and Turing were among them. In July of 1946, Newman had acquired a grant from the Royal Society and the group started work on a new computer. It was a stored program machine that used eight cathode ray Williams tubes for random access electrostatic memory coupled to a slower but larger memory (16,384 words of 40 bits) consisting of a magnetic drum.

At Cambridge, computer development was led by Maurice Wilkes, who constituted the staff of the "University Mathematical Laboratory," which was formed in 1937 to tend the university's Differential Analyzer. In May of 1946, Wilkes read von Neumann's "First Draft on the EDVAC," which he obtained from L. J. Comrie who had just visited the United States. Wilkes was invited to the Moore School 1946 summer lectures and returned home determined to build an electronic computer. Williams had not yet developed his electrostatic memory, so Wilkes used mercury delay lines. He called his machine the EDSAC (Electronic Delay Storage Automatic Calculator). Wilkes had acquired only enough funding for the initial design work, but a large British firm had heard of his work and hoped that the Wilkes machine could solve its accounting problems. This firm was J. Lyons and Co. Ltd., which supplied baked goods to small stores, cafes, and tea shops. The company had sent one of its executives to the United States to study the changes in office management induced by the war. The executive, T. R. Thompson, became very enthusiastic about the possibilities of electronic computing and decided to support Wilkes. The firm gave Wilkes some money and loaned him the services of one of their best engineers, E. H. Lennarts. The EDSAC was first demonstrated in 1949 and was the first full-scale stored program computer to function. The EDVAC was the first one to be designed and a very small version of stored programming was tested by the Manchester group in 1948, but the Cambridge machine was the first one to be operational.

EDSAC also led to the first computer to be used for commercial data processing. A re-engineered EDSAC was built for corporate use which was called LEO (Lyons Electronic Office) and which became operational in 1951.

In spite of these great achievements, the future course of digital computers was largely determined in the United States. The machines built immediately after the ENIAC that set this course were the EDVAC (1952), the IAS machine of von Neumann (1951), the SEAC (1950) and SWAC (1950) of the Bureau of Standards, the ERA 1101 (1950), the UNIVAC (1951), and Whirlwind (1951). Although the EDVAC was the last of this group to become operational, its design was well known and had a profound influence on all post-ENIAC machines, as has been already noted.

The Institute for Advanced Study machine was funded by the Army through the efforts of von Neumann and Goldstine. Von Neumann's great prestige and widespread contacts ensured that the IAS machine would be widely used. In fact, von Neumann's original plan was to distribute design plans to a number of institutions as they were developed so that the other organizations could rapidly build copies. A number of copies, with some alterations and improvements, were actually built at various institutions, including the Rand Corporation, the Los Alamos National Laboratory, the Argonne National Laboratory, and the University of Illinois. All of these were paid for by the United States government. Von Neumann's machines were not limited to the United States. Several versions were built abroad, including the SILLIAC in Australia.

The SEAC (Standards Eastern Automatic Computer) and the SWAC (Standards Western Automatic Computer) were built by the National Bureau of Standards in Washington and Los Angeles respectively. Both of these computers were designed to be completed rapidly in order to satisfy the computing needs of the National Bureau of Standards until either the IAS machine or the UNIVAC was completed. In May of 1950 the SEAC became the first post-ENIAC American computer to be fully operational. While it did not have the memory or the power of the larger machines being built, it did incorporate some innovations, the most important of which was that it used solid state elements. These were germanium diodes, 10,000 of which were used in SEAC. SWAC was completed by July 1950, at which time it was the fastest computer in the world. The SWAC construction was headed by Harry Huskey, who had returned from his sabbatical in England and was now with the NBS.

The ERA 1101 was the commercial version of a general purpose computer that Engineering Research Associates designed and built for the U.S. Navy. It was code-named ATLAS and became operational by the end of 1950. Only one 1101 was built, but a successor, the 1103, gave ERA the dominant market position in computers up to 1952. The 1103 also was a commercial version of a Navy machine, the ATLAS II.

Engineering Research Associates was founded right after the war by Harold Engstrom, William Norris, Ralph Meader, and John E. Parker as a firm to build special decoding equipment for the U.S. government, particularly the Navy. Engstrom and Norris were members of the Navy's code-breaking group, Meader had been head of the Naval Computing Machine Laboratory, and Parker was a Minnesota businessman who backed the others even though, because of security requirements, he did not have a clear idea of what they were doing. The Navy asked ERA to develop ATLAS in 1947 in an effort to decrease costs by replacing special purpose decoding equipment with a general purpose computer.

After leaving the Moore School, Eckert and Mauchly went into business for themselves as the Electronic Control Company, which was later incorporated under the name of the Eckert–Mauchly Computing Corporation. They were successful in getting a contract from the National Bureau of Standards, which was acting on behalf of the Census Bureau, to design and build a new machine which they called the UNIVAC (Universal Automatic Computer). Eckert and Mauchly were poorer businessmen, and their firm was in financial trouble from the beginning when they agreed to an inadequate NBS contract even though they knew it did not provide for enough money to build UNIVAC. In 1948 the firm received an infusion of money from Henry Straus, an engineer who had a monopoly on special electromechanical calculators to compute odds and winnings for parimutuel betting at race tracks. He invested $400,000 for 40 percent of the company, but this did not prove to be enough. Unfortunately, on October 25, 1949, just before Straus was to start on a major investment program for the company, he was killed in an air crash. The firm's financial position continued to deteriorate and on February 1, 1950, it sold out to Remington Rand.

In the meantime, Eckert and Mauchly had contracted to build a small version of EDVAC for Northrop Aviation, which needed a computer for in-flight control of the "Snark" missile. BINAC was completed in August of 1949. (The acronym stood for Binary Automatic Computer, since it contained two central processing units.) It was the first stored program computer to be operational in the United States and actually ran its first program a few months before the EDSAC and Mark I in England, although it was not fully completed until later.

The first UNIVAC was delivered to the Census Bureau on March 31, 1951.

All told, forty-six UNIVACs were built for governmental and commercial customers. Remington Rand became the first large-scale computer firm, and for a time was far in advance of anyone else in sales and in commercial computer technology. But the company could not take advantage of its lead and later rapidly lost ground to IBM.

The UNIVAC could store 1,000 numbers or characters in mercury delay lines. It was a serial bit machine but with a high pulse rate of 2.25 megahertz so that it did not need bit parallel operation for speed. It was the first machine to use magnetic tape input–output, which provided a mass storage that was much faster and more convenient than punched cards. The UNIVAC was an advanced machine that, for a time, was synonymous with electronic computer in the commercial world. Its fast and efficient input–output made it well suited for commercial data processing applications, whereas the bit parallel architecture of the IAS machine made it more useful for scientific work, for which fast CPUs were more important than fast input–output.

An example of the versatility of electronic computation and an indication of its future pervasiveness were provided by the 1952 presidential election. CBS had asked Remington Rand for the use of a number of typewriters and calculators on its televised coverage of election night, in return for which CBS would display the company's logo. Someone at Rand suggested that CBS could have UNIVAC try to predict the results of the election as the returns came in. This would certainly be exciting for the viewers. CBS agreed, although it regarded the entire exercise as only a bit of showmanship. The Eckert–Mauchly group in Philadelphia accordingly programmed UNIVAC using an algorithm based on the previous two presidential elections and regional voting patterns for the past twenty-seven years. The results were astonishing. With only 7 percent of the votes in, UNIVAC predicted the Eisenhower victory within one percent of the electoral votes. Unfortunately, CBS broadcasters could not bring themselves to believe this results, and it was not broadcast because it was contrary to all political expectations. The algorithms were changed to bring them more in line with expectations, but a victory was still predicted although with a smaller majority. Late in the evening a predicted result of a modest victory for Eisenhower was aired, but by 11:00 P.M. it was obvious that the earlier landslide prediction was correct.

Whirlwind had its origins in 1945 when the Navy sponsored a project at MIT to develop an analogue computer control mechanism in connection with flight simulators for training purposes. The project was based on a plan by Jay Forrester, who was a graduate student at the time. But analogue methods proved to be too slow for flight simulation, which required real time machine response. Forrester visited the Moore School to look into digital electronic techniques and returned to start a new design along the lines of EDVAC. By 1951 Whirlwind I was operating on a routine basis. In 1952 the electrostatic memory was replaced by ferrite core memory, which Forrester had conceived as early as 1950. The sponsorship of the Whirlwind project was taken over by the Air Force in 1950 and ultimately led to the AN/FSQ7, which was built by IBM for the SAGE (Semiautomatic Ground Environment) Air Defense System. The first AN/FSQ7 was operational in 1958, and twenty-four of them were installed in the SAGE system. It was the first production machine to use ferrite core memory.

COMPUTER COMPANIES

The growth of the computer industry was dominated by the growth of IBM beginning in the mid-1950s. The story of other firms' ventures into the field, both in the United States and abroad, has largely been the story of either finding a special market niche or of inadequate competition with IBMs broad product line.

International Business Machines was born of the need to mechanize the tabulation and computation of results of the U.S. Census. By the middle of the nineteenth century it was obvious that the traditional human-hand modes of dealing with the enormous amount of data gathered by the Census Bureau were totally inadequate. The first mechanization was the "Seaton Device," introduced in 1872 by Charles W. Seaton, Chief Clerk of the Census, to help in the tabulation of the census results of 1870. This was nothing more than a mechanical aid consisting of slots to line up columns and rows so that the tabulator did not have to hunt for the right place on his spreadsheet. It was a sign of desperation that even this simple technique was important. The beginning of real mechanization had to await the efforts of Herman Hollerith.

Upon graduation from the Columbia School of Mines in 1879, Hollerith had gone to work at the Census for one of his professors who was in charge of a survey of power machinery in manufacturing. While in Washington, Hollerith met Dr. John Shaw Billings and dated his daughter Kate Billings. Dr. Billings was an accomplished man who had been medical inspector for the Army of the Potomac during the Civil War, organized the Johns Hopkins Hospital in Baltimore, and was Director of the Hospital of the University of Pennsylvania. Upon retiring from Penn in 1896, he organized the New York Public Library. In 1881 he was responsible for the collection and tabulation of data on vital statistics for the 1890 Census.

Billings discussed the data problem with Hollerith and suggested that some mechanization would be possible. Hollerith took the conversation seriously and invented a tabulator that was based on holes punched in a roll of paper like that used in the Seaton Device. Each hole represented a particular data item. The holes were sensed by wire brushes making contact with a drum under the paper, thereby activating an electromagnet which turned a counter. There were serious problems with this machine, particularly with data retrieval, but Hollerith soon replaced this with the electric tabulator based on punched cards. He also developed a card sorter so that the data could be readily sorted and manipulated. The holes were sensed by electrical contacts in pools of mercury. At first he punched the cards by hand but later invented a card puncher. Later he redesigned the card to be the size of the dollar bill of the time. Apparently, he knew nothing of Babbage's work, although he could have known about the Jacquard loom since his brother-in-law was in the weaving business. Hollerith himself claimed to get the idea of the punched card from railroad tickets which were punched to code in passenger information.

Hollerith's machines were successfully used in the 1890 Census. By that time they had been adopted in a variety of places, including the health departments of New York and New Jersey. Hollerith provided improved machines for the 1900 Census, including 311 tabulators, 20 card sorters, and 1,021 card punchers.

In 1896, Hollerith incorporated under the name of the Tabulating Machine Company. He was the majority stockholder and Ferdinand W. Roebling was President. The Census was an important part of the firm's business, but it soon branched into the private sector by providing machines for railroads and manufacturing firms.

The 1910 census was not easy for Hollerith. Simon North, the new superintendent of the Census Bureau, was opposed to the use of the Hollerith machines and undertook to develop alternates. Also, Hollerith was in his fifties, his doctor had told him to slow down, and his company was short of capital. All of these factors made him receptive to the merger ideas of Charles R. Flint.

Flint believed that companies should be large, a view he held on strong philosophical grounds. He believed that large, efficient firms were necessary for the growth of the country, and he therefore dedicated himself to merging small companies together to make larger ones. He was largely responsible for the mergers that created U.S. Rubber and American Chicle. In 1900, Flint had formed the International Time Recording Company from a merger of small time clock companies, and in 1901 he formed the Computing Scale Company. These, along with the Bundy Manufacturing Company, were merged with the Tabulating Machine Company in 1911 to form CTR, the Computer-Tabulating-Recording Company. Flint raised the money to pay $450 per share for Tabulating Machine, and Hollerith retired with $2,210,500, although he continued to consult for the company.

On May 1, 1914, CTR hired Thomas Watson, who was a young salesman for National Cash Register. Watson took over at CTR, and in 1924 he changed its name to International Business Machines.

By the time the ENIAC project started, IBM was a major supplier of tabulating and accounting office equipment. As noted above, it had supported several large-scale computing projects and was aware of the Moore School work. But IBM's entry into the computer market was a bit late. In 1946 the company had started a research effort at the Watson Scientific Computing Laboratory at Columbia, which gave it some experience in digital circuit design. Also, Brian Havens, a senior IBM engineer, was working on very fast circuits with the intention of eventually building a state of the art machine.

The project received a boost in 1950 when the Navy supported the IBM construction of a powerful computer which became known as NORC (Naval Ordnance Research Calculator). The machine was completed in December of 1954. It had a Williams tube memory which could store 3,600 words of 16 decimal digits each, could execute 15,000 operations per second, and was so reliable that its down time was only 8 percent. Government support, not only for NORC but also in the form of subcontracts from MIT's Lincoln Laboratory for computers for the SAGE system, was important in getting IBM into the electronic computer business.

In 1951, while NORC was still being designed, IBM undertook to design and build a computer for the commercial market. This was the 701, which was on the market by 1953. It was a bit-parallel machine which was derived from the von Neumann IAS machine rather than the EDVAC and had a Williams electrostatic memory of 4,096 words of 36 bits each. All told, nineteen of the 701s were installed. The 701 was a good machine for scientific work but was hard to program for variable length character records. In 1953 IBM therefore announced the 702, which was a serial, variable word length machine. It was first installed in 1955. In the same year the IBM 650 was introduced, a machine with much less power than the 701 but that opened up important computational resources to a wide market. By 1960 these machines and their descendants put IBM in the forefront of the computer companies, but its continued leadership was far from certain. Other companies were also developing advanced computer technologies and successfully competing with IBM products.

It was the System 360 series of machines that confirmed IBM's position of pre-eminence. IBM in the early sixties was making six incompatible computer systems. The rapidly changing technology contributed to incompatibility, and the cost of software was becoming an ever greater fraction of the total cost of a computer system. There was a large barrier to the purchase of a new system because of the difficulties attending the creation and installation of new software. IBM had recognized the importance of software early, and in fact had given John Backus the time to develop FORTRAN while the 700 series was being designed. Now computer incompatibility was becoming an industry-wide problem and IBM responded with the 360 system. This was a series of machines with a similar basic architecture designed for software that could be used on all machines in the series. With this series of machines, customers could have a computer of the power that suited their current needs and be assured that when it became necessary to expand they could go into a larger computer without the need to redo all their software. Furthermore, the 360 was designed to be useful for either scientific or business applications. IBM even gave the 360 some compatibility with its other computers by supplying microprogrammed ROMs to emulate the earlier machines. The 360, along with the large array of peripherals and software that could be acquired for it, redefined the nature of the computer industry.

IBM's approach to the computer market was well managed and well supported by a strong research and development effort. Such was not the case with Remington Rand.

By the time 1952 was over, Remington Rand had a commanding position in the computer industry. UNIVAC was an unqualified success both from a technical and a commercial point of view, the firm was large and had access to capital and, above all, it had an impressive array of technical talent in the field of computers. In addition to the Eckert–Mauchly team, the firm had ERA, headed by William Norris. ERA had run into financial problems and was purchased by Remington Rand in 1952, which now had the best and largest group of computer engineers in the country. No other computer company could match it in size, resources, or expertise; IBM had not yet entered the field, and Remington Rand was far ahead of anyone else. All the ingredients were there for the firm to take advantage of one of the great business opportunities of the century. But, because of a combination of factors, the opportunity was never realized. The basic problem was that top management never really took control of the computer efforts or understood the sales, development, and managerial requirements of the new technology. The several computer groups in the firm were never coordinated, the sales effort was ineffectual, and not enough funds were provided for research and development. At IBM there was a highly placed product champion for electronic computers in the person of Thomas Watson, Jr., a large fraction of sales went into R&D, management was unified, and sales were carefully geared to customers' computing needs by people who thoroughly understood their customers' business. By the time of the merger with Sperry in 1955, Remington Rand was on the way to being eclipsed by IBM. In 1957 Norris, along with most of his former ERA people, left Sperry Rand to form Control Data, and in 1959 Mauchly left. The great days of Rand's dominance in the computer business were over.

Control Data Corporation attained success by producing large, fast computers for sophisticated users in the scientific and government community who needed maximum computer power but did not need too much continual help in running their systems. This was a market niche not well covered by IBM. One of CDC's important assets was Seymour Cray, who left to form his own firm in 1973 in order to build supercomputers.

Another market niche was penetrated by Digital Equipment Corporation, which was founded by Kenneth Olsen and Harlan Andersen in 1957. Olsen was a student at MIT in 1950 and had worked on the Whirlwind. He went to work for IBM and helped build the computers for the SAGE project. The market niche DEC started with was that of smaller, affordable computers that were easy to use. This was the PDP series (Programmed Data Processor) which were the first minicomputers. In 1977 DEC announced its first VAX, a machine that rivaled the performance of the smaller mainframes but was much cheaper. The VAX series became popular with research laboratories, especially at universities.

Scientific Data Systems was another firm that was successful in the minicomputer field, but it was purchased by Xerox in 1969, after which it declined. By 1975 Xerox was essentially out of the computer business except for automated office equipment and some peripherals.

In addition to the new firms that were founded on the basis of the technical expertise of specific individuals, a number of established companies other than IBM and Sperry Rand entered the computer field in the fifties and early sixties. These included Bendix Aviation, Burroughs, Marchant, Monroe, Minneapolis-Honeywell, National Cash Register, Hewlett–Packard, RCA, Raytheon, Underwood, Philco, and General Electric. These last three found that they could not compete with IBM, especially after the IBM 360 was announced, and by 1970 had all given up on general computers, although GE still maintains a computer services business and has an important interest in computer-based automation.

Of all the established firms that took up electronic computing, Burroughs has probably had the best continual success in competing with IBM. After merging with Sperry, the resulting firm, Unisys, became the second largest American computer company, a position that is now being successfully challenged by DEC.

American computer technology was born of the demands of war, and its early growth was fed by a steady diet of federally sponsored research along with purchases by government agencies. As the commercial market grew in importance, more firms tried to enter the market. Out of this IBM emerged as a giant with most of its successful competitors being new start-ups rather than long-established large firms. The government support, market competition, and availability of venture capital interacted with a large domestic market to give the United States pre-eminence in the computer industry. On a worldwide basis, this pre-eminence was absolute for many years in spite of great efforts in Europe and Japan to narrow the American lead. Today, the United States is still the leader, but Japan has succeeded in developing an important computer industry second only to that of the United States. Europe has been less successful and relies more on American and Japanese technology than on its own.

In 1950 the level of computer technology in Britain was at least equal to that in the United States. The several groups that built computers after the war had demonstrated a degree of competence of the highest rank. But several factors made it impossible for Britain to keep up. The most important of these was that the country had expended much of its resources on the war and the demands on capital for rebuilding were great. Also, the British governmental agencies, especially the military, did not take a very active role in computer development, although the Ministry of Supply had supported the Ferranti Mark I, which grew out of the Manchester work. The British market was not as large and did not accept computers as rapidly as that in the United States. Ferranti sold only six computers in Britain during the time period that forty six UNIVACs were sold in the United States.

By the late 1950s a number of British firms were producing computers.

Some of these were large producers of electrical equipment, such as Ferranti and English Electric, while others were office machine manufacturers. This latter group was weak technologically and received little government support while the electrical equipment manufacturers were totally oriented to military and scientific needs, so that as commercial demand accelerated in the early 1960s the country was not well positioned to compete with American imports. The English response was a national policy to consolidate the industry through government-sponsored mergers to take advantage of scale economies. This resulted in a single major computer firm, International Computers Limited, which was formed in 1968 and acquired by Standard Telephones and Cables in 1984. This concentration did not preserve the domestic industry. By 1974 only about a quarter of computer installations were from British sources.

The "national champion" policy was followed by other European countries, especially France. The French effort had begun with the work of Louis Couffignal, who had studied the design of binary electromechanical calculators in the 1930s, and of Machine Bull, a large office equipment manufacturer, and a small firm, Societe d'Electronique et d'Automatisme, which actually produced the first French computer. Couffignal had visited both the ENIAC and the Institute for Advanced study in 1946, and Bull engineers had toured both the United States and England to learn the latest computer technology. After a dismal period of governmental neglect and continual loss of market share to foreign firms, the French Government was moved to act when, in 1966, the United States Government denied a proposed export of a Control Data 6600 to slow the development of French nuclear weapons. The French response was to form a new firm, by the merger of the several firms in the computer business, to be the national champion. This firm was named the Campagnie International pour l'Informatique. But this did not increase the French share of its domestic market. In search of a larger unified market, CII cooperated with Philips of the Netherlands and Siemens of Germany to form a trans-European company in 1973 called Unidata. This did not work and in 1975 the French withdrew from Unidata. In the meantime, Bull had become Honeywell-Bull after being bought by American General Electric, who later sold their computer interest to Honeywell.

In 1975, the French decided that an agreement with Honeywell was a good idea, with the result that a new firm was formed which combined CII and Honeywell-Bull operations. This was ultimately 65.1 percent French-owned and called CII-Bull. France has pursued the policy of supporting a single large national champion more vigorously than any other European country.

Germany did not start building any electronic computers until 1955, when a ban imposed by the Allies was lifted. Before this, IBM was already marketing in Germany and in 1956 it was producing the 650 in Stuttgart. In the early 1960s four domestic firms were selling computers in Germany, but this was cut down to two as Zuse was bought by Siemens and Telefunken was merged into AEG to form AEG–Telefunken. But Siemens emerged as a kind of national champion since it was more successful than AEG–Telefunken, which had continuing financial problems. Siemens had developed a line of IBM compatible machines, but these did not include large mainframes. Siemens had relied on an agreement with RCA for these, and after RCA abandoned the computer business the company turned to Fujitsu. Siemens is now a major vendor of Fujitsu mainframes. Germany was somewhat more successful than France. The real success story in West Germany was that of Nixdorf which entered the small minicomputer market in 1968. This was a market segment that the American manufacturers had neglected, and Nixdorf became a major European supplier.

Japanese policy was different. Rather than designate a single firm as a national champion, Japan stressed technology transfer from the United States, government protection from foreign competition, government support of generic research, and government sponsorship of a limited number of firms that were encouraged to compete with each other.

Prewar work on calculating machinery had been pursued at the University of Tokyo and at the Electrotechnical Laboratory. ETL was a government laboratory run by the Communications Ministry. In 1948 the communications part of ETL was placed under Nippon Telephone and Telegraph, which was the new government telephone monopoly, and called the Electrical Communications Laboratory. In 1952, ETL became part of the Ministry of Trade and Industry (MITI), which had overall responsibility for national industrial policy.

Punched card equipment had been built by ETL in 1920, and in the late 1930s the laboratory extended some work on Boolean algebra that had been done at Nippon Electric (NEC). This later led to a relay-based computer which ETL built in the early fifties. At the University of Tokyo, Hideo Yamashita designed a binary statistical tabulator in the late 1930s, and two versions of his design were built after the war by NEC and Fujitsu. In 1951 Yamashita visited some of the computer projects in the United States and attended a UNESCO conference on an international computing facility, thereby bringing information to Japan on the latest computing technology.

Several computer projects were undertaken after the war, but the interest of government and industry in computers did not become serious until after the late fifties. The basis for the execution of policy already existed in the Electronics Industry Provisional Act, which was passed by the Japanese Diet in 1957. The Act identified electronics as a high priority field for the country's industrial development. It authorized direct research and development support, loans to companies, and special depreciation benefits. It even allowed MITI to exempt firms from antimonopoly laws and set up interfirm research and production programs. This strong government support was the first part of Japan's computer strategy.

The second part was to acquire foreign technology so as to become technologically sophisticated as fast as possible. In exchange for letting any

Japanese firm license its patents, the IBM wholly owned Japanese affiliate was allowed to manufacture and market its computers in Japan. This took place in 1961 and was a unique arrangement that was not repeated with any other foreign firm. However, a number of joint ventures between American and Japanese firms were concluded by the mid-sixties, which included Hitachi with RCA, Oki Electric with Sperry Rand, and Nippon Electric with Honeywell. These arrangements permitted the Japanese firms to manufacture and install the products of the American firms.

At the same time, imports were decreased by a tariff increase to 25 percent in 1961 and by tight control of foreign exchange by MITI. The result was that the non-Japanese share of the Japanese market decreased from 82 percent in 1961 to 27 percent in 1982, 22 percent of which was held by IBM.

To further encourage domestic production, MITI established the Japan Electronic Computer Corporation with the charter of buying Japanese computers and leasing them to domestic users. This greatly eased the financial burdens on the manufacturers. To be eligible for purchase by JECC, a product had to have a certain percentage of domestic content, which at first was 20 percent. As time passed this was gradually increased to 75 percent. JECC provided a powerful incentive for the Japanese firms to develop the domestic industry.

An important aspect of the Japanese policy was that no particular firm was singled out for exclusive favorable treatment from the beginning. MITI based its actions towards the companies on the basis of their performance so that the element of competition was retained even as government support became more massive.

In spite of this policy, the Japanese computer industry was in a dangerous position after the introduction of the IBM 360 and 370 series, particularly since Japan had agreed to abolish its import quotas by the mid-1970s in response to American pressure. With its American partners in disarray because of the new IBM machines and the prospect of a more open market, MITI responded by getting Fujitsu and Hitachi to work on a joint project to develop a product line that was IBM-compatible. They got an invaluable boost on this project from Gene M. Amdahl, who had left IBM in 1970 to start his own company. IN 1972 Amdahl needed capital and got Fujitsu to buy 24 percent of his stock by giving them the right to Amdahl's technical information. Since Amdahl had been a senior computer designer at IBM this gave Fujitsu access to the IBM technology, which now had the information needed to produce computers compatible with current IBM machines.

Of course, mere imitation of IBM was not enough. But this was coupled with large-scale support of semiconductor and integrated chip research. The end result was that three major main frame producers survived and did well; Fujitsu, Hitachi, and NEC.

The careful, long-range Japanese policy of importing technology, protecting the domestic market, providing government support, and fostering competition worked. The European policy of developing an indigenous technology with a single national champion did not.

EVOLUTION OF MODERN COMPUTER TECHNOLOGY

The structure of a modern stored program computer is rather straightforward. It is usually described as consisting of a central processing unit, a memory, and input–output equipment. The CPU is the device that performs the actual arithmetic and logical operations. Its speed is a major determinant of the power of the computer. The memory holds the data or instructions with which the CPU works, and the I–O equipment transfers data, instructions, and results between the CPU and memory and the human operators. Actually, there are several kinds of memory and some of it is part of the CPU. A somewhat more detailed specification of computer structure is needed to take this into account and to differentiate among arithmetic, logical, and control functions.

The CPU consists of a unit to perform arithmetic and logical operations, a control unit to direct the communication and transfer of data among the various parts of the machine, and a memory unit, called the main memory, to store the data and programs on which the arithmetic-logic unit operates. The arithmetic-logic unit (ALU) consists of very fast memory units called registers. Each register holds one word of data, a word being defined as a number of binary bits as determined by the design of the machine. For large computers, the most common word size is 32 bits. ALUs can contain several kinds of registers. These include accumulators, where the results of computation are placed, instruction registers, which hold the instruction being executed, and index or address registers for counting and for identifying parts of main memory. Data or instructions are transferred to and from the registers in the ALU by the control unit as needed.

Of course, registers are memory units. But differentiation between ALU registers and units of main memory recognizes the fact that great speed in the ALU is highly advantageous but that most data and instructions can reside in units whose access time is less than that in the ALU. Since very fast access and cycle times are expensive, this permits larger memories to be had for a given cost. In addition to registers and main memory, there is a need for mass storage, in which large amounts of information that is not currently being used can be held. This needs to be cheap since so much of it is needed to hold programs for future access and files that are the result of program operations. Mass storage is normally embodied in magnetic tape, drums, or disks.

The technical devices needed to physically realize this computer structure include sets of bistable devices to hold binary digits, logic circuits to enable instructions to be performed, mass data storage media, and devices to enter data into the machine and read it out. Engineering advances in these devices have been of great importance in driving the remarkable progress in computer technology ever since the ENIAC was first built.

Although the basic unit operations in a computer are simple, there are a large number of them in any useful program. The resulting logical structures are therefore complex, and the way the program is constructed, the methods of addressing main memory, the nature of indexing, and the instructions contained in the logic circuits all have an impact on the effectiveness of the computer. The factors that have to do with the logic and arrangement of the data and instructions constitute software technology, and its evolution also contributed much to increased computer performance.

Stages of computer development are often identified by the electronic technology which they incorporate. First generation machines were based on vacuum tubes. These included the ENIAC and the first group of post ENIAC machines. The second generation used transistors. Solid state devices were used as early as 1950 in the SEAC. These were germanium diodes acting as logic gates. But the first computer to include transistors as active elements was Philco's Transac S-2000 in 1958. Univac followed with its M460, and IBM modified its 709 into the solid state 7090 which was five times as fast as the 709. STRETCH, a supercomputer built for Los Alamos, was the first IBM computer to be designed for transistor electronics, but it was not completed until 1961.

Transistors were a major advance for computer technology because they decreased the size, power requirements, and costs of electronic components by large amounts. When Jack Kilby of Texas Instruments developed the integrated circuit, in which a number of solid state electronic elements were interconnected on the surface of a silicon chip, a trend toward further miniaturization was started that has been the main factor behind the rapidly increasing capabilities and decreasing costs of digital computers. The idea of integrated circuits had been proposed by the Briton W. A. Dummer, and Robert Noyce of Fairchild also contributed a great deal to integrated circuit technology. Kilby's chip was working in 1959 and Texas Instruments supplied a computer to the Air Force in 1961 that used integrated circuits, but the first commercial machines with integrated circuits were shipped by SDS and RCA in 1964. Machines using integrated circuits are called third generation computers. Examples of third generation computers include the highly successful IBM 360, the RCA Spectra 70, and the Univac 1108.

The trend towards larger scale integration has been steady and fast. From 1959 to 1979, there has been a regular exponential growth in the maximum number of components per chip of a factor of four every two years. There is a loose classification of fourth generation computers based on chip densities above 100,000 components per chip (VLSI, very large-scale integration). The fourth generation is more usefully thought of as having started in the mid 1970s when very large-scale integration was combined with other important advances to give the modern high performance computer.

Integrated circuits provided enormous improvements in capacity and speed for main memories. The ENIAC had no main memory and could store only twenty numbers in accumulators, although its function tables gave it a very limited additional memory capability. The UNIVAC could store 1,000 twelve-digit characters in acoustic mercury delay lines designed by Eckert. The electrostatic Williams tube did not have much greater capacity, although it facilitated bit parallel processing because of its random access character. The first breakthrough in main memory was the ferrite core, but it was not until the semiconductor memories made possible by the transistor and large-scale integration that memory became cheap. The memory capacity of even the personal computer is five orders of magnitude greater than those of the first generation computers.

Mass memory storage techniques have also improved but not at as fast a rate. The UNIVAC already had sophisticated tape storage, and rotating magnetic drums were being used in the early 1950s. These evolved into the magnetic input–output and the magnetic disks currently in use.

By the mid 1960s, many of the currently used major concepts of modern computer technology had been initiated and put into at least one machine. Later advances were of a more incremental nature in which the inherent potential of the technologies was realized. This was also true of software advances. The only major software languages to be introduced after 1965 were Pascal (1970) and C (1973). Time-sharing was first used in 1957 for SAGE and was introduced into commercial systems in 1962. Multiprocessor operating systems and virtual memory were used in the Ferranti/Manchester ATLAS in 1961. Even UNIX, which is considered to be a very new operating environment, was in use by 1973. Computer technology therefore took on its present form by the early seventies. A most important innovation of this time was the microprocessor, the so-called computer on a chip, which was introduced by Intel in 1971. Although this was a natural outgrowth of large-scale integration and in that sense was a derivative technology, the microprocessor was the beginning of a new wave of computer technology and computer diffusion. The microcomputer brought powerful computation to people, organizations, and functions that never had it before. It spurred the growth of logical, graphic, and linguistic applications and made it clear to everyone that the computer is not merely a super calculator but a device for manipulating symbolic systems.

A new round of radical and rapid technological advance can be started by new technologies for making chips, such as x-ray lithography to replace optical lithography; faster semiconductor devices such as Josephson junctions; new multiprocessing techniques such as true parallel processing, which goes beyond the array processing that is restricted to matrix or vector type computations; or new software developments such as artificial intelligence. At the moment, these are all in the early stages of development.

COMPUTER PERFORMANCE

The discontinuity in computer power represented by the ENIAC was the beginning of the evolution of the application of electronics to computing that has resulted in even more powerful machines. By any measure, the growth of the power of electronic computing has been the fastest sustained growth of any technological performance to date.

While the general notion of computer power is obvious, a precise measure that enables different machines at different times to be compared is difficult to construct. The overall performance of a computer depends on the basic hardware of memories, central processor, input–output devices, and the architectural arrangement of these parts. The software, ranging from operating systems to applications languages also affects computer performance. Given a system in which these factors are defined, it is easy enough to determine the speed of a specific arithmetical operation, but from the point of view of actual use this is insufficient. In the first place, different problems require a different mix of the basic arithmetic and logical operations, all of which have different times of execution. These include fixed point addition (and subtraction), floating point addition (and subtraction), multiplication, division, comparison, and branching or logical instructions such as "and" and "or." Furthermore, the time it takes to complete a problem depends not only on the time of execution in the central processor, but also on the time the central processor is idle while waiting for instructions and sending out results. The memory capacity is also important, since larger memories can handle large problems without breaking them up into smaller parts. All of this means that, even for a particular machine with a particular configuration, the performance is application-specific. The comparison of computer power for different machines clearly requires great care.

The best and most extensive comparison that permits a quantitative description of the evolution of computer power is that of Knight (1966, 1968, 1973, 1983), who defined a Figure of Merit for computer power as the product of speed and memory factors:

$$P = M^i S \tag{6-1}$$

M is the capacity of the internal memory, and S is a measure of the speed of execution of instructions. Both M and S are composite measures. M depends on word length and the total number of words in memory. Variable length words were weighted twice as much as fixed length words, and M was made dimensionless by making it unity for a memory containing 32,000 words of 7 bits each. The relative weights of the memory and speed components are given by the exponent i in Equation (6-1).

Unlike speed of execution, memory capacity cannot be directly related to performance using technical specifications, so Knight sought the opinion of forty-three engineers and programmers on the relative importance of memory. With this advice, he used the exponent $i = 1/2$ for scientific computing and $i = 1/4$ for commercial computing.

As noted above, the speed is different for different operations, so S must be a composite measure. Knight classified the different operations into five groups. These are:

1. Fixed add (and subtract) and compare instructions
2. Floating add (and subtract) instructions
3. Multiply instructions
4. Divide instructions
5. Other manipulation and logic instructions, including branch, logic, and load register instructions

For a particular applications program, the speed of execution can be determined if the frequency of each of these kinds of operations is known, along with its speed and the time that the central processor stands idle waiting for input–output to take place.

The time it takes to perform the operations for each of these kinds of instructions can be determined from the technical specifications of the particular computer system, but their frequencies and input–output times are program-specific. Knight recognized that scientific computation and commercial applications have very different mixes of basic operations and treated them separately. To determine the frequency of the different operations and the input–output time for scientific work, he ran about fifteen million operations in over 100 problems on an IBM 704 and IBM 7090. For commercial applications, he ran nine programs on an IBM 705. These included inventory control, billing, general accounting, payroll, and production planning. The runs were analyzed for the frequencies of different operations, which were used as weights to compute the execution time, t_c, for each computer in his study. The non-overlapping input–output time, $t_{i/o}$, was also determined, and the speed was defined as

$$S = 10^{12}/(t_c + t_{i/o}) \qquad (6\text{-}2)$$

where the times are in units of microseconds per million operations.

Using these definitions and the technical parameters for each computer in its most common configuration, Knight was able to construct a measure of computer power with the dimensions of operations per second that permitted a comparison of different machines and a description of the evolution of computer performance with time. This was done for 374 computers introduced in the United States from 1944 (the Harvard Mark I) to 1979. The fact that each computer system could be set up in a variety of configurations was met by choosing the most typical configuration. For machines that were

Table 6-1 Values of the Technological Progress Parameter $A(Y)$ From 1952 to 1962

Year	Scientific Computing	Commercial Computing
1950–52	6.469	5.973
1953–54	6.985	6.713
1955–56	7.307	6.980
1957–58	7.668	7.266
1959	7.898	7.311
1960	8.193	7.538
1961	8.605	7.803
1962	8.870	8.167

significantly changed during their product life cycle, the typical configuration when it was first introduced was used.

Along with the calculation of computer power P, the cost, or rather its reciprocal C in seconds per dollar of rental, was also recorded in order to analyze the relation of computer power to cost. In fact, Knight made the power–cost relation the basis of his analysis by showing that the data could be represented by an equation of the form

$$ln(C) = A(Y) - B\ ln(P) \tag{6-3}$$

where A is a constant for any given year but changes from one year to the next, and B is a constant for all years. For a given year, this equation gives a curve that connects cost and performance. This is called the technology curve for that year. The presumption is that there is a certain level of technological state of the art during a given year and that a variety of different computer systems of differing power and therefore differing costs are introduced during that year within the available technology. The technology curve is the cost-performance relation for that technological level. The changes in the constant $A(Y)$ represent technological advance as time passes. Of course, technical progress does not take place in annual steps, but some time grouping of the introduction of new systems must be made.

Actually, Knight performed the analysis for three different time periods and therefore got three different sets of results. For the period 1950–1962 he got the result that for scientific computing,

$$ln(C) = A(Y) - 0.510\ ln(P) \tag{6-4}$$

while for commercial computing,

$$ln(C) = A(Y) - 0.459\ ln(P) \tag{6-5}$$

with the constants $A(Y)$ for different years having the values shown in Table 6-1. (For early times, the data were grouped into multiple years since there would otherwise not be enough points for a significant statistical analysis.)

Table 6-2 Values of the Technological Progress Parameter $A(Y)$ From 1962 to 1966

Year	Scientific Computing	Commerical Computing
1962	6.823	7.441
1963	7.095	7.826
1964	7.238	8.164
1965	7.645	8.627
1966	7.811	8.991

For the period from 1962 to 1966 the results for scientific computing were

$$ln(C) = A(Y) - 0.322\ ln(P) \tag{6-6}$$

and for commercial computing,

$$ln(C) = A(Y) - 0.404\ ln(P) \tag{6-7}$$

with the constants for different years being given in Table 6-2.

The third time period (Cerveny and Knight 1983) covered the years from 1972 to 1979 but analyzed only scientific computing. The results were similar to those of the earlier studies except that the proportionality factor between the logarithm of cost and the logarithm of power in a given year was -0.668, indicating that the cost benefits of increasing computer power were not as great in these later years.

These data illuminate two important issues: the relation between power and cost at one period of time and the increase in computer performance at constant cost as a function of time. This is readily seen by rewriting Equation (6-3) in the form

$$P = K(1/C)^{1/B} \tag{6-8}$$

where the constant $K = \exp(A/B)$.

For the 1952 to 1962 data, $1/B$ is 1.96 for scientific computing and 2.17 for commercial computing. For the 1962 to 1966 data, $1/B$ is 3.11 for scientific computing and 2.48 for commercial computing. But for the period 1972 to 1979, $1/$B was only 1.5. This means that computer power at any given time increases much more rapidly than cost. For the early set of data, power varies as the square of the cost while the economies of scale were even higher in the 1962 to 1966 period. However, after 1972 economies of scale were less. These data approximately verify "Grosch's Law," a speculation advanced by H. Grosch in the late 1940s that computer power increased as the square of the cost. This speculation was not treated seriously, and it was even proposed that increased power was purchased at an increased cost per unit power.

It must be pointed out that the costs in these data are only the rental costs

Table 6-3 Annual Percentage Growth Rates in Average Computer Power at Constant Cost for Scientific and Commercial Computing

Year	Scientific Computing	Commercial Computing
1952.5	87.5	178.3
1954.5	44.0	92.3
1956.5	51.5	95.0
1958	38.0	85.3
1959.5	78.3	173.4
1960.5	124.3	184.0
1961.5	68.1	213.5
1962.5	132.7	189.2
1963.5	55.9	220.2
1964.5	253.9	246.2
1965.5	67.5	213.5

of the equipment. Large-scale systems are attended by high additional costs for maintenance, systems programmers, operators, consultants, and the entire administrative and operating structure of a computer center. These costs are not included in the Knight analysis. While there is clearly a substantial economy of scale in hardware costs, the data give no information on the degree of economy of scale when total costs are considered. One more technical point must be mentioned. At any particular time there is a limit on how powerful a computer can be built and as power is increased beyond a certain limit, further increases become very expensive. Thus, even for hardware costs alone, there is a limited range over which the power law can be expected to hold.

For two different years, labeled with indices 1 and 2, the relative change in computer power for constant cost is given from Equation (6-8) as

$$[P(2)-P(1)]/P(1) = \exp\left([A(2)-A(1)]/B\right) - 1 \qquad (6\text{-}9)$$

The growth rates in computer power for constant cost calculated from Equation (6–9) is given in Table 6–3.

The growth rates in Table 6–3 are remarkable; no other technology has progressed so rapidly. Over a period of a decade and a half the smallest annual growth rate was 44 percent and at times reached rates well over 200 percent per year. For all years, the rate of increase in computing power for commercial applications was substantially greater than for scientific applications. There is a major difference between scientific and commercial computing. Scientific computing is much more arithmetic intensive, while commercial computing requires much more input–output activity. In early machines, input–output was slow and its rate of improvement through the years analyzed by Knight has been very large.

These data are not accurate representations of the growth in the state of the art for computing. They deal with all computers and not just those that have important improvements. In any given year a variety of computers are built, many of which do not exceed the power of earlier computers. The "technology curves" represented by Equation (6–3) therefore describe average practice, not best practice. But a measure of the evolution of the state of the art can be obtained from Knight's data by examining those computers that have achieved a performance greater than any that have gone before. This was done for both scientific and commercial computing using the combined data of Knight from the ENIAC in 1946 to the CDC Cyber 203 in 1979 for scientific computing, and to the Philco 200–213 in 1966 for commercial computing. The results are shown in Figures 6–1 and 6–2.

The log plots show that the data fit exponential curves fairly well up until the mid 1960s, with a growth rate of maximum performance for scientific computing of 72 percent per year and for commercial computing of 63 percent per year. Note that the difference between scientific and commercial computing is much less for the state of the art than it was for the average performance data. In fact, for the state of the art, scientific computer power grew faster than commercial, whereas for average performance the reverse was true. For both scientific and commercial computing the average performance grew more rapidly than the state of the art, suggesting that as time went by more systems were being produced that had closer to maximum capabilities.

The data for scientific computing indicates that the rate of increase in state of the art computing power started to decrease after 1965 and, in fact, was much less than in the earlier years.

The evolution of the cost of state of the art computing from Knight's data is displayed in Figure 6-3, which shows the computer power per unit cost for scientific computing for successive machines whose performance is greater than any that have gone before. The graph shows that the best performance–cost ratio increased by an average of 61 percent per year for twenty years. Not only has the technical progress been phenomenally high, but the cost of computing has decreased at a corresponding rate.

Knight's work for the years after 1966 includes only three points on state of the art computing, and it would be useful to have more data. Unfortunately, there is no comparably detailed analysis of computer performance after 1967. However, some data collected in the 1980 "Science Indicators" of the National Science Foundation shed some useful light on the evolution of computer performance up to 1980. The first set of the NSF data lists the maximum addition rate available for different years. From this, the successive maximum addition rates, greater than any that went before, were selected and plotted in Figure 6-4. These data exhibit about a 47 percent growth rate up to 1967 but then level off to much smaller growth rates.

The second set of NSF data represents an attempt to measure growth in the state of the art of computing by constructing a performance index from a linear combination of speed, memory, and cost with weights assigned to

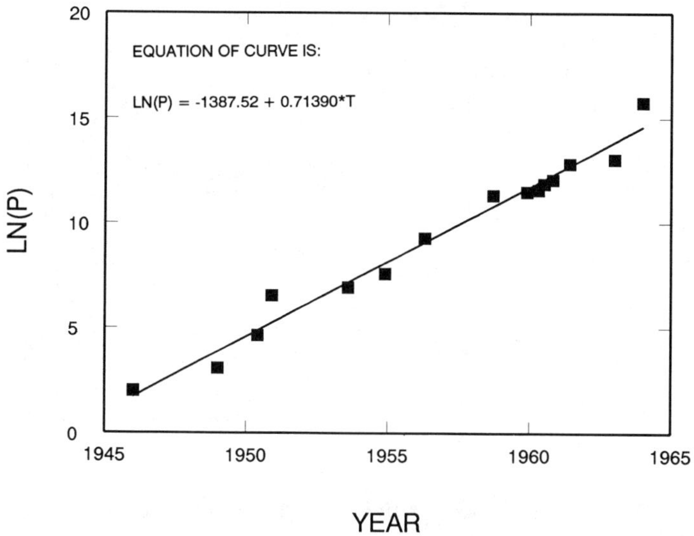

Figure 6-1 Maximum scientific computing power.

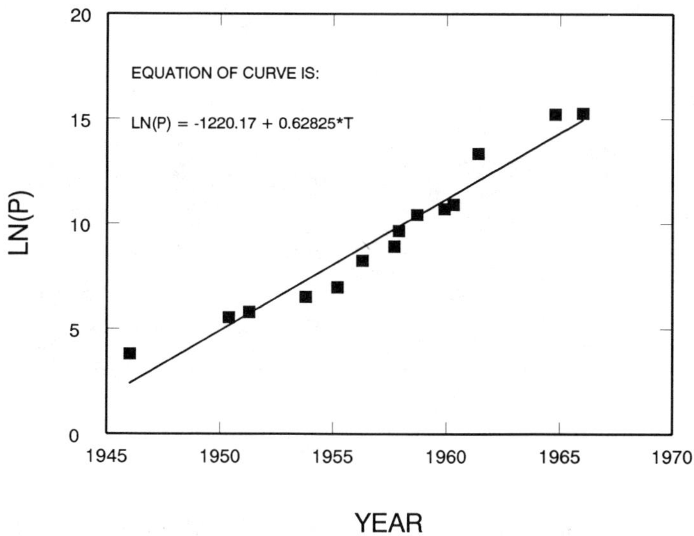

Figure 6-2 Maximum commercial computer power.

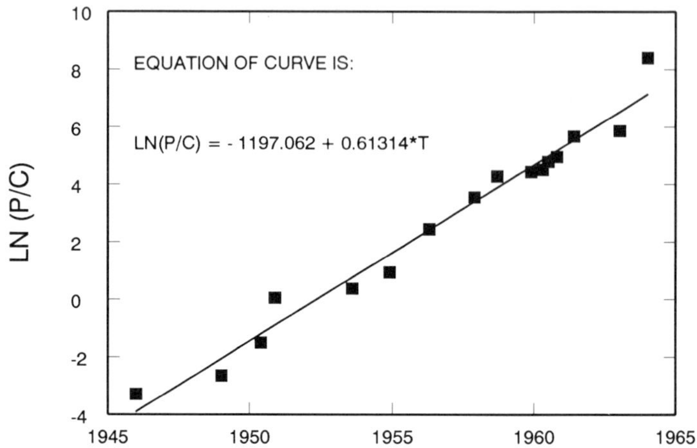

Figure 6-3 Computer power per unit: cost in operations per second per penny.

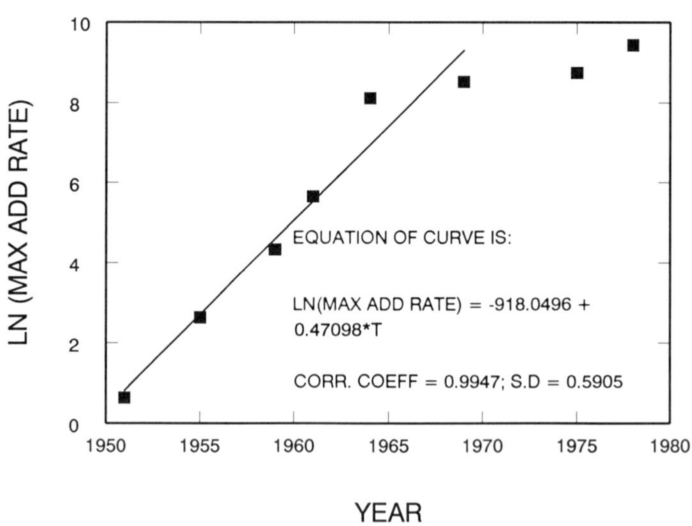

Figure 6-4 Maximum addition rate for computers with higher addition rates than any previous computer.

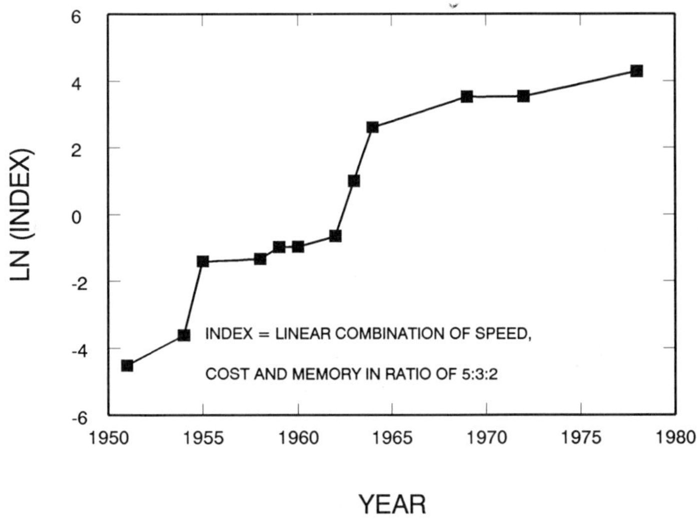

Figure 6-5 Successive maxima of computer power: NSF index.

these three components of 5:2:3. The speed itself was a composite measure for different operations, and the weights were chosen on the basis of expert opinion. A graph of these data is shown in Figure 6-5. This confirms the leveling off of performance after the mid 1960s.

There are two problems with the NSF performance index. The first is that it combines cost and technical performance data so that it is hard to consider these two factors separately. The second is that a linear combination is not always the best way to combine different performance parameters into a performance index. A linear combination is as good as any, providing a zero value for the parameter does not totally destroy the functionality. But for computers, either a zero speed or a zero memory makes the computer useless. A more appropriate measure is a logarithmic combination that takes this into account. I constructed such a logarithmic measure from the NSF data on speed and memory, giving the speed twice the weight of the memory. (This is completely equivalent to Knight's index, which was written as a product instead of a logarithm.) The results are shown in Figure 6–6. The results are similar to those of Figure 6-5 but more regular. It is of interest to consider the speed and memory separately, and this is done in Figure 6-7. This shows that the rate of increase in both speed and memory leveled off at about the same time in the mid-sixties, but while the speed increased regularly before then, memory capacity of state of the art computers was nearly constant from 1955 to the early sixties.

The improvements in costs also display a leveling off starting in the mid-sixties. This is shown in Figure 6-8, which is a logarithmic plot of the number

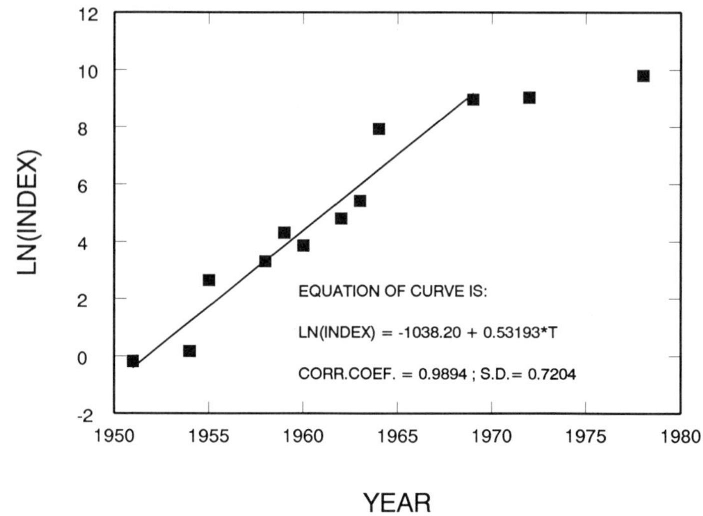

Figure 6-6 Successive maxima of computer power: L.A.G. index using NSF Data: $ln(\text{index}) = [2ln(\text{speed}) + ln(\text{memory})]/3$.

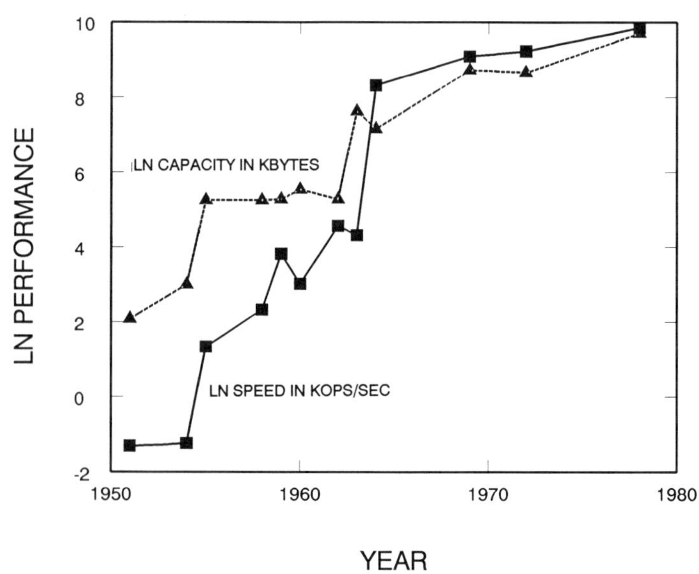

Figure 6-7 NSF data for memory and speed of computers.

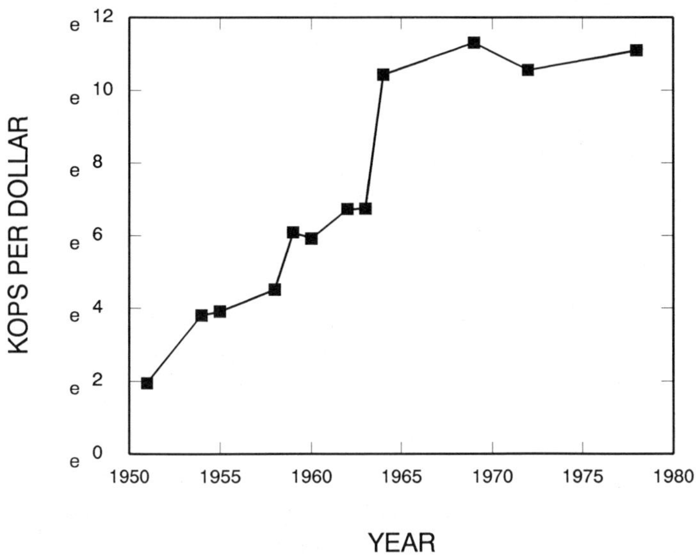

Figure 6-8 NSF data on computer cost for computers with maximum performance in KOPS per dollar.

of operations per unit cost for the state of the art machines defined by the NSF index.

It seems as if the general conclusion to be drawn is that the first twenty years of computer development was extremely rapid, but later the rate of improvement decreased, at least up to 1980. In fact this is true for the parameters illustrated, but these do not give a complete picture. The plots presented here represent the progress that has been driven by developments in electronics, particularly solid state microelectronics. But advances in software and architecture have also had dramatic effects. The latest indications on the future of parallel processing, for example, promise enormous improvements in the speed of problem solving. This is not simply the primitive parallelism of the von Neumann IAS machine, in which the bits forming a word were processed simultaneously, but involves the simultaneous use of a number of central processing units coordinated to work on different parts of a problem at the same time. Parallel architecture illustrates the great advances in performance that can be achieved by routes other than improved electronics.

DIFFUSION OF COMPUTERS

Many new technologies get their start in the market place by filling a particular niche in which the improved functionality makes the new technology

Table 6-4 Percentage Share of Computer Installations in the U.S. Economy

Year	Government	Education	Production	Services
1953	53.8	19.8	18.7	7.7
1959	22.9	6.1	43.1	27.9
1966	13.4	6.4	39.4	40.8
1969	12.1	6.3	39	42.6

Source: Phister 1979, p. 444.

worth its price, which is initially quite high. This was certainly the case with the computer, for which the first market niche was the government. It was the government, particularly the military, which had both the need and the resources to support computer development and then be the initial market for computers. This role of the Federal Government was developed during World War II and was not unique to computers. Radar, numerically controlled machine tools, aircraft design, communications satellites, semiconductors, and microelectronics are all examples of new technologies that were fostered by the Federal Government, which provided both R&D funds and an initial market.

All of the first set of post ENIAC computers built in the United States were either sponsored by or sold to agencies of the United States government. This includes those that went to universities. Even BINAC, which Eckert and Mauchly sold to Northrop Aviation, was designed for use in a bomber and was thought of by Northrop as a prelude to smaller computers that could act as guidance systems in missiles. The first UNIVAC was built for the Census Bureau, although a number of later copies were sold to industry. Also, IBM entered the computer market with orders for eighteen 701s by DOD customers.

The initial importance of government and the later growth of nongovernmental markets is shown in Table 6–4. This table gives the percentage of computer installations in various segments of the economy. The percentages in the second column refer only to computers installed in government agencies and do not include computers installed elsewhere but financed from government sources. Computers installed in educational institutions are listed separately from other services because most of them were funded by government agencies, especially in the early years. The total impact of government funding before 1960 is more closely represented by the sum of the government and educational installations than by the governmental installations alone. The great importance of the government market for the early growth of the computer industry is evident from Table 6–4.

The fourth column, labeled "Production," includes manufacturing, mining, construction, and agriculture; that is, the goods-producing sector of the economy. The last column represents all services (except education) including finance, insurance, real estate, wholesale and retail trade, transportation, and communications, as well as electric and gas utilities. The first penetration of computers into the nongovernment market was primarily into the production sector, although as time passed the service sector acquired a greater share of the installed computers. Of course, even in the production sector computers were used primarily to provide ancillary services such as accounting, cost control, and payroll. Even today these applications dominate, and only a small fraction of installed computer power is used to control production processes. The greatest recent growth in the use of computers directly related to the production process has been in computer-aided design.

The penetration of computers into the civilian sector was fairly rapid. By 1959 less than one-third of computer installations relied on government. By 1980, only 5.64 percent of all general purpose computers were in the hands of the Federal Government (Flamm 1988a, p. 110).

The quantitative analysis of the spread of computer usage is complicated by most of the problems that beset the study of the diffusion of far-reaching, basic technologies. The populations among which the diffusion of computers took place are multiple and heterogeneous. Government, industry, and academe all have different kinds of computer needs, different funding methods and economic constraints, and different modes of organization. Furthermore, not all institutions within a given classification are alike. The computer requirements of small firms are very different from those of large ones; service industries such as banking have different needs than manufacturing; research universities use computers differently than teaching colleges; the applications of computers in the Air Force are not the same as in the Census Bureau. These differences give rise to different determinants of the rates of diffusion, and a great deal is lost when these differences are ignored. Nevertheless, if some measure of the diffusion of computers in the entire country could be devised, an interesting picture of the rate of overall computerization would be obtained. But even this is complicated by the fact that computers themselves are not homogeneous products. The three main categories of computers are: mainframe general purpose computers, minicomputers, and microcomputers. Up until 1970, most computer installations were of the large general purpose type. But in 1965 the PDP-8 was introduced at a price of $50,000, which made it available to those that could not afford the high cost of mainframes, and after 1970 ever-increasing numbers of minicomputers were installed. By the early 1970s the microprocessor chip was available and this gave rise to the microcomputer. Clearly, mainframes, minicomputers, and microcomputers address different markets and different needs and therefore have different diffusion histories even though the remarkable advances in technology have blurred the lines of separation among these categories.

To illustrate the spread of computer technology, we have chosen to present the diffusion of mainframe general purpose systems. The reasons for this choice are that mainframes were the most important computers for many years, the right kind of data are available, and the performance data analyzed above refer to such systems. The diffusion data is taken from Phister (1979)

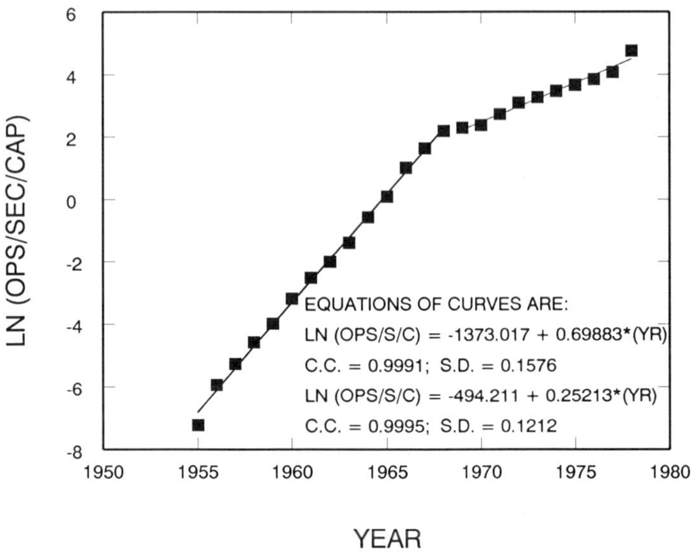

Figure 6-9 The diffusion of computer capacity in the United States in OPS per capita.

and is in terms of the total installed internal processor capacity in operations per second for general purpose systems. This is a better measure than the number of systems or the dollar value of systems or the number of institutions with computer systems. A logarithmic plot of the diffusion of computer capacity is given in Figure 6-9. The diffusion parameter was taken to be operations per second per capita of United States population. Actually, the rate of diffusion was so much higher than the rate of population growth that a plot of total ops/sec would have the same general characteristics as Figure 6-9. The diffusion rate is accurately described by two exponential curves, one up to 1968 with a growth rate of nearly 70 percent, and one from 1969 to 1978 with a growth rate of 25 percent. The break from a higher to a lower diffusion rate is very marked and occurs at approximately the same time as the decrease in the rate of improvement in computer performance and the decrease in the rate of cost effectiveness noted in the previous section.

SUMMARY

The development and growth of the electronic digital computer and its use represents one of the most remarkable examples of technological change in history. It is the prototypical twentieth-century technology, not only in its content but also in the driving forces of its evolution. It was scientifically based, involved extensive research and development, and had large-scale government support.

The computer was born of the exigencies of war, but it is important to note that before this a cadre of people with a profound interest in large-scale computation already existed. These people were research scientists and engineers with an academic orientation and an interest in solving scientific problems. Initially, they got only modest support for their work, but as the wartime needs for computation became evident the support became generous. World War II called forth the computer, but this was possible only because there were people who had been working on computation for purely academic reasons.

Although a number of people were working on the precursors of the modern computer both in the United States and Europe, it was a single event that gave the major impetus to the growth and adoption of computer technology. This was the completion of ENIAC by the Moore School team, which demonstrated both the feasibility and utility of electronic computing. The further development of computer technology was greatly facilitated by the efforts of highly respected and charismatic figures such as Turing and von Neumann, whose stature did much to generate interest and support for computers.

Starting with the ENIAC, there was a twenty-year period of rapid exponential growth in performance, cost effectiveness, and diffusion. All three of these exhibited concurrent rapid change, and all three started to level off to a lower growth rate at about the same time. This is an excellent example of the relations among performance, cost, and use. The data alone cannot yield the conclusion that decreases in the rate of improvement of computer performance and in the rate of cost effectiveness caused a decrease in the rate of diffusion, but there is no doubt that these three effects moved in parallel. The data given above refer only to large general purpose systems, and by the mid-sixties these had been adopted by most large firms. In 1967, nearly all firms with more than 1,000 employees had large computers. This might lead one to conclude that the simultaneous decreases in diffusion, performance, and cost effectiveness were coincidental. However, the diffusion parameter we have used is the installed processor capacity, not the number of firms with systems, and the history of corporate computer usage is one of continual expansion of computer power after the first computer is installed. The question of causal relations is therefore still open, but it is likely that the decrease in performance improvements and cost savings was at least partially responsible for the decrease in the diffusion rate.

The patterns of innovation and technological change for the computer are similar to those for many other important technologies. Precursors existed in terms of mechanical and electromechanical machines; a technological base that made electronic computing possible was in place; an important special event (the ENIAC) started a chain of further development; champions pressed for electronic computing in the face of skepticism and opposition (Eckert and Mauchly at the Moore School and Goldstine in the Army); diffusion started in a particular market niche (government) and then spread to other market segments; the diffusion was accompanied by major technological improvements and price decreases; complementary technologies (solid state electronics) played a major part in technological advances; differentiation of the technology took place for previously unanticipated applications (minis and micros, CAD–CAM, process control, robotics, telecommunications, and so on). This process is common to many large-scale, fundamental technologies such as the steam engine, electric lighting, and the airplane.

7
The Industrial Transition

MEASURES OF INDUSTRIALIZATION

The differences between industrialized and nonindustrialized societies are dramatic and extensive. In industrialized societies the per capita income, life expectancy, literacy rate, and education level are all much higher than in nonindustrial societies. The ease of communications, travel, and transport makes the population mobile and permits the widespread distribution of goods. Industrial societies devote a much smaller fraction of economic effort to agriculture and a larger fraction to consumer goods, health care, and education. Above all, industrial societies exhibit self-sustaining growth. That is, they enjoy a continuing increase in per capita wealth and income which results from the continual reinvestment of a portion of the national income to expand knowledge and to improve the technological base of the economy. An ever-increasing stream of inventions and innovations feeds into an expanding economy to provide goods and services at lower costs and to perform functions that were previously impossible. An industrial society is dynamic, not only in its technique but also in its social and institutional forms; it is continually recreating itself as technological change is reflected in, and complemented by, cultural and social changes.

In a nonindustrial society agriculture is the major economic activity and work is done primarily by human and animal muscle. The production of goods is carried out by hand and depends on craftsmanship. Energy sources and many important materials are of vegetable origin, wood is used both as a structural material and for fuel, textiles are handwoven from natural fibers, dyes are extracted from organic sources, and metals are used sparingly. The primary nonorganic materials used in any quantity are clay, stone, and glass.

In industrial societies, on the other hand, manufacturing is more important to the economy than agriculture. Mineral fuels provide cheap and abundant energy, metals, particularly iron and steel, are widely used, and mineral or synthetic materials ranging from dyes to textiles replace or supplement products of animal or vegetable origin.

A primary characteristic of an industrial society is that it uses massive amounts of energy and materials to produce large amounts of goods. The age-old economic problem of producing enough goods is largely solved in the sense that a known methodology exists to produce many goods in any quantity desired. One of the reasons that this is possible is that, in the process of industrialization, the productivity of agriculture is improved to such a great extent that it no longer absorbs the major effort of the population.

An overall result is that human beings acquire an ever greater control over their environment through technological means. Artificial lighting liberates people from the daily schedule imposed by nature; central heating and air conditioning provide climate control of indoor spaces; dams and irrigation mitigate and control the effects of natural water supplies; the effects of flood, drought, and earthquakes are eased in ways that are simply not possible for societies that do not have large-scale technological infrastructures.

There are many parameters that can be chosen to measure the degree of industrialization of a society, and a number of these will be used here. While there is no single parameter that can give an unequivocal measure of the extent to which a society has progressed, each of them gives some information and describes a specific aspect of the process by which it has moved from a preindustrial condition towards full industrialization. Actually, the choice of measures is, at least partially, a definition of what is meant by industrialization. It must therefore be kept in mind that no list of quantifiable measures can be complete, and that it should be supplemented by qualitative, descriptive factors that cannot be captured by numerical data.

The most obvious manifestation of the transition from an agrarian to an industrial society is the technological achievement. It is true that a frame of mind, a world view, must exist that leads people to develop technology; it is true that social and political institutions must foster innovation, or at least not inhibit it; it is true that the economy must attain some minimum level before any resources are available for investment in new technology; certainly a sufficient number of people must have an adequate level of technical and entrepreneurial skills. All these factors and more are critically important. But the industrial transition consists of the development and spread of new technology, and measures of technological change are central to the description of industrialization. Such measures would include the use of mechanical power, speed of travel and communications, extent of transport and communication systems such as railroads and telephones, use of synthetic materials, replacement of vegetable and animal products by mineral products, appearance of new products, and the increasing use of automation.

The most direct result of the growth of industrial technology is a change in the economic status of a society. Technology and economics are inseparable, and as the techniques of dealing with matter and energy change, so does the economy. For all societies that have been successfully industrialized, there has been a continuing increase in Gross National Product. This increase has outstripped the population increases that also attend industrialization, so that an increasing GNP per capita invariably attends a society in which new technology is being introduced and spread. Economic parameters such as GNP per capita, real wages, and the costs of goods are therefore an important part of the description of industrialization.

The avowed purpose of technology is to satisfy human wants and to contribute to the material well-being of the population. The extent to which this occurs can be determined by examining such parameters as consumption, longevity, causes of death, working hours, availability of leisure time, and educational level.

Clearly, a description of industrialization requires the analysis of technical, economic, social, and personal factors. Not all of these kinds of measures can be examined here. Instead, we will focus on relatively large-scale measures such as the overall extent of industrialization and the degree of mechanization.

There is an inherent tension in the introduction of new technology that has always made technological progress a rather discordant process. In the long run, industrialization and the introduction of new technology have proven to be beneficial, at least in material terms. Life spans have gone up, back-breaking labor has been eased, more kinds of goods at ever-decreasing real prices have been available, and material want has been overcome. But these benefits are not uniformly distributed even in the long term, and special political or social action must be taken to avoid the natural tendency for benefits to concentrate in ever fewer hands. Moreover, in the short run there are winners and losers, with the losers often far outnumbering the winners. Thus, a new technology may increase productivity, which ultimately benefits society by making goods cheaper. But in the short run the new technology produces unemployment among a particular group of workers. Even though technological advances create more jobs in the long run, they may hurt large numbers of people who might have to wait a long time for the benefits of technology to match their loss.

Technological unemployment is not the only part of technology's negative face. This is at least temporary and can be addressed with social and political programs that can attract widespread support. But there are two other kinds of effects that are harder to deal with. These are weapons developments and environmental effects. There is some evidence that we will be able to survive weapons technology. It is at least possible to contemplate a gradual growth in political maturity so that massive self-destruction is not inevitable. Because the threat is less immediate and less visible, it will actually be more difficult to control the environmental effects, although the technical means to do so are available.

Most of the description of industrialization presented here is of the "optimistic" kind, since it reflects a continual increase in those factors we all value. Indeed, the record of technological change is one of providing continuing benefits. There have always been dire warnings of the dangers of technology. In the short run it is indeed true that technological progress is often accompanied by terrible results, at least for some people. The suffering of the workers in the early British textile mills, the blight of company towns, the economic tragedy of worked out mining regions, and the recent loss of jobs in the manufacturing industries are all very real manifestations of the short-range dislocations induced by advancing technology. But cataclysmic predictions of technological doom have been proven wrong by history. It is hard to understand why such predictions have not simply been blown away by the evidence for the long-term beneficial effects of technology. Nevertheless it is important to keep in mind that we have an ever-increasing capacity to affect our environment. In fact, as more and more people are brought into the technological process, disastrous large-scale environmental effects become more probable. These negative factors are a part of industrialization and should be kept in mind when the effects of industrialization are considered.

WORLD INDUSTRIALIZATION

The rapid and continuing industrialization of the world since the middle of the eighteenth century is an obvious fact that has been the subject of extensive historical study. Some measure of the rate of this process would certainly illuminate our understanding of industrialization as a systemic phenomenon. But quantitative analysis of the process is extremely difficult and reliable estimates of the rates of industrialization are scarce. A number of statistical economic studies have been made that show how income, wealth, and national products have grown, but much less is known of the growth of industry itself, as distinct from its economic consequences.

A very useful analysis is that of Bairoch (1982), who computed the level of industrialization of a number of countries as a function of time from 1750 to 1980. To arrive at a sensible index, Bairoch started by defining industry to be "the totality of those activities whose object is to produce, or to transform, material goods, excluding all those activities described as agriculture (including everything up to the harvest)." He further defines "manufacturing industry" to be all of industry except agriculture, mining, building, electricity, gas, water, and sanitary services. Bairoch's index consisted of an aggregation of the output of his definition of manufacturing industry, each industry weighted by the unit prices of the finished products. The amount of work needed to construct such an index is certainly large and the limitations of existing data make substantial inaccuracy inevitable. Bairoch has estimated the possible margins of error. These errors range from about 10 percent to about 50 percent, the older data being more error prone. This is a surprisingly low uncertainty given the problems of collecting good data even in our own time.

The first point to note in Bairoch's analysis is that in 1750 the level of industrialization was about the same for all the countries of the world. The United Kingdom was in the lead even then, but not by much. For most of the world, the difference between the level of industrialization from that of

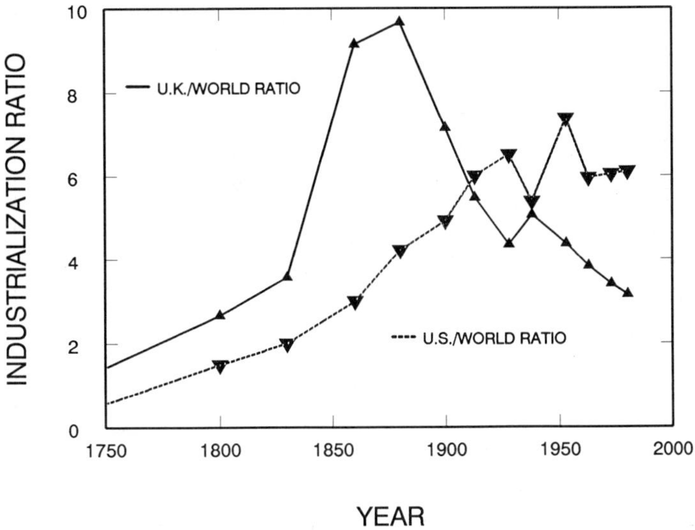

Figure 7-1 Ratio of industrial level of U.S. and U.K. to the rest of the world.

the United Kingdom was small and within the expected margins of error. The United States was the furthest behind, with an industrial level only 40 percent that of the U.K., a fact probably due to its colonial status as a supplier of raw materials and the overwhelming importance of agriculture. But this point must be stressed: In the middle of the eighteenth century, the different regions of the world were approximately equal with respect to industrial output per capita of population!

The rapid industrialization of the U.K. after 1750, followed by that of the other Western nations, radically changed this rough equality and totally changed the economic balance of the world. This is shown by the data for the industrial levels and their growth rates for a number of countries from 1750 to 1980. These are given in Tables 65 and 66 in Appendix 3. A graphical representation of the enormous shift in relative industrial positions is shown in Figure 7-1, which displays the ratio of industrial level of the United Kingdom and the United States to that of the rest of the world. These ratios were close to unity in 1750, but by 1900 the U.K. was ten times more industrialized than the world as a whole. The United States soon caught up, and in the twentieth century surpassed the United Kingdom as the world's greatest industrial power, achieving a position that was six to seven times that of the world as a whole. But even this does not illustrate the dramatic shift of industrialization with sufficient clarity because, in referring the ratio to the industrial level as a whole, the level of industrialization of the advanced countries is included in the denominator. Figure 7-2 shows the ratio of the

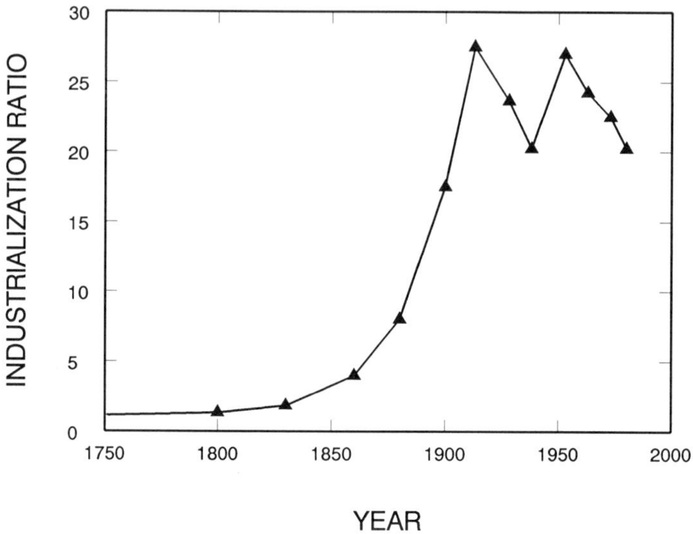

Figure 7-2 Ratio of industrial level of developed to underdeveloped countries.

level of industrialization of developed to underdeveloped countries (using modern definitions of which are the developed and underdeveloped countries). This graph shows that the gap between developed and underdeveloped countries started to increase rapidly in 1825 and became very large by 1925. Since 1925, the ratio of industrial level of rich to poor has leveled off and may even be decreasing. This conclusion is supported by other detailed economic studies (Kravis, Heston, and Summers 1982). The data for Figure 7-2 were computed from Bairoch's tables and are listed in Table 67 of Appendix 3.

It is remarkable that the growth of industrial level over the long term can be described by simple exponential curves. We computed the best exponential fit to Bairoch's data by the least squares method applied to the logarithms of industrial level versus time. The results are given in Table 68 of Appendix 3. In all cases but two, the fit to an exponential was quite good. For Japan and the Soviet Union, two exponential curves were needed to represent the data.

Figure 7-3 shows the data and the fitted curves for the United Kingdom and the United States. In the United Kingdom industrialization grew at a rate of 1.6 percent per year from 1750 to 1980, while in the United States it grew at an annual percentage rate of 2.3 percent per year, a rate sufficiently greater than 1.6 to allow the U.S. to overtake the U.K. by 1925. Germany and France did not start to increase their industrial levels until after 1800, but then they also started on a course of exponential growth.

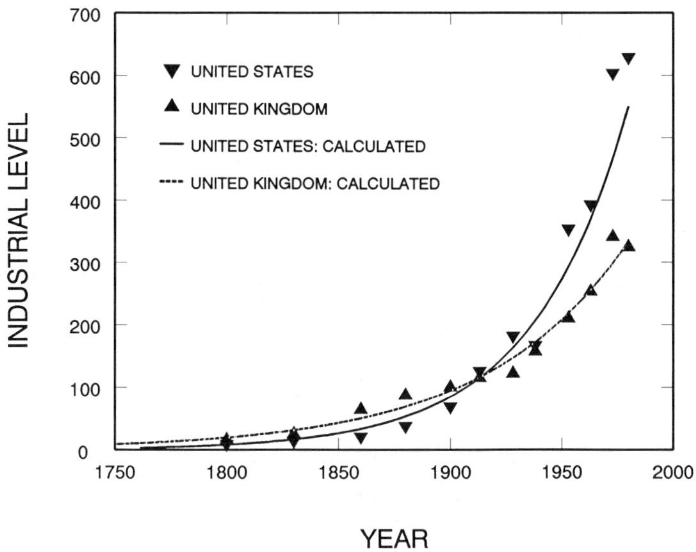

Figure 7-3 Industrial level of the United States and the United Kingdom as a function of time.

Both Japan and Russia show somewhat different courses of development. The exponential growth of Japanese industrial level began about 1870, after the Meiji Restoration of 1868. The new government, formed under the authority of the Emperor Meiji, dismantled the feudal system, introduced money reform with a centralized currency, built modern infrastructures, including railroads, and fostered the development of modern industry. The result was an exponential growth rate of 2.5 percent per year as shown in the early part of the graph of Figure 7-4. This growth was interrupted by Japan's defeat in World War II. But after the war, growth resumed at the incredible rate of 8.4 percent up to the mid-1970s, making Japan one of the world economic superpowers. There is evidence that this rate slowed down in the eighties.

The growth of industrial level in Russia (up to 1913) and the Soviet Union (1928 to 1980) also exhibits two exponential regimes. A growth rate of 1.7 percent per year was achieved from 1860 to 1913, but after the war the growth rate was 5 percent per year. The growth rate tripled under the Communist government relative to that under the Imperial Government, even when including the less industrialized non-Russian parts of the Soviet Union in the comparison. Again, recent evidence shows that there are significant decreases in the rate of growth.

It is remarkable that the industrialization of the world as a whole also follows an exponential curve, as shown in Figure 7-5. The growth rate for

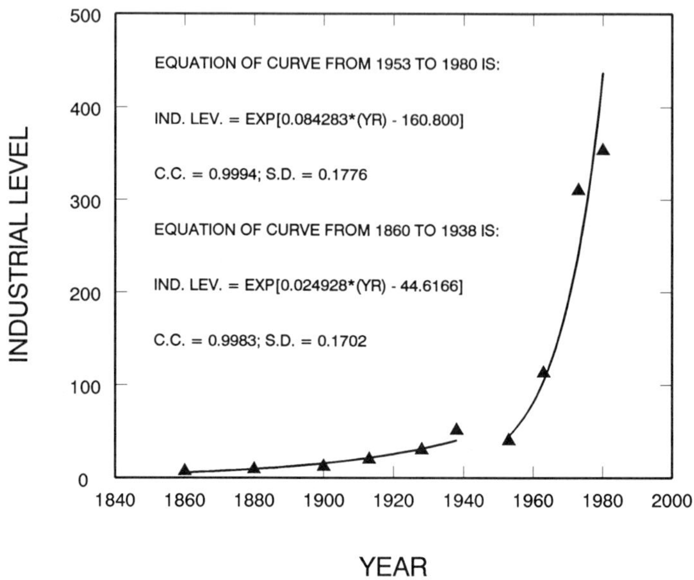

Figure 7-4 Industrial level of Japan as a function of time.

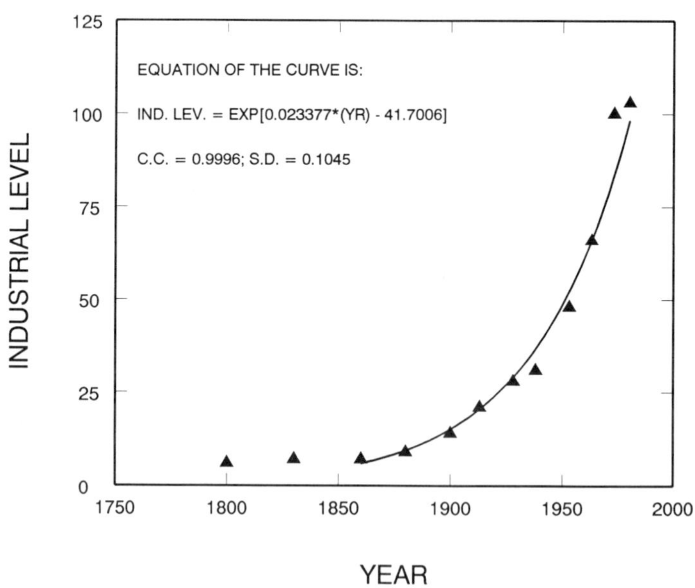

Figure 7-5 Industrial level of the world as a function of time.

the world since 1860 has been 2.3 percent per year. This is a result of both the continuing increase in industrial level for industrialized countries and the increase in the number of countries in which industrial growth has been initiated.

The exponential growth of industrial level described by Bairoch's data represents a fundamental, systemic phenomenon. When modern industrialization begins, it grows exponentially. But it must be emphasized that this is a large-scale description of industrial growth. Bairoch's data points are far apart and cover long time spans. The temporal resolution of the data and the calculated curves is poor. On a finer time scale it would be clear that the course of industrialization is much more irregular, with significant fluctuations both above and below the large-scale curves as the techno-economic system is affected by such factors as wars, depressions, rapid technological change, and governmental economic policies. Nevertheless, beneath all the chaotic activities of an industrial economy and all the ups and downs that so profoundly affect economic well-being, there is an inexorable and regular overall movement of increasing industrial output.

The continual increase in industrial output is the direct result of technological change. If technology were to remain static, there would be a maximum possible output consistent with the established techniques, and the growth of industrial level would not be possible. (Actually, this is not quite true, since industrial output could increase even for a fixed technology by shifting labor and capital from other economic sectors, such as agriculture or services, to manufacturing. But even so, the increase that could be achieved this way is limited and eventually growth would stop.) Technological advances yield increases in productivity which permit and induce higher levels of industrialization.

A number of economic historians have pointed out that the rate of economic growth, as measured by Gross National Product per capita, is generally faster for countries that start industrialization later. Rostow (1980) has summarized and extended this work, using historical GNP per capita data for fourteen countries. The rational for this is that latecomers to industrialization have the earlier results to build on. The later the takeoff date, the greater the store of technology, the larger the markets, and the more capital there is in existence. Thus, a country has more to start with if it industrializes late.

Although Bairoch's data is for industrialization, not income or GNP, it is of interest to test this idea with his data. A graph of the growth rate of industrial level versus the date of takeoff is shown in Figure 7-6. The growth rates are those for the exponential fits to Bairoch's data as listed in Table 68, Appendix 3. Until the twentieth century, the growth rates do not show an increase with takeoff dates. In the twentieth century, however, there is a definite and substantial rise with time.

Economic historians have not merely proposed that the later a country industrializes the greater its rate of progress. Their thesis is that while this

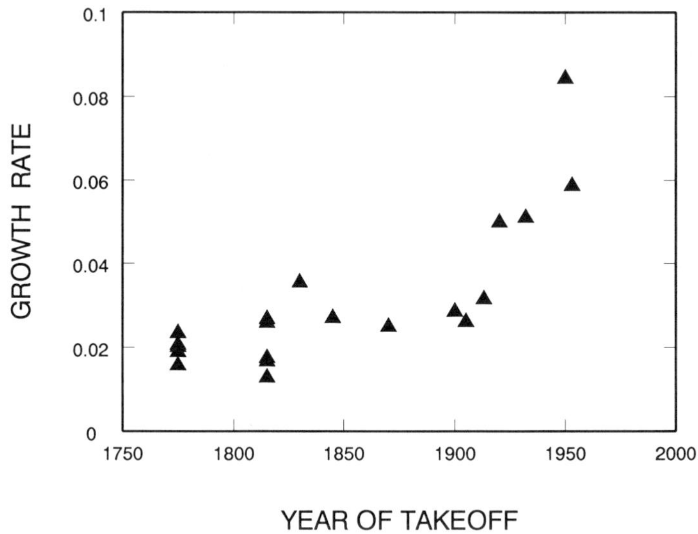

Figure 7-6 Growth rate of industrial level versus date of industrial takeoff.

is true, it is also true that the growth rate increases at the beginnings of industrialization, rises to a maximum, and then decreases. At different times different countries will therefore be the world economic leaders, the gap between rich and poor continually narrows, and ultimately all nations that industrialize will have a high and approximately equal per capita income.

The picture that emerges from this is that the long-range growth of a technological economy is best represented by an S-shaped curve rather than an exponential. The S-curve may level off and approach a maximum or continue to rise, but growth does not remain exponential indefinitely. This, of course, is meant to be a systemic statement in that it is valid in the large over long time scales and is often perturbed by special circumstances that may be specific to individual countries.

The relationship between economic growth and industrial growth is very strong but very complex. The detailed impact of industrialization on national income depends on a great many factors. If, for example, agriculture is a major component of the economic sector, growth of national product will lag behind industrial growth until agriculture absorbs new technology in sufficient amounts to yield large increases in productivity. Or, if a country's economy depends on raw material exports, it could have a high per capita income even in the absence of a high level of modern industry. In such a case, a country would not have achieved the condition of self-sustaining growth, but this would be masked by raw material income. But these are idiosyncratic factors; in general, rising industrialization induces rising gross national product, and rising GNP can only be sustained by increasing industrialization sustained by continually improving technology.

ECONOMIC GROWTH AND INDUSTRIALIZATION

The impact of technological change on economics is dramatically illustrated by the fact that before the Industrial Revolution per capita wealth grew slowly when it grew at all. After the Industrial Revolution, growth was considered to be the norm, and now a country's economic well-being is regarded as being defined by the growth rate of wealth per person, not merely by the level of that wealth. Technology has redefined the economic perspective.

In his book on the relation of population to economics, Schultz (1981) points out that until the Industrial Revolution the Malthusian principle was indeed operative. When population grew, there were more people than could be supported by the land, wages went down, hardships increased, birth rates went down, and death rates went up. When population decreased sufficiently, wages went up, life eased up a bit, and the population started to rise again. This was the result of the classical Malthusian dilemma in which population could increase geometrically while increases in agricultural output were limited by land availability which, even in favorable cases, could only increase arithmetically. The result was a continual oscillation between poverty and relative sufficiency. Self-sustaining growth was not even a dream.

Of course, the Malthusian principle was not completely valid when applied over time spans of centuries. For millennia there had been gradual improvements in agriculture and continuing technical innovations, and in the Middle Ages there was an acceleration of technical advance. It was these improvements that created the wealth and the conditions that made the Industrial Revolution possible when it finally did take place. But the improvement was very gradual and barely perceptible to the general population. Now, of course, growth is measured, and expected to be positive, on an annual basis.

Rostow (1980) has studied the systemic relation between economic and technological growth in broad terms. He identified a critical date for each of a number of countries, called "take-off," when a country industrializes sufficiently to enter the regime of self-sustaining growth, and defined a time period, called the "drive to technological maturity," during which a country effectively transforms to a modern technological industrial base. From an analysis of earlier work and a detailed consideration of data on real GNP per capita which he assembled in an appendix to his book, Rostow drew the following conclusions:

1. The later a country starts on the road to industrialization, the faster the growth of GNP per capita.
2. During the course of industrialization, the rate of growth of GNP per capita is significantly lower in the later years, after the country is tech-

> nologically mature, than in the earlier stages when it is striving towards complete industrialization.

These are important statements and they deserve a critical analysis. Both statements can be supported by an attractive rationale. If a country starts late, it can take advantage of the experience gained by others. It does not have to invent and develop the technological processes and devices from scratch. For the late starter, technology is available at a much lower cost than for the pioneers. Also, examples exist for its application, and for the organization, financing, engineering, and marketing that must accompany successful technological growth.

A point that is often overlooked is that latecomers can find it easier to market technological products. Such products can only be purchased if sufficient wealth exists above the subsistence level. Wealth of this sort is found in industrialized nations. Latecomers therefore can find ready-made markets in which they can compete because their labor costs are generally lower than those in developed countries. The early starters, on the other hand, had to create markets so that their growth was necessarily slower.

The rationale for the ultimate slowing of growth in later stages of industrialization is that success creates conditions that do not favor rapid change. Large-scale technological substructures are difficult to modify; pervasive technologies are accompanied by large institutions with many people who have vested interests in the status quo and therefore resist change; wealth and the ease it brings are not conducive to risk taking; incentives for new technologies are much greater for the poor than for the rich nations. In short, people in a wealthy society are not interested in changes which may upset their lives and threaten their positions.

It is, of course, the systemic growth rates we are interested in here. Gross National Product often changes erratically on a year-to-year basis but shows quite regular trends when examined over long time periods. This is an example of short-term fluctuations, which are determined by continually changing short-term conditions superimposed on a systemic growth which is determined by the fundamental characteristics of the technosociological system. The fluctuations sometimes make it difficult to interpret the growth rates reported by economists, who often report rates for specific time periods without specifying the methods of obtaining these rates. Fortunately, Rostow's tables of GNP per capita for different countries enabled me to perform calculations of my own to test the hypotheses that latecomers grow faster and that older technological societies slow down. These tables listed the GNP per capita in terms of 1967 U.S. dollars and different countries could therefore be compared to one another.

Since I was searching for systemic growth rates, I plotted the data for each country as a function of time, using both linear and logarithmic coordinates. In this way, it was possible to test whether or not any of the data followed an exponential law from which long-range growth rates could be computed. The result was that for every country long periods of time could be found

Table 7-1 Systemic Growth Rates of GNP per Capita From Data of Rostow

Country	Take-Off Date	Time Frame	Annual Growth Rate (2)
United Kingdom	1783	1790–1900	1.50
		1947–1973	2.30
France	1830	1826–1913	1.99
		1946–1971	4.22
Germany	1840	1851–1912	1.61
		1950–1971	5.66
United States	1843	1789–1860	1.36
		1861–1929	1.92
		1947–1973	2.18
Sweden	1868	1870–1930	2.06
		1932–1971	3.61
Japan	1885	1886–1939	1.75
		1946–1972	8.56
Russia-USSR	1890	1880–1946	1.40
		1951–1972	4.21
Italy	1895	1905–1940	1.02
		1946–1971	5.19
Canada	1896	1871–1928	1.53
		1935–1971	2.76
Australia		1862–1930	0.58
		1931–1971	1.96
Argentina	1933	1901–1971	1.08
Brazil	1933	1931–1961	2.67
		1962–1972	3.05
Turkey	1933	1934–1953	1.13
		1954–1972	3.03
Mexico	1940	1896–1928	1.31
		1933–1970	3.24
Taiwan	1953		3.32
		1961–1972	6.71
South Korea	1961	1954–1964	2.43
		1965–1972	8.36

for which the GNP per capita was well described by an exponential curve. In fact, it was a general rule that most of the data for any country displayed systemic exponential growth, although it was often true that it took more than one exponential to fit the data over different time periods. For the twentieth century, there was a large perturbation of the systemic trends resulting from the depression of the thirties and World War II. Even so, by the 1950s the GNP per capita for most countries had settled down to an exponential trend that could be closely extrapolated from the trend before 1930. The results of fitting data in various time periods are shown in Table

7-1. These time periods were chosen because the exponential curves fit them with a correlation coefficient of at least 0.9999 and with smaller standard deviations than other possible choices of time periods. The take-off dates in this table are those of Rostow (1980).

The time periods listed in the third column are those regimes in which it makes sense to regard the GNP per capita as growing in a systemic way according to an exponential growth law. The times that are omitted are periods in which fluctuations were so large that no regular trends could be discerned. The period from 1930 to 1946 for the United States, for example, was dominated first by the Great Depression and then by World War II, so that there was a significant drop in GNP per capita, followed by a large rise as the country went to a war economy. The depression and the war were such large-scale disturbances that systemic economic trends were obliterated. However, it is interesting to note that after the war the GNP per capita settled down to a level and growth rate that was close to what it would have had if neither the depression nor the war had taken place.

Many countries had GNP curves that fluctuated greatly during the period from 1930 to the mid-forties, since the depression and the war were worldwide events. However, different countries had different time periods over which the nonsystemic fluctuations took place. Great Britain, for example went through a half century (1900–1947) during which there was considerable stagnation, growth was sporadic, and systemic trends were not evident. Sweden, on the other hand, had minimal disruption in its systemic trends even during the depression and the war.

The remarkable result of this table is that it does not support Rostow's theses. There is no discernible trend of increasing growth rates with time of take-off, and there is no evidence that growth rates decrease when a country becomes highly industrialized and technologically mature. On the contrary, the data show that the growth rate for *all* countries is higher in later than in earlier years. In fact, the most conclusive statement that can be made from these data is that growth rates throughout the world increased after World War II. Also, the data after the mid-forties show a regular, systemic growth on a worldwide basis, the fluctuations being minor in comparison to the major disturbances of the Great Depression and the war.

POSTWAR WORLD ECONOMIC GROWTH

The increased growth rates for later years that emerged from the analysis given above suggests that World War II marked a significant change in the world economy. This is indeed the case. For two centuries high technology and its attendant sustained economic growth was restricted primarily to Western societies. That is, Europe and some of the countries arising from European colonization had a monopoly on advanced technologies that brought them ever increasing living standards while the rest of the world grew slowly or not at all. The only exception to this was Japan, which had made a national decision to emulate the West and to industrialize. The modern technological and economic successes of Japan have their roots in national Japanese policy established in the nineteenth century, and Japan was an industrial country well before the War. But the rest of the non-Western world was still economically and technologically backward.

The years since the war have seen a dramatic change in this state of affairs. Many non-Western nations are developing a technological and industrial base and are undergoing rapid economic growth as a result. The most important study documenting this fact is that of the United Nations International Comparison Project directed by Kravis, Heston, and Summers (1982). The 1982 report contains their basic data, which was used by Kravis and Lipsey (1984) and Lipsey and Kravis (1987) to analyze world growth. Recently, the data have been updated and refined by Summers and Heston (1984, 1988), who also made the data available in machine readable form (for spreadsheets) on three diskettes. The result is that a measure of real income per capita is now available for each country in the world for the period 1950 to 1985. This is a monumental statistical achievement and will act as a base for studying the world economy for many years.

The importance of the International Comparisons Project is that, for the first time, it provides a sound comparison of the per capita income in different countries by determining what can be purchased in that country by a unit of the local currency. This was done by the use of extensive lists of goods and services. In each country, the unit price of each item on the list was obtained in the local currency. Using a weighted average, the real costs of goods and services to the residents could be compared among countries. This is far superior to the use of official currency exchange rates to convert prices from one country to another. Exchange rates are influenced by currency speculations, international and domestic political factors, world trade, and economic policies. As a result they sometimes change rapidly and are not a reliable measure of comparative purchasing power. The ICP provides a stable, realistic measure of national income based on the amount of goods and services that can actually be purchased with the local currency.

The measure used in the ICP was Gross Domestic Product. The Gross Domestic product includes the total economic output generated within the territorial boundaries of the country, whether it is produced by residents or nonresidents. This is different from the Gross National Product, which is the output generated by residents whether the output is produced at home or abroad. For the large, advanced countries, the difference between these two measures is small, but it can be significant for smaller, lower income countries.

Growth rates of the GDP per capita for the periods after World War II have been computed from the ICP data and reported by Kravis and Lipsey (1984) and Lipsey and Kravis (1987). To compare countries at different levels of development, they used the following definitions:

Industrial countries: United States, Canada, Australia, New Zealand, South Africa, Japan, Austria, Belgium, Denmark, Finland, France, Germany, Iceland, Ireland, Italy, Luxembourg, Netherlands, Norway, Spain, Sweden, Switzerland, United Kingdom.

Centrally planned economies: Bulgaria, Peoples Republic of China, Czechoslovakia, German Democratic Republic, Hungary, Poland, Romania, Soviet Union.

Developing market economies: These were too diverse to consider as a group so they were further classified as follows:

1. Extremely low income countries: Benin, Burundi, Central African Republic, Chad, Ethiopia, Gambia, Guinea, Lesotha, Madagascar, Malawi, Mali, Niger, Rwanda, Sierra Leone, Somalia, Togo, Uganda, United Republic of Tanzania, Upper Volta, Zaire, Afghanistan, Bangledesh, Burma, Nepal, Haiti. (These are the 25 countries with the lowest GDP per capita.)
2. Oil exporting countries: Algeria, Indonesia, Iran, Iraq, Kuwait, Libya, Nigeria, Saudi Arabia, Venezuela.
3. Relatively industrialized countries: Hong Kong, Philippines, Rhodesia, Swaziland, Singapore, Syria, Malta, Portugal, Yugoslavia, Argentina, Brazil, Chile, Dominican Republic, Mexico, Nicaragua, Peru, Puerto Rico, Uruguay. (These derived at least 20 percent of their 1975 GPD from manufacturing.)
4. Other middle income countries with developing market economies that did not fit under the above classifications.

A classification based completely on per capita Gross Domestic Product would lump together all the developing market economies except the 25 lowest (item 1) as middle income market economies.

The Kravis–Lipsey analysis verifies the conclusion we made from Rostow's data in that, as a whole, economic growth rates were significantly greater after World War II than at any other time. On page 9 of The Conference Board Report (1987), Lipsey and Kravis present a table comparing real growth rates of industrialized countries, using the data of Maddison (1982) for the periods before 1950. A simplified version of this, in which growth rates of real GDP per capita after 1950 are compared to those for the first quarter of the century, is shown in Table 7-2. The last column is the ratio of the later to the earlier growth rates. Comparisons of the postwar period to prewar periods other than 1900–1925 lead to the same conclusions.

That postwar growth was not restricted to the industrialized nations is shown by the data in Table 7-3, taken from Kravis and Lipsey (1984).

In view of the fact that until this century, and indeed until World War II, self-sustaining economic growth was restricted primarily to a few countries of European origin, this table displays an astonishing result. Only 7 percent of the countries (with only one percent of the world population) had negative growth. Only twenty-four countries, or 5 percent of the population, had growth less than one percent. More than half of the world's people (54.7

Table 7-2 Growth Rates of GDP per Capita for Selected Industrialized Countries (in Percent per Annum)

Country	1900–1925	1950–1984	Ratio
United States	1.73	1.91	1.10
Canada	1.03	2.41	2.34
Japan	1.03	5.73	5.56
Belgium	0.90	2.72	3.02
Denmark	1.51	2.65	1.75
Finland	1.40	3.60	2.57
France	1.07	3.33	3.11
Germany	1.15	4.19	3.64
Italy	1.42	3.75	2.64
Netherlands	1.19	2.62	2.20
Norway	1.70	3.10	1.82
Sweden	2.17	2.50	1.15
United Kingdom	0.54	1.94	3.59

Source: Lipsey and Kravis 1987.

percent) were caught up in growing economies that exhibited more than 3 percent per year growth.

The widespread and large increases in growth rates indicate that the world economy experienced a major change after the war. Part of this is certainly due to the new technologies of transport and communications whose developments were accelerated in response to wartime needs. The electronic computer and microelectronics were particularly important technologies, which transformed the nature of business by enabling large amounts of information to be processed. The number and complexity of transactions that could be handled was thereby greatly increased and business decisions could be made with a wealth of information that was not previously available.

An idea of the difficulty of determining systemic growth rates can be obtained by comparing the postwar growth rates computed from Rostow's

Table 7-3 Distribution of Growth Rates per Annum of GDP per Capita for Countries of the World and for World Population

Growth Rate in Percent	Percent of Countries	Percent of Population
Negative	7	0.9
0–0.99	17	4.1
1–1.99	26	27.1
2–2.99	31	13.2
3–3.99	24	44.5
>4	21	10.2

data (Table 7-1) with those given by Lipsey and Kravis (Table 7-2). In general, there is agreement on the ordering of rates of growth among nations, but the Lipsey–Kravis rates are consistently lower, being from 67 to 88 percent of the Rostow rates. Part of this is due to the fact that the Rostow data only went to the early seventies while the Lipsey–Kravis data extended to 1984. Also, the Rostow data were for Gross National Product while the Lipsey–Kravis data were based on Gross Domestic Product. But the most important origin of the difference is probably the fact that the Lipsey–Kravis data were determined from a comparison of actual domestic prices, in domestic currency, of a wide range of goods and services. The International Comparison Project provides the best available data for computing and comparing national economic growth rates. The conclusion that a major increase in economic growth rates took place throughout most of the world after World War II is born out by both sets of data.

The variability of the computed growth rates according to the time period chosen for the calculation was investigated by Kravis and Lipsey (1984, 1987). They concluded that there were indeed significant variations. In the group of industrial countries, for example, the computed annual growth rates of the per capita income GDP for the periods 1950–1963, 1963–1973, 1973–1984 were 3.1, 3.95, and 1.74 percent respectively. Throughout the world, the growth rates during the 1950s and 1960s were larger than in the 1970s and 1980s. But beyond this variation, there was no obvious period in which growth was stagnant while being extremely high in other periods.

CATCH-UP

At the end of World War II the United States stood supreme. Throughout the twentieth century American technology, industry, and business was dominant, and the American economy was the most advanced and richest in the world. The war solidified the position of the United States as the pre-eminent economic, and therefore political, power in the world. But after the war, the new phenomenon of worldwide growth brought most countries into the era of self-sustaining progress. Furthermore, the growth rates were not the same everywhere and many countries experienced growth rates that were higher than that of the United States. The result was that there is now greater economic equality among the nations of the world than at any time since the Industrial Revolution created the great disparity between the rich and the poor nations. The quantitative aspects of this remarkable transformation of the world economy have been greatly illuminated by the data from the International Comparison Project. The analysis presented here is based on that data.

There are two separate aspects of the catch-up phenomenon. The first is that the industrialized countries are catching up to the United States in per capita income. The second is that the developing countries are catching up to the industrialized countries.

Table 7-4 Catch-up of Some Industrial Countries to the United States

Country	1950	1984
Italy	23.9	66.2
Germany	37.6	88.9
Netherlands	45.1	75.8
France	47.5	83.2
Belgium	54.0	81.2
Denmark	55.9	89.4
U.K.	57.4	73.4
Japan	17.1	76.3

Source: Lipsey-Kravis data; real GDP per capita; U.S. = 100.

The first point is evident in Table 7-2 and even more obvious from the data in Table 7-4. (Lipsey and Kravis, 1987). This table lists an index of the real GDP per capita for some industrialized countries relative to that of the United States in 1950 and 1984. For both years, the index is chosen so as to assign a value of 100 to the per capita GDP of the United States.

The differential growth rates given in Table 7-2 led to the dramatic changes shown in Table 7-4. In 1950, Denmark was the best off relative to the United States, having a per capita GDP 55.9 percent of the U.S. value. The GDP per capita in Italy was only 23.9 percent that of the U.S. By 1984, Italy had advanced to 66.2 percent of the U.S. position. The other industrial countries did even better. Keep in mind that, in constant 1982 dollars, the per capita GNP in the United States went from $7,935 to $14,771, an increase of 86 percent. The figures in Table 7–4 therefore represent an enormous increase in the wealth of the industrial countries.

A measure of the relative economic positions of the nations of the world can be constructed from the data of Lipsey and Kravis (1987), who listed growth rates for the GDP per capita for the periods 1950–1963, 1963–1973, and 1973–1984. By assuming that the GDP per capita was a smooth exponential function of time during each of these periods, I computed the ratio of GDP per capita in 1984 to that in 1950. This gave a measure of the improvement that was not strongly dependent on annual fluctuations and permitted the comparison of different economies. The results of this calculation are given in Table 7-5 for national groupings and in Table 7-6 for some individual countries.

According to Table 7-5 only the centrally planned economies are catching up to the industrialized market economies, since they increased their per capita GDP by a factor of 3.54 whereas the industrialized market economies increased theirs by a factor of only 2.65. The oil exporting nations were the only other group to match or surpass the performance of the industrialized market economies, with a ratio of 2.68. The picture that emerges from Table 7-5 is that the advanced countries have, by and large, maintained their lead

Table 7-5 Ratio of per Capita GDP in 1984 to That in 1950

Group	Ratio
Industrial	2.65
Low income	1.61
Middle income	2.44
Oil exporters	2.68
Relatively industrialized	2.49
Other	2.25
Centrally planned	3.54

Source: Computed from data of Lipsey and Kravis (1987) for national groupings.

with only the centrally planned economies showing any tendency to catch up. The poorest countries did show a 60 percent improvement, but this is much lower than that of the richest countries so that the gap between the richest and the poorest continues to grow.

But the data in Table 7-5 are highly aggregated and a great deal of variation among the countries within each group is hidden. These data were therefore supplemented by that for some individual countries and listed in Table 7-6.

The performance of the advanced market economies was quite varied, and if the German and Japanese results are set aside, the growth of many of the developing economies compares favorably with that of the advanced industrial societies. Thus, while several advanced nations have improved

Table 7-6 Ratio of Per Capita GDP in 1984 to That in 1950

Country	Ratio
U.S.	1.92
Canada	2.27
France	3.11
Germany	4.18
Italy	3.59
U.K.	1.93
Norway	2.87
Sweden	2.34
Japan	7.07
India	1.77
Brazil	3.33
Korea	5.82
Mexico	2.60
China	5.68
USSR	3.14

Source: Computed from data of Lipsey and Kravis (1987) for some countries.

their relative position, so have large numbers of developing nations. This is confirmed by the data in Table 7-3, which shows that large numbers of countries with a large portion of the world population were caught up in rapid economic growth.

THE FIRST TRANSITION

The eighteenth century saw the beginnings of the greatest transformation of the human condition the world has ever seen. Only the transition to settled agriculture even remotely compares to the Industrial Revolution in depth and extent. Earlier societies certainly achieved a degree of wealth based on agriculture and commerce and reached great heights of cultural and artistic expression. But these achievements were based on gross inequalities, by which the mass of the population lived at subsistence levels while the higher aspects of culture and comfort were the property of a few. Of course, there were periods of relative prosperity. But these could not be sustained in societies that could not produce beyond the abilities of preindustrial technology.

It is remarkable that of all the great civilizations that emerged over a period of about ten millennia since the spread of agriculture, only one became a high technology society. At first sight, it is surprising that this started in Great Britain, an island nation of about six million people. Historians have advanced a number of reasons why the Industrial Revolution took place in Britain rather than elsewhere. Some of the more important reasons that have been proposed are:

1. Britain, particularly England, was already relatively rich in the seventeenth and eighteenth centuries even before the Industrial Revolution. Compared to the rest of the world, and even to Western Europe, Britain was economically well off. There was a sufficient surplus for an important export business and enough people lived above the subsistence level to yield a significant national savings. In fact, the preindustrial per capita income of Britain was considerably larger than that of some modern underdeveloped nations.
2. Britain had been favored by a long period of good harvests, which resulted in an increased income, more jobs, and greater demand for manufactured goods.
3. Britain was a single economic unit with few barriers to internal trade. France, on the other hand, had six different economic areas, while in Eastern Europe every little principality was an independent economic power with its own taxes, tariffs, and regulations.
4. Contracts and the honoring of contracts were more highly developed in Britain than elsewhere. Arbitrary seizures of property by nobility and clergy was much more restricted in Britain. Also, industry and commerce were viewed as legitimate and even desirable pursuits in Britain. This

was in strong contrast with Continental Europe, where government or military service, or the Church, were considered the only worthy occupations. All this gave Britain an environment in which industrial advance was welcome.

5. Britain's coastline and waterways provided a natural transportation system by which raw materials and goods could be transported cheaply, thereby making expanded markets possible. This, along with the availability of natural resources such as coal and metallic ores, was effective in the integration of local economies into a national economy.
6. Her colonies not only gave Britain a source of wealth that could be invested in industrialization, but also provided large markets for her manufactured goods.
7. There was a considerable number of skilled workers in Britain who were familiar with mechanical devices, knew how to extract and process metals, work with glass and leather, and were accustomed to industrial work. These included miners, smiths, joiners, millwrights, tanners, and clock makers. There was a well-developed textile industry whose workers were familiar with simple machines. In short, there was a cadre of people that made rapid industrialization possible. The numbers and abilities of these skilled workers were enhanced by both the demands of industry and a surprisingly widespread availability of technical education.
8. The seventeenth and eighteenth centuries were ages of an optimistic scientific rationalism in Britain, as exemplified by the career of Isaac Newton. The intellectual climate was right for a spurt of technology and entrepreneurship.

These are among the conditions that have been cited as being the causes of the Industrial Revolution. As always, historical causality is very hard to sort out and I leave you to do it yourself, starting with the references in the bibliography. If you do this, you will find that the Industrial Revolution was aided by these factors and also that its roots go back to at least the Middle Ages.

It is easier to note the changes that accompanied, and indeed constituted, the Industrial Revolution. Landes (1969) has chosen three general categories of change as being essential. These were the use of mineral energy sources along with the invention of heat engines, the use of new mineral raw materials, which replaced many materials of organic origin, and substitution of human handwork by machines. The most important of these categories was the development of new energy sources that were concentrated and plentiful. In the beginning, this was coal, which was later followed by oil, natural gas, and nuclear power. Before the advent of mineral energy sources, animal and human muscles were the main sources of motive power and wood was the main source of heat. Land transport was animal-powered, farming was done by men and animals with little mechanical help, animals were used in mills, factories, and mines where they shared the labor with human beings.

Windmills and water power were well developed but were limited in both amount and flexibility. Decent sites were at a premium, and no river that could generate power escaped the installation of water wheels along its length. There was just not enough rapidly flowing water to satisfy the needs of an expanding industry. The iron industry, for example, was continually short of waterpower for its bellows, hammers, and mills. The average iron works only operated for about 40 weeks per year and much less than this in times of drought. Other industries were similarly constrained by the limited availability of power. A man turning a crank could generate 0.08 horsepower and a horse about 0.5 horsepower (at a walk). The average eighteenth-century water wheel could generate about 2 to 5 horsepower; the best of them seldom could put out more than 10 horsepower. It was coal and the invention of the steam engine that changed this. By 1790 there were steam engines that generated nearly 50 horsepower, and by 1900 engines of 1,000 horsepower and more were in use in electric power stations.[1] Furthermore, the location of industrial activity was often dependent on the location of adequate water power until coal-fired steam engines came into use. For the first time a highly concentrated source of energy could be transported so that industries could locate according to the dictates of market and business factors, not according to the availability of power.

The steam engine was first developed in response to a specific problem. The use of coal as a heat source had greatly increased even before the 1700s because of a growing shortage of wood, so coal mining was already an important industry before the use of steam engines. But the flooding of mines by underground water was a serious barrier to the efficient exploitation of coal that could only be overcome by pumping out the water. In fact, not only coal, but other mines were plagued by the presence of water that had to be pumped out. This was a major problem in the Cornish copper and tin mines, which became prime markets for steam engines. So steam engines were first used for pumping water from mines. But as the engines became more efficient, cheaper, and smaller (per horsepower), and as their utility became evident to more people, their use spread to other industries. At first, it was common to use the engines to pump water to a height from which it could be used to drive water wheels. This was a conservative step in which the steam engine continued to be used as a pump and the industrial power mechanism continued to be that of water power. But with the invention of the parallel bar mechanism, Watt's engine provided a simple means of transforming the reciprocating action of the piston to rotary motion. This aided the direct use of steam power, which then was applied in many industries and in both land and sea transportation. The diffusion of steam followed a typical course for major technologies in that it first penetrated a specific

---

[1]The numbers in this paragraph were taken from "Power to 1850" by R. J. Forbes, pages 148–167 in Singer et al., *A History of Technology*, 1958.

market niche for which it was essential. Then, as it proved its worth and became more efficient, it diffused into other markets until it was in widespread use. The steam engine also illustrates the close connection between diffusion and technical performance. Early steam engines had low efficiencies and were restricted to reciprocating motion. The success of the Watt engine as a power source for pumping water from mines encouraged, and indeed enabled, further improvements, the most important of which were increases in efficiency and rotary motion. By 1800, the firm of Boulton and Watt had produced at least 490 engines. Watt made his first two engines in 1776, so there was clearly a rapid increase in the use of mechanical power in the last quarter of the eighteenth century. Not all of this was in mines. The diffusion of steam into other applications spurred further developments such as the steam ship, the steam locomotive, and the steam turbine. To this day, steam is a critical element in prime movers, particularly for the generation of electricity, although, for transport, it has largely been supplanted by heat engines burning petroleum products.

Minerals transformed not only the nature of energy sources but also the materials that were used by industry. The most celebrated example of this was the growing use of iron and steel. The important preindustrial materials for machines and construction were wood, stone, and clay. But there could not be any great degree of industrialization that depended on such materials only. In 1770 British iron production was only 32,000 tons. By 1800 this had expanded to 281,000 tons.

The steam engines, machinery, factory buildings, bridges, and rails that were the obvious material realizations of the Industrial Revolution could not have been made without low cost, abundant iron. But the use of new mineral materials was not restricted to structures and machines. The textile industry, for example, was transformed by the invention of machines, but also by the development of inorganic materials for bleaching, cleaning, and dying. Sulfuric acid and hydrochloric acid were particularly important. Chlorine-based bleaches were far superior to sunshine or buttermilk. As another example, industrial alkali, both in the sodium and potassium varieties, were obtained from the ashes of plants; potassium from wood ash and sodium from saltwort or kelp. Such a source of supply could not begin to keep up with the growing demand. But the Leblanc process of converting salt to sodium carbonate made alkali available from mineral sources. The production of soaps and glass could then be expanded at a low cost. These early beginnings of industrial chemistry ultimately grew to the point where practically none of the materials in use in an industrial society have a direct animal or vegetable origin. The major exceptions are textiles and leather, and even these constitute only a small fraction of textiles and leather-like goods.

The increasing availability of mineral energy sources and materials was accompanied by an increased use of machines that replaced hand labor. Initially, the greatest impact of this mechanization was felt in textiles, especially the cotton industry. The flying shuttle and the power loom in weaving, the water frame, the jenny, and the mule in spinning, the carding machine, the cotton gin, and engraved copper roller printing completely revolutionized textile production. The resulting ability to meet an exploding demand for textiles with the cheap production of ever-improving quality goods was a cornerstone of the enormous economic benefits accruing to Great Britain during the seventeenth and eighteenth centuries. Note that the early machine revolution was not necessarily a power revolution nor a materials revolution. The early textile machines were designed to run with the power sources then available, which were water, animals, and human beings, and they were made largely of wood. The essence of the change to mechanization was that mechanical devices replaced hand work. Regardless of the source of power, the introduction of mechanisms increased both the rate of production and the quality of the products. Machines were faster than people, they did not get tired, and they were capable of far greater production uniformity than the human hand. They greatly increased not only production capacity but also quality. Before mechanization, for example, the cotton industry in Britain was weak and could not compete with cotton fabrics from India in either quality or cost. As machines were applied to cotton fabric production, the position was reversed. British cottons became far superior and cheaper than those that could be produced by the hand labor of India. Of course, the full potential of mechanization was achieved only with the introduction of the intensely concentrated and almost limitless power based on mineral energy sources and the construction of machines from iron and steel.

Along with the technical factors of materials, power, and mechanization, a restructuring of modes of production and indeed of all society took place. Home crafts were replaced by the factory system; industry grew in importance at the expense of agriculture; urban populations expanded.

Britain was the first nation to undergo the industrial transition. The result was that it became extremely wealthy compared to the rest of the world, and was the dominant world power for over a century and a half. It showed the way for others who later matched and then surpassed its accomplishments.

THE AMERICAN TRANSITION

The American colonies declared their independence in the same year that Watt installed his first steam engines; one at the Bloomfield mines and the other at Wilkinson's blast furnace in Shropshire. The United States was caught up in the Industrial Revolution quite early, as would be expected from its close relationship with England and the requirements of a dynamic, expanding society.

As early as 1790, a mechanized textile factory was established in Providence, Rhode Island, by Samuel Slater, who installed carding machines, roving and winding machines, draw heads, and a spinning machine. Slater was an immigrant from England who had worked in a cotton spinning plant, and the machines he set up in Providence were copies of Arkwright's machines in England. Initially, the machinery was operated by hand, but later it was driven by water power. By 1804 a factory in Pittsburgh was building a variety of textile machinery, including drawing frames, carding machines, roving frames, mules, and water frames.

The steam engine was introduced to the United States at an early date. Fenichel (1964) cites evidence that the first steam engines were installed before 1790 and perhaps even before 1780. By 1812, ten of Oliver Evans' engines were in use in five Eastern states, and by 1827 at least 163 engines, totaling 4,500 horsepower, were in use throughout the country (Fenichel 1964).

Steam and textile machinery were only two examples of the rapid adoption of new technology by the Americans. The enthusiasm with which they adopted, adapted, and created new industrial technology during the nineteenth century was remarkable and, as shown in Section B, the United States soon matched and then surpassed the nations of Europe in the level of its industrialization.

The foundations of American industrialization were laid during the time from Independence to the Civil War. Along with the great territorial expansion, there was a dramatic growth in the development of technology that was the basis for the transition from a rural, agricultural society to the most modern of industrialized nations.

A number of the important technical developments were already in place or well on their way by 1860. The cotton gin had been invented in 1793, Fulton had steamed the "Clermont" up the Hudson, McCormick had produced his reaper, John Pitts had developed a combined thresher–separator in 1834, Colt was producing his revolver in 1835, Goodyear vulcanized rubber in 1841, Morse set up a telegraph line between Baltimore and Washington in 1844, two firms were producing sewing machines by 1851, Otis built his first safety elevator in New York in 1854, and Drake drilled the first oil well in 1859.

The first steam locomotive in the United States was a Stephenson engine, the "Stourbridge Lion," which was imported from England for the Delaware and Hudson Canal Railroad, but by 1831 Americans were building locomotives themselves. In that year, the "Best Friend of Clinton" ran in South Carolina on the first line to be designed for steam traction from its inception, and the "Dewitt Clinton" had a successful run on the Mohawk and Hudson Railroad. American locomotive engineering and production soon surpassed that of Europe, and by 1838 locomotives were being exported to other countries, including England. The growth of railroads was so rapid that the first transcontinental rail link was made on May 10, 1869.

A most important development was the growing ability to manufacture metal products with interchangeable parts. Although it did not come to full fruition until the twentieth century, interchangeability was well on the road to reality before the Civil War. The interest in interchangeability had its roots in the military, where the advantage of having firearms whose parts were interchangeable was obvious. The origin of this idea went back as far as 1765, when General Jean-Baptiste de Gribeauval tried to introduce standardized weapons into the French Army. The idea was taken up by the armories at Springfield and Harper's Ferry, as well as by private American manufacturers. In 1813 Simeon North was given a contract by the War Department to make pistols with interchangeable parts. The initial approach was to use a model piece against which all gauges and jigs were calibrated and to make parts that were identical to those in the model. This was the beginning of the advances in parts manufacture that led to mass production. The gunmakers were the first to succeed, and the method, called the "American System" of manufacture, quickly diffused to the production of other products including sewing machines, typewriters, farm implements, and bicycles. Diffusion of the American System was effected by the machinists in the gun industry who changed jobs and went into other types of manufacturing enterprises. From 1830 to the Civil War, Springfield Armory was a clearing house for information on interchangeability and the latest metalworking methods. It was also a major source of trained people for other metal-based manufacturing industry.

In short, there were continual and growing technological developments that embraced the Industrial Revolution from the very beginnings of the United States and dramatically changed the nature of the country even before the Civil War. In 1790, essentially all prime movers were animals or men, with a small contribution from water and wind. By 1860, there were over five million horsepower of inanimate prime movers installed, practically all of it being steam power and about 40 percent being for railroads. The railroad trackage being operated in 1860 was over 30,000 miles and growing rapidly. The railroads were part of a transportation system that grew from the most primitive beginnings to include a network of highways, canals, and railways that tied together a rapidly growing nation.

The concomitant social and political changes were no less dramatic. From a land area of less than 900,000 square miles in 1790, the United States encompassed nearly three million square miles on the eve of the Civil War. Furthermore, the nature of economic activity and living conditions were also changing. In 1790 only a bit more than 5 percent of the population lived in urban locations with more than 2,500 people. By 1860, increasing urbanization had raised this to nearly 20 percent. Over the same time period, there was an increase in the manufacturing labor force. From 1810 to 1860 the percentage of the labor force in manufacturing went from 3.2 to 13.8 while the percentage in agriculture decreased from 73.7 to 67.

The increasing urbanization and industrialization was accompanied by a growth in at least the beginnings of the amenities of modern life. By the late 1820s, both Boston and New York had gas lighting, and indoor plumbing with bathrooms had been installed, at least in hotels. Stoves and then central heating were replacing fireplaces, and running water started to be extensively available in the 1830s. Homespun was being replaced by manufactured clothing, carpets were replacing sanded floors, dishes and cutlery were being used by ordinary people, and canned food was being produced and sold. Before the Civil War, the United States was well on its way to modernization.

Innovation and industrialization continued at an increasing pace after the Civil War. In 1860, coke was used for the first time in Pittsburgh for the production of iron and, with the adoption of the Bessemer process, American production of steel soared. By the end of the nineteenth century, the focus of inventive activity and technological innovation had shifted from Europe to the United States. Some of the more noteworthy American events were the construction of the first universal milling machine by Joseph Brown, the development and production of celluloid by John W. Wyatt in 1868, Bell's patenting of the telephone in 1876, Edison's inventions, ranging from the microphone in 1877 to the first electric light power station in New York and London in 1882, the Hall process for producing aluminum in 1888, the Kodak camera and celluloid-based film in 1888, the gasoline tractor in 1892, the phonograph in 1896, and cellulose acetate in 1902.

The trend continued into the twentieth century. Up to the Civil War, industrialization was dominated by steam and mechanical technologies. Although these continued to be developed, electrical technologies became the leading edges of innovation in the latter part of the nineteenth century. In the early part of the twentieth century, these were joined by automobile, aviation, and chemical technologies based on petroleum. After the war, the leading edge technologies became plastics and polymers, electronics, and jet engine aircraft. While the older technologies continued to be improved, there was a continual shift to new areas as the older ones became mature.

Three basic measures of industrialization are consumption of iron and steel, energy usage, and the degree of mechanization. The consumption of energy was discussed in Chapter 5 and shown in Figures 5-2 and 5-3. The per capita consumption of energy was roughly constant until 1900. Then, except for the decade of the thirties, it grew dramatically until the 1970s, after which it leveled off and actually decreased. The drop in energy consumption in the 1970s is of a different character than that during the 1930s. In the earlier decade, all economic activity was low because of the Great Depression, and this was reflected in a lower energy consumption. There was no comparable economic disaster in the seventies. The decreased consumption was the result of more efficient energy usage. Of course, efficiency had been increasing continually since colonial times, but there was a spurt in the rate as a result of the OPEC oil embargoes.

The constant energy consumption up to 1900 is misleading. There were large increases in efficiency of energy use as a result of switching to mineral fuels, the mechanization of industry, and technical improvements in prime movers, such as steam engines. This means that the effective use of energy had been increasing continually long before 1900. The substitution of min-

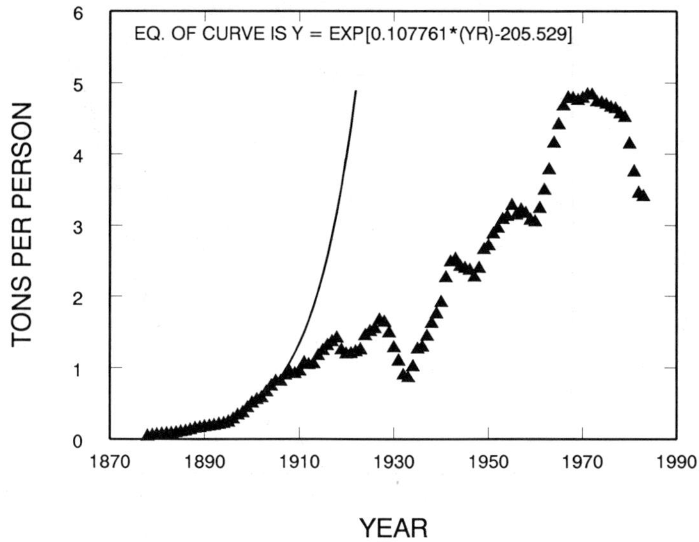

Figure 7-7 Consumption of steel per capita in the United States: five-year moving average.

eral fuels for wood was particularly important. As shown in Figure 5-3, by 1882 half of the energy consumed in the United States came from mineral sources, primarily coal.

The consumption of steel is shown in Figure 7-7, which is a log plot of the per capita consumption as a function of time. The data is divided into three regimes; the first displays an exponential growth from 1878 to 1909 with the remarkable growth rate of 10.5 percent per year. The second regime runs from 1910 to 1965 and, except for the decade of the Depression, shows a growth rate of 2.5 percent per year. After 1965, the consumption levels off and even decreases.

The general picture of the use of energy and materials in the process of American industrialization is that there is an initial rapid increase in per capita consumption but that the rate of consumption ultimately slows down. In the advanced stages of industrialization, the per capita consumption becomes constant and even decreases. The total consumption shows a similar slowing trend in the later years since the rate of population growth also decreases. These data directly contradict the popular Malthusian thesis that an industrial society requires amounts of materials and energy that continually grow exponentially. These neo-Malthusians properly point out that exponential growth of consumption cannot go on indefinitely, but then assume that it will only be checked by a collapse of technological society arising from severe shortages. The data simply do not support such ideas. Con-

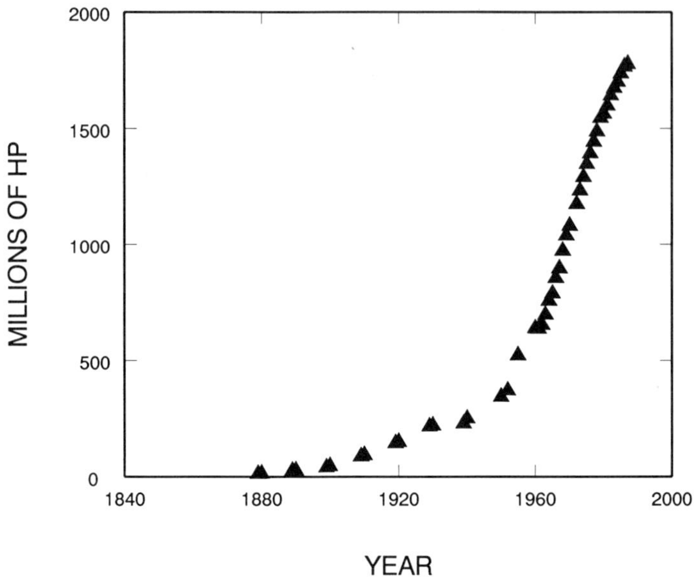

Figure 7-8 Installed nonautomotive horsepower in the United States.

sumption does not continue to grow exponentially. It eventually levels off, at least on a per capita basis, and most probably ultimately even on a total basis. The reasons for this are several. In the first place, the needs that these resources satisfy do not grow forever. When there are bridges across all the rivers, and skyscrapers in all the cities, when rail networks fill the country, and every household has every major appliance, levels of saturation are reached. Less materials and energy are needed to maintain a completed industrial social structure than to create it. In the second place, the efficiency of use for both energy and materials increases with advancing industrialization. This was pointed out for energy consumption in Chapter 5. There also were corresponding improvements in the use of materials. As a technological society advances, it needs less energy and materials for a given application.

The mechanization of the United States is represented by the graph of Figure 7-8. This shows the total horsepower installed in the United States in mechanical prime movers. It excludes wind, water, and muscle sources of power. The graph includes horsepower installed in factories, mines, electric utilities, railroads, ships, and aircraft, but does not include automotive transport. There was a steady growth of 3.9 percent until 1920. After a slowdown because of the Depression, growth resumed until the decade of the eighties, where there is the beginnings of a saturation effect similar to that for steel consumption.

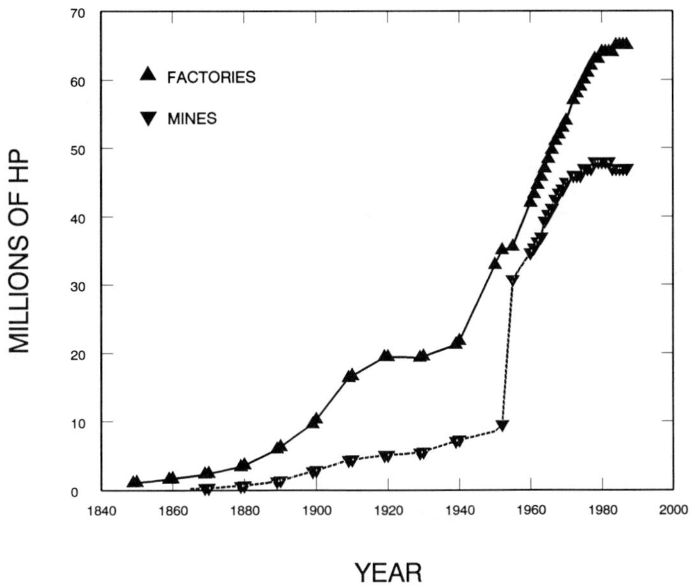

Figure 7-9 Installed horsepower in factories and mines in the United States.

If it were not for the continuing growth of the horsepower installed in aircraft and electric utilities, the leveling off in the later years would be more pronounced. It is in fact instructive to look at the separate components of mechanization. When this is done, it is apparent that there is a saturation of the mechanization of factories, mines, and farms. For railroad horsepower there is even a decrease in recent years. The data for factories and mines are shown in Figure 7-9 and for farms in Figure 7-10. The saturation effect in the decade of the eighties is clearly apparent. The data for farms can be approximately represented by a logistic with an initial growth rate of 8.6 percent per year and a maximum of 360 million horsepower. Incidentally, the total horsepower on farms is over half that in electric generating plants and more than three times that in factories and mines combined. This shows how highly mechanized and energy-intensive American agriculture has become.

Although the installed horsepower for electrical generation has not yet saturated, it certainly has started to slow down. This is apparent in Figure 7-11, which is a logarithmic plot of horsepower versus time for generating plants. There was a rapid growth period from 1890 to 1930 (11.7 percent per year), after which growth was much slower. In the eighties, growth is starting to decrease towards zero. The beginning of a saturation effect is also clearly shown in Figure 5-19 (Chapter 5; "The Dynamics of Energy Usage"),

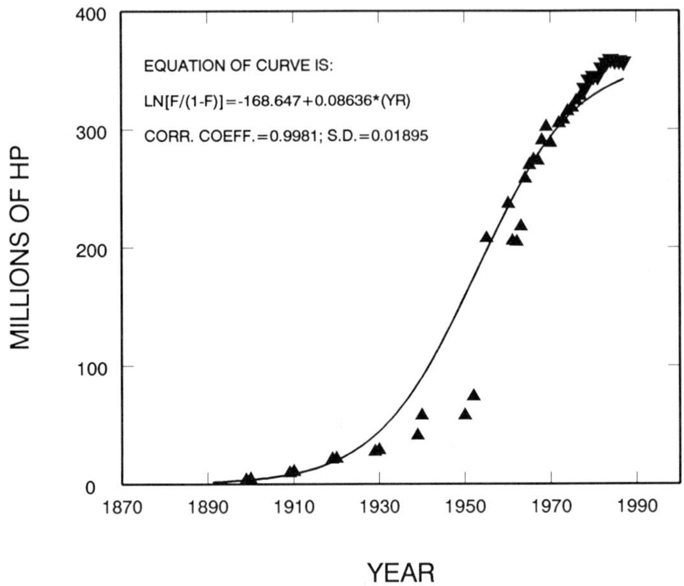

Figure 7-10 Installed horsepower on United States farms.

in which the production of electricity in the United States is seen to be leveling off in the 1980s.

Automotive transport was separated out from these data because it represents a quite different phenomenon. Personal transportation and high mobility are obviously very highly valued. To be able to go anywhere, to any distance at any time by means under personal control is a much desired and quite remarkable result of modern technology. The total amount of horsepower in automotive transport swamps that in all other installations combined. In 1987, of a total of 33.3 billion installed horsepower, 31.5 billion was in automotive vehicles and most of this was in passenger cars. But even here there is evidence of a coming saturation effect. Figure 7-12 shows that the per capita automotive horsepower started to level off in the mid-eighties. This is consistent with Figure 5-23 in Chapter 5, which shows a decreasing rate of annual automobile registrations in recent years.

From the information given here, along with that in Table 4-6 of Chapter 4 and the graphs already referred to in Chapter 5, a broad outline of the American transition to industrialization can be summarized as follows. From the very birth of the nation, the United States was caught up in the Industrial Revolution, and by the time of the Civil War had laid the foundations for the growth which was to make it the premier technological country of the world. The great modern technological systems had their origins in developments that became ever more rapid after the Civil War. First canals, steam,

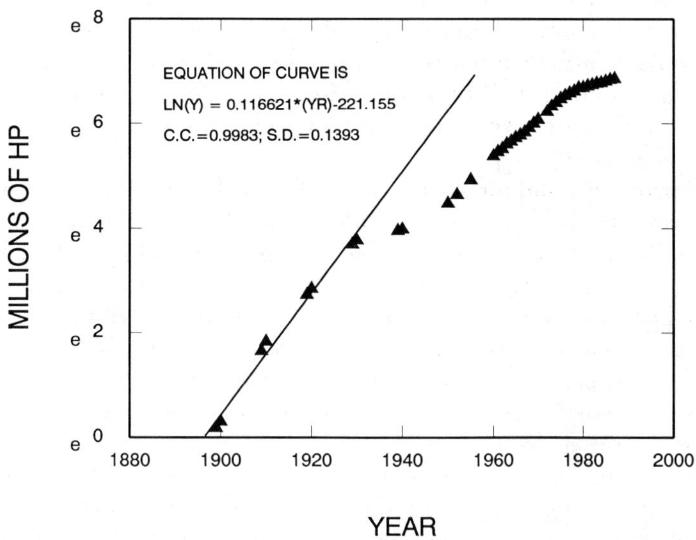

Figure 7-11 Installed horsepower in electric generating plants.

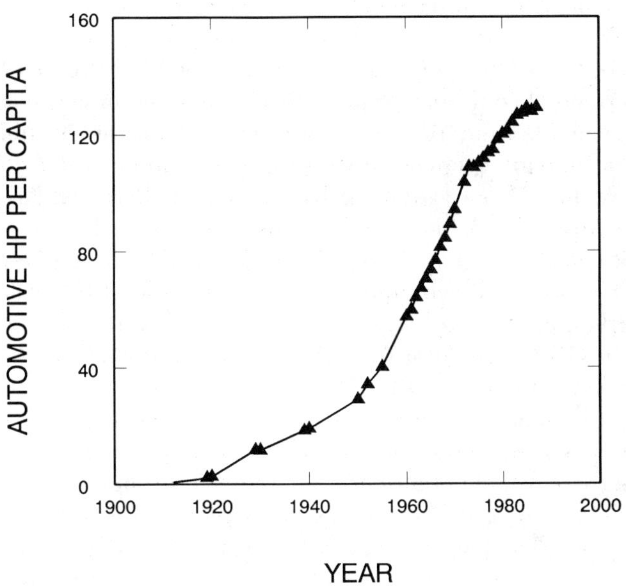

Figure 7-12 Automotive horsepower per capita in the United States.

metal working, and railroads; then the telegraph, telephone, and steel; then mechanization, tractors, electricity, and automobiles; then synthetic textiles, radio, and aircraft; then plastics, jet engines, and electronics. Each of these groups had rapid growth periods in succession that drove the direction of inventive and innovative activity and provided the basis for a continually growing economy.

The periods of rapid industrial growth for the United States are over. This is clear from the data, which show saturation effects for the major industrial technologies. There is a sense in which the problems of providing food and goods are solved. That is, there is a method by which agricultural and industrial production has been increased by large amounts and can be further increased so that there is no technical barrier to meeting the needs of large numbers of people for food and goods. The problems of providing food and goods are no longer technical, or even economic, but political and social. Also, the large-scale technological systems of production, transportation, and communications are in place. A new kind of technological and economic growth is now taking over. This is based on technologies such as computers, communications, robotics, quantum optics, and biotechnology. These are not nearly as energy- and material-intensive as the industrial technologies and are much more directed towards services such as finance, health, education, and entertainment than towards production. Of course, productivity will continue to increase for industrial production just as it has for agriculture. But the major share of employment and Gross National Product is no longer in these sectors. The American industrial transition is over. The future will be driven by postindustrial technologies.

The growth of American technology has made the United States the richest country in the world, on both an absolute and a per capita basis. Except for the decades of the 1930s and 1940s, the Gross National Product has shown a long-term trend of continued growth. Real GNP data for each year in 1958 dollars are available from 1890 to the present in Government publications. Also, five-year moving averages of the GNP per capita in 1967 dollars from 1789 to 1973 have been estimated by Thomas S. Berry and reported by Rostow (see Appendix 3 for data). Because one set of data is in terms of annual GNP and the other is a five-year average, they cannot be reduced to a single set by the usual technique of finding the ratio of the deflators from data for overlapping years. Both sets of data were therefore used to examine the growth in GNP per capita from 1789 to the present. Graphs for these data are shown in Figures 7-13 and 7-14.

The first point to note about these figures is that they display a remarkably steady growth except for the years of the Great Depression and World War II. Of course there were fluctuations up and down throughout the entire past two centuries. Even so, it is clear that the Great Depression was a unique event superimposed on long-term growth. The rise in GNP per capita starting in 1940 is obviously associated with the war. After the war, the GNP per capita fell again until it met the long-term trend curve and then continued

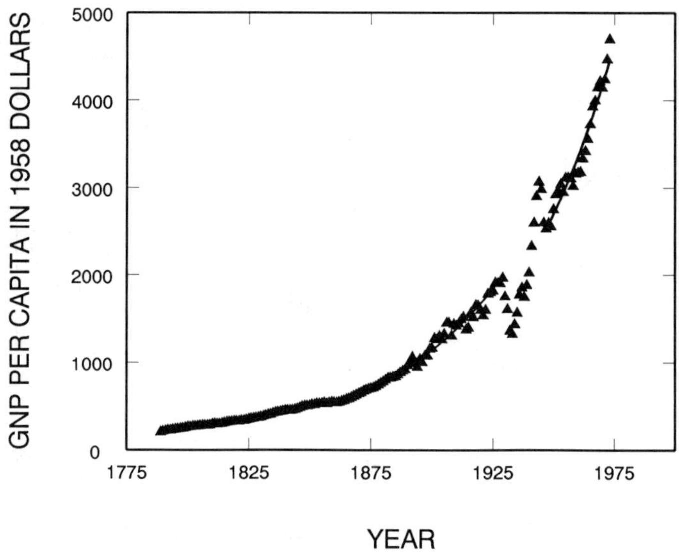

Figure 7-13 GNP per capita from Berry/Rostow data: five-year moving average.

its growth. The curves show that, except for the decade and a half starting in 1930, the economy was driven by largely systemic factors with relatively small idiosyncratic fluctuations for two centuries. An exception is the time period from 1853 to 1859, during which there was no growth. But even the massive disturbances of the Depression and the war had only a temporary effect. Once these incidents passed, there was a self-healing process that brought the economy back to its systemic progression.

The data can be used to separate out the systemic growth in GNP per capita because they are sufficiently regular to be well described by exponential functions. Visual observation of both sets of data shows that for the Berry data there are three regimes that are linear on a logarithmic scale. These are: 1789 to 1848, 1860 to 1929, and 1947 to 1973. For the Government data, there were two logarithmically linear regimes: from 1890 to 1929 and 1947 to 1987. I calculated exponential curve fits for a variety of time periods to verify that these regimes actually gave the best fits to the data. For these regimes, Table 7-7 shows the curve fitting parameters for the equation

$$G = \exp(At - B) \tag{7-1}$$

where G is the GNP per capita and t is the year A and B define the exponential curve, A being the growth rate. The correlation coefficients (C.C.) and standard deviations (S.D.) are also given in Table 7-7.

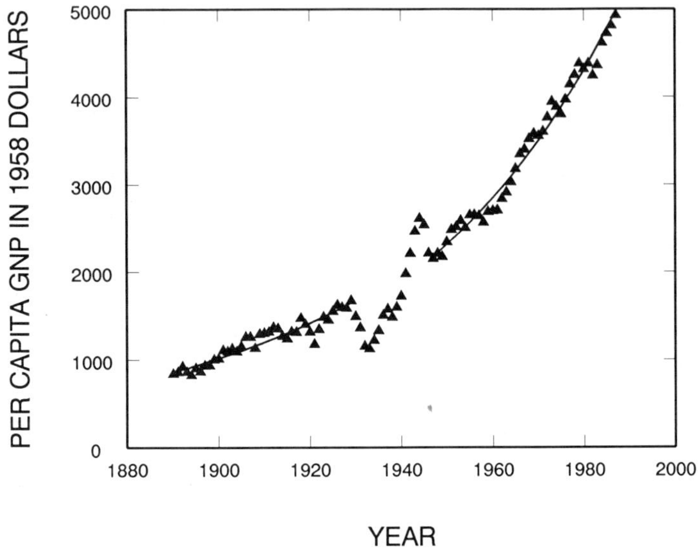

Figure 7-14 GNP per capita from government data.

Taken together, the data in Table 7-7 show that there was a systemic growth rate of 1.4 percent per year from 1789 to the decade before the Civil War. From the Civil War to the present the systemic growth rate according to the Berry data was about 2 percent per year. The government data also show a growth rate of about 2 percent per year for the period from 1947 to 1987, but from the Civil War to 1929 it shows a growth of only 1.7 percent. At any rate, it seems safe to conclude that the systemic growth rate before the Civil War was just under about 1.5 percent per year, and that this increased to about 2 percent per year after 1860. This increase coincided with the accelerating pace of industrialization after the Civil War.

Unlike the degree of mechanization and the consumption of energy and

Table 7-7 Curve Fitting Parameters for Real GNP per Capita in the United States

Data Set	A	B	C.C.	S.D.
Berry 1789–1848	0.013949	19.58406	0.99999	0.01992
Berry 1860–1929	0.019195	29.43040	0.99998	0.04091
Berry 1947–1973	0.021869	34.7453	0.99999	0.03776
Gov't. 1890–1929	0.01668	27.3955	0.99996	0.06445
Gov't. 1947–1987	0.020668	32.5550	0.99999	0.03423

materials, the growth in GNP shows no signs of leveling off. This supports the idea that the postindustrial technologies are taking over as the leading sectors driving economic growth. The American industrial transition started with the Industrial Revolution and took two centuries to complete. It is now an accomplished fact and can provide the base for continued growth using new technologies. For most of its history, the United States could rely on its domestic markets and had little technological competition. With the industrialization of non-Western societies and the increasing globalization of national economies, this is no longer true. Worldwide technological competition is intense and will become more so. The industrial transition is in various stages for societies throughout the world and the number of roughly comparable technological competitors is increasing.

MATERIAL RESOURCES

An industrial society uses concentrated energy sources to transform natural resources into useful products. It is characterized by the intensive use of energy and materials and by their major importance in economic growth, the critical technologies being those related to power and the large-scale production of goods. Industrial society would cease to exist without sufficient quantities of raw materials and sufficient energy to process them. And these quantities are immense. The consumption of the more commonly used metals, iron and aluminum, is measured in units of hundreds of millions of tons per year in the United States alone. In the aggregate, Americans use about 40 tons of materials per year per person. Economic growth was based on, and accompanied by, a continual growth in the quantity of natural resources extracted from the earth and turned to human ends. The need for such resources will persist as long as large-scale technological societies exist so that questions of patterns of use, growth rates of consumption, and availability of supplies are of great importance.

The availability of resources is determined by an intimate mix of economic and technological factors that change with time. It is this dynamic feature of the resource base, which is so often overlooked, that has sustained technological society so far and promises to do so in the future. But even a proper recognition of the interacting roles of economics and technology is inadequate for a practical definition of resources. The materials we take out of the earth are not useful simply because they exist; we value them for the functions that they perform; they constitute resources only to the extent that they can perform these functions and as long as we are willing to have them do so. A particular material is used for specific purposes in end products. The techno-economic interaction is therefore driven, modified, and supplemented by a tangle of human preferences, objectives, and actions, which are expressed in a variety of ways ranging from individual consumer choices to government policies. While this is a complex situation, it can usually be understood in terms of a generalized form of the idea of comparable utility.

This idea simply states that people must make choices of what they can have and that they will choose that which they deem most useful to them. Making a choice means that something else must be given up; the choice will therefore be made on the basis of a combination of need and price.

Material resources are used in such huge amounts that there has always been a great deal of worry about their future availability. A technological society requires a large supply of the right kind of materials. At more immediate levels, the degree of access to natural resources has strong effects on the power and economies of nations, the fortunes of corporations, and the well-being of consumers. Resource projections are therefore of great interest in spite of their inherent uncertainties. The nature of such projections depends on the objectives to be served, and their validity rests on the validity of the assumptions on which they are based. Failure to realize this has led to a long string of projections which predicted a degree of resource exhaustion that would lead to catastrophic consequences. To date, such predictions have been consistently wrong.

These predictions were wrong because they were based on two incorrect ideas. First, they treated resources as if they were a fixed amount of well defined materials residing in a warehouse. Second, the exponential increase of consumption that existed in the past was assumed to hold true for the indefinite future. If a definite amount of material residing on warehouse shelves is being depleted at an exponentially increasing rate, then it is easy to determine when the shelves will be empty. Also, the amount of material existing on the shelves is usually assumed to be some multiple of proven reserves, as if multiplying the amount of proven reserves by a factor of two to five is a generous estimate of the total amount of a resource. Using such assumptions, a prediction of doom is inevitable.

It is instructive to recall the work of Meadows et al. (1974), who computed the effects of exponential growth. They were aware that exponential resource growth could not continue so they also did calculations using a bell-shaped curve for resource production. They still predicted catastrophic collapse because of resource shortages and/or excessive pollution sometime in the 21st century. I differ with their approach in that it seems clear from the data that a decreasing rate of resource use per capita and a decreasing rate of population growth are systemic in a technological society. Slowing of consumption and population growth are both natural results.

In considering the role of natural resources in a technological society the most important point to recognize is that resources do not consist of a static stock of material in nature analogous to the stock of supplies in a storehouse. Rather, a resource should be regarded as a flow, whose volume and rate is determined by a host of technical, economic, political, and social conditions. This was pointed out in Chapter 5 where we described an analysis of the energy inventory. The McKelvey diagram and the Hubbert–Hewett model are important tools for studying the availability of both materials and energy resources and, in fact, were first applied to metallic minerals.

Mining regions are notoriously unstable. When deposits are discovered they attract substantial investment and large numbers of people, rapidly creating boom conditions and a period of local prosperity. But as the ores are depleted, production decreases and ultimately ceases altogether. The result is the decay and even abandonment of formerly dynamic communities. It was this cycle of instability that led Hewett, and later Hubbert, to a quantitative statement of mineral production.

The Hewett–Hubbert model has been found to be quite general. The model simply states that the rate of production of any resource as a function of time is a curve that starts out slowly, grows with increasing rapidity until it reaches a maximum, and then decreases to zero. In its simplest form the curve is bell-shaped and can be represented by the derivative of the logistic equation as was done for energy resources in Chapter 5. This model is quite reasonable. After a deposit is discovered and decisions are made to mine it, some time is required to begin and increase production. As facilities are installed, additional excavations are made, transportation needs are met, new customers are found, and production grows at an increasing rate. But as the mines extend deeper into the earth, as metal content decreases because of the geological gradients, and as the deposit is depleted, it becomes more difficult to maintain high production rates. Ultimately, the deposit becomes exhausted or the remaining ore cannot be extracted economically.

An analysis of the limits of resource recovery by Earl Cook (1976) provides some insight into the specific factors that affect resource production rates. A basic consideration is the geologic structure of the deposit. A typical porphyry copper deposit, for example, consists of a body of ore that is pipelike and oval with dimensions of 1,000 by 2,000 meters near the surface and a depth up to 3,000 meters. But the pipe narrows continually with depth. Mineralization is in concentric circles, the copper content at the edges being much less than at the center. At the Cuajone mine in Peru, the deposit forms a truncated, inverted cone of mineral containing copper in amounts ranging from 1.3 to 0.2 percent. The copper concentration decreases rapidly as the grade drops below 1 percent, and the deposit has a rather sharp geochemical boundary. An important point is that the total amount of copper in the low grade parts of the deposit is much less than in the high grade regions. Also, the energy to mine and mill a unit of copper rises dramatically as the grade of ore becomes very low. The data for copper content and energy requirements versus grade are shown in Table 7-8 for the Cuajone mine (Cook 1976).

Practically all of the copper at Cuajone is contained in ore of grades greater than 0.3 percent. The energy requirements rise gradually and are inversely proportional to grade until the grade decreases below 0.3 percent. At and below 0.2 percent the energy needed to mine and mill the ore is so great that the small amount of copper remaining cannot be recovered economically under present conditions.

The Comstock Lode in Nevada is an excellent example of a deposit with

Table 7-8 Ore Grade, Copper Content and Energy Costs at Caujone[a]

Grade (%)	Amount of Ore	Amount of Copper	Energy Cost
1.32	20,000	264	0.76
0.99	430,000	4,237	1.01
0.32	102,000	326	3.13
0.20	1,057,000	67	>16,000

[a]Energy is in units of the energy needed to mine and mill a ton of 1% grade ore; amount of ore and copper content are in thousands of short tons.

a sharply defined geochemical boundary. Its production cycles are shown in Figure 7-15 (Cook 1976). Starting about 1860, the discovery of the Ophir, Gould, Curry Bonanza resulted in a large and rapidly increasing annual production of silver. This started to decrease before the decade was over, but the discovery of the Great Bonanza started the production over again until this too was depleted. Both these Bonanzas consisted of high grade ore. But technical advance permitted the exploitation of lower grade ores, thereby giving another production cycle, although it was considerably smaller than the previous ones. Finally, the cyanide extraction method was developed to the point where it was economical to extract silver from some very low grade

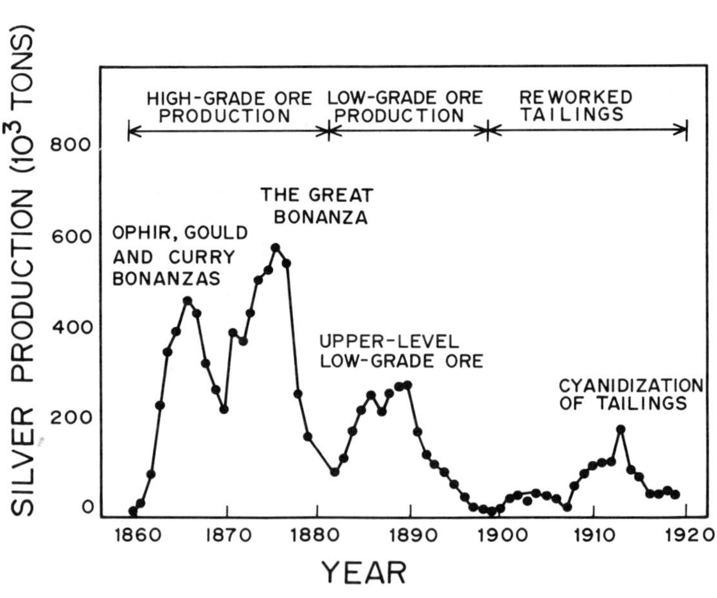

Figure 7-15 Silver production cycles of the Comstock lode.

material that was regarded as waste up to that time. The dollar yield per unit of metal decreased substantially as lower grade ores were worked.

The production history of the Comstock Lode is a series of bell-shaped curves of the type to be expected from the Hewett model. Each of these curves is different and illustrates some of the conditions that define the model. In the first place, a specific Hewett curve refers to a specific body of ore as shown by the superposition of the Great Bonanza production curve on the later parts of the first cycle. In the second place, a specific curve refers to specific types of ores and a specific technology. When technical progress had reached the point at which low grade ores could be profitably mined, a new production cycle was initiated. This does not mean that there is no technological advance during a particular cycle. We know that quite the contrary is the case. But innovations during the cycle for a particular ore are normally of an incremental nature and contribute to the efficiency of extraction in a continuous manner. It is when technology reaches a stage permitting the exploitation of a different kind of material with different properties that a new cycle is started.

A particularly important case of this phenomenon is the development of taconite technology. Since the latter part of the nineteenth century the Lake Superior district has been the major source of high grade iron ore in North America. This ore was hematite and was mined in ever-increasing amounts until about 1945. But most of it was gone by 1970, and if it had not been for the development of a new technology, the district would have been exhausted by 1980. Of course, there have been continual innovations in all aspects of iron mining. Important improvements in drilling and blasting, hoisting, loading, hauling, crushing, grinding, and transport have been going on throughout the 100-year life cycle of Lake Superior hematite. But there was only so much hematite. It was the development of methods to use the abundant deposits of taconite that extended the life of the district.

The major problems with taconite are that it is very hard and its iron content is low (30 percent or less). Hematite is softer and of higher grade (about 50 percent); therefore it is easier to mine and requires minimal beneficiation relative to taconite. But research on taconite had been going on since E. W. Davis started his work in 1913 (see Kakela, 1981; and Gold, Pierce, Rosegger and Perlman, 1984). By 1950 the technology reached the point where, under the war-induced fears of supply shortages, it was applied to the huge taconite deposits of the Lake Superior district.

The problem of breaking up the hard taconite rock was solved with the advent of Linde's jet piercing method in which a kerosene-oxygen flame jet heats the rock. A water spray rapidly cools it, thereby breaking up the rock which is then carried to the surface by the water stream. Breaking and grinding the taconite is straightforward, but it must be ground to the consistency of flour so that the iron-rich magnetite can be separated from the waste by magnetic means. Such fine powder is useless for a blast furnace, and a major bottleneck was the lack of a suitable method for agglomerating the powder into a form suitable for reduction to iron. The method that was finally developed produced pellets by rolling damp taconite powder (containing clay to act as a binder) in a ball drum. The pellets were then fired to harden them. The result was a superior feedstock for the blast furnace. The uniformity of the product and its improved permeability almost doubled blast furnace productivity. The pellets also reduced the coke requirements because they distributed heat uniformly. Thus, although the costs of mining, beneficiating, and pelletizing were high relative to the costs of hematite, the savings at the blast furnace were higher.

The development of taconite technology was accelerated by government policies as a result of wartime concerns, which were renewed by the Korean conflict, of the adequacy of domestic supplies. Kakela (1981) has summarized the federal and state policies relevant to taconite and traced their effect on the emergence of taconite as a new and important resource.

The production history of the Lake Superior district is instructive (Cook 1976). The region was productive for a hundred years, and the mining of hematite shows the general features of a Hewett–Hubbert cycle. The area started to go into decline during World War II when increasing amounts of low grade ore were being used. But taconite has started the district on a new production cycle. The production curve not only shows the rejuvenating effect of a new technology, but also displays the powerful effect economic conditions have on resource production. There are a number of minor short-term fluctuations in production, but the most dramatic feature is the large fall in production during the Great Depression, making the curve bimodal so it does not have the simple bell shape of the Hewett model. This is quite a different situation from the multiple maxima resulting from changes in type of ore or dramatic technical improvements. The Hewett model clearly requires that a number of factors need to be constant or at least change in a regular, continuous manner in order for the model to be applicable. The state of the economy is one of these.

There are two important points to extract from this analysis. The first is that, for a given level of technical and economic conditions, the extraction of minerals from a particular source goes through a life cycle. The second is that the life cycles can be renewed or extended by technological advances. Both of these conclusions are important for short and intermediate term strategic considerations.

Paradoxically, long-range global projections are easier to make once assumptions are made on population size and the degree of industrialization of the various portions of the population. But even here it is assumed that certain broad patterns of evolution of the past will continue into the future.

MATERIALS AND WORLD INDUSTRIALIZATION

How long can the industrialized world go on using material resources at ever-increasing rates before they are exhausted? This question has been at the core of a long-standing debate concerning the ultimate viability of industrial societies. I have addressed the relation of industrialization to resource availability before (Girifalco 1983) and have drawn on this earlier work for the material presented here.

The tone of the debate was set during the eighteenth century. Scientific discovery, technical progress, and growing affluence gave rise to optimistic views on the future of humanity. The growth of knowledge and a rational approach to all aspects of human activity was thought by many to ensure that there was no limit to how much the human condition could be improved. Malthus' "An Essay on the Principle of Population as It Affects the Future Improvement of Society" was a response to this optimism. This essay appeared in 1798 and has had a profound influence on succeeding generations to the present day.

Malthus' thesis was simple. Left to itself, population grows geometrically so that the number of people that need to be fed grows very rapidly. But geometric growth of food production is not possible because the amount of arable land is finite. The best that can be hoped for is an arithmetic increase in agricultural production. This means that sooner or later the food supply is insufficient to support the growing population. The result is early death from hunger and illness until the population is brought into balance with the food supply. The economic history of society will therefore always be dominated by an oscillation between relative happiness and relative misery. When the food supply is large enough to support the population in relative comfort, the population will automatically increase until the food supply becomes inadequate. Agriculture struggles to keep up but it cannot, and the result is misery and death. A balance is thereby reestablished and the cycle starts over again. Malthus was an excellent mathematician, knew how to treat data, and collected historical data which supported his thesis. Modern economic historians have concluded that up to Malthus' time there indeed was a series of fluctuations of the kind described in his essay.

The intellectual descendants of Malthus apply his logic to the exhaustion of natural resources and conclude that industrial societies must die. The exponential growth of population and the concomitant exponential growth in consumption must lead to disaster. This line of thought has made terms such as "the population bomb" and "limits to growth" a part of modern thought. It is the exponential law that frightens the projectors of catastrophe. Indeed, the calculations in Appendix 1 show the destructive power of exponential growth. Even a resource that would last ten thousand years if consumed at a constant rate would vanish in less than two centuries if the consumption increases at the rate of 3 percent per year.

The opponents of this catastrophic view of resource utilization agree that Malthus was correct for the eras that preceded him. But by the beginning of the nineteenth century the Industrial Revolution was well under way, and continuing technological change had redefined future possibilities. The Malthusian thesis is right only if technology is relatively static. But ever-increasing agricultural productivity and the continual development of new resources are the natural results of advancing technology.

A proper resolution of the issue of future resource availability hinges on the following question. Will exponential growth continue until it is checked by a catastrophic failure, or will growth slow down gradually as a result of systemic factors in the technological society? I have already shown that there is strong evidence that the per capita consumption of both materials and energy levels off as an industrial society matures. The question therefore resolves itself into the expected growth of population.

A detailed model for the projection of resource use would start with a projection of the per capita consumption and of the population for each country of the world. Such projections would require assumptions about the future growth laws for both population and per capita consumption for each resource of interest. The cumulative consumption computed from this model would then be compared with an inventory of available resources to determine how long they would last. But such a detailed calculation can be short-circuited to give a lower limit on the resource exhaustion times because there is strong evidence that world population will eventually level off. If this is true, then a calculation can be made of how long a resource would last if the ultimately stable population existed right now and maintained its number into the indefinite future. All that is needed is a figure for the per capita consumption of the resource. If the per capita consumption is taken to be that for a mature industrial society, this procedure will give an overestimate for consumption and an underestimate for the length of time our resource base can support a fully industrialized world. The steps in the calculation are:

1. Determine the amount of resources available.
2. Project the future population of the world.
3. Determine per capita annual usage of resources.
4. Multiply per capita use by population to get total consumption rate.
5. From total consumption and amount of resource, determine the time at which the resource will be exhausted.

A convenient summary of the mineral resource inventory has been published by Goeller and Weinberg (1976). They computed the number of years it would take to exhaust a resource if it were consumed at the rate of world consumption in 1968. Their results are for the entire resource inventory and include speculative as well as identified resources. Their numbers are given in Table 7-9. The last six entries, starting with titanium, are my own numbers computed from data in the Department of the Interior Geological Survey Professional Paper 820 entitled "United States Mineral Resources" by D. O. Brobst and W. P. Pratt, Eds.

The significance of these numbers for industrialization of the world depends on future population growth. Not too many years ago it was believed

Table 7-9 Resource Inventory of the Most Commonly Used Elements Referred to World Consumption in 1986

Resource	Amount[a]
CHx	2500
C (oxidized)	4×10^6
Si	5×10^6
Ca	4×10^6
H	10^{10}
Fe	4.5×10^6
N	10^8
Na	3×10^8
O	3.5×10^7
S	3×10^7
Cl	4×10^8
P	1300
K	4×10^7
Al	2×10^8
Mg	4×10^8
Mn	13000
Ar	2×10^8
Br	6×10^8
Ni	1.4×10^6
TI	4×10^5
Zn	10^3
Co	450
Cu	170
Pb	450
Sn	200

[a]Years to exhaustion at 1968 use rate.

that population would continue to grow at explosive rates. Modern demographic studies, however, project quite a different picture. The general scholarly consensus is that the world is in the midst of a demographic transition, a phenomenon which is observed in many biological populations. When the food supply (or resource base) for a population is constant, it adjusts to a constant size determined by resource availability. If the resource base (or food supply) increases, there will be a corresponding population increase to a new, constant level in keeping with the larger size of the resource base. A large, rapid increase in resources will give rise to a rapid increase in population to a higher but constant level. This is the demographic transition. Industrialization gave us a new method of generating resources to levels that were previously impossible. The natural result was an increase of the human population. From an initially stable condition in which birth and death rates balance, there is a decrease in the death rate which results in population growth. Initially, the birth rate remains high, but later it starts to decrease and eventually matches the death rate. The result is that the population levels off to a new stable condition with a higher value.

There is strong evidence that world population is about halfway through such a demographic transition. The Industrial Revolution ushered in an era of unprecedented population growth. It took 1,800 years for the population to triple from its value in 1 A.D., but in the succeeding 190 years it has more than quadrupled. Of course this cannot continue. Population growth must stop sometime. The only question is will population be checked by a Malthusian catastrophe from a lack of resources, or in a more orderly manner by forces internal to the industrial sociotechnological system.

The annual death rate before the Industrial Revolution was about 40 per thousand in Western countries (Schultz 1981). This decreased steadily and leveled off to about 10 per thousand with increasing industrialization (Keyfitz 1982). For the more developed countries as a whole the birth rate has been decreasing to a level that is now nearly matching the death rates, so that their populations are relatively constant. The fully industrialized societies have therefore completed the demographic transition. The lesser developed countries are leaving the early stages of the demographic transition. Fertility in the LDCs has decreased by percentages ranging from 10 percent to 70 percent, the rate of childbearing has decreased, and birth rates have gone down steadily since 1950. Detailed analysis of current trends leads to the conclusion that the world as a whole is indeed in the midst of an orderly demographic transition and that the world population will level off at something over 10 billion sometime around the year 2100 (Coale 1983).

Some important details of the demographic analyses need to be noted. The first is that while the population is decreasing in some of the more developed countries, in others there is a slight increase. The result is a low but positive population growth for the industrialized countries as a whole. The second point is that the population trends in these more developed countries seem to be the result of conscious choices. There certainly are no general Malthusian imperatives to limit population in the advanced countries. In fact, a survey by Berelson reported in "Limits to Growth" (Meadows et al. 1972) shows a strong correlation between desired family size and GNP. As the GNP increases, the number of couples wanting four or more children decreases dramatically, levels off, and then exhibits a slight rise. The final point is that projections cannot distinguish between a demographic transition that results in a completely stable population or one that is slowly growing. But the evidence suggests that a slow, non-zero growth rate would be by choice, not by the unthinking urge to breed assumed by Malthus.

Detailed projections of the population of all countries and of the world as a whole to the year 2100 have been published by the World Bank (Zacharia and Vu 1988). These were based on the best available demographic data. They indeed show a demographic transition with a leveling off in world population at about 11 billion. Figure 7-16 shows the world population from 1750 to 2150, the points from 1990 on being taken from the World Bank projections. The leveling off at 2100 is clearly evident.

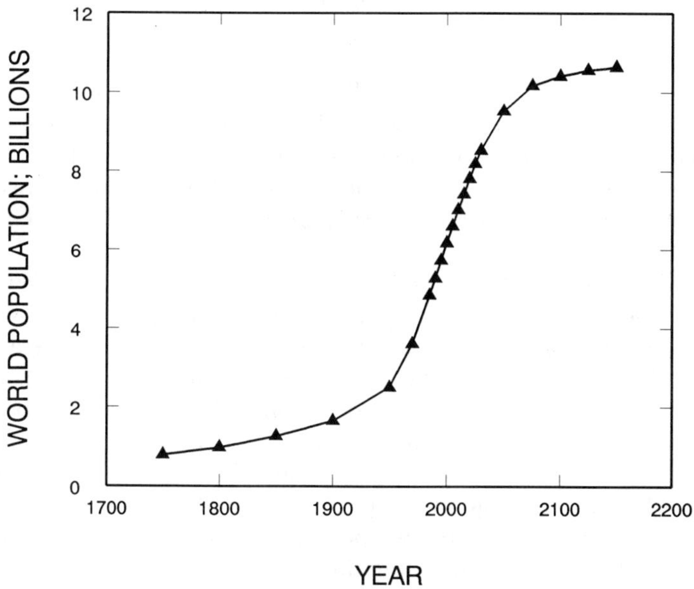

Figure 7-16 The demographic transition.

This projection implies that the entire world will be industrialized after 2100. This is a necessary conclusion because the demographic transition is driven by the rising affluence of industrialization.

Because of the projected constant population, along with the constant per capita resource use of a mature industrial society, there will be a constant rate of resource depletion after the demographic transition. Now the "short circuit" calculation, in which we pretend that the high consumption rate after the transition starts right now, can be done as follows:

If A is the total amount of a resource, and C_p is its per capita rate of current consumption and P_p is the population doing the consuming, then the time it would take to exhaust the resource if both the population and the per capita consumption were constant is

$$T_p = A/C_pP_p \qquad (7\text{-}2)$$

This is just the linear exhaustion time referred to current consumption.

If the resource were being consumed at a constant rate characteristic of the conditions after the demographic transition, then the linear exhaustion time would be

$$T_L = A/C_xP_x \qquad (7\text{-}3)$$

where C_x and P_x are the consumption rate per capita and the population after the demographic transition. Taking the ratio of Equations (7-2) and (7-3) gives,

$$T_L = T_p(C_p/C_x)\ (P_p/P_x) \qquad (7\text{-}4)$$

Estimates are available for all factors in this equation. For the exhaustion time referred to current consumption, just take the values in Table 7-9. These refer to 1968, but a few decades don't matter because the exhaustion times are much larger than this and are certainly not known with an accuracy of a quarter century.

We will assume that the per capita consumption is the same after the transition for the entire population as it is now for the population of the advanced industrial countries. The advanced countries currently use about 90 percent of all resources; within the accuracy of the calculation, all resource usage can be allocated to the developed nations. If P_p is taken to be the population of the industrialized nations, then the per capita consumption ratio in Equation (7-4) is unity. The population of the more developed countries in 1985 was 1.18 billion. If the population after the transition will be 11 billion, then the ratio of the two populations is 0.11. The exhaustion time referred to the steady state after the transition is therefore given in terms of that referred to current consumption by

$$T_L = 0.11T_p \qquad (7\text{-}5)$$

This equation states that the times to exhaustion of resources for the population after the transition are about one-tenth of the values shown in Table 7-8.

The important conclusion from this calculation is that *even if we immediately started using resources at a rate characteristic of a fully industrialized world, many resources are in almost infinite supply.*

The table also shows that some of the metals we use are in rather short supply. These latter include cobalt, copper, lead, and tin, which are industrially important metals. Copper and tin, for example, would only last a few decades in a fully industrialized world, while zinc would last about a century. Nevertheless, Goeller and Weinberg contend that the relatively small resource base for these metals cannot lead to any kind of neo-Malthusian catastrophe. They properly point out that these metals are valued for their functions and that substitutes for them can be found. Goeller performed a detailed analysis of the extent to which cadmium, zinc, lead, copper, tin, and mercury could be replaced by materials that have a much larger resource base. Copper, for example, is most widely used to conduct electricity. But this function can be performed by aluminum, which is in much greater supply. In fact, optical fibers are rapidly replacing copper in telecommunications, and there is no copper shortage. The only functions of mercury that cannot be replaced with known technologies are the production of pharmaceuticals containing mercury and high performance mercury batteries. Goeller and Weinberg point out that this is hardly a major threat to civilization. The point is that nearly

95 percent of the worldwide use of metals consists of iron and aluminum, both in essentially infinite supply, and substitutes can be found for those metals that will become scarce in the future.

Of course, these optimistic estimates depend on an adequate energy supply. As lower grade ores are utilized, more energy will be required to process them. Goeller and Weinberg have examined the energy budget required to do this. While more energy per unit of metal will indeed be needed, the amounts are not extremely large; the per capita energy use would be greater than the current United States consumption by as much as a third. This ignores the fact that the efficiency of energy use has been increasing and is likely to continue to do so in the future. Referring to Chapter 5, it can be concluded that the availability of energy will not be a physical limitation on a fully industrialized world.

Hydrocarbons are a scarce resource. At current consumption rates they will last only a few millennia. After the demographic transition, they will last only a few centuries. Alternative energy sources to hydrocarbons will have to be developed if the world is to continue on its course of technological development.

Table 7-8 indicates that phosphorous is in even shorter supply than hydrocarbons. This is distressing because phosphorous is an essential element to sustain life. If, after the demographic transition, we only have a 130-year supply, we will be in serious trouble. But the numbers in Table 7-8 are all underestimates. Hypothetical and speculative resources are generally larger than usually reported. For phosphorous in particular, the amount reported in Table 7-8 does not include resources that are poorly understood but, according to the U.S. Geological Survey, are probably very large. Depending on the amount in these speculative resources, phosphorous may have to be recycled, or large amounts of energy will need to be expended to get it from very low grade sources. Neither of these options are impossible, and neither would bring industrial society to a halt.

It is worthwhile to point out that recycling can extend the life of a resource base considerably, but only if a lot of the resource is recycled. Recycling extends the resource base by a factor of $(1 - r)^{-1}$, where r is the fraction of the annual use of the resource that is recycled. Thus, 20 percent per year recycling extends the resource life by only 25 percent, 50 percent recycling doubles the resource lifetime, but if 90 percent of the resource is recycled its life is extended by a factor of ten. Recycling can be important but it must be done efficiently.

Of course, if enough energy is available the scarce resources can be extracted from many common rocks in which the resource concentration is small. The fact that the scarce resources (excluding hydrocarbons) comprise only a small fraction of our total material resource consumption is important. This means that even if the energy needed to extract them becomes larger, the economic result would not be catastrophic. Goeller and Weinberg have examined the energy budget for extracting scarce resources. The conclusion that a per capita increase of energy consumption by one-third would be needed to provide material resources indefinitely is theirs.

The detailed evolution of world societies certainly cannot be inferred from calculations such as these. Rather, they are an attempt to specify the conditions of resource availability within which industrialization must evolve. Within the uncertainties of data, the calculations provide an answer to the question of how soon shortages of material resources will pose a serious threat to industrial society. The uncertainties are surely not trivial; both population and resource availability are notoriously hard to predict. The critical resource is energy, and some of its components can be predicted with fair accuracy while others cannot. The amounts of hydrocarbons and uranium are probably known within a factor of two or three, but this degree of inaccuracy is not important for long-range projections. Hydrocarbons are in such short supply that they cannot provide a long-range energy source, while uranium is an essentially infinite source if used in breeder reactors. These conclusions are not changed by the uncertainties in our knowledge of their amounts.

Even given the large errors inherent in these calculations, it is still possible to conclude that materials and energy shortages in themselves are not likely to cause the collapse of civilization. But there is no cause for complacency either. If the entire world is to industrialize, there will have to be major large-scale changes during the next century. A ten-fold increase in materials and energy use cannot take place without extensive change. Some industries will vanish, others will grow, continuing technical innovation at an accelerating rate will be necessary, international competition will become more intense, and pollution problems will be much greater than at present. An industrialized world population of 10 to 12 billion will require a high order of human ingenuity. The Malthusian disaster is not the only reason for anticipating trouble on the way to industrialization. But at least the resources seem to be there. If industrial society collapses, it will be because of non-Malthusian forms of self-destruction. With sufficient wisdom, the standard of living and quality of life can continue to grow even as population increases.

APPENDIX 1

Mathematical Analysis of Technological Growth

Technological change is largely a set of interrelated growth phenomena. To come to grips with it in a quantitative manner, we must have in mind some model, a way of thinking about technology in a structured way. Technological change is effected through a large number of individuals and institutions, it is enormously complex, and its progress is the result of a great many interacting influences. It is remarkable that it exhibits any regularities at all; yet many regularities exist which are very useful for providing insights into the reasons for technological change, the forces that sustain it, its results, and its probable future course.

Let's start by listing some of the questions we would like to have illuminated by quantitative studies of technological change. The first of these might be of a macroscopic nature, which address the large-scale effects of technological progress. These would include:

1. What is the effect of technology on life span, on birth rates, on death rates, and on population growth?
2. How is technology related to the Gross National Product, the standard of living, and the relative distribution of income?
3. Are natural resources sufficient to maintain technological growth in the future? Is there a danger that they will be exhausted?
4. Are less-developed countries catching up to the West technologically? To what extent is technological change responsible for differing national growth rates?

There are many other large-scale issues that could be considered, but these are illustrative of questions whose quantitative analysis can give results of global importance. At a somewhat more refined level we could ask the following:

5. What is the time lag between invention (the conception of a technological advance) and innovation (putting it into practice)?
6. What is the frequency of invention and innovation? Do these frequencies change with time?
7. What is the rate at which innovations spread among nations, among corporations, or among consumers?
8. What is the rate at which a new technology replaces an older one? What is the lifetime of a given technological process or product type?
9. What are the limits to improvement of particular types of technological products or processes? How rapidly are these limits reached?
10. What fraction of inventions result in commercial applications (innovations)?

These are representative of the questions that deal most directly with the process of technological change itself. A further group of questions arises in connection with technology based corporations. Corporations are the institutions through which most technological innovation is implemented in the Western societies, so they have a particular importance when we are trying to understand technological change. We therefore consider the following:

11. How profitable is innovation to individual corporations relative to society as a whole?
12. What is the relation between the life cycle of a given technology and the limits to the markets for the products of that technology?
13. How is the growth rate of a corporation related to the level and rate of innovation?
14. What is the effect of differing growth rates on the relative positions of corporations in the absence of direct competition? In the presence of direct competition?

Technology has its most important effects on individuals, and it is for these effects that we value or criticize technology. A great many questions relating to these can be investigated quantitatively. Among them are:

15. What is the rate of change of environmental pollution?
16. What is the rate of increase of personal mobility and personal communication?
17. How is technological change related to job creation or technological unemployment?

18. What is the trend in prices to consumers of basic necessities, of technology-based products, of luxuries?
19. What is the rate of change in hours of labor required to maintain the average standard of living?
20. What are the trends in personal hazards in the work place, at home, in general? To what extent are these the result of technological change?

Technology so pervades modern life that these lists could go on indefinitely. But this is sufficient for the purpose here, which is to have something specific in mind as I present the general model by which the structure of technological change is analyzed.

The basic unit in the model is the technological enterprise, that is, any organization or group that effects or contributes to technological change. Examples are R&D project teams, R&D laboratories, corporations, entire industries, university science or engineering departments, technical institutes, professional organizations, consumer groups, law-making bodies, government agencies, and even the nation or the world as a whole. Some of these units have a direct responsibility for technological change while others do not. But each of them has a direct or indirect impact on technological change as a result of its activities.

Each of these technological enterprises is a social system in the sense that it contains people who are organized into groups that have some functions of their own and also carry on activities that contribute to the function of the larger enterprise. That is, a technological enterprise is a system that consists of subsystems. The subsystems may be more or less tightly coupled to each other. In some corporations, for example, all the divisions are made to follow a specific set of objectives and policies so that the work of one division is strongly linked to that of the others. In universities, the work in one department is often not strongly influenced by that of the others. In the nation as a whole some of the subsystems are closely coupled while others have no direct connection with each other at all.

With this in mind let us consider a technology-based corporation as an example of a technological enterprise. Its growth is determined by three different kinds of factors. The first set of factors arises from the internal dynamics of the corporation itself. These include the corporation's strategic posture, the allocation of resources, the nature of its marketing effort, its R&D expenditures, its proclivity for risk or for risk aversion, and so on. That is, the corporation has a set of characteristics that strongly influence its growth and would completely determine its growth if it were not for the other factors. But these other factors do exist, and they arise from the actions of the outside world that affect the corporation. There are two kinds of external influences, regular and irregular. The regular factors are those that are consistent, are relatively constant, or change smoothly with time, and can therefore be explicitly included in the corporation's response to the outside world. Such factors might be the existence of competitors, degree of saturation of the market, taxes, the state of the art of relevant technology, and regulatory restrictions. The irregular factors are those that do not have this kind of consistency. They might include sudden changes in regulations, taxes or interest rates, international political events such as the oil embargo, or the sudden emergence of an unexpected technology.

In our model, the world contains a large number of technological enterprises of a wide variety of types, sizes, and functions. Each of these has a degree of autonomy arising from its internal nature that is the equivalent of immanent forces that drive its evolution in time. But this evolution is not determined solely by internal factors. The enterprise is subject to external forces because it is coupled to other units whose functions and evolution interact with the enterprise. This is analogous to a simple input–output model, in which the enterprise generates an output as a result of its internal activities but also receives inputs from other units which influence its own internal functions, thereby affecting its output. The inputs include both those that are deterministic, or systemic, in the sense that they vary smoothly in time and can be anticipated, and random, or idiosyncratic, factors, in the sense that they are not predictable far enough in advance to permit an optimum response. The terms "deterministic" and "systemic" are meant to imply that there are certain characteristics of our culture, economy, and social structure that are relatively constant in time and that are major determinants of technological change. The terms "random" and "idiosyncratic" are meant to recognize the fact that natural events and human activities are seldom completely predictable and that there is a degree of uniqueness to every event and every human action.

The most important feature of this model is that it is partially systemic. It does not view the evolution of technology as the result of totally inevitable laws, nor does it treat technological change as a series of mere historical or accidental events. It assumes that the actual state of affairs is somewhere between these two extremes and that the mix of determinism and indeterminism is different in different circumstances and at different times. The task of technological dynamics is to sort out any determinism that is operative in technological enterprises. It cannot say much for the nondeterministic or random factors which, by nature, are not amenable to simple mathematics and are better dealt with by historical analysis. The distinction between causal and random factors is not always clear, and a random event can be absorbed into the system and become part of its deterministic structure. The sharp rise in oil prices, for example, qualifies as an irregular, idiosyncratic factor because it was the result of conscious political action and was not inherent in the market forces up to that time. But once the price rise took place, it was gradually absorbed into the market and the role of OPEC became an intrinsic part of the system.

I have emphasized this description of technological change several times because it is the essential viewpoint of this book that such a description is basic to understanding the quantitative aspects of the evolution of technology.

Of course, historical analysis has been and will continue to be a powerful method of learning about all aspects of technological change, including those that are systemic. There is no substitute for the detailed knowledge of individual decisions, social interactions, institutional structures, and special events that are revealed by a good historical analysis or case study. But we are after something different. We want to know to what extent quantitative regularities exist in the evolution of technology and what fundamental causes give rise to these regularities. There is no doubt that this is of value in its own right and that it also can inform and guide historical studies.

A correct mathematical description of our model would obviously be very complex. In fact, the mathematical analysis of technological change is in a rudimentary state and a great many basic studies remain to be done. Accordingly, we adopt a rather straightforward approach by isolating the streams of data that show a regular change with time and subject them to analysis using simple growth laws. Each stream of data refers to a particular enterprise and is the result of all the forces, both internal and external, that act on that enterprise. A correct mathematical description of our model would consist of a large number of coupled equations to represent these data. Technological change is not yet understood in sufficient detail to support such a large-scale detailed analysis, but considerable progress can be made with a more modest approach in which growth laws are applied to one or a small number of time series.

A variety of equations have been used to describe growth of various kinds. In this Appendix, I summarize the ones that have been most commonly used and that are representative of the kinds of growth observed in technology.

FORMS OF GROWTH LAWS

Growth is essentially a mathematical concept which describes the change of some quantity with time. Normally, the quantity is measured at regular time intervals; the Gross National Product, for example, is reported annually, and the sign and magnitude of its change from year to year is taken as a measure of economic health.

The analysis of growth starts with a time series of some quantity (such as production rates, GNP, or population). At a series of times given by,

$$t_1, t_2, t_3, \ldots t_j, \ldots t_N \tag{1}$$

the results of a measurement of a quantity x are

$$x_1, x_2, x_3, \ldots x_j, \ldots x_N \tag{2}$$

The growth rate for a given time, t_j, is defined by

$$a_j = \frac{x_j - x_{j-1}}{x_{j-1}(t_j - t_{j-1})} \tag{3}$$

That is, we take the difference in the values of x at two succeeding times divided by its value at the earlier time and the time interval. If the time intervals are all equal, we can set them to unity (i.e., one year) and Equation (3) takes the simpler form

$$a_j = \frac{x_j - x_{j-1}}{x_{j-1}} \tag{4}$$

and succeeding values of x are related through the growth rates by

$$x_j = (1 + a_j)x_{j-1} \tag{5}$$

In analyzing growth it is assumed that there is some underlying regularity in the time series; that is, some mathematical law exists that describes the sequence of x's. Otherwise the x's would just be a set of random numbers from which not much of interest could be learned. In its simplest form, the growth law is a continuous function of time.

$$x = x(t) \tag{6}$$

with x_j being the measured values of the function $x(t)$ at particular times.

But we must be careful here. When a growth law is assumed, a kind of causality is assumed along with it. The precise nature of the causality may be in doubt, or even totally unknown, but the mathematical law implies that causes exist from which it flows. There are two problems that must be addressed in relating the time series to the growth law. The first is common to all empirical work and concerns the accuracy of the measurement itself. Even in the best of controlled laboratory conditions there is scatter in the data; in the analysis of economic, social, demographic, or historical trends the problems of accurate measurement and data collection are extreme, and significant scatter can be expected as a result. This is a serious problem, but it can be taken into account by being careful in the data selection process, by recognizing the limitations of the data, and by statistical analysis. The second problem, however, arises from the meaning of causality itself in complex systems. For example, we may expect a regular increase in the GNP with time as a result of continuing capital investment, technological advancement, and population increase. These are all factors that increase GNP, and if they are constant then the GNP can be expected to increase smoothly. But wars, political acts such as the oil embargo, severe weather, regulatory changes, and sudden changes in consumer confidence can all have serious effects on the actual GNP. Growth laws are gross oversimplifications of the actual dynamics of real systems that imply a single causal factor or a small set of related causal factors as being the major driving force for the growth rate. They assume that certain conditions continue to prevail over the time period being studied or that these conditions change slowly and regularly so that they are smoothly reflected in the growth law itself.

If a growth law exists, we therefore adopt the following posture towards the quantity defined by the growth law and the corresponding quantity observed in real systems: The observed quantity is the sum of the effect of the causal factors underlying the growth law, the inherent inaccuracies in the data, and the irregular factors defined in our model of the technological enterprise that are unpredictable from the growth law. To formalize this, we write the growth law as a continuous function of time,

$$y = y(t) \tag{7}$$

and relate the observed quantities x to those in the growth law by

$$x_j = y(t_j) + d_j \tag{8}$$

The quantity d_j is the deviation of the actual data from that defined by the growth law. A major problem of the statistical analysis of time series for growth is the extraction of the functional form $y = y(t)$ from the observed numbers x_j.

Growth laws are mathematical functions that describe the change in time of some parameter of interest. They are a special class of functions in that their first derivatives depend on the value of the parameter itself. This is an important distinction. If, for example, a population's entire increase takes place by immigration then we define this as accretion rather than growth. The key definition of growth is that the rate of growth depends on how much growth has already occurred. The biological increase of population is an example of growth, since the growth rate of the population at a given time clearly depends on the size of the population at that time. Growth laws therefore have the form.

$$dy/dt = F(y) \tag{9}$$

where $F(y)$ is some function of the growing parameter y. If $F(y)$ is positive, Equation (9) describes self-sustaining growth. If it is negative, y is in decline (negative growth).

An important class of growth laws can be obtained by expanding the function $F(y)$ in a Taylor series to get

$$dy/dt = K + ay + by^2 + cy^3 \ldots \tag{10}$$

The shape of the growth curve depends on the values of the coefficients $K, a, b, c, \ldots,$. The simplest case is that for which all coefficients are zero except K, thereby giving a constant rate and a linear growth law. This is seldom observed in real systems.

If all coefficients are zero except a, then (10) yields the familiar exponential rate equation:

$$dy/dt = ay \tag{11}$$

For positive values of a, this integrates to the exponential rise, while for a less than zero it gives the exponential decay. We will be primarily interested in the exponential rise.

Next consider the case for which only the first two constants, K and a, are nonzero in Equation (10) and let K be positive but a be negative. Then (10) can be written as

$$dy/dt = k(Y - y) \tag{12}$$

where k and Y are positive constants and Y is always greater than y.

This is the rate equation for the decaying exponential rise. Note that there is an upper limit on the growth rate, which is a maximum at $y = 0$ and then decreases to zero as y approaches Y. This is quite different from the rate for the exponential, which can increase indefinitely.

Another important special case of Equation (10) is that for which a is positive, b is negative, and all other constants are zero. We then write (10) as

$$dy/dt = ky(Y - y) \tag{13}$$

with $k > 0$ and $Y > y$.

This is the logistic rate equation and represents a wide variety of growth phenomena. For small values of y, it starts out as an exponential rise but, as y approaches its limiting value of Y, the rate goes to zero.

There are two equations involving the logarithms of the growth variable that are important for technological change. The first is the Gompertz rate equation which is

$$dy/dt = ky(lnY - lny) \tag{14}$$

which approaches zero as y approaches Y, and the second is the super acceleration equation

$$dy/dt = kylny \tag{15}$$

which has no upper limit and increases very rapidly with increasing y.

Each of these rate equations implies something different for the forces that drive the growth of the variable y, and it is worthwhile to consider the general nature of the causality implicit in them.

The exponential is the prime example of self-sustained growth. Equation (11) states that the rate at which y increases is directly proportional to y so that doubling the value of y doubles the rate of its increase (assuming that a is positive). This is the simplest kind of feedback in which an increase in y increases the rate of its increase which further increases y and so on ad infinitum. It is this feedback which gives this growth law its remarkably expansive power.

Equation (11) represents a law whose causality is easily understood at least at a superficial level. If a population is increasing exponentially this means that the increase depends directly on the size of the population. If the GNP of an industrialized society increases exponentially this means that it continually reinvests a portion of its growing wealth. Of course, this only scratches the surface of the causality in such situations, but the mathematical form provides a base for an analysis of the deeper deterministic factors.

The logistic equation is an example of self-sustained growth, but its causal implications are different from those of the exponential. In addition to being driven by how much has already been achieved, the logistic is also limited by a finite potential of achievement. This is the meaning of the term $(Y - y)$ in Equation (13). The logistic assumes that there are limits to growth which may be the result of limitations in the environment or inherent in the growth process itself. The exponential assumes that there are no limits and that growth can continue indefinitely.

The binary adoption model is an example of a mechanism represented by the logistic equation. In this model a new innovation is adopted by individuals or groups as a result of coming into contact with those that have already adopted the innovation. At each contact, there is a constant probability that the innovation will be adopted. Then the rate of adoption will be proportional to the number that have adopted, times the number that have not, and the rate constant is the frequency of binary contacts.

Another way in which an innovation spreads is through the dissemination of information from a constant source such as radio or television. In this case, the rate of adoption is proportional only to the number that have not yet adopted. The rate is then given by Equation (12), the rising exponential decay. The mechanism represented by (12) is therefore one in which the cause is a constantly operating source and does not depend on binary contacts.

The Gompertz rate equation is one in which growth is determined by percentages rather than absolute values. To see this, rewrite (14) as

$$dlny/dt = k(lnY - lny) \tag{16}$$

This shows that the change in the logarithm of y in an increment of time is proportional to the difference between the logarithms of the current value of y and its ultimate value. When y is small, the rate is exponential, but as y approaches its limit, Y, the rate becomes proportional to the percentage difference between Y and y.

The mechanism for the superacceleration equation is similar to that for the exponential in that the rate increases with increasing value of the variable, but it is much more powerful. The presence of the logarithm greatly enhances the positive feedback beyond what it would be if only the variable y appeared in Equation (15). One interpretation of the superacceleration equation is that the growth rate increases not only because of the value of y, but also because of the cumulative effect of past growth.

PROPERTIES OF THE EXPONENTIAL

Integrating Equation (11) gives the exponential growth equation:

$$y = y(0)\exp(at) \tag{17}$$

where $y(0)$ is the value of y at an initial time $t = 0$.

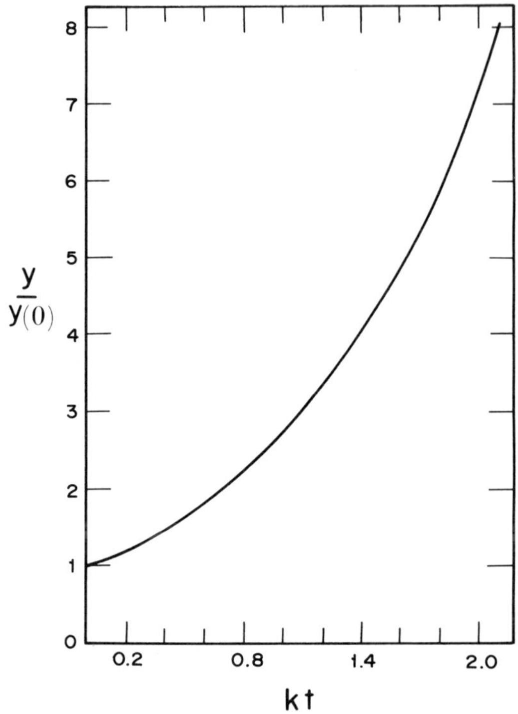

Figure A1-1 The exponential curve.

Note that if (17) is substituted into (11) the result is

$$dy/dt = ay(0)\exp(at) \tag{18}$$

so that the rate of change of y as well as y itself increases exponentially.

The general properties of the exponential function are shown in Figure A1-1. It starts off slowly but soon accelerates and rapidly reaches high values. Note that there is no upper limit to the growth. Just wait long enough and the growth variable can reach any value no matter how large. It is this mathematical fact that so frightens the proponents of policies for limiting economic or population expansion.

The exponential is often used to analyze two different kinds of phenomena. The first is typified by the unrestricted growth of a population and the second by the consumption of natural resources. For the first case, let $P(t)$ be the population at time t of an exponentially growing population so that

$$dP/dt = aP \tag{19}$$

$$P(t) = P(0)\exp(at) \tag{20}$$

where a is the rate constant for the population growth and $P(0)$ is the population at time $t = 0$.

Table A1-1 Times for Increases by Factors of 2, 10, and 100 for Various Rates of Exponential Growth

Growth Rate = a	$t(2)$	$t(10)$	$t(100)$
0.01	69.3	230.5	460.5
0.02	34.7	115.1	230.3
0.03	23.1	76.8	153.5
0.04	17.3	57.6	115.1
0.05	13.9	46.1	92.1
0.06	11.6	38.4	76.8
0.07	9.9	32.9	65.8
0.08	8.7	28.8	57.6
0.09	7.7	25.6	51.2
0.10	6.9	23.1	46.1

We want to relate the growth constant a to the time it takes to increase the population to some multiple of its initial value. To do this, take the value of (20) at two times t and t' and divide one by the other to get

$$\frac{P(t')}{P(t)} = \frac{\exp(at')}{\exp(at)} \tag{21}$$

from which

$$(t' - t) = \frac{1}{a} ln\left[\frac{P(t')}{P(t)}\right] \tag{22}$$

Usually, the effect of exponential growth is described in terms of the time it takes to double the population. In addition to this, we will also examine the time it takes to increase the population by a factor of ten and a factor of 100. That is we compute the times for $P(t')/P(t)$ to assume the values 2, 10, and 100. From (21), these times are

$$t(2) = (ln\ 2)/a = 0.6931/a$$

$$t(10) = (ln\ 10)/a = 2.3025/a$$

$$t(100) = (ln\ 100)/a = 4.6051/a$$

The results of calculations from these formulas for growth rates ranging from 1 to 10 percent per year are given in Table A1-1.

The results in Table A1-1 illustrates the power of the exponential. At a growth rate of only 3 percent per year, a population doubles in 23 years, increases tenfold in less than 77 years, and one hundred times in a little more than a century and a half. Furthermore, rather small increases in the growth constant have profound effects. At a 5 percent annual growth rate it only takes 92 years for the population to increase a hundredfold. No wonder there is so much concern about world population growth. Of course, whether or not real populations grow exponentially over long periods of time is another story. In fact they do not. Sooner or later limiting factors come into play.

Incidentally, the relation between the rate constant and the doubling time, $a = 0.6931/t(2)$, permits the exponential to be written in a universal form by defining a reduced time t_r by

$$t_r = t/t(2) \tag{23}$$

so that

$$at = 0.6931t/t(2) = 0.6931t_r \tag{24}$$

The exponential then takes the form

$$y = y(0)\exp(0.6931t_r) \tag{25}$$

This describes all exponential growth in terms of the ratio of the time to the doubling time.

The analysis of the exponential for resources starts with the assumption that a nonrenewable natural resource is consumed at a rate given by

$$dR/dt = bR \tag{26}$$

so that

$$R(t) = R(0)\exp(bt) \tag{27}$$

where $R(0)$ is the amount of the resource that has been consumed at time $t = 0$, $R(t)$ is the amount consumed at time t, and b is the annual growth constant for consumption.

Let the total amount of the resource available at time $t = 0$ be $R(\text{avail})$. If the rate of consumption were to remain constant at the value it has at time $t = 0$, then the number of years the resource would last would be the total amount available divided by the consumption rate. Since the consumption rate at $t = 0$ is given by Equation (26) as $bR(0)$, the time it would take to exhaust the resource if this rate were to continue is

$$t_L = R(\text{avail})/bR(0) \tag{28}$$

We have called this time t_L to indicate that it is a linear extrapolation assuming the consumption at $t = 0$ does not change. This time is of interest because reserves of resources are often reported in terms of the number of years they would last at current rates of consumption, which is just t_L.

If the exponential law holds, however, then the time to exhaust the resource is obtained by computing the time it takes to use up the amount $R(\text{avail})$. Let this time be denoted by t_x. The amount of resource available is then the amount used up by time t_x minus the amount used up at time zero. That is,

Table A1-2 Exponential Exhaustion Times for Various Growth Rates and Linear Exhaustion Times

Growth Rate	Linear Exhaustion Time (Years)					
	10^2	10^3	10^4	10^5	10^6	10^7
0.001	95	693	2400	4620	6910	9210
0.002	91	549	1520	2560	3800	4950
0.003	87	462	1140	1900	2670	3460
0.005	81	358	786	1240	1700	2160
0.010	69	238	461	690	921	1150
0.015	61	185	334	487	641	794
0.020	55	152	265	380	495	610
0.030	56	114	190	267	343	420
0.050	35	78	124	170	216	262

$$R(\text{avail}) = R(t_x) - R(0)$$

Using Equation (27) this becomes

$$R(\text{avail}) = R(0)\exp(bt_x) - R(0) \tag{29}$$

From the above two equations, (28) and (29), we readily see that the linear and exponential exhaustion times are related by

$$t_x = ln(1 + bt_L)/b \tag{30}$$

Table A1-2 shows the exponential exhaustion times for different linear times and growth rates as computed from this equation.

Table A1-2 shows that even very large amounts of existing resources are used up rapidly if their use increases exponentially. For example, a resource that would last ten thousand years if consumed at current levels would last only 461 years if the consumption rate increases by one percent per year. Even a resource that would last ten million years at current rates of consumption would be exhausted in 262 years if consumption increases at the rate of 5 percent per year. It is clearly important to determine if the consumption of natural resources actually does increase exponentially.

Clearly, the exponential generates large numbers in relatively short times. Just as clearly, in most real situations the immense growth implicit in the exponential cannot occur. When an exponential growth is observed in real data, the first question that must be asked is if there are intrinsic physical limits to the indefinite growth. Population growth, for example, cannot continue to the point where every square foot of land is occupied by human beings.

If there are physical limits, then the question arises as to whether the exponential is an intrinsic property of the growing system or if it is an approximate representation of the data to date. That is, we would like to know if the data represent a general law of the system. If it does, and if the consequences of the law are undesirable, then we can think about ways of changing it. Very often, however, observed exponentials are the early stages of a growth phenomenon that has ultimate limits.

PROPERTIES OF THE LOGISTIC

It is often convenient to write the logistic equation in reduced form by defining the ratio

$$f = y/Y \tag{31}$$

which is just the fraction of its limiting value that y has reached at time t. In terms of f, Equation (13) becomes

$$df/dt = bf(1 - f) \tag{32}$$

where $b = kY$.

This equation is easily integrated by using partial fractions to convert it to the form

$$\frac{df}{f} + \frac{df}{(1 - f)} = bdt \tag{33}$$

which integrates to

$$f = \frac{C \exp(bt)}{1 + C \exp(bt)} \tag{34}$$

where C is an integration constant.

It is useful to absorb the integration constant into the exponential by defining a new constant $t_{1/2}$ and rewriting (34) as

$$f = \frac{1}{1 + \exp[b(t_{1/2} - t)]} \tag{35}$$

Note that when $t = t_{1/2}$, $f = 1/2$. That is, $t_{1/2}$ is the time at which growth has reached half its limiting value. Accordingly, $t_{1/2}$ is called the half-time of the growth process.

An interesting feature of this equation is that it assigns an upper limit to the possible values of y. As the time increases, y goes up but it can never exceed the value Y. This is easily seen from Equation (35), since, as t becomes very large, the exponential in the denominator goes to zero. Then f goes to unity. That is, y approaches Y as t approaches infinity.

When f is small, it can be neglected relative to unity in Equation (30), which then reduces to the exponential rate equation. At early times, therefore, the logistic starts out growing exponentially as shown in Figure A1-2. As time goes on, f can no longer be neglected and the growth starts to slow down under the influence of the factor $(1 - f)$. The result is that f increases

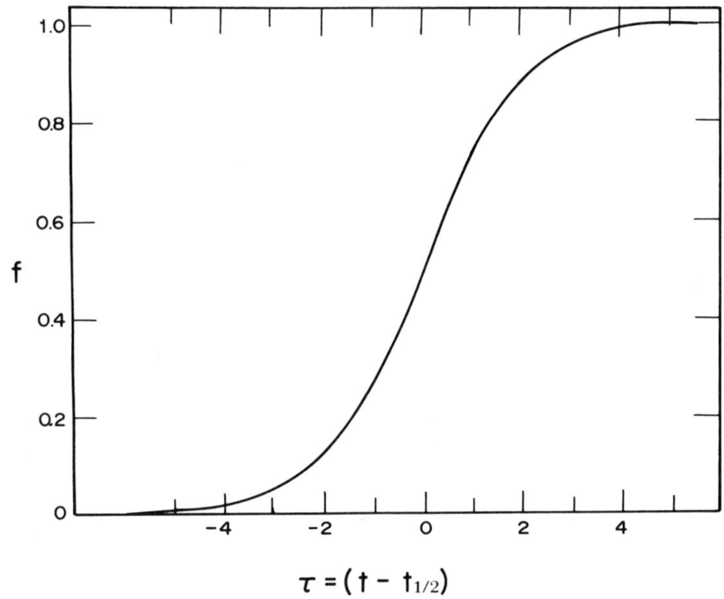

Figure A1-2 The logistic curve.

more and more slowly, asymptotically approaching its limiting value of unity. The maximum value of the growth rate occurs when $f = 1/2$ at a time equal to the half-time $t_{1/2}$.

Figure A1-3 shows the form of the growth rate as a function of time; this is the derivative of Equation (33).

The maximum in the growth rate at $t = t_{1/2}$ corresponds to the inflection point in Figure A1-2.

Equation (35) is often written in the form

$$\frac{1}{1 - f} = \exp[b(t - t_{1/2})] \tag{36}$$

so that a linear plot can be obtained from the equation

$$ln[f/(1 - f)] = bt - bt_{1/2} \tag{37}$$

Plotting empirical data according to this equation yields a straight line if the growth is logistic. The growth constant b and the half time $t_{1/2}$ can then be computed from the slope and intercept of the line.

It is often of interest to have a measure of how long the growth takes to run its course. However, the logistic equation extends indefinitely into the past and never reaches its final value, although it approaches it asymptotically. The best we can do is to define a time period over which most of the growth takes place. The conventional definition is that of the takeover time

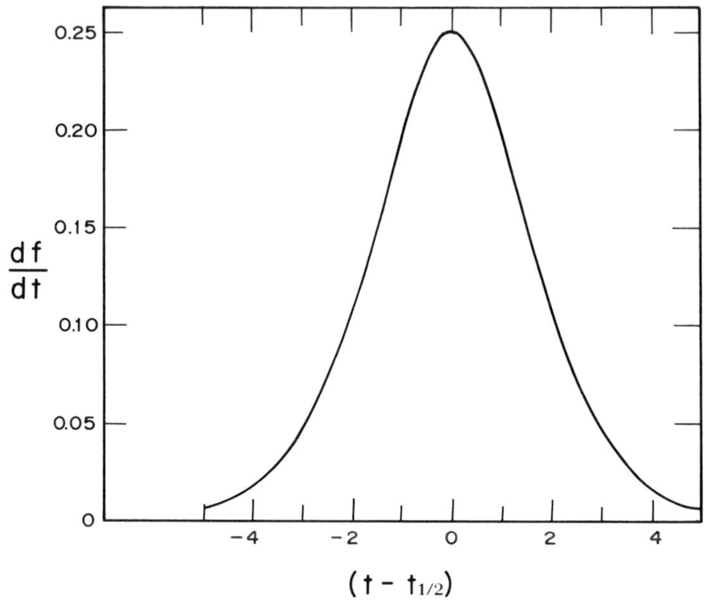

Figure A1-3 The rate curve for the logistic.

which is the time required for the process to go from 10 percent to 90 percent completion. This is obtained by writing (35) for the times $t(.1)$ and $t(.9)$ when f equals 0.1 and 0.9 respectively.

$$ln(0.1/(1 - 0.1)) = -2.197 = bt(.1) - bt_{1/2} \tag{38}$$

$$ln(0.9/(1 - 0.9)) = 2.197 = bt(.9) - bt_{1/2} \tag{39}$$

Subtracting (38) from (39) gives the takeover time $T_{.9}$ as

$$T_{.9} = t(.9) - t(.1) = 4.394/b \tag{40}$$

Note that if $4.394/b$ from (40) is substituted into (36) the result is

$$ln[f/(1 - f)] = 4.394(t - t_{1/2})/T_{.9} \tag{41}$$

which is a universal logistic equation for the fraction of completed growth as a function of the single parameter $\phi = (t - t_{1/2})/T_{.9}$.

It is usually more realistic to use the logistic equation to analyze the consumption of natural resources rather than the exponential. The increase in consumption cannot continue indefinitely. Eventually, the consumption rate of any resource must slow down either because the supply of the resource is limited or because the demand for it falls or levels off.

The case of supply-limited resource use assumes that all that can be produced will be used and that a constant or increasing demand will spur improvements in extraction techniques until most of the resource is consumed. In this case the use of the resource starts at a low rate but increases rapidly until its increasing scarcity and expense of extraction forces the production rate down. The production curve then has a bell shape, in which the rate of production is small at early and late times and a maximum sometime between. The logistic differential equation is usually applied to this situation so that the production rate of a supply-limited resource is

$$\frac{dR}{dt} = \alpha \frac{1}{R_M}(R_M - R) \tag{42}$$

where R_M is the total amount of the resource in existence before it first started to be extracted and α is the rate constant. The rate of use is illustrated by Figure A1-3. The total amount of resource produced at a time t is just the integral of (42), which is

$$R = \frac{R_M}{1 + \exp[\alpha(t_{1/2} - t)]} \tag{43}$$

Proceeding as in the case of the exponential use of resources in which the linear and exponential times to exhaustion were related, the half-time $t_{1/2}$ can be related to the linear production time as follows:

If the production rate were to stay constant at its value for $t = t_0$, then the time to complete depletion would be

$$t_L = \frac{(R_M - R_0)}{(dR/dt)_0}$$

where R_0 is the rate of production at time t_0. Taking $t^0 = 0$, combining this with (42) gives

$$t_L = R_M/\alpha R_0$$

Using Equation (43), this gives

$$t_{1/2} = \frac{1}{\alpha} ln(\alpha t_L - 1) \tag{44}$$

PROPERTIES OF THE DECAYING EXPONENTIAL RISE

Integration of (12) yields

$$y = Y - [Y - y(0)]\{1 - \exp[-k(t - t_0)]\} \tag{45}$$

where $y(0)$ is the initial value at the initial time t_0. If y and t are taken to be zero at the initial time, then Equation (45) becomes

$$y/Y = 1 - \exp(-kt) \tag{46}$$

Differentiation gives the rate as a function of time:

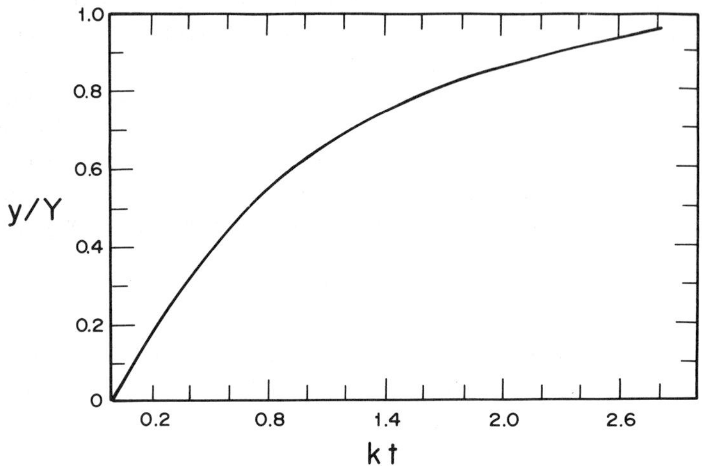

Figure A1-4 The decaying exponential rise.

$$dy/dt = Yk\exp(-kt) \tag{47}$$

which shows that the rate for this process decays exponentially from an initial value, hence its name.

Figure A1-4 shows a curve for the decaying exponential rise in which y/Y is plotted against kt according to Equation (46). The initial rise is rapid but becomes slower as time passes because the slope decreases as given by Equation (47). The rate itself starts out at a high value but decreases rapidly as shown in Figure A1-5. The first point to note is that the initial part of the growth is not concave upwards as is the case for both the exponential and the logistic. As a result, the shape of the growth is quite different from that of the logistic, even though both approach a limiting value asymptotically for long times. Also, the decaying exponential rise starts at time zero with a finite value of the growth variable. In Equation (46) this value is zero. For the logistic, however, the growth variable continually gets smaller as we go back in time, and there is no time at which it equals zero.

It is of interest to compare times for the decaying exponential rise and the logistic. To do this write Equation (46) as

$$t = -[ln(1 - y/Y)]/k \tag{48}$$

and compute the time it takes for y/Y to go from zero to 90 percent of its limiting value of unity, which is

$$t(.9) = -(ln(.1))/k = 2.303/k \tag{49}$$

Comparing this with the result for the takeover time of logistic growth given by Equation (40) shows that, for equal values of the rate constant,

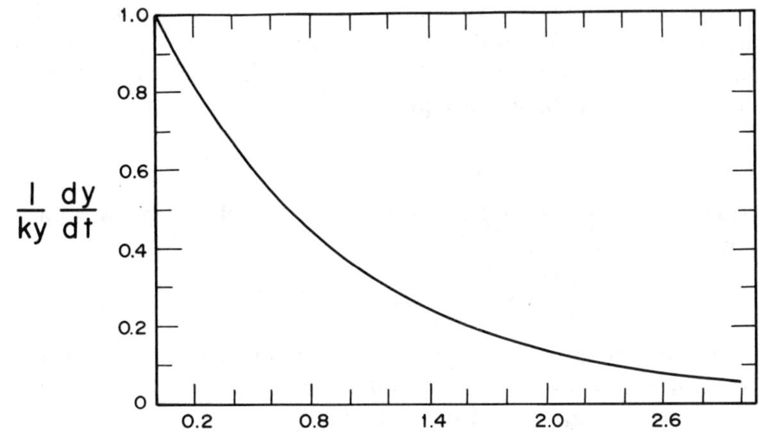

Figure A1-5 The rate for the decaying exponential rise.

takeover is faster for the decaying exponential rise. However, it must be kept in mind that the rate constants have different units and different meanings in the two equations. In the decaying exponential rise the rate constant is analogous to the frequency of individuals receiving messages; that is, it is a frequency per individual. In the logistic the rate constant is analogous to a binary collision; that is, it is a frequency per pair of individuals. A more compatible comparison would therefore multiply the logistic rate constant by the value of the growth variable. Even taking this into account, however, the decaying rise takes over more quickly for comparable rate constants because its initial growth is higher. This presumes that the limiting values are the same for both the logistic and the decaying exponential rise representations.

THE GOMPERTZ EQUATION

Using the reduced variable $f = y/Y$, the Gompertz rate equation, (14), becomes

$$d\ln f/dt = -k\ln f \tag{50}$$

The integration is readily performed by treating $\ln f$ as the variable of integration with the result that

$$\ln f = C\exp(-kt) \tag{51}$$

where C is a positive integration constant. The Gompertz equation is usually written in the form

$$f = \exp(-C\exp(-kt)) \tag{52}$$

This yields an S-shaped curve roughly similar to the logistic but with some important differences. The inflection point, for example, does not occur at the value $f = 1/2$. To get the inflection point, just differentiate the rate equation (50) and set the result equal to zero. The result is

$$lnf(T_M) + 1 = 0 \tag{53}$$

where $f(T_M)$ is the value of f at time T_M for which the growth rate is a maximum. From (53),

$$f(T_M) = 1/e = 0.36788 \tag{54}$$

Thus, the growth rate is a maximum when the growth is 36.8 percent complete rather than 50 percent complete as for the logistic. The relation between the rate constants and the time T for the maximum rate is readily seen to be, from (51),

$$T_M = lnC/k \tag{55}$$

Figures A1-6 and A1-7 shows that the Gompertz curve and its derivative are not symmetric about the inflection point as is the case for the logistic. This is often cited as an advantage of the Gompertz curve in fitting empirical data since data do not necessarily have the strict symmetry forced by the logistic function. However, the Gompertz has a symmetry of its own since the inflection point must always occur at the value $f = 1/e$.

SUPERACCELERATION

The curve for superacceleration is the exponential of an exponential, as is evident from integrating Equation (15) which gives

$$y = Y\exp(\exp kt) \tag{56}$$

With increasing time, this curve increases very rapidly, even for very small values of the rate constant k. This is shown in Figure A1-8 in which y/Y is plotted against kt.

The rate as a function of time is obtained by differentiating Equation (56) to get

$$dy/dt = Yk\exp(kt)\exp(\exp(kt)) \tag{57}$$

so that the rate increases even more rapidly than the growth variable itself.

For the exponential growth function, we saw that the growth variable doubled for every increase in time given by $t(2) = 0.6931/k$. For superacceleration, it is the logarithm that doubles for each such time change. This means that y can start to grow very slowly, but as time passes it picks up speed at an ever-increasing rate until the growth of y becomes almost ex-

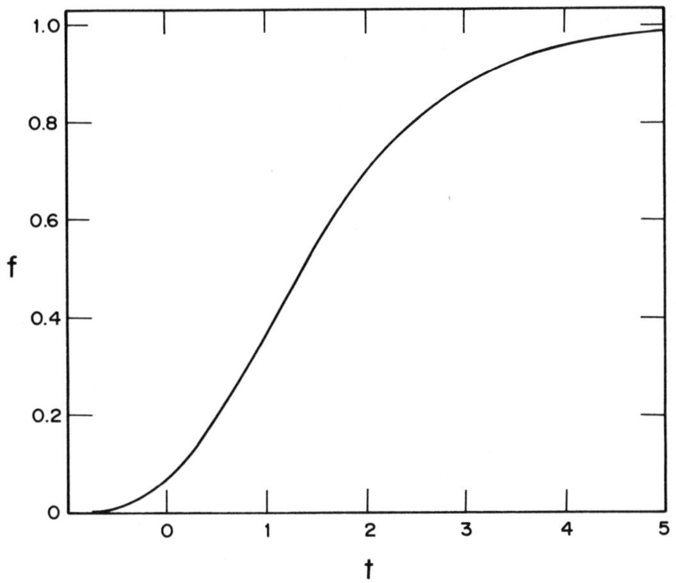

Figure A1-6 The Gompertz curve.

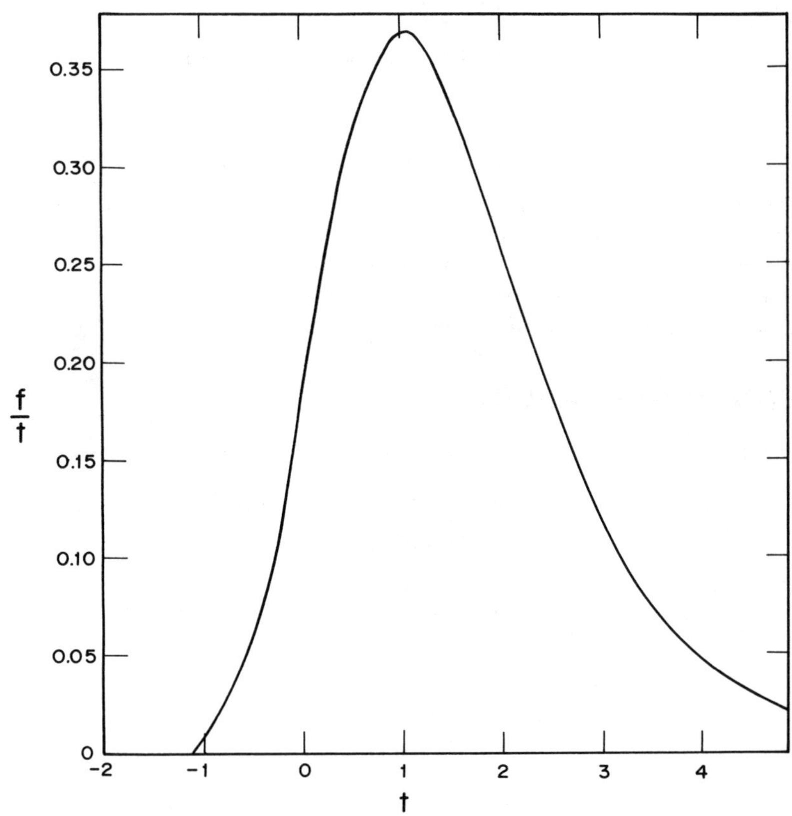

Figure A1-7 The rate for the Gompertz curve.

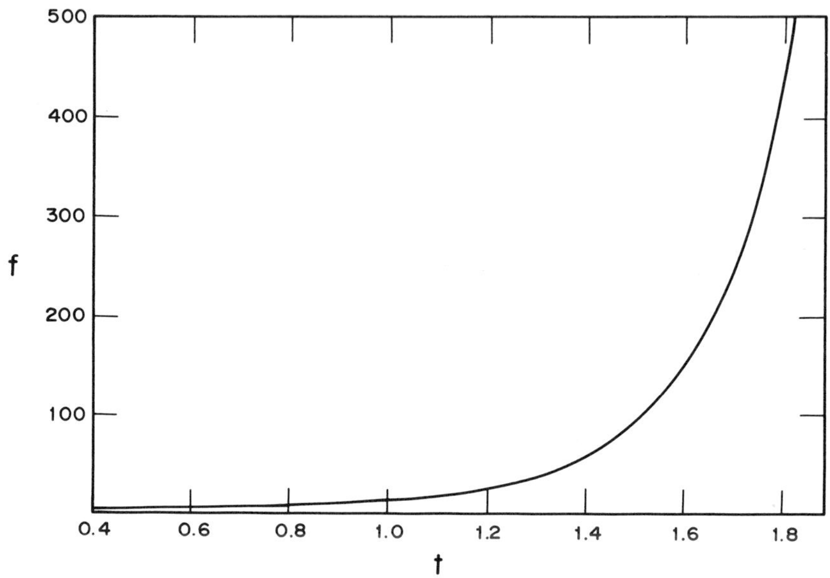

Figure A1-8 The superacceleration curve.

plosive. Because of this property, the superacceleration equation has been used to describe the long-range development of technology. For several million years, technological growth was very slow by today's standards, while the rate of advance since the Industrial Revolution has been so great that it makes previous ages seem static. This is precisely the kind of situation that the superacceleration equation describes.

COMPETITIVE GROWTH

Consider two exponential growth processes which follow the equations

$$x = x(0)\exp(at) \tag{58}$$

$$y = y(0)\exp(bt) \tag{59}$$

These equations may represent, for example, the growth of two subgroups of a population or the growth of the income of two individuals or the growth in sales of two corporations.

If (58) and (59) are solved for the time t and the results set equal so that the time is eliminated, the following equation is obtained:

$$x = Cy^g \tag{60}$$

where $C = x(0)^a/y(0)^b$ and $g = b/a$.

Another form of (60) is

$$lnx = glny + lnC \tag{61}$$

Equation (60), or (61), is called the allometric equation and describes the relation between the magnitudes of two growing quantities. It has wide application and, for example, describes the relation between the sizes of different organs in a growing organism. Note that if a is greater than b then x grows more rapidly and ultimately becomes very much larger than y.

An important point is that the allometric equation holds whether or not there is any connection whatsoever between x and y. The allometric equation would hold even if one of the exponentials described the growth of bacteria in a Petri dish while the other described the GNP of Brazil. The time can always be eliminated between two growth equations to yield a relation between two growing quantities. This is why it is dangerous to ascribe causal relations from the mere fact that two parameters are correlated. Such a causal relation may or may not exist, and this can be determined only from considerations that go beyond the fact that two data sets are correlated.

The allometric equation has important consequences for the relative positions of two growing entities whether or not the growth rates affect each other since the entity growing at the faster rate will eventually completely dominate. To see this, let x and y represent the size of two similar entities (for example, two populations, two corporations, or the income of two individuals) and compute the fraction

$$f(x) = x/(x + y) \tag{62}$$

which is the share obtained by the entity x at time t.

Using Equations (58) and (59) in (60) and rearranging gives

$$f(x) = \frac{1}{1 + A\exp(Bt)} \tag{63}$$

where $A = y(0)/x(0)$ and $B = (b - a)$.

But Equation (63) is just the logistic equation so that the share of x will increase towards unity while that of y will approach zero.

This result has some interesting consequences. If two corporations continue to grow exponentially, the one with the faster growth rate will eventually have a much greater fraction of sales and will economically totally dominate the other even though the slower corporation continues to get larger in absolute terms. Of course these results carry over when many competing units are considered. Those countries with faster growth rates get increasingly larger shares of the Gross World Product and those individuals with the fastest growing incomes get larger and larger shares of the wealth even though everyone's lot is improving. This is the fundamental reason that the distribution of wealth in an open market is uneven. If we grant the simplifying assumption that individual income increases are uneven in a market economy because of the varying abilities of people to command these increases, then an uneven distribution of wealth is inevitable. Furthermore, the distribution becomes more uneven as time goes by until the few with the highest growth rates acquire the preponderant share. The factors that mitigate this result in the real world are the fact that relative growth does not stay constant and the fact that governments take action in the form of progressive taxation or regulations or social welfare programs to redistribute income.

The time can be eliminated between any two growth equations, no matter what their form, to give a relation between the two growth variables. If this is done for two logistic growth functions f_1 and f_2 with rate constants a and b and half-times T_1 and T_2, the result is

$$ln\left(\frac{f_1}{1 - f_1}\right) = Kmln\left(\frac{f_2}{1 - f_2}\right) \tag{64}$$

where $K = \exp[a(T_1 - T_2)]$ and $m = a/b$.

Of course, competition can take place directly between two growing entities so that the growth of one is affected by the other. As an example, consider two companies that are competing for the same fixed market. Then the limiting market size is the same for each company, and the growth rate for each company will be proportional to the product of the sales already achieved and the total possible sales that remain. But the remaining possible sales are the result of subtracting the sales to date of both companies. The rate of growth equations for the two companies is therefore

$$dx/dt = ax(1 - x - y) \tag{65}$$

$$dy/dt = by(1 - x - y) \tag{66}$$

where x and y are the fraction of cumulative sales relative to total possible sales and a and b are the rate constants for the two companies. These are two coupled rate equations since the rate of change of x depends on the value of y as well as the value of x. At least one effect of the competition is easily apparent from these equations. Because of the presence of y in Equation (65), x will not approach the limiting value of unity. That is, the total sales that company x might achieve are decreased by the activity of company y and vice-versa.

Equations (62) and (63) lead to the allometric equation just as if they were two independent exponentials. This is readily seen by rearranging them to give

$$dx/axdt = 1 - x - y \tag{67}$$

$$dy/bydt = 1 - x - y \tag{68}$$

Equating the left-hand sides of (67) and (68) and rearranging gives

$$dlnx = gdlny$$

which integrates to

$$x/x(0) = [y/y(0)]^g \tag{70}$$

where $g = a/b$ and $x(0)$ and $y(0)$ are the initial values of x and y. This is just the allometric equation and this exercise emphasizes the result that an allometric relation says nothing about possible causal connections between the two variables. Such a relation may or may not exist. In this case it does because the allometric is the result of combining the two coupled equations (65) and (66), but in the case of combining two independent exponentials or logistic equations there was no causal connection between the variables.

MULTILEVEL SUBSTITUTION

Binary technological substitution, in which an old technology is displaced by a new, has a simple mathematical representation which was given in Chapter 4. Multilevel substitution, in which there is displacement of a sequence of technologies, is slightly more complex and is presented here.

There are two alternate approaches that can be used for multilevel substitution, as described in Chapter 4.

To develop the first approach, we label a sequence of innovations by an index $j = 1,2,3, \ldots N$ such that any innovation in the sequence substitutes for all the previous ones as a group. Therefore if n_j is the share of the jth technology, then the quantities that satisfy the binary substitution equations are:

$$\begin{aligned} f_2 &= n_2/(n_1 + n_2) \\ f_3 &= n_3/(n_1 + n_2 + n_3) \\ &\vdots \\ f_N &= n_N \end{aligned} \tag{71}$$

with

$$n_1 + n_2 + n_3 + \cdots + n_N = 1 \tag{72}$$

The first of Equations (71) states that technology number 2 is substituting for technology number 1; the second equation states that technology number 3 is substituting for technologies 1 and 2; and so on.

We call f_j the partial fraction for the jth technology. These are the quantities that are fitted to empirical data according to binary substitution equations.

The quantities of interest for comparison of the model to empirical data are the n's, so it is necessary to solve Equations (71) for the n's in terms of the f's. This is readily done by rewriting Equations (71) in the following form:

$$\begin{aligned} f_N &= n_N \\ f_{N-1} &= n_{N-1}/(1 - n_N) \\ f_{N-2} &= n_{N-2}/(1 - n_N - n_{N-1}) \\ &\cdots \end{aligned} \tag{73}$$

from which it follows that

$$
\begin{aligned}
n_N &= f_N \\
n_{N-1} &= f_{N-1}(1 - f_N) \\
n_{N-2} &= f_{N-2}(1 - f_{N-1})(1 - f_N) \\
n_{N-3} &= f_{N-3}(1 - f_{N-2})(1 - f_{N-1})(1 - f_N) \\
&\quad \cdot \quad \cdot \quad \cdot \quad \cdot \\
n_2 &= f_2(1 - f_3)(1 - f_4) \ldots (1 - f_N)
\end{aligned}
\tag{74}
$$

Note that f_1 is not defined and that n_1 is obtained from the normalization condition of Equation (72). The analytic procedure is now simple. Binary substitution equations such as the logistic are applied to each of the partial fractions defined by Equations (73). The actual fractional share of each technology is obtained from the partial fractions from Equations (74). The results give the evolution in time of the share of each technology.

In the second approach to multilevel substitution, the competition is assumed to be pairwise so the quantities that satisfy the binary substitution equations are:

$$p_{ji} = n_j/(n_i + n_j) \tag{75}$$

This states that the technology j displaces the technology i independently of all the others. We call the p_{ji} the binary fraction for j replacing i. It is convenient to order the subscripts to make $j > i$.

To convert the binary fractions to total shares n_j, rewrite Equation (75) as

$$n_i = n_j[1 - p_{ji}]/p_{ji} \tag{76}$$

Summing (76) on i up to $i = (j - 1)$ and using the normalization of the n's (Equation 72), gives the market shares as

$$n_N = \left[1 + \sum(1 - p_{Ni})/p_{Ni}\right]^{-1} \tag{77}$$

$$n_{N-1} = (1 - n_N)\left[1 + \sum(1 - p_{(N-1)i})/p_{(N-1)i}\right]^{-1} \tag{78}$$

$$n_{N-2} = (1 - n_N - n_{N-1})\left[1 + \sum(1 - p_{(N-2)i})/p_{(N-2)i}\right]^{-1} \tag{79}$$

The procedure for applying these equations is similar to that for the first approach. The general shapes of the curves obtained by the two approaches are usually similar but can differ in important details.

THE LOGNORMAL DISTRIBUTION

Analysis of allometric relations shows that the more rapidly an entity grows the larger its size becomes relative to other entities. This idea can be given a more precise form by computing the distribution function for entity size. That is, we seek to obtain a formula for the fraction of entities whose size is in a given range. To do this, let's assume that we have a large number of

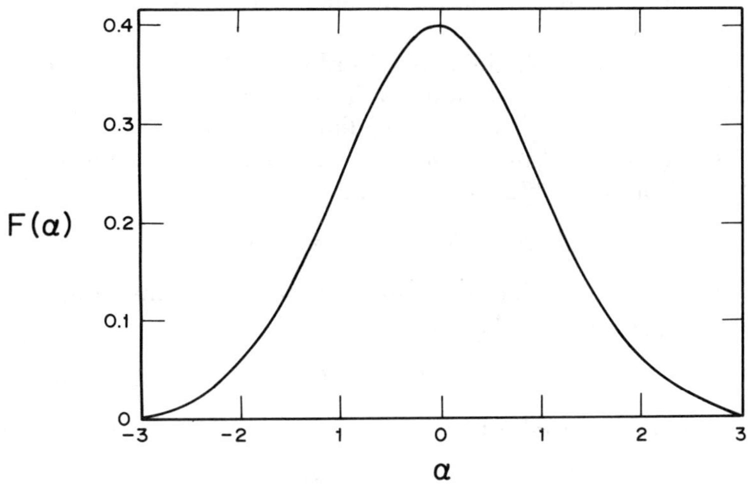

Figure A1-9 The normal distribution.

entities with growth variables increasing at different rates. These variables may be the incomes of individuals, the population of cities, the membership of trade unions, or the size of corporations. To be concrete, we will refer to corporations but the results will be applicable to many other types of entities.

Business firms are subject to a large number of forces that affect their growth. These forces affect different firms differently and the firms' responses can be highly variable. It is therefore no surprise that there is such a wide variety of rates of corporate growth. We assume that our set of firms is very large so that we can use continuum analysis, and we assume that the growth rates are distributed among them randomly so that we can use the normal distribution. This means that if we choose a range of growth rates $d\alpha$, centered on the growth rate α, then the fraction of the firms with growth rates in this range, $F(\alpha)\,d\alpha$, is given by

$$F(\alpha)d\alpha = \frac{1}{\sigma(2\pi)^{1/2}} \exp[-(\alpha - \langle\alpha\rangle)^2/2\sigma^2]d\alpha \tag{80}$$

where $\langle\alpha\rangle$ is the average growth rate for all firms and σ is the standard deviation of the growth rates about the average.

Equation (80) yields the familiar bell-shaped distribution curve which is shown in Figure A1-9. Note that the curve is symmetric so that there are just as many firms with growth rates below the average by a given amount as there are above the average by the same amount. Also, the average growth rate occurs at the maximum in the curve, which means that the average growth rate is the most probable. The width of the curve is determined by the standard deviation; the larger the standard deviation, the greater the spread of growth rates.

We are interested in obtaining the distribution of firm size, given that the growth rates are distributed normally. To do this, we assume that all firms started out at the same size Y at some time $t = 0$, and that their sizes are related to each other through the allometric equation. That is, if we label each firm by an index 1, 2, 3, . . . , then,

$$ln(y_1/Y)/\alpha_1 = ln(y_2/Y)/\alpha_2 = ln(y_3/Y)/\alpha_3 \ldots \tag{81}$$

Each term in this equation must equal the same constant because (81) is true for independent y_j's. Therefore,

$$ln(y_j/Y) = K\alpha_j \tag{82}$$

Averaging this over all firms gives

$$\langle lnz \rangle = K\langle \alpha \rangle \tag{83}$$

where $z_j = y_j/Y$.

The average growth rate is defined by

$$\langle \alpha \rangle = [\alpha_1 + \alpha_2 + \alpha_3 + \ldots]/N_T \tag{84}$$

where N_T is the total number of firms. Similarly,

$$\langle lnz \rangle = [lnz_1 + lnz_2 + lnz_3 + \ldots]/N_T \tag{85}$$

Eliminating the constant K between Equations (82) and (83) gives

$$\langle \alpha \rangle lnz_j = \alpha_j \langle lnz \rangle \equiv \alpha_1 \mu$$

where $\mu = \langle lnz \rangle$ is a characteristic value of the set of growing firms that depends only on time.

Adopting the continuum notation, which is equivalent to assuming that the number of firms is very large, we rewrite this as

$$\langle \alpha \rangle ln(y/Y) = \alpha \langle lnz \rangle = \alpha\mu \tag{86}$$

which is the general relationship between size and growth rate that follows from the allometric equation.

Now the distribution of size can be related to the distribution of growth rates by recognizing that the number of firms in a given range must be the same whether the range is measured by size or growth rate. That is, if $F(y)dy$ is the fraction of firms in the size range dy, then

$$F(y)dy = F(\alpha)d\alpha \tag{87}$$

or,

$$F(y) = F(\alpha)\, d\alpha/dy \tag{88}$$

The derivative in this equation is easily obtained from Equation (86);

$$d\alpha/dy = \langle\alpha\rangle/\mu y \tag{89}$$

from which

$$F(y) = (\langle a\rangle/\mu y)\, F(\alpha) \tag{90}$$

Putting this in Equation (80) and using the allometric equation to replace the growth rates by sizes gives the size distribution as

$$F(Y) = \frac{1}{y\sigma_L(2\pi)^{1/2}} \exp[-(lny - \mu)^2/2\sigma_L{}^2] \tag{91}$$

where σ_L is defined by

$$\sigma_L = \sigma\mu/\langle\alpha\rangle \tag{92}$$

This is the lognormal distribution function with mean μ and standard deviation σ_L. It describes a set of variables whose logarithms follow the normal distribution curve. The parameter μ is the average of the logarithms of the size. To get the average of the size itself, we use the definition of the average, which is

$$\langle y\rangle = \int yF(y)dy \tag{93}$$

Putting Equation (91) into this and defining $x = y/\langle y\rangle$, the integral reduces to a well-known form with the result that the average size is given by

$$\langle y\rangle = \exp(\mu + \sigma_L{}^2/2) \tag{94}$$

Using this, the formula for the distribution function, Equation (91), can be put in the form

$$F(x) = \frac{1}{x\sigma_L(2\pi)^{1/2}} \exp[-(lnx + \sigma_L{}^2/2)^2/2\sigma_L{}^2] \tag{95}$$

This is a form of the lognormal distribution that is convenient for calculation. It describes the distribution in terms of the average size through the variable x and the standard deviation σ_L. Note that as time goes by, the standard deviation increases. This is clear from Equation (92) since the sizes are larger at later times, and μ increases with increasing size.

Figure A1-10 is a plot of the lognormal distribution with standard deviation equal to unity. It is quite different from the normal distribution in that it is not symmetric, but is skewed to the left, and the most probable value is less than the average value. It is clear from this curve that, in an economy where growth rates differ, there are few firms of large size and many more of moderate or small size. The larger the value of the standard deviation, the more pronounced this effect.

Another way to examine the distribution of size is to compute the fraction of firms that have sizes less than some specified amount. This is easily done

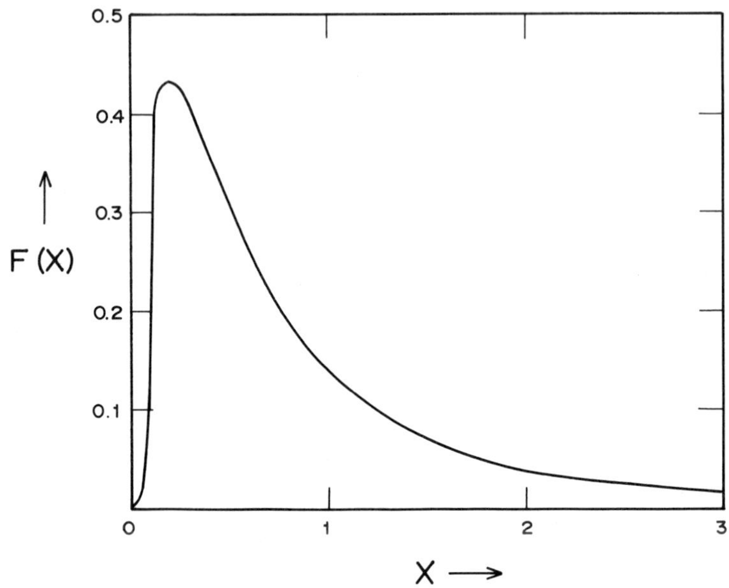

Figure A1-10 The lognormal distribution.

by integrating the distribution function up to various values of the size. That is, the fraction of firms with size less than X is given by

$$G(X) = \int_0^X F(x)dx \tag{96}$$

This is readily put in a form that permits it to be computed from tables of the integral of the normal distribution by defining the variable u,

$$u = (lnx + \sigma_L^2/2)/r_L \tag{97}$$

With this definition, Equation (96) becomes

$$G(X) = P(U) \tag{98}$$

where $P(U)$ is the normal integral given by

$$P(U) = \frac{1}{(2\pi)^{1/2}} \int^U \exp(-u^2/2)du \tag{99}$$

and U is related to X by

$$X = \exp\ (U\sigma_L = \sigma_L^2/2) \tag{100}$$

so that a curve of $G(X)$ versus X can be constructed from tables of the normal integral. Figure A1-11 shows such a curve for standard deviation equal to

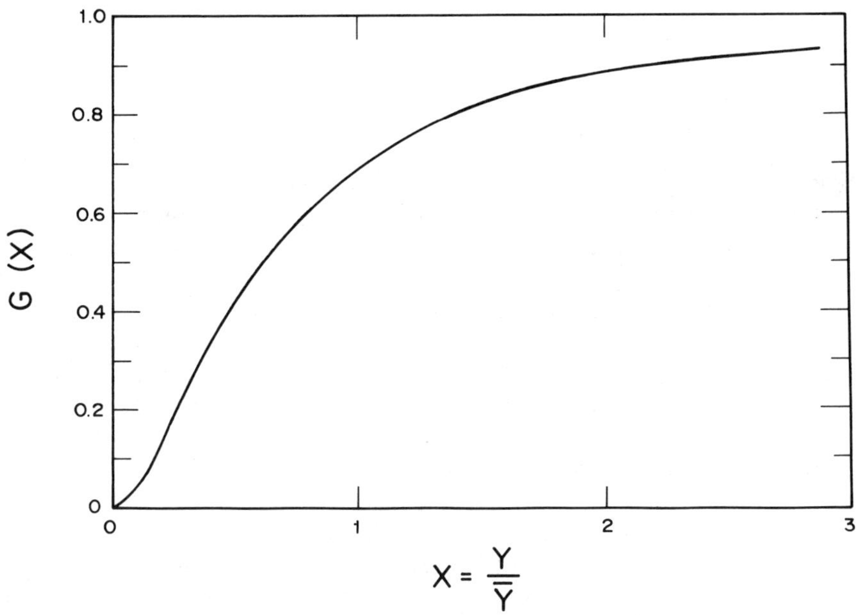

Figure A1-11 The lognormal integral.

unity. It graphically demonstrates the unequal size distribution resulting from unequal growth rates. In our example, nearly 70 percent of the firms are less than average size and only 11 percent of the firms have a size more than twice the average, while more than 40 percent have size less than half the average.

It is well to remind ourselves of the simplifying assumptions in arriving at the result that sizes are distributed lognormally. These are: the growth rates are distributed normally; no new entities are formed and none die; growth starts for all entities from the same initial size at the same initial time; the sizes of the entities are related through the allometric equation. These assumptions can be relaxed to some degree and still generate a distribution that is at least approximately lognormal. It has also been assumed that the growth rates are constant in time; however, it can be shown that this is not a necessary condition for the lognormal distribution to be valid. It is sufficient to have the distribution of the *average* growth rates be Gaussian, where the averages are taken from time zero to t for each growing entity. This will lead to the lognormal distribution for sizes.

In spite of the simplifying assumptions, the lognormal is often found to give a good approximate representation of the size distribution of a variety of entities, including personal incomes, crushed particles, city populations, and business firms. It is remarkable that the lognormal representation of sizes can be derived using only the allometric equation and the normal distribution of growth rates without detailed knowledge of the determinants of the growth.

HORIZONTAL ANALYSIS

The primary method of studying the relation between two variables is to select a system and to compare the two variables at different times. Thus, in searching for a connection between energy consumption and GNP, for example, a particular country is chosen and the two quantities for each year are obtained. The relation between them is then studied by plotting one against the other, each data pair being for a different year. This is called longitudinal analysis since it compares a particular pair of variables over a period of time. The allometric equations are expressions of longitudinal analysis.

Sometimes data is not available over sufficient periods of time for this kind of longitudinal study. Another approach is then taken in which a particular time is chosen and then a correlation is sought between the two variables at that time for a number of different systems. In the example of energy versus GNP, these two quantities would be obtained for a number of different countries for a given year and then plotted against each other. This is called *horizontal analysis* or *cross-sectional analysis.* Such a procedure has some rather strong assumptions built into it. It assumes that the systems are all sufficiently alike so that the horizontal correlation has some legitimate meaning, and it assumes that allometric growth rate constants are the same in the various systems.

Let's assume that each of a set of systems exhibits an allometric relation between two variables given by

$$lnx_j = g_j lny_j + lnC_j \tag{101}$$

where (101) is the equation for the jth system. It is clear that a horizontal analysis is legitimate only if the constants g_j and c_j are the same for all systems. Since these constants contain the rate constants and initial values for the growth of x and y, horizontal analysis places stringent requirements on the systems. Of course, if there is a causal relation between the two variables so powerful that it overides the differences among the systems, then horizontal analysis is valid. In fact, a successful horizontal correlation is often taken to be evidence for the existence of such a causal relation. However, scatter in the data is often sufficiently large to make such a conclusion dangerous. In practice, horizontal analysis can be useful but it must be interpreted with care.

LEARNING CURVES AND TECHNOLOGICAL PROGRESS

The power law we met as the allometric equation has a variety of applications. For the study of technological dynamics, an important application is that of the learning curve.

Experimental studies of learning show that the rate at which something new is learned follows a power law up to a point, after which learning stops. An example would be learning to give correct responses to words written upside down or the number of correct turns taken in a maze. In all such learning experiments it is found that the number of correct responses is a power function of the number of trials in the test.

The power law seems to be a good representation of learning situations. Animals, individual human beings, groups, and organizations all exhibit a power law relation between the efficiency of doing a task and the number of repetitions. For our purposes, the important example is the relation between productivity or efficiency and the cumulative production associated with an industrial process.

In general, if E is a measure of efficiency in a production process and N is the total volume of production since the process was initiated, then

$$E = AN^b \qquad \text{for } N = 0 \text{ to } N_M \tag{102}$$

$$E = \text{constant} \qquad \text{for } N > N_M \tag{103}$$

where N_M is some maximum after which there is no further improvement.

The constants A and b are usually obtained empirically. The measure of efficiency might be labor productivity or the reciprocal of the time needed to complete an operation. Note that A is the initial efficiency; that is, the efficiency in producing the first unit. Figure A1-12 shows a typical learning curve for $b = 0.5$. Empirical studies show that the exponent b is generally less than unity, making the learning curve concave downwards.

There is evidence that the maximum efficiency, reached at N_M, is a measure of incentive and that it can be increased by increasing the incentive to learn.

Clearly, the cost of production is related to its efficiency, so that if C is the cost of labor per unit of some product it will go down as the number of units produced goes up according to

$$C = C(1)N^{-b} \tag{104}$$

until some minimum cost is achieved. $C(1)$ is the cost of producing the first unit (providing, of course, that wages do not change). A labor cost curve computed from (104) is shown in Figure A1-13.

The learning curve is often applied in logarithmic form as

$$lnE = blnN + lnA \tag{105}$$

for the efficiency of performing the task, or in terms of cost as

$$lnC(N) = -blnN + lnC(1) \tag{106}$$

where $C(N)$ is the labor cost of the Nth unit produced.

The learning curve has the characteristic that an increase in the number of trials (or units produced) by a given ratio always results in the same ratio of improvement regardless of the total number of trials. To see this, write Equation (93) for two values of N and subtract the two equations to get:

$$ln(E_2/E_1) = bln(N_2/N_1) \tag{107}$$

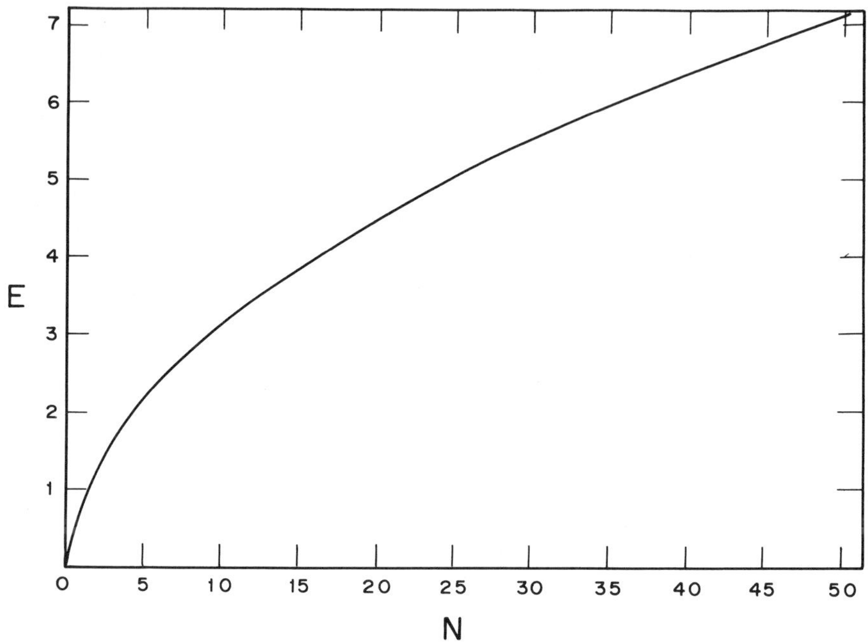

Figure A1-12 The learning curve.

where E_2 is the efficiency after N_2 trials and E_1 is the efficiency after N_1 trials. Thus, if the ratio (N_2/N_1) is constant, so is the ratio $E_2/E_1)$.

It is common to express the results of industrial learning curve studies in terms of the fractional increase for a doubling of the number of trials or units produced. For a doubling of N, Equation (107) gives $ln(E_2/E_1) = bln2 = 0.69315b$ for the improvement ratio for each doubling of the number of trials. For the learning curve expressed in terms of labor costs, $p(2)$ is the fractional decrease in labor cost per unit for a doubling of the number of units produced and is given by $p(2) = \exp(-0.69315b)$.

Empirical studies show that a number of industrial processes exhibit efficiency improvements such that a doubling of the number of units produced results in a 20 percent productivity improvement. This is called the 80 percent rule and corresponds to a value of $b = 0.32$. The 80 percent rule is far from universal, but it is useful for illustrative purposes. The curve in Figure 13 was calculated using $b = 0.32$.

The learning curve is often used to project labor costs for production processes in which the labor content is repetitive and high, such as assembly. It is then of interest to have the average labor cost as a function of the number of units produced. This is easily obtained from the definition of the average cost which is

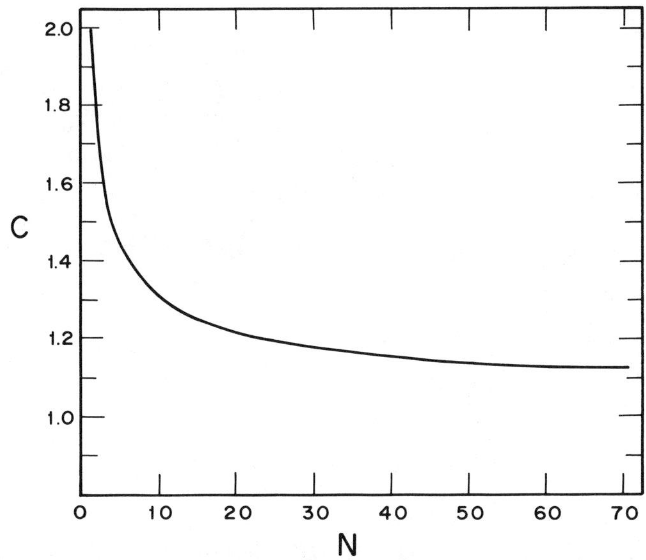

Figure A1-13 The cost reduction curve.

$$\langle C \rangle = \frac{1}{N} \Sigma_n C(n) \tag{108}$$

where $C(n)$ is the cost for the nth unit and the summation is taken over all values of n from 1 to N. Using Equation (104) this becomes

$$\langle C \rangle = \frac{C(1)}{N} \Sigma_n n^{-b} \tag{109}$$

If the number of units produced is very large, the sum can be approximated by an integral so that

$$\langle C \rangle = \frac{1}{N} \int_0^N Cdn = \frac{C(1)}{N} \int_0^N n^{-b}\, dn \tag{110}$$

or

$$\langle C \rangle = \frac{C1}{(1 - b)} [N^{-b} - 1/N] \tag{111}$$

For large values of N, the average cost curve has the same slope as the unit cost curve, when plotted logarithmically.

To show the error that can be expected if the integral form is used instead of the summation, Table A1-3 lists the values of $\langle C \rangle / C(1)$ calculated from Equations (109) and (111), along with the percentage error if the integral

Table A1-3 Comparison of Integral and Summation Forms of the Learning Curve

N	$\langle C\rangle/C(1)$ Sum	$\langle C\rangle/C(1)$ Integral	Percent Error
1	1	0	100
5	0.749	0.585	21.9
10	0.633	0.557	12.0
20	0.526	0.490	6.8
30	0.469	0.446	4.9
40	0.432	0.415	3.7
50	0.405	0.391	3.4
100	0.329	0.321	2.4
150	0.290	0.286	1.4
200	0.2656	0.2625	1.2

form is used. The 80 percent rule was adopted for these calculations so that $b = 0.32$. The table shows that the integral formula can be used with about one percent error if the number of units produced is above 200. Even for a value of N of 30 the error is only 5 percent. Since the scatter in observed data is usually at least this high, the integral formula is useful for all but rather small values of N.

After a new product is introduced, it usually undergoes steady modification and improvement for a considerable time. Thus, not only does the product become cheaper with time, it also improves technologically. This has led to the suggestion that the total cumulative production is also a measure of technological change, such that parameters that describe the level of a technology improve according to a power law similar to that for production efficiency. The use of the learning curve equation in this way combines all factors that contribute to increased efficiency. Sometimes it gives a good representation of such data but often it does not.

DATA SMOOTHING

The time series of data subjected to analysis for the study of technological change is often rather "noisy." While definite trends may be discernible there is usually a considerable amount of scatter in the data. Standard statistical regression techniques are used to extract as much information from the data as possible, but other approaches are sometimes used by operating on the raw data before analysis. A common technique is to use moving time averages of the data in a time series rather than the individual data points themselves. In a time series of annual data, for example, a five-year moving average would be constructed by adding the data for five consecutive years, dividing by five, and associating the result with the third year of the set. The next point would be obtained by dropping the first point of the set and adding the point directly after the fifth. This has the effect of smoothing out random fluctuations, since there are likely to be both positive and negative deviations from the trend line. Thus if we go back to the time series of Equation (1), its five-year moving average would consist of terms of the form $(x_j + x_{j+1} + x_{j+2} + x_{j+3} + x_{j+4})/5$.

In relating the moving time average to a growth law as in Equation (9), the deviation of the moving average would then be an average of five deviations, and there would be a better chance that the data points represent the growth law.

Another technique is to use accumulated points whenever possible. A lot of data, such as production figures, are given as annual rates. As we have seen, growth laws can be written in differential or integrated form. In their integrated form, they often represent the value of the reported data accumulated over time. Using the sum of the annual figures up to a given time, as a function of time, gives plots with less scatter than plots of the annual figures. However, the decrease in scatter is not only because the random fluctuations are averaged out as they get summed for succeeding years. The cumulative sum gets ever larger as time passes, so that any deviations from regularity become a smaller percentage of the total even though their absolute values do not change. The major effect of cumulative data on fluctuations is to hide them rather than to smooth them out.

Often, absolute values of a growth variable exhibit large fluctuations while relative values are rather smooth. Annual production of steel by a given process, for example, shows large fluctuations from year to year. This is to be expected since steel production is highly sensitive to economic and political conditions. However, if the fractional share of annual production by each production method is computed, it is found to be a remarkably smooth function of time.

Of course, something is lost whenever data are massaged by such methods. In the case of the moving average, the time resolution of the data is decreased. For the accumulated data, direct rate information is lost and the early points are more subject to fluctuations than the later ones. Using relative values gives no information on the absolute growth. Nevertheless, these methods allow us to study growth in a more precise way than would otherwise be possible.

This completes our introduction to the equations behind the quantitative study of growth. Only a general introduction to their properties has been given, and a number of growth equations have been used that are not mentioned here. However, the most commonly used equations and their most salient properties have been presented. Greater detail is most appropriately given in connection with specific applications.

APPENDIX 2

Chronology of Innovation

A temporal panorama of innovation is extremely useful in illustrating the dynamic flow of technological progress. The systemic analysis of technological change can be fully appreciated only in conjunction with some knowledge of the actual innovations that have determined the course of technology and shaped the modern world. I therefore offer the following chronology, which consists of a list of innovations to which dates and brief descriptions have been attached.

There are three major problems that need to be met in constructing such a list. The first is to decide where to start, the second is to choose which innovations to include, and the third is to assign dates to the innovations.

With respect to the first issue, I decided to start at the year 1700. My interest is in technological change in the modern West and, in spite of the fact that the Industrial Revolution had its roots in medieval times, I thought Savery's steam engine was a good place to begin. As to which innovations to include, this was dictated primarily by the flow of my studies and by my subjective judgment as to what was important. While no list can even pretend to completeness, I believe that I have presented a good overview of the important innovations that represent the evolution of modern technology.

Choosing dates for the innovations was the most difficult task. Invention and innovation are part of a continuous process; there are false starts, duplications, and conflicts; sometimes initial commercialization fails; sometimes earlier achievements are forgotten and must be repeated. Nevertheless, there are events that are sufficiently important that they rise above the continuum progress displays when analyzed on a fine scale. Whenever possible, I have tried to choose a unique date of innovation as the date of successful commercialization. When I could not do this, that fact is evident from the text describing the innovation.

1700 Atmospheric Engine (Steam Engine)

Thomas Savery patented his "Atmospheric Engine" in 1698 and first advertised it for sale in 1702. It consisted of a boiler and two oval vessels with appropriate piping and valves. Steam from the boiler was led to one of the vessels, which was full of water. The steam expelled the water to the other vessel. The first vessel was then cooled by pouring cold water on it from outside, thereby creating a vacuum which enabled water to be drawn up from a lower level. Savery tried to use high pressure steam (up to 8–10 atmospheres) but the materials available could not stand high pressures.

Savery's engine was built to raise water from mines. His written description of it appeared in a pamphlet called "The Miner's Friend."

1709 Coke Blast (Iron and Steel)

Abraham Darby developed the first successful use of coke in a blast furnace. The secret of his success was careful choice of both coal and iron ores. The major disadvantage of coke relative to charcoal is the high S and P content of coal. These embrittling agents drastically lower the quality of iron and steel. Darby used low sulfur coal and selected ores but even so the resulting pig iron was brittle because of high P content.

The advantages of coke were that it was in much greater supply than charcoal, it was less frangible, and could stand a stronger blast. The greater strength meant that larger blast furnaces could be used without having the fuel become crushed and block the blast. Also, the lower frangibility made transport of coke easier.

The higher temperature attainable with coke gave an iron with higher silicon content. This, with the high phosphorous, gave a product of lower melting point, which enabled Darby to make better castings. His cast iron pots and vessels were thin and competed well with copper.

The diffusion of the use of coke in blast furnaces was slow, only six or so other coke furnaces being built before mid-century. But by 1760 there were seventeen, by 1775 there were thirty-one, and in 1790 there were eighty-one coke-fired blast furnaces out of a total of 106. This was the result of a series of improvements, a most important one being the remelting of the pig in a reverberatory foundry furnace, which removed much of the impurities and gave a more homogeneous product. Also, the introduction of blowing cylinders in 1762 (John Smeaton) to get the stronger blast needed for coke accelerated its substitution for charcoal.

1711 Newcomen Engine (Steam Engine)

Thomas Newcomen built his first engine in 1705 with the help of his assistant John Cawley and started to sell it in 1711. The engine consisted of a boiler, a cylinder containing a piston, and a heavy counterpoise connected to a pump. Steam from the boiler entered the cylinder, enabling the piston to be raised by the counterpoise. Cooling the cylinder created a vacuum, causing the counterpoise to move back, thereby working the pump. At first, the cylinder was cooled externally, but later it was cooled by injecting cold water into the cylinder. The engine worked at atmospheric pressure so it was safer and more durable than Savery's engine. The piston-cylinder, internal cooling, and production of translational motion were important improvements over the atmospheric engine. Furthermore, the pumping height was not limited by the steam pressure as was the case in Savery's engine. It was almost exclusively used to raise water, even though the motion of the beam made it suitable for other purposes.

Newcomen brought Savery in as a partner because Savery's patent was written so broadly that Newcomen could not get around it. They erected their first engine in 1712 at the Dudley Castle Colliery. It generated about 5.5 HP. Savery's original engine generated 1 HP.

Cylinders could not be bored accurately if the diameter was greater than 7″. But Newcomen's engines needed large cylinders. One in 1725 had a cylinder of 29″, and another in 1765 had a diameter of 74″. The lack of tight fit contributed to the low efficiency.

1731 Seed Drill (Agriculture)

Jethro Tull published his book, *Horse-Hoing Husbandry*, in which his invention of the seed drill in 1701 was made public for the first time. It was the first practical drilling and seeding machine and the first step towards farm automation.

1733 Flying Shuttle (Textiles)

John Kay invented the flying shuttle, which greatly accelerated weaving. He put the shuttle on wheels, had a hammer to drive the shuttle through the warp, and returned the shuttle by a cord. This increased the width of cloth that could be woven since previously the weaver had to push the shuttle from hand to hand. But it was not widely used until the 1760s, primarily because the bottleneck was spinning. One weaver could use the output of five or six spinners. The flying shuttle made matters worse because in addition to permitting larger widths to be made, it also increased the weaving speed. The pressure to remove the spinning bottleneck was great.

1738 Carding Machine (Textiles)

Lewis Paul invented a carding machine. The wool or cotton was passed between two rollers revolving at different speeds, thereby drawing out the fibers before passing them to a spindle which imparted a twist. In 1740 Paul and Wyatt opened a mill using these machines but the firm failed because, according to Paul, the workers would not work at the regular pace required by the machines.

1746 Sulfuric Acid (Lead Chamber) (Chemicals)

John Roebuck, in partnership with Samuel Garbett, erected the first lead chamber plant in which saltpeter and sulfur were burned together to produce sulfuric acid. The reaction between sulfur and saltpeter was commonly used previously and had been used by Joshua Ward in 1737. But this was in glass chambers, and Roebuck's process made large-scale cheap production possible. After 1830 sodium nitrate was substituted for saltpeter, and after 1840 sulfur was replaced by iron pyrites. This further decreased the price of sulfuric acid.

The availability of cheap sulfuric acid was extremely important for the Industrial Revolution. It had many uses in the metals industries, such as pickling and cleaning, and as a replacement for sour milk in the bleaching of textiles. The effect of this on the textile industry can be gauged from the fact that sulfuric acid was twenty-five times faster than sour milk in the bleaching process. Even this speed was insufficient to keep up with the rapidly expanding industry. It was the use of chlorine that eventually solved the textile bleaching problem.

1750 Screw-driven Lathe (Machine Tools)

Antoine Thiout (France) devised a lathe with a tool holding carriage that was screw-driven, thereby attaining a degree of accuracy not previously available.

1751 Slide Lathe (Machine Tools)

The first known metal lathe was built by Jacques de Vaucanson about 1751.

1751 Crucible Process (Iron and Steel)

Benjamin Huntsman commercialized the crucible process, which he had started to develop about 1740. In this process, steel made by cementation was heated by a coke fire surrounding a vitrified fire clay crucible. A variation was to use a charge of iron and charcoal. This was the first time that a temperature high enough to melt steel was reached (2900°F). Carbon was removed by CO formation, the reaction being judged complete when bubbling stopped. Si content went up from reaction with the silica from the crucible. Maximum charges were 100 lbs., since this was what a man could lift out of the crucible. The process took 6–8 hours per batch.

The steel was of good quality, suitable for watch springs and cutlery. In fact, Huntsman was a clock maker and developed the process to get better steel for his clocks and watches.

1752 Copper Plate Printing (Textiles)

Printing on cotton by engraved copper plates was invented near Dublin by Francis Nixon in 1752. He brought the technique to England over the next four years. Copperplate printing permitted a fineness of detail not possible with wooden block printing. England and Ireland had a monopoly on the new technique for the next twenty years. The designs were almost always monochrome, although there was some two-color printing.

1770 Screw-cutting Lathe (Machine Tools)

Jesse Ramsden built a lathe for cutting screws that had the rod to be cut parallel to the guide bench and turned by gears attached to the driving crank of the lead screw. A model screw was made by hand from a geometric pattern and used to guide the lathe drive.

1771 (1769?) Water Frame (Textiles)

Richard Arkwright patented the water frame; the first successful machine to automate spinning. In spite of its name, it was originally designed to be powered by a horse, although it was later driven by water power. This was the first textile machine to require much more power than could be supplied by human muscle and was a prime instrument in moving the textile industry from cottages to factories.

In the water frame, rovings made of raw cotton which had been cleaned, carded, and slightly twisted, were held on horizontal bobbins. The roving was drawn through pairs of rollers, one of which moved faster than the other. The roving was thereby compacted and stretched. After leaving the rollers, the yarn passed through a flyer and was wound on a bobbin on a spindle.

In 1781 Arkwright brought suit against other manufacturers for patent infringements. He lost and appealed. During the appeal, the water frame was attributed to Thomas Highs. Arkwright lost his case and was not recognized as the inventor, but as an enterprising businessman.

1772 (1770?) Spinning Jenny (Textiles)

James Hargreaves patented the spinning jenny. It was more suitable for weft, whereas Arkwright's machine was better for warp and strong hosiery yarn. The spinning jenny was small and could be powered by one person. It was suitable and cheap enough for cottage use. The first jenny had eight spindles; the one that was patented had sixteen spindles. By 1788 there were about 20,000 of the Hargreaves' jennys in use. By 1800 they held 120 spindles.

In the jenny, the roving was drawn from bobbins at the bottom by a bar which was pulled back and forth on the top and wound on spindles. Both the bar and the spindles were hand-driven.

The jenny broke the spinning bottleneck. Even with the simplest jenny one spinner could produce all the weft that one weaver could use. The weavers became the new bottleneck.

1775 Carding Machine (Textiles)

Richard Arkwright took out his second patent for a carding machine. In addition to Paul's attempt in 1738, at least two other tries were made previously. Arkwright and D. Bourne had taken out patents in 1748. But it was the 1775 machine that was definitive; it had the features that are still in use. The mechanization of carding was necessary so that it would not become the new bottleneck when spinning was mechanized.

1775 Boring Mill (Machine Tools)

John Wilkinson built a boring mill for boring cannon. He kept the piece stationary and rotated the tool, which was a boring bar keyed to a drive wheel at one end and supported on bearings at the other. It achieved an unprecedented degree of accuracy.

1776 Watt Engine (Steam Engine)

James Watt started work on his engine in 1765, but his first two engines were not installed until 1776. His great improvement was to introduce a condenser so that the cylinder and piston were not cooled between strokes. This greatly improved the efficiency.

The delay from conception to production was the result of inadequate materials, skilled labor, and machining techniques. The boring of accurate cylinders was a major problem, since Watt required tight fits. It was John Wilkinson's boring mill for cannon, developed in 1774, that made accurate cylinders possible. Wilkinson made the cylinders for Watt's engines, one of the first two being used to supply air to one of Wilkinson's blast furnaces.

In 1774 Watt joined with Matthew Boulton, a leading manufacturer. This consolidation provided funds, business acumen, and political influence. The firm of Boulton and Watt was very successful.

In 1756 Watt got a job as instrument maker at the University of Glasgow. In this capacity, he was asked to repair an atmospheric engine which could produce only a few strokes. Watt found that there was nothing mechanically wrong with the engine. It was in searching for the reasons why it did not perform that he was led to the idea of the condenser, so that the cylinder was not cooled between strokes. His approach was quantitative and analytic.

Watt was well acquainted with Joseph Black, Professor of Chemistry at Glasgow. He also knew prominent British scientists of the time through his membership in the Lunar Society, which included Joseph Priestley, William Small, Erasmus Darwin, and Thomas Wedgewood.

1776 Steam-powered Blast (Iron and Steel)

John Wilkinson used a Watt steam engine to produce the blast for a blast furnace. This enabled the coke to burn more easily and reach higher temperatures. Also, the blast furnace could now work continuously, since it depended only on the supply of coal and iron and not on the seasonally variable water power that was previously used to drive the blast.

By 1780 Wilkinson had four engines at work for furnace blast, and by 1800 the British iron industry was using twenty-four engines for blast air.

1779 Spinning Mule (Textiles)

The spinning mule was designed and built by Samuel Crompton. It was a hybrid of the water frame and the jenny, which resulted in finer and stronger yarn than hitherto available. The mule was hand-driven. Its thread was suitable for both warp and weft. The product was of high quality and much finer than anything handmade. It was twice as fine as the best Indian cloth.

In the mule, the roving was passed through rollers just as in the water frame, but they were placed at the far end of the frame. There was a carriage, just as in the jenny, which was drawn back at the same rate as the yarn left the rollers. But before the bar had traveled all the way, the rollers were stopped, while the bar went back more slowly. The spindles, mounted on the bar, continued to twist. After the bar's complete traverse, the spindles were turned back a bit to free the yarn. The bar was then pushed back and a wire "faller" guided the yarn to the spindles.

Spinning machines solved the old problem of getting uniform yarn. Hand spinning produced yarn that was variable in thickness and strength, and it was always a problem to assemble uniform batches.

1782 Seed Drill (Agriculture)

James Cooke brought out the seed drill that was the forerunner of modern seed drills. An important feature was that it replaced belts and chains in the seed-dropping mechanism with gears.

1783 Double Acting Engine (Steam Engine)

By 1783 Watt built a double acting engine, introduced the governor, and used the natural expansion of the steam near the end of the stroke to decrease mechanical shock and increase efficiency. He had also invented the parallel bar mechanism. The steam engine was now suitable for rotary motion, which was not possible with the single acting engine.

1783 Copper Plate Roller Printing (Textiles)

Thomas Bell invented multicolor roller printing using copper plates. A roller could do the work of forty hand printers.

1784 Cort Puddling Process (Iron and Steel)

Henry Cort established the puddling process of making wrought iron. The process consisted of melting pig iron in a reverberatory coal furnace and having it continually stirred (by hand). As carbon is oxidized, pure iron, with a higher melting point, clumps and is collected into lumps, which are placed in a forge to hammer out slag. A major innovation then was to reheat and pass the iron through rollers to remove slag. This greatly speeded up the process and gave a good quality wrought iron. It previously took twelve hours to make a ton of wrought iron. Fifteen tons could be made by Cort's process in the same time.

1785 Steam-powered Spinning (Textiles)

The first application of a Watt steam engine to spinning took place in 1785. By 1835, 35,000 HP of steam and 10,000 HP of water power were being used in the cotton industry. This grew to 71,000 HP of steam and 11,000 HP of water by 1850. Power use increased by a factor of 15 by 1907.

1785 Javel Water (Chemicals)

C. L. Berthollet in France showed that a solution made by passing chlorine through potash had a very strong bleaching action.

1786 Threshing Machine (Agriculture)

The first successful threshing machine that could replace the flail was built by Andrew Meikle of Scotland in 1786 and was starting to come into general use by 1805. It consisted of a drum and a concave (semi-cylinder). The drum, driven by horses, water, or even by hand, rotated against the concave as grain was passed through it. It was, in effect, a kind of continuous flail.

1787 Power Loom (Textiles)

Edmund Cartwright introduced the power loom in 1787. But a quarter of a century later there were still only 2,400 in the country, since there was no shortage of hand looms or weavers.

1791 Leblanc Soda Process (Chemicals)

In 1791 Nicholas Leblanc patented a process for making soda ash (sodium carbonate) in France in which common salt was treated with suphuric acid to make salt cake (sodium sulphate), which was then roasted with charcoal and limestone. The sodium carbonate was then extracted with water and evaporated to dryness in open pans.

Development of this process was critical to the evolution of the chemical industry. Alkali was an essential material for making glass, soap, saltpeter, alum, and bleaches. Before its synthesis from salt, alkali was obtained from organic sources, primarily from wood ash (potash) and kelp (soda ash). These sources were so limited that the industrial shortage was intense. There had been earlier attempts to produce alkali from salt, some of them quite similar to the Leblanc process, but it was the Leblanc process that took hold and provided industry with all the alkali it needed.

1793 Cotton Gin (Textiles)

Eli Whitney invented the cotton gin, which mechanized the removal of seeds from cotton bolls and greatly increased the supply and decreased the cost of raw cotton. It made possible the use of short staple cotton which could be grown in the inland southern United States.

1796 Cotton Gin Improvement (Textiles)

Hogden Holmes improved the gin by replacing Whitney's hooked cylinder by toothed saws and flat metal ribs for the slotted bar.

1796 Water-powered Printing (Textiles)

Water-powered printing by engraved copper rollers on cotton was introduced.

1799 Bleaching Powder (Chemicals)

Charles Tennant started manufacture of bleaching powder by passing chlorine over lime. He was aware of Berthollet's work, having been informed by Watt, who learned about it on a trip to Paris. This development was very important for the textile industry, which required large amounts of bleaching agents. The chlorine was obtained by treating hydrochloric acid with manganese dioxide. The acid was available in abundance as a waste product from the Leblanc process.

1800 Self-acting Mule (Textiles)

John Kennedy started experiments on using steam power to drive the spinning mule in 1793. By 1800 he was successful, although his machine still required skilled workers. Further developments, particularly by Richard Roberts, made the power mule fully automatic, and it came into general use after 1830.

1800 Screw-cutting Lathe (Machine Tools)

Henry Maudslay produced his famous screw-cutting lathe, which was very precise. He was a major figure in the improvement and development of machine tools. He used all metal machines for greater rigidity, invented the slide rest, and worked with the use of accurate standards.

1801 Jacquard Loom (Textiles)

J. M. Jacquard used punched cards to control the weaving of patterns. The holes in the cards let the proper threads through. This automated the weaving of complex patterns by having needles pass through the perforations. The needle lifted a hook to which a thread was attached. Operation of the threadle wove the proper threads and advanced the card to another pattern.

1801 Steam Car (Transport)

Richard Trevithick made his first steam carriage in 1801. This was a great improvement over earlier attempts (Nicholas Cugnot; 1796; France) (William Murdock; 1785; Cornwall). The cylinder was vertical and partially enclosed in the boiler to keep it hot. The wheels were driven by a connecting rod to the piston.

1802 High Pressure Engine (Steam)

Richard Trevithick built the first successful English high pressure steam engine. This was a double acting engine. One he built in 1802 operated at a pressure of 10 atm. The boiler was made of 3.5″ thick cast iron and the cylinder was only 7″ in diameter with a stroke of 3 ft.

Independently, Oliver Evans was working on a high pressure engine in the United States, and by 1804 he was using it to grind plaster of Paris and saw marble in Philadelphia.

1806 Coal Gas (Lighting)

In 1806–1807 gas light was installed at a Salford cotton mill. Over 900 gas lights illuminated the factory, a private road, and a private residence. The system was designed by William Murdock and was installed by Boulton & Watt. Watt had sent his son Gregory to Paris to look into the work of Phillipe Lebon, who was working on gas lighting using gas distilled from wood. Murdock was interested in materials to preserve the bottom of ships and worked on the making of tar by distilling coal. In 1792 he used gas generated from coal to light a house. In 1798 he went back to Boulton & Watt and developed coal gas illumination.

1807 Paddle Wheel Steamer (Transport)

Robert Fulton successfully sailed the *Clermont* up the Hudson River on August 17, 1807. This date is recognized as the beginning of the commercial use of water stream travel. Actually, some thirty steamboats had been built before this, three of which had been in practical use.

As early as 1873 the Marquis Jouffry d'Abbans ascended part of the Saone River near Lyons in the *Pyroscaphe*, a 182-ton paddle wheel steamer.

In 1784, James Rumsey of Virginia built a steam-powered, jet action boat which was run on the Thames by his friends after Rumsey's death.

In 1790, John Fitch had built a wheel steamer which went between Philadelphia and Trenton on a regular schedule, but there were not enough passengers to cover costs.

In 1802, John Stevens built the *Little Julian*, which was equipped with a high pressure engine and twin screw propellers.

1808 Woodworking Automation (Machine Tools)

Samuel Bentham automated the manufacture of pulley blocks for the Royal Navy which used a 30 HP steam engine, turned out 130,000 blocks a year, and cut the labor force from 110 skilled to ten unskilled men. In 1793 he took out a comprehensive patent, which included veneer cutting, mortising, molding, conical cutters for dovetail joints, and rotary cutters for the simultaneous cutting of several sides at once. But Marc I. Brunel approached Bentham in 1801 with methods that were better than Bentham's, who recommended that the Admiralty adopt Brunel's proposals. The block-making machines were designed by Brunel and built by Henry Maudslay.

1808 Plough (Agriculture)

Robert Ransome took out a patent for an iron plough frame to which standardized replaceable iron parts were fitted. This completed the evolution of the plough design for the horse-drawn plough, which afterwards was virtually unchanged. Earlier (patented in 1785) Ransome had developed a method for hardening the under surface of a cast iron share (so that it was self-sharpening) by more rapid cooling of the under surface. This accelerated the replacement of wooden by iron ploughs.

1813 Interchangeability of Parts (Metal Working)

The U.S. War Department had a long-standing interest in producing arms with interchangeable parts. This interest originated from French influences. As early as 1765, General Jean-Baptiste de Gribeauval was trying to introduce standardized weapons into the French Army that permitted interchangeability of parts. In 1786, Honore Blanc was given funds to start an armory to produce muskets with interchangeable parts, but this project was aborted by the French Revolution. Thomas Jefferson was enthusiastic about Blanc's work and tried to get it adopted in the United States. Also, French officers who stayed in the United States after the American Revolution were a source of the drive for interchangeability. One of the most important of these was Louis Tousard, who published a book on artillery in 1809.

In 1813, Simeon North was given a contract to provide twenty thousand horse pistols to the War Department, with the proviso that their parts be completely interchangeable. He did this by making all parts fit a single model piece. He also pioneered the use of special purpose machinery and installed the first milling machine in America in 1816.

1814 Cylinder Press (Mechanical Printing)

On November 29, 1814, the *Times* issued the first paper to be printed on the cylindrical press that was developed by Friedrich Konig, a German who came to England in 1806. It was steam-driven and had a capacity of 1,100 impressions per hour. The hand-inked press which preceded it had a capacity of three hundred impressions per hour. Konig patented his press in 1810, and it was used by Thomas Bensley in 1811 to print a part of the *Annual Register*. This first machine had a capacity of 800 impressions per hour.

1817 Planing Machine (Machine Tools)

Richard Roberts, who had worked for Wilkinson, made the first metal planing machine in 1812.

1820 Quinine (Pharmaceuticals)

Pierre Joseph Pelletier and Joseph B. Caventou were professors of chemistry at the Paris School of Pharmacy. They isolated quinine and other alkaloids (brucine, chinchonine, and strychnine). This marked the beginning of definite compounds in pharmacology rather than infusions of vegetable origin.

1824 Portland Cement (Construction)

Portland cement is made from a mixture of limestone and clay (or shale) heated to 1500°C and ground to a fine gray powder. On being mixed with water, it sets and hardens. Mortars based on slaked lime or gypsums were used since ancient times, but they were not always satisfactory. John Smeaton experimented with mortars when commissioned to rebuild the Eddystone Lighthouse off Cornwall and found that the best results were obtained with limestone containing a lot of clay. In 1796 James Parker patented a hard burned lime. John Aspdin patented a process for making Portland cement in 1824. He named the product because of a similarity to Portland stone, a limestone quarried at Portland.

1824 Rifle (Firearms)

John Hall invented a breech loading rifle which he tried to patent in 1811. The commissioner of patents, William Thornton, claimed that he had invented such a rifle in 1792! Hall was forced to include Thornton as co-inventor. In 1824, Hall delivered 1,000 breech loading rifles to the War Department under a contract which was let in 1819. The importance of the rifle was not only that it replaced the muzzle loader, but also that the rifles were made with truly interchangeable parts, using special purpose machines. Hall drew on the work of North (1813), who made pistols with interchangeable parts, and of Blanchard (1822), who developed machinery for the manufacture of gun stocks. The entire movement towards mechanized manufacture of products with interchangeable parts started with small arms production and then spread to other industries.

Hall produced his rifles at Harper's Ferry Armory. His work strongly influenced the efforts at Springfield Armory, which successfully produced muskets with interchangeable parts in the 1830s and 1840s. The goal of complete interchangeability was achieved at Springfield in 1849 or 1850 with the United States Model 1841 rifle. The Springfield Armory also was the major force in the diffusion of "armory practice," and acted as a center for the reception and dissemination of information on modern machining.

1825 Steam Locomotive (Transport)

George Stephenson demonstrated the *Locomotion* on a 27-mile track between Stockton and Darlington. It attracted enough attention to encourage more ambitious projects. By this time a number of locomotives had already been built. Richard Trevithick had equipped a steam engine with wheels to pull five trucks at an iron mine in South Wales in 1804, and in 1805 he built a second locomotive for a colliery near Newcastle. In London in 1808 he exhibited a steam locomotive that pulled a carriage full of passengers around a circular track for a fee. By 1820 there were a number of steam-powered industrial railways in Britain.

1829 Hot Blast (Iron and Steel)

James Beaumont Nielson preheated the air in the blast before it entered the blast furnace. This allowed higher temperatures to be reached and led to a great increase in iron production. It also permitted the use of anthracite for smelting. This work was done in South Wales and in Pennsylvania, where anthracite was plentiful.

1830 Steam Railroad (Transport)

George Stephenson opened the Manchester and Liverpool railway, which was the first railway to carry both goods and people solely by steam locomotion. The engine used was the famous *Rocket*. The steam pressure was 40 lbs. The tubular boiler and direct drive piston set the pattern for future developments.

1831 Reaper (Agriculture)

Cyrus McCormick produced the first successful reaper. It used a knife and cutter bar and was pulled by horses. Before 1831 there were two French, one German, thirty-three English, and twenty-two American patents for reapers. Fifteen hundred of McCormick's machines were sold in the United States by the end of 1850. Obed Hussey had another design in 1833, but it was not as commercially successful as McCormick's.

The McCormick reaper established the definitive reaper design. It was pulled, not pushed; the cutting was done by a horizontal knife set at right angles to the machine; a triangular divider separated the swath from the standing crop; a cutting bar above the knife had projecting fingers that held the straws vertically while being cut; a reel bent the grain heads to the knife and pushed the cut stalks to a platform behind the cutter bar and knife; a "bull wheel" carried and moved the machine and transmitted power to the reel and cutter by cogs and bands.

The reaper solved the problem that reaping was the most labor-intensive of farming operations. Before the invention of the reaper, a farmer had the means to plant twice as much as he could harvest. This is an example of a production limitation calling forth an invention to break the bottleneck. John Lane and John Deere had designed steel plows that scoured even in the black, sticky soil of the American prairies, and during the Civil War a horse-drawn, two-wheeled sulky plow permitted plowmen to ride rather than walk along the field while plowing. These developments, along with seed drills, speeded up planting. A method of reaping at a corresponding efficiency was needed and provided by the reaper.

This development was the beginning of the extensive mechanization of agriculture.

1834 Thresher (Agriculture)

A successful thresher had been developed by Andrew Meikle of Scotland in the 1780s, but the grain, chaff, and straw still had to be separated. In 1834, Hiram and John Pitts of Maine developed a combined thresher–separator that became the basis for future designs.

1835 Revolver (Firearms)

Samuel Colt formed the Patent Arms Manufacturing Co. of Paterson, New Jersey, in 1835, the year his patent for the revolver was granted. The product at Paterson was of poor quality, and the company stopped production in 1842. Ultimately, Colt moved his operation to Hartford, Connecticut. The Hartford plant did not achieve interchangeability but was a progressive leader in mechanization.

1836 Daniell Cell (Electric Power)

John F. Daniell (England) developed a nonpolarizing cell consisting of zinc anode in zinc sulfate, magnesium sulfate, or dilute sulfuric acid and a copper cathode. This gave a constant open circuit voltage of 1.08 volts and was used for telegraphy. The Voltaic cell (1800) generated hydrogen and was subject to polarization. It was thus not suitable for industrial application, but much electrical research was carried out by the Cruickshank version of Volta's cell.

1837 Telegraph (Electric Communication)

William F. Cooke and Charles Wheatstone installed the first telegraph line. It was a bit more than one mile long, between Euston and Camden in London. It was a "five needle" system, in which the magnetic needles were moved by magnetic induction and a code was used to convert the needle positions to letters of the alphabet.

In 1838 they installed a telegraph on the Great Western Railway between Paddington and West Drayton. It was a "five needle" design. It was extended in 1842 and again in 1843, using a simpler "two needle" system.

1837 Daguerreotype (Photography)

L. J. M. Daguerre succeeded in making clear and permanent photographs using a relatively short exposure (about 15 minutes in sunshine).

The plate was polished silvered copper, which had been exposed to iodine to make a thin layer of silver iodide. After exposure, the plate was developed by exposure to mercury.

Daguerre had gone into partnership with Joseph N. Niepce, who had obtained photographic images on silver chloride sensitized paper as early as 1816 and took the world's first photograph in 1826 using a bitumen that hardened and became insoluble in white petroleum on exposure to light. The exposure took 8 hours.

1838 Screw-driven Steamship (Transport)

The first successful screw-driven steamship, the *Archimedes*, was constructed in Britain. The screw propeller was invented in 1836 by John Ericsson, a Swede who came to the United States in 1839 and built a screw-driven ship for the Navy. It was Ericsson who designed the ironclad *Monitor*, which was built in 1861.

1841 Vulcanized Rubber (Chemicals)

Natural rubber was being used from the early 1900s, but had deficiencies, chief of which was that it was easily affected by temperature. When cold it was brittle and when warm it was tacky. Charles Goodyear, a hardware merchant in Philadelphia, solved this problem by heating sulfur with natural rubber. The product was not soluble in organic solvents and was relatively unaffected by temperature changes. By 1841, Goodyear was making continuous sheets of uniform quality.

1844 Telegraph (Electric Communication)

Samuel Morse erected a telegraph line between Baltimore and Washington in 1844. From the beginning, Morse's idea was to use a code controlled by an intermittent current. He was helped on the batteries by Professor Gale, a chemist, and on manufacture of equipment by Alfred Vail, who actually devised the Morse Code. On May 24, 1844, Morse, from the Supreme Court, sent the message "What hath God wrought" over the 40 miles of wire connecting Washington and Baltimore. By 1850, virtually all of the United States east of the Mississippi was interconnected by the telegraph.

1845 Turret Lathe (Machine Tools)

Stephen Fitch developed the vertical turret lathe.

1846 Ether (Drugs)

Ether had been tried as an anesthetic for minor surgery in 1842 by Dr. Crawford W. Long, an American surgeon, but this fact was not well known. The use of ether did not spread until it was used by W. T. G Morton during a tooth extraction at Massachusetts General Hospital in September of 1846 and to anesthetize a patient of J. C. Welles in October for surgery. Morton was a former student of Horace Wells, a dentist who had one of his teeth extracted while under the influence of nitrous oxide in 1842 and later used the gas on his patients. Later in 1846, ether was used by Robert Liston, a surgeon in London. After 1846, the spread of the use of ether as an anesthetic was rapid.

1847 Rotary Press (Mechanical Printing)

The first successful rotary press was that of Richard Roe, who patented his machine in 1845. It was installed by the *Philadelphia Public Ledger* in 1847.

The Roe press was imported into England to print *Lloyd's Weekly Newspaper* and was adopted by the *Times* in 1858. In 1866 the *Times* installed a rotary press, sometimes called the Walters press after John Walter III, publisher of the *Times*, which was the prototype of the modern press. It printed both sides of a continuous web of paper at a rate of 25,000 impressions per hour.

1851 Sewing Machine (Appliances)

Two firms started producing sewing machines in 1851: the Wheeler and Wilson Manufacturing Company and the Singer Manufacturing Company. These, along with Wilcox & Gibbs, were the major firms until Singer, in the late 1860s, became the dominant manufacturer.

Isaac M. Singer produced an excellent practical home sewing machine and contributed a number of critical inventions that made it possible. But other inventors also made critical contributions. The first sewing machine was invented by Elias Howe, who worked in a factory producing cotton machinery. Howe came up with the critical design of having the eye of the needle at its point and using a lockstitch. Howe got his patent in 1846 but was unable to commercialize his invention either here or in England. His machine could only sew straight seams of limited length. Singer's machine had a straight needle, an adjustable presser-foot to hold the cloth in place, and was operated by a foot treadle. The cloth-feeding mechanism was invented by Allen B. Wilson. There was a great deal of patent litigation, which threatened to destroy the industry until the formation of a patent pool in 1856 called the Great Sewing Machine Combination.

1851 Wet Collodion Process (Photography)

Frederick Scott Archer, an English sculptor, introduced the wet collodion process in 1851. This became the dominant photographic process for the next two decades and was used by the photographers who covered the Civil War.

In this process, a glass plate coated with nitrocellulose (collodion) solution containing potassium iodide was immersed in a silver nitrate solution (in the dark). The plate was exposed while wet and developed immediately upon exposure. It was a negative-positive process and, despite its obvious disadvantages, was widely used because of the high quality of the negatives.

1851 Converter Steel (Iron and Steel)

William Kelly found that pig iron could be converted to steel by a blast of air on the surface of the molten metal. He built a series of converters from 1851 on, obtained a patent in 1857, but went bankrupt the same year.

1854 Elevator (Construction)

Elisha G. Otis built the first safety elevator in 1852 for a new factory and demonstrated it in New York in 1854, showing that it was safe even if the cable was severed. Otis' contribution was a pawl, spring, and ratchet mechanism that made the elevator really practical for the first time. This innovation was essential for the development of tall buildings.

1855 Electromagnet in Generator (Electric Power)

Soren Hjorth of Denmark was granted a British patent for a generator that used electromagnets instead of permanent magnets. He came close to having a fully self-excited machine. There is no evidence that this machine was ever made.

1856 Converter Steel (Iron and Steel)

In 1856, Henry Bessemer read his paper "On the Manufacture of Malleable Iron and Steel Without Fuel" to the British Association for the Advancement of Science. It is hard to overestimate the importance of this development. Until then steel was an expensive commodity, since it was made by the cementation or cupola process, both of which were slow and restricted to small batches.

In the Bessemer process, air is played on the surface of molten iron. This oxidizes the carbon, manganese, phosphorous, sulfur, and silicon in the pig iron and generates enough heat to keep the metal molten. The reaction is rapid, large batches being converted in 15 to 30 minutes.

Early converters were of the acid type. That is, the converter lining was silica, which produces an acid slag. This removes carbon, silicon, and manganese but not phosphorous or sulfur. Bessemer's initial tries were successful because he used a low phosphorous ore, and he soon found that high phosphorous ores gave steel of poorer quality.

The converter produced a highly useful mild steel, intermediate in carbon content between the carbon free wrought iron and the high carbon crucible steel. A problem was that carbon steels have high melting points and furnaces of the time were unable to melt them.

1856 Aniline Purple (Chemicals)

W. H. Perkin was a student of August Wilhelm von Hoffman, a German chemist who became the first superintendent of the Royal College of Chemistry. Hoffman was among the first to realize that coal tar could be the source of many interesting compounds. Perkin tried to make quinine. In the process, he found that oxidizing aniline resulted in some purple crystals which were suitable for dying cloth a brilliant mauve. This was the beginning of the organic chemical industry. Perkin himself made enough money from producing the dye to retire at age thirty-five to do chemical research. In 1869, he synthesized alizarin, the red coloring matter of the madder root.

1858 Generator Driven Arc Lamp (Lighting)

Frederick Hale Holmes installed an arc lamp for trial at South Foreland. The lamp was powered by a steam-driven generator, using rotating coils and fixed magnets, of the type designed by F. Nollet in about 1850.

1858 Telegraph (Electric Communication)

On August 16, 1858, Queen Victoria cabled greetings to President Buchanan over the first trans-Atlantic telegraph. The cable was laid by two warships, the American *Niagara* and the British *Agememnon,* after Congress authorized it in 1857. But the signals started to fade in October 1858, and repairs were delayed by the Civil War. Two new cables were laid in 1866 and 1868, and trans-Atlantic telegraphy was resumed. In 1869 France laid a cable between Paris and New York via Nova Scotia. For the next half century, undersea telegraph cable was the chief international communication technology.

1859 Lead Acid Battery (Electric Power)

Raimond Louis Gaston Plante constructed the lead acid battery. It was not very useful as a storage battery, even though it could be charged and recharged many times, because no good primary source was available.

1859 First Oil Well (Petroleum)

Edwin L. Drake drilled for and struck oil at Oil Creek, Pennsylvania.

George Bissel was interested in lighting and had Benjamin Silliman, who was Professor of Chemistry at Yale, analyze petroleum. Silliman separated the components by fractional distillation and identified petroleum as a mixture of hydrocarbons. He recognized the uses of the different fractions for lighting and lubrication. Bissel was on the verge of failure when Drake came along and realized that oil should be drilled for rather than just letting it bubble up through the ground.

1860 Repeating Rifle (Firearms)

The development of the metallic cartridge by Horace Smith and Daniel B. Wesson (1852–1855) paved the way for modern repeating rifles. In 1860 both Christopher Spencer and B. Tyler Henry patented repeaters. The Spencer had a magazine in the butt stock while the Henry had a tubular magazine under the barrel. Both rifles saw service in the Civil War, although the Spencer was much more widely used.

1862 Milling Machine (Machine Tools)

Joseph Brown at the American firm of Brown & Sharpe built the first universal milling machine. It was first designed to cut helical grooves in twist-drills, but Brown recognized its wide applicability. It was widely used in the United States, although milling machines were not in general use in Britain until after 1900.

1863 Open Hearth—Martins-Siemens (Iron and Steel)

In 1863, Pierre and Emil Martin, in France, used a regenerative furnace to melt steel bars and scrap. The regenerative principle allowed higher temperatures to be reached. The furnace had been built by Siemens' engineers.

William and Friedrich Siemens erected a plant at Birmingham, and by 1868 had developed the open hearth process. Its features were that it used gaseous fuel and a regenerative checkerwork to pre-heat both air and fuel. The lining was silica. (The first basic open hearth, dolomite lining, was erected by J. H. Darby at Brymbo, Wales.) Pig iron was refined by adding ore or iron oxide to oxidize the impurities. The process was slower than the Bessemer process and enabled easier control.

1863 Solvay Process (Chemicals)

Ernest and Alfred Solvay in Belgium built a plant at Couillet to produce sodium carbonate, using a soda-ammonium process which they patented in 1861. In their process, saturated brine is saturated with ammonia and then carbonated with carbon dioxide. Sodium bicarbonate precipitates out and is converted to sodium carbonate by heating. This had a number of advantages over the Leblanc process, which generated large amounts of hydrochloric acid and "galligu," a sulfur containing foul waste. The Solvay process also had the advantage of being able to be made continuous, with the ammonia and carbon dioxide being recycled.

1866 Trans-Atlantic Cable (Electric Communication)

William Thomson was knighted for his contributions to telegraphy. He was scientific advisor to the Atlantic Telegraph Company, which laid the first successful trans-Atlantic cable. A first try, in 1857, failed, and 300 miles of cable were lost. After a second failure, there was a successful linkage in 1858, but the cable broke down and was unserviceable after a few weeks.

1867 Dynamo (Electric Power)

S. A. Varley, E. W. von Siemens, and Charles Wheatstone each described or demonstrated DC machines using self-exciting electromagnets. They became known as dynamos. A number of people had been working on generators and motors since Faraday's disc generator in 1831, but the first dynamos that were industrially useful were those of Zenobe T. Gramme, who built a generator for alternating current in 1867 and one for direct current in 1869. Self-excitation was a major step in generators, since electromagnets could be much stronger than permanent magnets.

1867 Dynamite (Explosives)

Alfred Nobel of Sweden patented dynamite in England in 1867 and in the United States in 1868. Nitroglycerine had been discovered in 1846 by an Italian chemist, Ascanio Sobrero, but it was so sensitive to shock and temperature that its manufacture was banned in several countries after some serious accidents. Doctor C. F. Schonbein, in 1845, invented guncotton, but his product was also unstable, and it was not until 1866 that Frederick Abel showed that it would be stable if all traces of sulfuric and nitric acid were removed.

Nobel's important contribution was to absorb the nitroglycerine in kieselguhr.

1868 Carbon-Zinc Cell (Electric Power)

Georges Leclanche invented the carbon-ammonium sulfate-zinc cell which used a manganese oxide depolariser. This became the familiar dry cell when the electrolyte was made into a paste. It was known as a primary cell, in contrast to the lead acid storage cell, which was known as a secondary cell. The Leclanche cell was useful because it could be used intermittently over long periods of time without attention, but the voltage dropped with use. The Daniell cell was therefore used for telegraphy until the advent of the chromic acid cell in the 1870s.

1870 Gramme Generator (Electric Power)

Zenobe Theophile Gramme, a Belgian working in Paris, developed the ring armature and the first really practical electric power generator. (Actually, the ring armature had been made by Antonio Panciotti about 10 years earlier, but his work attracted little attention.) Many Gramme generators were sold during the 1870s, mainly for lighting. These were DC machines.

Many people contributed to the development of electric generators after Faraday's disc generator of 1831. Some of these were:

In 1832 Antoine Hippolyte Pixii in Paris used a coil, therefore having the first assembly of elements of the modern generator; a coil and a field magnet rotating relative to each other. Faraday's was a DC machine while Pixii's was AC.

William Sturgeon was probably the first to develop the metallic commutator (before 1840).

Charles Wheatstone made "magnetos" for his telegraph. In 1841 he patented a 5-armature machine which gave a relatively smooth DC output.

John Stephen Woolrich in 1842 patented an electroplating method using magnetos.

Frederick Hale Holmes installed a steam-driven generator to provide power for an arc light at the South Foreland lighthouse.

By 1866 several self-excitation generators were conceived and developed. Soren Hjorth, a Dane, obtained British patents in 1854 and 1855 for devices that almost constituted full self-excitation. In 1866, C. F. Varley, Werner von Siemens, Charles Wheatstone, Henry Wilde, and Moses Farmer all patented, disclosed, or built self-excited machines.

1872 DC Motor (Electric Power)

An electric motor is simply a generator run backwards. In 1873, Gramme demonstrated a reversible combination of two machines, one acting as a generator and the other as a motor.

1873 Celluloid (Chemicals)

John Wesley Hyatt of Newark, New Jersey, started to manufacture Celluloid (nitrocellulose plasticized with camphor). He used it to make solid objects. Celluloid was invented by Alexander Parkes in 1861, but was successfully developed by Wyatt in 1868.

Hyatt made the first successful plastic by dissolving nitrocellulose under pressure. He organized the Albany Dental Plate Company in 1870 and the Celluloid Manufacturing Company in 1871. Celluloid is a homogeneous colloidal dispersion of camphor in cellulose nitrate. Its first use was to make dental plate blanks, but it was later used for other applications, such as automobile side curtains and telephones. The first intended use was for billiard balls. Phelan and Collander, who manufactured ivory billiard balls, had offered a $10,000 prize to anyone who could find a good substitute for ivory. John Hyatt and his brother Isaiha described methods for making the balls from Celluloid in patent disclosures from 1870 to 1872.

In 1888 Hyatt produced clear sheets of celluloid with a uniform thickness of 0.01", which were suitable for a base for photographic film.

1873 Dry Plate Processor (Photography)

B. J. Sayce and W. B. Bolton invented the collodion dry plate process in 1864, and prepared plates were on sale in 1873. Photographers could store the plates, expose them without preparation, and develop them later. It was no longer necessary for the photographer to carry a dark room lab with him. The major disadvantage relative to the wet collodion process was that the dry plates were less sensitive and needed longer exposures. But in 1871 a plate using gelatin instead of collodion was invented by R. L. Maddox, which solved this problem. The gelatin dry plates were on sale by 1876.

1873 Typewriter (Mechanical Printing)

In 1873 the Remington Company started production of a typewriter designed by C. S. Sholes, an American printer. There had been other attempts at constructing a typewriter, in which the letters were brought to position by moving a semicircle of metal on which they were mounted. A French printer had built a typewriter in which each character was mounted on a separate bar. But Sholes' was the first practical typewriter. It had separate bars for each character, and used a horizontal platen with an inked ribbon. Sholes was the one to introduce the QWERTY keyboard, which solved the problem of mechanical interference of the bars from keystrokes that were too rapid. The QWERTY keyboard keeps letters apart that frequently occur together.

1876 Telephone (Electric Communication)

Alexander Graham Bell patented his telephone. The sound–electricity conversion was done electromagnetically, using an electromagnet and a diaphragm. Initially, it was conceived as a pairwise device, but by 1877–1888 exchanges were built that permitted one-to-many communication.

1877 Microphone (Electric Communication)

Edison invented the carbon resistance microphone. The resistance of a compact of carbon particles depends on the pressure exerted by the sound waves. This was an important development for telephony, since the electromagnetic transducer was neither strong nor accurate enough.

In the same year, Emile Berliner utilized the same variable contact principle, but his transducer used metallic electrodes.

1878 Storage Battery (Electric Power)

R. L. G. Plante in France demonstrated a lead storage battery which came into use about two years later. Plante's design was improved by others, especially Sir Joseph Swan, who invented the cellular lead plate in 1881. By the early 1880s lead storage batteries were being used for lighting railway carriages and propelling road vehicles. The constant voltage and rechargability were a great advantage over previous cells.

1878 Converter Steel—Thomas-Gilchrist (Iron and Steel)

Sidney Gilchrist Thomas and his cousin Percy Gilchrist developed a basic converter process using a bottom blown converter, a basic lining, a basic flux, and an afterblow. This removed carbon, manganese, silicon, phosphorous, and sulfur, which made the use of phosphoric ores for good quality steels practical. This was particularly important for western Europe.

1878 Otto Cycle (Internal Combustion Engine)

Nickolaus August Otto introduced the four-cycle engine, which he had patented in 1787. In a few years, the firm of Otto & Langen sold 35,000 engines. A total of 50,000 were sold over a period of seventeen years. The fuel was gas; petroleum products were not used for fuel until later.

The four-cycle engine had been patented by a Frenchman, Alphonse Beau de Roches in 1862. De Roches worked out the essential theory of the internal combustion engine, but did not commercialize it.

Other pioneers included Ettiene Lenoir, who built a gas engine that resembled a double acting steam engine. Ignition was by a battery driven induction coil spark.

1880 AC Generator (Electric Power)

Baron M. de Meritens manufactured an AC generator using permanent magnets which was widely adopted in France.

1881 Incandescent Lamp (Lighting)

Joseph Swan started production of his carbon filament lamp. Swan had been working on carbon filament lamps since 1848, but gave it up in 1860 because he realized that a vacuum was needed that was beyond existing capabilities.

In 1877 he took up the problem again after reading about Sir William Crooke's research using the mercury vacuum pump invented by Hermann Sprengel in 1865.

In the meantime, Edison was also working on filament lamps and made them using carbonized bamboo (1879). Swan did not patent his work, but Edison patented everything, including metallic filament lamps.

By 1910 tungsten lamps were in production, which were developed by the General Electric Laboratories, AEG, the Auer Gesselschaft, and Siemens & Halske.

1882 Incandescent Lighting (Lighting)

In April 1882 and September 1882, electric lighting was turned on in London and New York respectively at the Holburn Viaduct and Pearl Street stations. Both were run by Edison companies.

1884 Compound Turbine (Steam Turbine)

Charles Parsons patented and built the first modern compound steam turbine in 1884. It ran at a speed of 18,000 rpm. In 1887 he made a compound reaction turbine with high and low pressure stages, and in 1891 he built the first condensing engine, which was more efficient than a conventional steam engine of the same capacity.

Parsons said that his work on the steam turbine was initiated by calculations based on the analogy of water in a water turbine under small pressure differences. These calculations indicated the steam turbine was feasible.

The first use of steam turbines to generate electric power was in 1890, when the firm of Clark, Chapman, Parsons & Co. supplied two 75-kw turbogenerators for the Firth Banks Power Station of the Newcastle and District Electric Light Co.

1885 Transformer (Electric Power)

The Hungarians Max Deri, Otto Blathy, and Karl Zipernowski invented a transformer with closed iron cores. It was the first practical commercial transformer and was used to light 1,067 lamps at the Hungarian National Exposition in 1885.

1885 Machine Gun (Firearms)

Continuous automatic loading and firing of cartridges was achieved by Sir Hiram Maxim, who demonstrated and started to manufacture the maxim machine gun by 1885. This was much better than the Gatling gun (1862), which worked on a revolving multiple barrel principle. The machine gun was one of the major technological advances that contributed to the enormous casualties of World War I.

1887 AC Generator (Electric Power)

Henry Wilde built the self-exciting AC generator.

1887 Aluminum (Hall-Heroult Process)

The first commercial production of aluminum by the electrolytic process was in Neuhausen, Switzerland, by the Swiss Metallurgical Company using Heroult cells. The electrolytic process was discovered independently and simultaneously by Charles Martin Hall in the United States and Paul Louis Touissaint Heroult in France. Alumina is so stable that direct reduction methods are not feasible. Earlier production of the metal was achieved by reacting the aluminum chloride with molten sodium. The chloride was made from alumina from bauxite.

The electrolytic process consists of dissolving the alumina in molten cryolite (a double fluoride of sodium and aluminum) and passing a current through the melt. Aluminum is deposited and the cryolite is reused.

American production was first commercialized by the Pittsburgh Reduction Company in 1888. In Britain, the British Aluminum Company first produced the metal in 1896, but for many years the Swiss were the main producers.

1888 AC Motor (Electric Power)

Nikola Tesla was a strong advocate of alternating current for the generation, transmission, and use of electrical power. While at Westinghouse, he developed the modern alternating current motor.

1888 Photography Flexible Film

John Carbutt, a plate manufacturer in Philadelphia, marketed celluloid sheets (obtained from Hyatt) coated with photographic emulsion.

Hannibal Goodwin of Newark had applied for a roll film patent in 1887, but it was not granted until 1898. In the meantime, the Eastman Company was marketing a roll film invented by H. M. Reichenbach and patented by Eastman in 1879.

1888 Kodak Camera (Photography)

Eastman introduced the Kodak camera, which so simplified photography that it became open to amateurs. The camera required only three operations to make an exposure, weighed only 2 lbs 3 oz, and had a built-in roll film mechanism holding 100 exposures. After exposure, the camera was given to the Eastman Processing service, which developed and printed the film and returned a freshly loaded camera to the customer. This was the beginning of the rapid spread of photography.

1888 Pneumatic Tire (Transport)

John Boyd Dunlop, a veterinary in Belfast, developed a pneumatic tire to improve his son's bicycle. A pneumatic tire had been patented in 1845 by R. W. Thomson ("a hollow belt of India rubber or gutta percha and inflated with air . . ."). Some were fitted to a carriage and successfully tested but achieved no commercial success. The tire was ahead of its time.

1889 Smokeless Powder (Explosives)

Nobel produced ballistite, the first smokeless gunpowder. In the late 1840s, C. F. Schonbein, a German-Swiss chemist, had discovered nitrocellulose (guncotton) and tried to make a smokeless powder from it. But it was too unstable, and the factories kept exploding. In 1889, Sir F. A. Abel and Sir James Dewar invented and developed cordite, a smokeless powder made from a mixture of nitrocellulose and nitroglycerine. Nobel sued them, but lost.

1889 Gasoline Engine (Internal Combustion Engine)

Gottlieb Daimler of Wurtemburg patented a four-stroke two-cylinder V type engine which ran on volatile petroleum fractions. It included a carburetor which he had patented in 1885. It was built in considerable numbers and used for motor cars and motor launches as well as a stationary engine.

1891 Rayon (Textiles)

The Count Hillaire de Chardonnet developed the first commercially successful process for producing rayon. He produced his first fiber from a nitrocellulose solution in 1884 and went into production in 1891. The filaments were coagulated by heated air.

A French chemist, Louis Henri Despeissis, developed the cuprammonium process in 1890, but it was too expensive until stretch spinning, which had been patented by Edmund Theile, permitted the manufacture of very fine filaments in 1919. Purified cellulose is immersed in copper sulfate solution, giving a product that is soluble in ammonia. The solution is spun into an acid bath which neutralizes the ammonia and removes the copper sulfate. Cuprammonium rayon is used for fine-textured fabrics.

The viscose process was patented in 1892 by C. F. Cross and E. J. Bevan. It was put into production in 1905. Purified cellulose is first treated with NaOH and aged, after which CS_2 is added to form cellulose xanthate, which is then dissolved in NaOH. The solution, viscose, is forced through spinnerettes, washed, desulfurized, bleached, washed again, oiled, and dried.

At first, the raw material was cotton linters (the short fibers left on seeds after ginning), but now cellulose from soft wood is used.

1892 Gasoline Tractor (Agriculture)

John Froelich of Iowa mounted a single cylinder van Duzen engine on a tractor, which completed a 50-day threshing run. It became the forerunner of the John Deere tractors.

1893 Motor Car (Transport)

Karl Benz, of Mannheim, built a four-wheeled vehicle with a horizontal crankshaft and a two-speed belt drive using an engine of his own design. The car was successful and many hundreds were built. This was the beginning of the automobile industry.

Gottlieb Daimler's successful model appeared in 1894 as a product of the Cannstatt works. The Cannstatt-Daimler cars later became the Mercedes.

Other inventors in other countries were also active at this time. Charles and J. Frank Duryea and Henry Ford were building experimental models of gasoline-powered cars as early as 1892, and R. E. Olds built a two-passenger car in 1896.

1894 Electric Motor Car (Transport)

The Electric Carriage and Wagon Company introduced electric taxicabs into New York City in 1894 and was operating about 100 electric cabs by 1898.

Among the first electric automobiles was one built in England in 1888 by J. K. Starley. At the same time, Fred M. Kimball was building one in the United States. Another American electric auto pioneer was William Morrison, who finished his car in 1891.

Although it was eventually eclipsed by the internal combustion powered auto, the electric car was widely used for a time. By 1922, Studebaker was producing five models of electric autos and sold twenty of them that year. It had a maximum speed of 13 miles per hour. In 1904, one-third of the cars were electrics in New York, Chicago, and Boston. By 1912, there were over 20,000 electric cars in the United States, and 10,000 electric vehicles were produced that year. But the disadvantages of low speed, short range, heavy batteries, and low battery life were great, and by 1930 Detroit Electric was the only manufacturer of electrics left in the United States. Currently, most electric vehicles are industrial trucks, golf carts, and delivery vehicles.

1895 Motion Pictures (Photography)

In Lyons, Auguste and Louis Lumiere demonstrated their movie camera and projector. They had seen Edison's Kinetoscope in New York. Edison had patented a movie camera in 1881 (the kineograph). His system was too expensive for mass viewing, being suitable for one viewer at a time. Edison invented the 35 mm. perforated movie film with four perforations per frame which is still standard.

The Lumieres completed the final step in a complex development that made movies commercially feasible.

1896 Phonograph (Sound Reproduction)

The Berliner Gramaphone Co. was building and selling phonographs by 1896. Edison had patented a phonograph in 1877 which used helical grooves on a tinfoil cylindrical surface, but soon got too busy with electric lighting to pursue it much further. Alexander Graham Bell, with Chichester Bell and Charles Tainter, developed the wax cylinder, which they patented in 1885. Bell and Tainter organized a company to sell their machines in 1887.

Emile Berliner, a German-born American, made the important contribution of using a flat disc in which the grooves varied from side to side. He first proposed to use hard rubber for the disc but soon found a company that provided him with thermoplastic material. Berliner got the help of Eldridge Johnson of Camden, who developed a spring-driven motor with a friction governor. He made the entire mechanism for Berliner and made a number of improvements for the gramaphone. Johnson and Berliner formed the Victor Talking machine Co. in 1901. The Bell and Tainter patents became the basis for the Columbia Phonograph Co.

1897 Diesel Engine (Internal Combustion Engine)

Rudolf Diesel was a German engineer born in Paris. He took out his first English patent for the diesel engine in 1892. Diesel consciously sought high efficiency by introducing fuel only after a high compression of air. The compression heated the air sufficiently to ignite the fuel oil spray. It worked well with a variety of oils, and its efficiency was 11 percent greater than any other prime mover of the time. The diesel engine was first successfully manufactured in 1897.

1897 Radio (Electric Communication)

Guglielmo Marconi established the first radio station at the Isle of Wight, from which he made a number of firsts: first maritime transmission (18 miles), first paid for wireless telegram, first 200-mile transmission (1901).

His transmitter was a spark gap apparatus, similar to that used by Hertz to demonstrate the existence of electromagnetic waves. The spark was produced by a spark coil (a transformer that converted a continually interrupted low voltage from a battery into a high voltage). The spark jumped across to balls or plates. They acted as a condenser when the current overshot. This produced a rapidly decaying spectrum of waves. Marconi's detector was a coherer, a device consisting of metal filings in a tube whose resistance varied when exposed to electromagnetic waves. A "trembler" had to be used to shake loose the filings after they had responded to the electromagnetic wave so they could respond to the next signal. These constituted a wireless telegraphy system, which transmitted and received Morse code by varying the time of spark pulses by using a tapping key.

1902 Cellulose Acetate (Textiles)

In 1902, Arthur D. Little, William H. Walker, and Harry S. Mork were granted the first U.S. patent for spinning a man-made fiber. This led to the first successful manufacture of acetate rayon. Little had kept abreast of English experiments. Lauren Nadin and Paul Schutzenberger first made cellulose acetate in 1865. C. F. Cross and E. J. Bevan patented a production process for it in 1894.

In 1903, G. W. Miles discovered that partial saponification of cellulose (with NaOH) gave a product that was soluble in acetone. Unsaponified cellulose acetate had to be dissolved in chloroform.

1903 Electrocardiograph (Medicine)

The first electrocardiograph was built by Willem Einthoven, a professor at Leyden. It weighed 660 pounds.

1903 Airplane (Transport)

Orville Wright made the first successful powered flight in a plane built by himself and his brother Wilbur. It used a 12-HP gasoline engine that they designed and made themselves. There had been a considerable number of efforts at powered flight, including those of Lillienthal, Langley, and Cayley, but these were not successful because they did not solve the problem of control. The Wrights achieved control by having a device to warp the wing tips and by including a rudder.

The Wright brothers had been experimenting with flight for some years and flew three biplane gliders at Kitty Hawk, North Carolina, from 1900 to 1902. They saw control during flight as the major problem and used glider experiments to study control techniques. Their first attempt at powered flight was on December 14, 1903, but it failed. The vehicle was their 1902 glider, which they fitted with the 12-HP engine and two chain-driven push propellers. They tried again on December 17, making four successful flights that day. The first flight lasted 12 seconds. The fourth flight lasted 59 seconds and traveled 852 feet.

Of the earlier efforts, those of Samuel Pierpont Langley probably came closest to matching the Wright's achievement. Langley was Secretary of the Smithsonian Institution and was an excellent scientist and engineer. He experimented with small models for a number of years and flew one in a 3/4 mile circle in 1896. He finally built a full-scale machine, using an engine designed and built by his assistant, Charles Manley. Manley's engine was a good design and actually had a horsepower to weight ratio that exceeded that used by the Wright brothers. Langley tried to launch his plane from a catapult on top of a houseboat on the Potomac, but on each occasion the plane was fouled by the catapult equipment and crashed into the river. His second and last try was on December 8, 1903, just nine days before the first successful flight by the Wright brothers.

1904 Glass (Automation)

The first fully automatic machine for making bottles, designed by M. J. Owens, went into production in 1904. By 1913, 164 such machines were in operation. The device was soon superseded by feeder-fed machines in which the glass gobs were formed by gravity feed and a plunger rather than by suction. By 1920, two billion bottles a year were being made by feeder-fed machines.

1904 Diode (Electric Communication)

Sir John Ambrose Fleming, in his capacity as a consultant to the British Marconi company, discovered that the "Edison Effect" could be used in a diode as a rectifying detector.

1906 Triode (Electric Communication)

Lee de Forest invented the triode. By means of a battery and a grid between the anode and cathode, electric currents could be amplified. This was important to detection as well, because the diode in its original form was no better than the crystal detector in detecting radio waves.

1907 Bakelite (Plastics)

L. H. Baekeland in the United States and James Swinburne in England independently and almost simultaneously succeeded in making a thermosetting plastic based on phenol-formaldehyde. Baekeland's patent was filed one day before Swinburne's. Baekeland received his patents in 1909 and formed the Bakelite Corporation in 1910. The material became known as Bakelite. Eventually, the British and American interests merged.

1911 Gyrocompass (Transportation)

The gyrocompass of E. A. Sperry and the three-gyro compass of Hermann Anschutz-Kaempfe and Max Schuler were both given sea trials in 1911. Sperry's gyrocompass was immediately adopted by the U.S. Navy. The German Navy was already using a single-gyro compass which Anschutz-Kaempfe and Schuler had submitted for sea trials in 1908. The three-gyro compass was a great improvement.

Anschutz-Kaempfe became interested in a nonmagnetic compass because he wanted to explore the North Pole in a submarine and a magnetic compass was not suitable for this. By 1903 several experimental models were built. Schuler joined him in 1906 and made important contributions to later designs.

1912 Gyrostabilizer (Transportation)

The U.S. Navy installed its first gyrostabilizer on a ship in 1912. The ship was the U.S.S. *Worden*, a 433-ton torpedo boat destroyer. The gyrostabilizer was developed by Elmer Sperry. It was not the first such device. Gyrostabilizers had been tried by both the British and German navies. However, these devices were large and sluggish. Sperry's methods of feedback control made gyrostabilizers practical.

1912 Stainless Steel (Iron and Steel)

The three types of stainless steel are the martensitic (10–18 percent chromium, 0.08–1.1 percent carbon), the ferritic (15–30 percent chromium, 0.08–0.2 percent carbon), and the austenitic (16–22 percent chromium, 6–12 percent nickel). The austenitic stainless was developed by Edward Maurer and Benno Strauss at Krupp's, which produced it in 1912.

The martensitic stainless was first produced in 1914 by Thomas Firth and Sons Ltd. in England, after being developed by Harry Brearley, who was head of a laboratory operated jointly by Firth and another steel firm, John Brown & Co. Ltd.

The ferritic stainless was developed by Elwood Haynes, an independent American inventor who first applied for patents on his steel in 1912, but was refused. It was 1919 before the Patent Office issued him a patent. A steel syndicate then bought his patents.

1913 Thermal Cracking (Petroleum)

The first successful thermal cracking process was developed by William M. Burton of Standard Oil. He started his work in 1909, and it was first used commercially in 1913. Commercial use of Burton's process was resisted by the Standard Oil Board of Directors and was made possible only after Burton became a member of the Board of Standard Oil of Indiana, a company newly formed as a result of Federal anti-trust action. The Burton process cracked the large molecules in the heavier fraction of petroleum into the smaller molecules that constituted gasoline by using high temperature and pressure. Until 1920 it was the only available method for producing gasoline from the heavier fractions of petroleum.

It should be noted that the cracking effect of high temperatures was known as early as 1861, and that the production of lower boiling fractions by pressure and temperature was observed by James Dewar and Boverton Redwood in 1889.

From 1920 to 1923, four new thermal cracking processes were developed and used. A major improvement embodied by these was that they were continuous rather than batch processes.

1915 Haber Ammonia Process (Chemicals)

BASF opened the first plant, at Oppau, to produce ammonia by the Haber process. It had a capacity of 8,700 tons per year.

In 1918 Fritz Haber was awarded the Nobel Prize for his process, which consisted of the direct combination of hydrogen and nitrogen in the presence of a catalyst at temperatures of 600°C and pressures of 200 atm. The first catalysts were uranium carbide or osmium, but later a much cheaper activated iron compound was used.

The Haber process provided cheap ammonia for the manufacture of fertilizers and explosives.

1917 Gasoline Tractor (Agriculture)

The Fordson was first marketed in 1917. This was the tractor that initiated the rapid diffusion of tractors on American farms. It had the general form that came to dominate tractor technology.

1917 Cellophane (Plastics)

Cellophane was first produced by the French firm, La Cellophane, which was formed by the Comptoir de Textiles Artificiels for this purpose. Cellophane was developed by Jacques E. Brandenberger, a Swiss-born French Chemist. Comptoir de Textile Artificiels, the largest rayon producer in France, financed Brandenberger and hired him.

Comptoir was a partner with Du Pont in a jointly owned Fibersilk Company. Du Pont acquired the patents and know-how from La Cellophane in 1923.

The first cellophane was produced in the U.S. in 1924, but it was not moisture proof. In 1926, William Hale Charch and Karl Edwin Prindle found ways to apply waterproof coatings to both sides of the film and thereby assured the success of cellophane as a major product.

Sylvania Industrial Corporation started production of cellophane in 1935 under license from Du Pont.

1922 Insulin (Pharmaceuticals)

In 1922 the first human patients were treated with insulin at the Toronto General Hospital after Frederick G. Banting found a method of extracting insulin from pancreas.

Insulin exists in the islet tissue of the pancreas. Previous attempts at isolation failed because a substance in the surrounding pancreas deactivated the insulin. Some insulin had been isolated by tying off the main pancreatic gland, which then deteriorated leaving the islet intact, but the method was much too slow and only very small amounts could be produced.

Banting succeeded by working at low temperatures and using ethyl and methyl alcohol, which slowed up the deactivation process.

The process was widely licensed in the United States and abroad, the Eli Lilly Company doing a lot of the important process development in the United States.

1924 35 mm Camera (Photography)

The 35 millimeter camera was invented in 1914 by Oskar Barnack and produced by Leitz in 1924. It was called the Leica and started a trend that revolutionized photography.

1926 Liquid Fueled Engine (Rockets)

The first successful liquid-fueled rocket was that of Robert H. Goddard in 1926 at Auburn, Massachusetts. In 1935 he had succeeded in reaching an altitude of two kilometers. The early developments of rocketry, however, were dominated by Germany, where Hermann Oberth had been working on rockets since the 1920s. By 1942, the A-4, a liquid-fueled rocket with a range of nearly 200 miles, had been developed at Peenemunde under the leadership of Wernher von Braun.

1929 Electroencephalograph (Medicine)

Hans Berger, Professor of Physiology at Jena, built the first EEG in 1929. He was supported by the Carl Zeiss firm, which saw the value of the device when very few others were interested.

1931 Neoprene (Rubber)

Neoprene was first manufactured by Du Pont in 1931 after successful research led by Edward K. Bolton, with some critical help from Wallace C. Carothers and Arthur M. Collins. Bolton had suggested that Du Pont look for a synthetic rubber in 1925. But little progress was made until he heard a lecture at an American Chemical Society meeting by Father Julius A. Nieuwland. Nieuwland was a priest from Belgium who had obtained his doctorate from Notre Dame in 1904 and had decided to study acetylene. By 1921 he had polymerized acetylene. His divinyl acetylene was the starting point for the development of neoprene.

The elastomers from Nieuwland's work were too soft. Collins investigated the impurities of divinyl acetylene, for entirely different reasons, and found that one of them contained chlorine. Carothers suggested that the reaction between hydrogen chloride and monovinylacetylene be investigated. This gave a new compound, chloroprene, which spontaneously polymerizes to form an elastomer similar to vulcanized rubber.

1932 Crease Resistant Fabrics (Textiles)

In 1932 Tootal Broadhurst Lee Co. Ltd., a medium size Lancashire textile firm, announced that a crease resistant fabric was ready for commercial production. The firm had set up a research group immediately after World War I to find ways to reduce the creasing of cotton fabrics, which was a serious defect of cotton cloths. The project took 14 years. The problem was solved by polymerizing a resin inside the fibers after the fabric had been soaked in monomer solutions and catalyst.

An unpredicted benefit was that the treatment also greatly increased the strength of rayon, which is weaker than cotton or linen, especially when wet.

This was a case of corporate directed research conducted over the long range that had outstanding success.

1934 Diesel Locomotive (Transport)

General Motors built the version of the diesel locomotive that ensured its commercial success. The critical factor was the method of transmission of power to the wheels. In the GM diesel–electric, the diesel engine was linked to an electric generator which drove the wheels. This overcame the problem that the diesel does not respond rapidly to varying loads.

1935 Radar (Electric Communication)

Radar detection up to a distance of 19.5 miles was demonstrated in England. The British development of radar was a military project led by Robert Watson-Watt, who made major contributions to its development and clearly laid out the basic principles of aircraft radar detection. John Logie Baird, who had pioneered British research in television, was a critical figure in radar developments.

Radar requires the emission of intense short waves, made possible by the magnetron, invented by Albert Hall of General Electric in 1921. The much more powerful cavity magnetron was developed in Britain in 1939.

1935 Kodachrome (Photography)

Kodak placed Kodachrome on the market in 1935. It was the work of two independent inventors, Leo Godowski, Jr., and Leopold Mannes, who had produced a crude color picture using a three-layer process as early as 1923. Godowski and Mannes were concert musicians. They accepted an offer from Kodak and worked on the commercial development of their ideas at Kodak for ten years.

1935 Methyl Methacrylate (Plastics)

Methyl methacrylate polymers were produced by Rohm and Haas under the name of Plexiglass in 1935 and under the name of Pontalite,which was later changed to Lucite, by Du Pont in 1936.

The first patent for the polymerization of methacrylates was obtained by William Chalmers as a result of his work at McGill University, while a graduate student in 1930–1931. Early in 1931, he approached Rohm and Haas and Imperial Chemical Industries, both of which took up the idea. The Du Pont product was produced under license from ICI.

1936 Television (Electric Communication)

The BBC began regular television for two hours a day, six days a week, from London in 1936. In the same year RCA began its first demonstration transmission from the Empire State Building; its first public service was started in 1939.

The British broadcasts used the mechanical system of John Logie Baird, which was based on the rotating disk of P. Nipkow. Holes were arranged in a spiral pattern on the disk so that, when light shone through the holes while the disk rotated, a beam completely scanned the target. This was later abandoned in favor of electronic methods.

The American systems were electronic from the start. Electronic television was developed in the United States primarily by RCA, which recognized that the iconoscope, patented by V. K. Zworykin in 1925, made electronic television possible.

1937 Catalytic Cracking (Petroleum)

In June of 1936, Socony produced a small commercial catalytic cracking unit based on the work of Eugene J. Houdry. This was followed by much larger units installed by Sun Oil in April of 1937 at Marcus Hook, Pennsylvania. These developments were the result of a joint venture involving Houdry, Sun Oil, and Socony.

Houdry was a French inventor and industrialist who became interested in better fuels because of his interest in auto racing. He started work on catalytic cracking in 1922. His successful catalyst consisted of aluminum–silicon oxides. After using much of his own capital and being unable to interest French sources in his projects, he went to the United States and ultimately joined with Sun and Socony.

The major advantage of the catalytic over the thermal cracking was that the quality of the catalytic product was better and the financial return per barrel of oil was better.

1938 Nylon (Textiles)

Du Pont first produced nylon commercially in 1938. It had been first obtained at the Du Pont research laboratory in 1935 by Wallace H. Carothers.

1939 Turbojet (Transport)

The first jet-powered plane flew in Germany. It was the Heinkel HE-178, using an engine designed by Hans von Ohain. Independently, Frank Whittle had been working on a turbojet in England, and the first English flight was in 1941.

The turbojet works by taking in air from the atmosphere, using a centrifugal (axial) compressor to compress it to 3 to 12 times its original pressure and adding fuel. The mixture burns to reach temperatures of 1200 to 1700°F. The hot gas drives a turbine which drives the compressor. The discharge at the compressor is at a pressure high enough to drive the aircraft forward (about twice atmospheric pressure).

The advantage of the jet engine is that it has a high power to weight ratio. Much more power can therefore be used for the aircraft. Also, the efficiency of the jet is high at high speeds, while that of the propeller is lower at high than at low speeds. The absence of a propeller permits the jet plane to be more streamlined and aerodynamically efficient than the propeller-driven plane. Also, the speed of the propeller is higher than that of the plane. Thus, the propeller tips reach sonic velocities before the plane does. The resulting wave-making losses at the sonic velocities limits the practical speeds of propeller planes to not much more than 400 mph.

1940 Frequency Modulation (Electric Communication)

Commercial FM broadcasts were authorized by the Federal Communications Commission and assigned a band from 42 to 50 MHz. By that time there were already twenty-two experimental FM stations in existence.

FM was a response to the difficulty of having static free broadcasting with AM. By its very nature, amplitude modulation is highly susceptible to interference. Any electromagnetic source easily affects the amplitude of the radio waves, and since the audio information is carried by amplitude variations, it is easily distorted.

Edwin F. Armstrong started to attack the problem of static in 1923 with the backing of RCA. He was successful by 1933, and in 1934 he had an experimental transmitter on top of the Empire State Building that started broadcasting on June 16th. However, RCA had decided to commit all its resources to television and closed the station in March of 1935. It was not until the FCC approval in 1940 that FM broadcasting began to grow.

1941 Helicopter (Aviation)

Development of the helicopter has a long history. One capable of flight was built by J. C. Ellehammer in 1912, and in 1936 Heinrich Focke flew a helicopter of his own design. But it was only with the work of Igor Sikorsky at the United Aircraft Corporation that a combination of technical advances and industrial backing ensured practical success. By 1941 the Sikorsky design was flying and military production was started.

1942 Atomic Pile (Atomic Power)

On December 2, 1942, the first controlled self-sustaining chain reaction was achieved in a pile, designed by a group led by Enrico Fermi.

1942 V-1 and V-2 (Rockets)

The Germans opened their main rocket research station at Peenemunde in 1935 and launched their first successful test of the V-2 in 1942.

1943 Penicillin (Pharmaceuticals)

In 1943, production of penicillin was started in the United States.

In 1928, Alexander Fleming, at the University of London, had noted the antibacterial action of the mold Penicillium Notatum. He determined that it was nontoxic to animals and white corpuscles and showed that it was effective against a number of bacteria but not against others. To go further, methods of extracting penicillin from the mold, stabilizing it, and increasing the yield had to be worked out. These were pursued by Prof. H. Raistrick at the London School of Hygiene and Tropical Medicine. The later development of penicillin was due to H. W. Florey and E. B. Chain in Oxford. In 1939, they established its chemotherapeutic properties, and in 1941 enlisted the help of American companies to manufacture it.

The development to a marketable product started with work at the U.S. Department of Agriculture in Peoria, where deep culture fermentation (which permitted large-scale production) and the use of corn steep liquor (which increased the yield) were introduced.

1944 Titanium (Metals)

In 1944 the first plant was built to produce titanium at Boulder City, Nevada. Its capacity was 100 pounds per week.

The process that made the production of titanium feasible was invented by W. J. Kroll, a metallurgist with a private laboratory in Luxembourg. He started work on the process in 1930. By 1938 he had succeeded and went to the United States to sell the process. In 1940 he emigrated to the United States.

Further development of the Kroll Process by the Bureau of Mines in the early forties led to its production at the Boulder City plant.

The Kroll process depended on the reduction of titanium tetrachloride by calcium. Later, as the result of work at Imperial Chemical Industries, sodium was used. ICI built a plant in 1955 with a capacity of 1,500 tons per day.

1945 Atomic Bomb (Explosives)

On July 16, 1945, the first atomic bomb was exploded at Alamagordo.

1946 ENIAC (Computers)

In 1946, the first electronic computer was completed at the Moore School of the University of Pennsylvania under contract to the Army.

1947 Transistor (Electronics)

John Bardeen, Walter Brattain, and William Shockley invented the transistor at the Bell Labs. Semiconductors had been used in radar because they were much more sensitive for the detection of microwaves than vacuum tubes, but could not be used for amplifiers until the Bell Labs' work.

1948 Instant Photography (Photography)

The Polaroid black and white instant camera and film was developed and marketed by Edwin Land in 1948. In 1963, instant photography was extended to color.

1948 Long Playing Record (Sound Recording)

The Columbia Records Company introduced the long playing record in 1948. This was the work of Peter Goldmark, who was head of the Columbia Broadcasting System research laboratories.

Goldmark decided on discs rather than tape or wire because discs were in common use and were cheaper. The vinyl he used permitted finer grooves to be cut in the disc and also reduced surface noise.

The important features of the LP are: slow rotation (33-1/3 rpm), finer grooves, vinyl plastic, and light weight pickup.

RCA tried to counter this development with the 45-rpm record, which was reasonably successful for popular music but did not seriously hurt the LP. The 45 found its own niche in singles and popular music, while the LP is used for albums and classical music.

It is interesting to note that RCA had tried to introduce a 33-1/3 rpm as early as 1931. This effort failed because proper players were not available and the records were made of soft plastic. The records wore out quickly under the heavy load of the tone arms of the time.

1950 Xerography (Mechanical Printing)

The xerox copying method was introduced into the market by the Haloid Corp., which later became Xerox. Xerography was invented by Chester Carlson, who started on it in 1934 and had the basic concepts worked out by 1937.

Carlson was a patent attorney for P. R. Mallory & Co. and worked on his project at home in his spare time. His work for Mallory brought him into contact with the Battelle Memorial Institute, which undertook the development of xerography. By 1946 Battelle had developed the process to the point where commercialization was possible. Haloid had learned about the process through a 1944 article and took a sub-license from Battelle in 1946.

1951 Nuclear Reactor (Atomic Energy)

The first reactor to produce useful electricity was built by the Argonne National Laboratory in 1951.

1952 Hydrogen Bomb (Explosives)

On November 1, 1952, the first hydrogen bomb was exploded in the Pacific.

1954 Color Television (Electric Communication)

The first demonstration of color television was that of J. L. Baird in 1928 using mechanical scanning. H. E. Ives, Peter Goldmark, and others worked on color television in the United States, but it was not until 1953 that standards were adopted that made color television compatible with the existing black and white system. Compatible broadcasting started in 1954.

1954 Transistor Radio (Electronics)

The first transistor radio was placed on the market in 1954. This was an American product called the Regency. In 1955 Sony introduced the TR-55, which was the first radio that was fully transistorized.

1955 Transatlantic Cable (Electric Communication)

Until the TAT–1 (Trans Atlantic Cable No. 1) was placed in service in September of 1955, overseas telephony depended on radio. The cable had great advantages over radio since transmission was not degraded by radio interference. TAT–1 had a capacity of only forty-eight conversations, but later cables increased this to thousands. Optical fiber cable TAT–8 (1988) can handle over 17,000 circuits simultaneously and can be upgraded to carry 40,000 channels.

1957 Sputnik (Satellites)

The Soviet Union launched the first artificial satellite on October 3, 1954. One month later they launched a dog (named Laika), which survived the launch and made several orbits. In 1960 the Soviets launched two dogs, which were recovered unharmed, and in 1961 Yuri Gagarin successfully orbited earth and returned safely. The United States reaction was intense and culminated in the moon landing of Neil Armstrong and Edwin Aldren on July 21, 1969. On October 4, the Soviet Union launched the first man-made earth-orbiting satellites.

1960 Laser (Optics)

C. H. Townes, with Arthur Schawlow, had suggested the possibility of light amplification by stimulated emission in analogy with the maser. In 1960 the first laser was built by T. H. Maiman.

1960 Echo I (Satellites)

Echo I was a passive reflector satellite. The reflector was a large gas-filled balloon with an aluminized surface. It was launched in 1960.

1961 Integrated Circuits (Electronics)

Texas Instruments and Fairchild Semiconductor both offered integrated circuits commercially in 1961. The idea of an integrated circuit on a chip had occurred to others and appears in a paper by W. A. Dummer, a British radar specialist. But Jack Kilby of Texas Instruments was the first to implement the idea. At Fairchild, it was Robert Noyce, who later was a co-founder of Intel, who developed integrated circuits. Integrated circuits revolutionized electronics and made the microprocessor possible.

1962 Telstar 1 (Satellites)

On July 10, 1962, the United States launched the first active communications satellite, which relayed television messages across the Atlantic.

1971 Microprocessor (Computers)

Ted Hoff and Stan Mazer of Intel developed the 4004, the first commercially available microprocessor in 1971. It contained about 1000 transistors, used four-bit words, and was capable of doing arithmetic. The 4004 was the result of an effort to make an inexpensive hand-held calculator.

1974 Personal Computer (Computers)

The first personal computer was offered for sale as a hobbyist's kit in 1974 by MITS (Micro Instrumentation Telemetry Systems), a manufacturer of calculators that was having financial trouble. It was called the Altair, after a planet in the Star Trek series. After a cover story on the kit by *Popular Mechanics*, MITS was swamped with orders. Steve Wozniak and Steve Jobs started Apple Computer, and in 1976 sold 200 Apples from Job's garage by mail order. Other companies were attracted to the market, and in 1977 Commodore had the Pet and Tandy was selling the TRS–80. These were fully assembled systems.

1977 Optical Fibers (Optics)

Optical fibers were first used for communications systems in 1977 when a 1.5-mile telephone and video link using a 24-fiber cable was installed in Chicago.

Glass fibers had been used as light guides much earlier. In the 1950s, highly transparent glass fibers were being used for applications over very short distances in which signal losses were not important. These include illumination of automobile dashboards and the "fiberscope," which was used to examine airplane engine components and for internal examinations in medicine.

Upon its development in the 1960s, the laser was recognized as an ideal light source for optical communication. The continuous, room temperature semiconductor laser developed in the early seventies made the development of fiber optics practical. Development of the first fiber system prototypes was started by AT&T in 1974.

APPENDIX 3

Data Tables

GENERAL COMMENTS

The data given in this appendix were collected from a variety of sources, most of which are secondary sources and many of which are government publications. In many cases, I have performed elementary calculations on the data in these sources and presented the results here. When these are of a trivial and obvious sort, such as dividing by population figures to get a per capita result, or converting from one set of units to another, I will not explicitly mention this fact in these notes. When the manipulation of the data are more extensive, an explanation of what was done is given below. The source of the data and any brief comments are given in the list of tables.

An important fact is that, by the very nature of the data and the means available to acquire it, many of the numbers are of quite limited accuracy. Thus while official speed records are certainly accurate to at least the nearest mile per hour, quantities such as energy consumption, dwellings with electric service, and even population figures can be very inaccurate. Usually, this does not have any serious effect on the conclusions drawn in this book from the data because any conclusions that depend on doubtful accuracy are avoided.

It is convenient to abbreviate certain references that are used repeatedly. These are:

Historical Statistics of the United States: Colonial Times to 1970, U.S. Department of Commerce (1975), hereafter abbreviated as *HSUS.*

Statistical Abstracts of the United States, U.S. Department of Commerce. These are annual compilations which update much of the data in HSUS. They will be abbreviated by initials followed by the year of reference. For example, the volume for 1986 will be referred to as *SAUS–86.*

Annual Energy Review, Department of Energy, various years. These are annual compilations of energy statistics and will be referred to (for the 1984 edition, for example) as *DOE–84.*

Science Indicators, National Science Board, various years. These are compilations of data on science activities published by the National Science Foundation. They will be abbreviated as *NSF–XX.* Note that after 1986, the volumes are called *Science & Engineering Indicators.*

GROSS NATIONAL PRODUCT

A GNP series in constant ("real") dollars is constructed by dividing a series in current dollars by a deflator. This deflator is the ratio of a price index for the current year, divided by a price index for some base year. A variety of "constant dollar" years can therefore be constructed depending on which year is chosen as the base year. For any two constant dollar series, based on different years, the ratio of any two values for the same year is the ratio of the price indices for the two different years. If two series are available for two different base years, which cover different ranges but have some overlap, a series can be constructed that covers the sum of the ranges of the two series. This is done by finding the ratio of the price indices for the two base years and using this as a multiplier to extend either of the two series. The data presented in Table 3 were constructed from the GNP series in 1958 dollars given in *HSUS*, page 8, and an overlapping series given in 1972 dollars in *NSF–85*, pages 216 and 217. The two series overlapped from 1960 to 1970. In this range the ratio of the 1972 to the 1958 values was 1.5. The 1972 series was therefore divided by 1.5 to get values of GNP in 1958 dollars for the years after 1970.

ENERGY CONSUMPTION IN THE UNITED STATES

Although the amount of data published on the production and consumption of energy is large, it is not a trivial task to construct consistent series of data over an extended time period. Published data by different agencies and governments often differ, and definitions sometimes vary among sources of data. In constructing the energy tables in this book, I have tried to achieve a consistency of 5 percent using easily available sources. An effort much beyond this is not justified because an accuracy better than 5 percent is too much to hope for. I simply do not believe that barrels of oil, tons of coal, and cords of wood can be counted with such accuracy. A certain amount of energy consumption goes unrecorded as does a certain amount of production. The inherent inaccuracies must be kept in mind when using such data. However, the series presented here are reasonably consistent and serve the purposes of this book.

The consumption of energy in the United States was computed from the sum of energy from coal, petroleum, natural gas, nuclear fuel, wood, water, and geothermal energy. The consumption of hydrocarbon fuels is generally reported as total consumption, including non-fuel uses such as for chemical feed stocks and lubricants. To get energy consumption, non-fuel uses of hydrocarbon minerals must be subtracted from the total consumption.

For the consumption of coal up to 1950, the data were taken as the sum of bituminous and anthracite consumption reported in *HSUS*, page 587. The data for 1951 and later were taken from *DOE–80* and *DOE–85*. For 1800 to 1845, the data are for coal production, not consumption, since coal consumption was not reported. I assumed that consumption was close enough to production to include these figures in the table. The non-fuel use of coal was subtracted even though it was less than 1 percent, as shown on page 702 of HAUS–86 and on page 71 of *Energy Perspectives–2 (1976).* There are no data on non-fuel use of coal before 1947. For these early years, I have arbitrarily used 0.5 percent for the non-fuel use of coal.

Petroleum consumption was defined to include natural gas liquids. Up to 1948, data were taken from pages 587 and 588 of *HSUS.* This included natural gas liquids, net imports, and domestic production. After 1948, data were taken from *DOE–86.* Non-fuel uses of petroleum from 1947 to 1974 were taken from *Energy Perspectives–2 (1976)* by adding up the non-fuel industrial use and asphalt use for the household and commercial sectors. For 1975, and for 1879 to 1984, non-fuel use of petroleum was computed from data in *HAUS–86,* page 702. Before 1947, it was assumed that 5 percent of petroleum was for non-fuel use. For 1976, 1977, and 1978, it was assumed that non-fuel use was 11 percent, which is an interpolated value. For the early years (before 1947) I arbitrarily used 5 percent for non-fuel use of petroleum. There is always an asphalt residuum that can be as high as several percent. At any rate, the early figures are probably not accurate to 5 percent.

Natural gas consumption up to 1970 was taken from the *HSUS* series on pages 587–588 (natural gas, dry). From 1970 to 1984, consumption was computed by adding losses from Table 60 of *DOE–86* to consumption in Table 3. The SCF to quads conversion was effected by using the thermal conversion factors in *DOE–80.* Non-fuel use of natural gas was taken from *Energy Perspectives–2 (1976).* No reliable data before 1947 were found. It is surprising that 11 percent of natural gas is reported to be for non-fuel use in 1947. But before 1947, the dominant fuel was coal, so errors in non-fuel use of gas are not serious. I arbitrarily took non-fuel use of gas before 1947 to be 10 percent. The error introduced by inaccuracies in this assumption are certainly less than a few percent for the total consumption of energy.

The total error introduced from lack of data on non-fuel use in the early years varies from 0 to 0.2 quads, or 0 to 2 percent.

Consumption of nuclear energy is taken from *DOE–84* and is the total heat liberated by fission at nuclear plants.

Wood energy use up through 1970 was taken from *HSUS,* pages 587–588. Data for 1980–1983 are from *DOE–84.* Consumption was steady in the late sixties but took a big jump in 1980 in reaction to high fuel prices. Rose (1986, page 173) estimates that there was a 40 percent increase in the number of households burning wood from 1978 to 1980. If 0.4 quads is subtracted as the 1960's baseline and the 40 percent increase is applied to 2.0 quads, this gives 1.8 quads for 1978 and 2.2 quads for 1979. I therefore assumed that wood energy consumption remained constant at 0.31 quads from 1971 to 1973 and then increased gradually to 0.6, 0.7, 1.0, and 1.3 quads up to 1978, when it jumped to 1.8 quads. These assumptions cannot be off by more than 0.5 quads, which would give an error of less than 1 percent for total energy use.

Data for water power were taken from *HSUS*, pages 587–588, up through 1970 and from *DOE–84* after 1970. There is always an element of arbitrary definition in defining the consumption of energy from water, which can range from the maximum mechanical energy of flowing water to the energy actually delivered at water wheels or turbines. I have adopted the usual convention of defining water energy at the "prevailing central station equivalent." That is, consumption of water energy is defined as the consumption of fossil fuel that would be needed to produce the energy actually delivered at water wheels or turbines.

Geothermal energy is reported as the amount of electricity actually produced. The amount of geothermal energy consumed is so small that it is not even worth correcting.

Consumption figures computed as described above were used to construct Tables 5 to 13, Table 17, and Table 22.

WORLD ENERGY CONSUMPTION

It is more difficult to construct a series of energy consumption for the world than for the United States. In many countries there is little or no reliable data collection. Often, in the developing and lesser developed countries only "commercial" energy is counted. That is, only fuels that are bought and sold through recognized channels can be captured in the statistics. Reported energy consumption figures are therefore probably significantly less than actual consumption. Refinements such as accounting for non-fuel uses and differentiating between production and consumption could not significantly improve accuracy.

World energy consumption for 1925, 1938, 1950, and 1955 were taken from Darmstadter, Teitlebaum, and Polach (1975), *Energy in the World Economy, Resources for the Future*, Johns Hopkins University Press. Their data were in millions of metric tons. This was converted to quads by multiplying by the conversion factor 0.0278. From 1960 to 1973 the data were taken from *Energy Perspectives–2*, U.S. Dept. of the Interior (1976). From 1974 to 1983, a production series was used taken from *DOE–84*. These results are given in Table 4.

FRACTION OF FUELS CONSUMED TO GENERATE ELECTRICITY IN THE UNITED STATES

HSUS, page 820, gave the net production of electricity up through 1970. This was multiplied by the heat rate on page 826 to get quads of fuel consumed to generate electricity. Data after 1970 were computed from *DOE–80* and *DOE–84* data. Heat rates after 1970 were taken from *DOE–84*. The calculation includes industrial as well as utilities' production of electricity and also includes the fossil fuel equivalent of water power. Energy consumption as computed above was used as the divisor to get the fraction of energy used to generate electricity. The results are in Table 22.

PRODUCTIVITY OF UNDERGROUND COAL MINING

The data for productivity of underground bituminous coal mining was based on the data in Gold, Pierce, Rosegger and Perlman (1984), page 194, and on data in *DOE–86*, page 177. As stated on pages 232–233, the data listed in Table 9-1 of Gold et al. are not consistent. Starting in 1932, payroll records were used to determine the number of days worked, but up to 1932, days the tipple operated was used. Also, starting in 1934, one day was seven hours. Before 1932, the workday was eight hours, I therefore corrected the data first by reducing them to tons per man-hour and then by fitting two exponentials, one up to 1931 and the other from 1932 to 1949. These two curves did not meet, so a constant was then subtracted from the data from 1900 to 1931 to make the two curves match at 1931. This constant was 0.285 tons per man-hour. This procedure gave a corrected set of data up to 1963. *DOE-86* data were used for the years from 1964 to 1985, but these also had to be corrected since DOE assumed that a workday was eight hours. The result of the corrections was a consistent set of data that agreed well where there was overlap between the *DOE* and the Gold et al. compilations. The data exhibited four regimes in which the time dependence of the productivity was different. From 1900 to 1950 the data were well described by an exponential of the form $\mathrm{Pr} = \exp(0.017272t - 33.8953)$ with a correlation coefficient of 0.9993 and a standard deviation (for the logarithm) of 0.02639. From 1951 to 1968 the data fit the equation: $\mathrm{Pr} = \exp(0.055652t - 108.6676)$ with a correlation coefficient of 0.9976 and a standard deviation of 0.03299.

From 1969 to 1978 there was a steady decrease in productivity as a result of the Federal Coal Mine Health and Safety Act of 1969 and of environmental concerns. From 1978 to 1985, there was a resumption of increasing productivity which fit the equation $\mathrm{Pr} = \exp(0.081777t - 161.5916)$ with a correlation coefficient of 0.9987 and a standard deviation of 0.02517.

The productivity in these equations is in tons per man-hour worked, and the time t is the year.

In Figure 2-13, the curves were plotted from the three exponential equations given above.

List of Tables in Appendix 3

A3-1	World Population from Prehistoric Times to the Present. Data to 1975 from Schultz (1981), p. 36. Data from 1975 to 1934 from *SAUS* (1986), p. 833.	420
A3-2	Population of the United States. Data to 1970 from *HSUS*, p. 8. Data after 1970 from *HAUS*–86, p. 5.	421
A3-3	Gross National Product per Capita of the United States.	422
A3-4	Total and per Capita Consumption of Energy Worldwide.	423
A3-5	Total and per Capita Energy Consumption in the United States.	424
A3-6	Energy Consumption in the United States by Source.	425
A3-7	Non-fuel and Total United States Consumption of Mineral Fuels.	428
A3-8	Disposition of U.S. Energy Consumption by End Use.	429
A3-9	Energy from Mineral Sources Consumed in the United States.	430
A3-10	Energy from Nonmineral Sources Consumed in the United States.	432
A3-11	Fraction of Energy Consumed from Different Sources in the United States.	434
A3-12	Fraction of Energy from Mineral and Nonmineral Sources Consumed in the United States	436
A3-13	Total and per Capita Petroleum Consumption in the United States. Data through 1970 from *HSUS*, pp. 587–588, by adding crude oil consumption and petroleum product imports. Data after 1970 from *DOE–84*, pp. 111.	437
A3-14	World oil production. Data from *DOE–84*.	439
A3-15	United States Crude Oil Production. Data from *HSUS*, pp. 587–588 and *DOE–80* and *DOE–84*.	439
A3-16	United States Petroleum Proven Reserves. Data from *HSUS*, pp. 593–594 and *DOE–80* and *DOE–84*.	441
A3-17	Ratio of Energy Consumed to Gross National Product for the United States.	443
A3-18	U.S. Cost of Fossil Fuels as a Fraction of Gross National Product.	444
A3-19	Fraction of United States Gross National Product Spent on Fuels and Electricity. Data from *DOE–84*, p. 24 and *DOE–86*, p. 27.	444
A3-20	Production of Electricity in the United States by Utilities and Firms.	445
A3-21	Total and per Capita Production of Electricity in the United States.	446
A3-22	Fraction of U.S. Consumption of Primary Energy Used to Generate Electricity.	447
A3-23	Generation of Electricity by Nuclear Power in the United States. Data from *DOE–86*, p. 7.	448
A3-24	Maximum Addition Rate for Computers Introduced from 1951 to 1979. Data from *NSF-80*.	448

A3-25 Successive Maximum Performance Indices for U.S. Computers. Linear and logarithmic performance indices computed from data in *NSF–80*. See text. 449

A3-26 Successive Most Powerful Computers Using Knight's Data and Definition of Computer Power for Scientific and Commercial Computation. Data extracted from data of Knight (1966, 1968, 1983). 450

A3-27 Diffusion of Computer Capacity. Data from Flamm (1987), p. 216. Flamm collected the data from the Phister compilation (Phister 1979). 450

A3-28 Worldwide Cumulated Production of Tractors. Data from Sahal (1981), p. 341. 451

A3-29 Tractor Efficiencies. Data from Sahal (1981), p. 341. 452

A3-30 Successive Maximum Efficiencies of Farm Tractors as a Function of Cumulative Production. Data from Sahal (1981), p. 341. 453

A3-31 Successive Maximum Tractor Efficiencies. Data extracted from that of Sahal (1981), p. 341. 453

A3-32 Labor Productivity of Underground Coal Mining for the United States. 454

A3-33 Diffusion of the Use of Continuous Mining Machines for the Underground Mining of Coal in the Untied States. Data from Gold, Pierce, Rosegger, and Perlman (1984), pp. 169–170. 455

A3-34 Cruising Speed, Passenger Capacity, and Passenger-Miles per Hour of Commercial Aircraft. Data from Martino (1983), pp. 338–339. 455

A3-35 Maximum Speeds of U.S. Military Aircraft Compared with Official World Aircraft Speed Records. Data from Martino (1983), pp. 337 and 361. 457

A3-36 Horsepower to Weight Ratio of Reciprocating Aircraft Engines. Data from Martino (1983), p. 367. 458

A3-37 Number of Locomotives in Service in the United States. Data from *HSUS*, pp. 727–728. Up to 1916, data were reported as a mid-year count. From 1916 to 1970, data were reported as an end-of-year count. For consistency, I have therefore recorded that data up to 1916 at the half year and the later data at the following year. 459

A3-38 Fraction of Diesel and Steam Locomotives in Service in the United States. Data recorded here are fractions of the sum of diesel and steam locomotives. The small number of other types is not included. 461

A3-39 Average Traction of Locomotives in Service in the United States. Data from *HSUS*, pp. 727–728. 462

A3-40 Fuel Efficiency of Railroad Freight Transport. Data in this table were computed from the fuel received by U.S. railroads and from ton-miles of freight carried. Up through 1970, data were from *HSUS*, pp. 732–733 for the ton-miles and p. 737 for the fuel received. After 1970, data of Rose (1986), pp. 187, were used. An error is introduced because the fuel received included that for passenger service, but this is small as evidenced by the fact that the number of freight cars is very much larger than the number of passenger cars. In 1970, there were over 1.4 million freight cars and less than 11.5 thousand passenger cars. **463**

A3-41 First Adoptions of Diesel Locomotives by American Railroads. Data from Mansfield (1968), p. 175. **464**

A3-42 Total and per Capita Miles of Railroad Track Operated and Miles of Track Built in the United States. Data from *HSUS*, pp. 727–728. **464**

A3-43 Percentage of U.S. Dwellings with Electric Service. Data from *HSUS*, p. 827. **468**

A3-44 United States Households with Radios. Data from *HSUS*, p. 796. **469**

A3-45 Aluminum Production Capacity. Data from *HSUS*, pp. 597–698. **470**

A3-46 Number of Incandescent Lamps Produced in the United States. Data from *HSUS*, p. 696. **471**

A3-47 Total and per Capita Number of Registered Automobiles in the United States. Data from *HSUS*, p. 716, and *SAUS–86*, p. 597. **472**

A3-48 Telephones in Service in the United States. Data through 1970 from *HSUS*, p. 783. After 1970, data from *SAUS*, p. 545. **473**

A3-49 Amount of Rayon and Acetate Yarns Available to United States Mills, Including Domestic Shipments and Imports. Data from *HSUS*, pp. 691–692. **474**

A3-50 Steel-making Capacity in the United States. Data from *HSUS*, p. 698. **474**

A3-51 Diffusion of Telegraphy in the United States. Data to 1970 from *HSUS*, pp. 787–788. Up to 1916 only Western Union data are available, so the table is actually the number of messages handled by Western Union. Data only to 1910 are given here since the growth of telegraphy was confined to the earlier years. **475**

A3-52 Diffusion of Internal Combustion Farm Tractors in the United States. Data from *HSUS*, p. 469. **476**

A3-53 Diffusion of Television in United States Households. Data from *HSUS*, p. 796, and *SAUS*, p. 545. **477**

A3-54 Diffusion of Color Television in the United States. Data from *HSUS*, p. 796, and *SAUS*, p. 545. **477**

A3-55 Substitution of Power for Sail in the United States Merchant Fleet. Data from *HSUS*, p. 751. **478**

A3-56 Learning Curve Data for a Pressman on a New Two-color Press. Data from Levy (1965), p. B-152. **481**

A3-57 Efficiency of Incandescent and Fluorescent Lamps. Data from Martino (1983), p. 363. **482**

A3-58 Efficiency of Artificial Light Sources. Data from Martino (1983), p. 340. **482**

A3-59 Efficiency of U.S. Generation of Electricity by Fossil Fuel Generating Plants. To 1970, data were computed from *HSUS*, p. 827, data on the Btu's per KWH, taking the electrical equivalent of heat to be 3412 Btu/KWH. Data after 1970 taken from Energy Information Administration report DOE/EIA-0455(84). **483**

A3-60 Steam Engine Efficiencies in Mechanical Energy Output. Data from Martino (1983), p. 344. **484**

A3-61 Average Mileage of Passenger Autos in the United States. Through 1970, data were computed from data in *HSUS*, pp. 716 and 718, which gave the total amount of motor fuel used and miles driven for passenger vehicles. Dividing these gave a good estimate of miles per gallon for passenger autos because by far most of passenger travel is by private auto. **484**

A3-62 Distribution of Research Funds at 91 United States Universities in 1985. These data were computed from a National Science Foundation fact sheet which gave the research budgets for American research universities in 1985. The third column is the number of universities with research budgets in the range that starts in column 1 and whose center is in column 2. The last column is the probability that a university has a research budget in the range defined in columns 1 and 2. **485**

A3-63 Income Distribution in England in 1688. Data from Lindert and Williamson (1982). **485**

A3-64 Size distribution of U.S. Cities in 1984. Data were computed from data in *SAUS*, pp. 16–17, by counting the number of cities in the given size ranges. The second column is the number of cities with population greater than that listed in the first column. The third column is the percentage of cities with population greater than that in the first column. The fourth column is the probability that a city has a population between the value in the first column and its preceding value. **486**

A3-65 National and World Levels of Industrialization from 1750 to 1980. Data from tables of Bairoch (1982), pp. 292 and 302. Levels of industrialization were defined by Bairoch to be the total industrial output per capita of population, the output being defined as the volume of each commodity weighted by its relative price. The index used by Bairoch took 100 to be the level of industrialization of the United Kingdom in 1900. The category of developed countries includes all Europe except Turkey, along with the United States, Canada, South Africa, Australia, New Zealand, and Japan. The remaining countries constitute the third world. Bairoch's paper should be consulted for details of the definitions he used. **487**

A3-66 Annual Percentage Growth Rates of Levels of Industrialization. The average growth rates for the periods between the years for which Bairoch (1982) reported levels of industrialization were computed from his data by assuming an exponential growth during these time periods. That is, the exponential average rate was calculated by taking the ratio of the level of industrialization at one reported year to that in the previous reported year, taking the logarithm of this ratio and then dividing by the difference between the reported years. Multiplying this result by 100 gave the average exponential growth rate defined in Table A3-66. **488**

A3-67 Ratio of Levels of Industrialization of Rich to Poor Countries. These ratios were computed from the data in Table A3-65. **489**

A3-68 The Takeoff to Exponential Growth of Industrialization. The data in Table A3-65 all exhibited the property that the industrial level was an exponential function of time after some date before which the level was either constant, fluctuated slightly, or actually decreased. The data after this date were used to compute a best fit to an exponential growth curve. The parameters for these curves are shown in Table A3-68. The first takeoff date is the date of the first data point used in computing the parameters for the exponential curves. But the data points as given by Bairoch are far apart and the takeoff point cannot be determined with precision. The second takeoff point is my estimate of a reasonable takeoff point from my examination of the data. **489**

Table A3-1 World Population from Prehistoric Times to the Present (millions)

Year	Population	Year	Population
−500000	2	1960	2986
−5000	5	1970	3610
0	300	1975	4033
1000	310	1978	4284
1500	500	1979	4378
1650	545	1980	4453
1750	791	1981	4530
1800	978	1982	4607
1850	1262	1983	4685
1900	1650	1984	4760
1920	1860	1985	4837
1930	2069	1986	4917
1940	2295	1987	5055
1950	2501	1988	5143

Table A3-2 Population of the United States (millions)

Year	Population	Year	Population	Year	Population
1790	3.929	1915	100.546	1952	156.954
1795	4.607	1916	101.961	1953	159.565
1800	5.297	1917	103.268	1954	162.391
1805	6.258	1918	103.208	1955	165.275
1810	7.224	1919	104.514	1956	168.221
1815	8.419	1920	106.461	1957	171.274
1820	9.618	1921	108.538	1958	174.141
1825	11.252	1922	110.049	1959	177.830
1830	12.901	1923	111.947	1960	180.671
1835	15.003	1924	114.109	1961	183.691
1840	17.120	1925	115.829	1962	186.538
1845	20.182	1926	117.397	1963	189.242
1850	23.261	1927	119.030	1964	191.889
1855	27.386	1928	120.509	1965	194.303
1860	31.513	1929	121.769	1966	196.560
1865	35.701	1930	123.188	1967	198.712
1870	39.905	1931	124.149	1968	200.706
1875	45.073	1932	124.949	1969	202.677
1880	50.262	1933	125.690	1970	205.052
1885	56.658	1934	126.489	1971	207.661
1890	63.056	1935	127.362	1972	209.896
1895	69.580	1936	128.181	1973	211.909
1900	76.094	1937	128.961	1974	213.854
1901	77.584	1938	129.969	1975	215.973
1902	79.163	1939	131.028	1976	218.035
1903	80.632	1940	132.122	1977	220.239
1904	82.166	1941	133.402	1978	222.585
1905	83.822	1942	134.860	1979	225.055
1906	85.450	1943	136.739	1980	227.757
1907	87.008	1944	138.397	1981	230.138
1908	88.710	1945	139.928	1982	232.520
1909	90.490	1946	141.389	1983	234.799
1910	92.407	1947	144.126	1984	237.001
1911	93.863	1948	146.631	1985	239.729
1912	95.335	1949	149.188	1986	241.613
1913	97.225	1950	151.684	1987	243.915
1914	99.111	1951	154.287	1988	246.113

Table A3-3 Gross National Product per Capita of the United States (1958 dollars)

Year	GNP/cap	Year	GNP/cap	Year	GNP/cap
1890	836	1922	1345	1954	2506
1891	856	1923	1482	1955	2650
1892	920	1924	1450	1956	2652
1893	859	1925	1549	1957	2642
1894	819	1926	1619	1958	2569
1895	900	1927	1594	1959	2688
1896	865	1928	1584	1960	2699
1897	930	1929	1671	1961	2706
1898	933	1930	1490	1962	2840
1899	1000	1931	1364	1963	2912
1900	1011	1932	1154	1964	3028
1901	1105	1933	1126	1965	3180
1902	1093	1934	1220	1966	3348
1903	1126	1935	1331	1967	3398
1904	1092	1936	1506	1968	3521
1905	1149	1937	1576	1969	3580
1906	1258	1938	1484	1970	3555
1907	1255	1939	1598	1971	3603
1908	1130	1940	1720	1972	3767
1909	1290	1941	1977	1973	3946
1910	1299	1942	2208	1974	3885
1911	1313	1943	2465	1975	3802
1912	1366	1944	2611	1976	3969
1913	1351	1945	2538	1977	4146
1914	1267	1946	2211	1978	4251
1915	1238	1947	2150	1979	4383
1916	1317	1948	2208	1980	4319
1917	1310	1949	2172	1981	4382
1918	1471	1950	2342	1982	4247
1919	1401	1951	2485	1983	4362
1920	1315	1952	2517	1984	4617
1921	1177	1953	2587		

Table A3-4 Total and per Capita Consumption of Energy Worldwide

Year	Total Consumption (Quads)	Per Capita Consumption (Quads/million)
1925	41	0.0211
1938	50	0.0225
1950	74	0.0285
1955	93	0.0332
1960	131.5	0.0433
1961	132.0	0.0427
1962	138.9	0.0442
1963	146.6	0.0458
1964	155.9	0.0479
1965	163.5	0.0493
1966	171.1	0.0507
1967	177.6	0.0517
1968	190.9	0.0545
1969	202.8	0.0568
1970	216.8	0.0596
1971	225.5	0.0608
1972	235.8	0.0624
1973	244.8	0.0635
1974	248.8	0.0632
1975	246.0	0.0613
1976	260.4	0.0636
1977	269.1	0.0644
1978	274.4	0.0643
1979	288.8	0.0662
1980	287.6	0.0646
1981	281.8	0.0619
1982	280.2	0.0602
1983	280.1	0.0589

Table A3-5 Total and per Capita Energy Consumption in the United States (total energy in Quads and per capita energy in trillion BTU's per person)

Year	Total	Per Capita	Year	Total	Per Capita
1800	0.003	0.00056	1927	21.379	0.1796
1805	0.004	0.00064	1928	21.899	0.1817
1810	0.005	0.00069	1929	23.208	0.1906
1815	0.007	0.00083	1930	23.183	0.1882
1820	0.009	0.00093	1931	18.313	0.1475
1825	0.012	0.00106	1932	15.949	0.1276
1830	0.023	0.00177	1933	16.443	0.1308
1835	0.047	0.00312	1934	17.454	0.1380
1840	0.064	0.00372	1935	19.970	0.1568
1845	0.121	0.00601	1936	20.820	0.1624
1850	2.356	0.1013	1937	22.101	0.1714
1855	2.808	0.1025	1938	19.252	0.1481
1860	3.159	0.1003	1939	20.944	0.1598
1865	3.405	0.0954	1940	24.544	0.1858
1870	3.946	0.0989	1941	25.840	0.1937
1875	4.315	0.0957	1942	27.107	0.2010
1880	4.986	0.0992	1943	29.579	0.2163
1885	5.621	0.0992	1944	30.874	0.2231
1890	6.914	0.1097	1945	31.819	0.2274
1895	7.613	0.1094	1946	30.728	0.2173
1900	9.516	0.1251	1947	32.907	0.2283
1901	8.093	0.1043	1948	33.753	0.2302
1902	8.628	0.1090	1949	31.175	0.2090
1903	10.145	0.1258	1950	33.642	0.2218
1904	10.066	0.1225	1951	35.888	0.2326
1905	13.094	0.1562	1952	35.638	0.2271
1906	11.799	0.1381	1953	36.549	0.2291
1907	13.688	0.1573	1954	35.376	0.2178
1908	12.102	0.1364	1955	38.663	0.2339
1909	13.379	0.1479	1956	40.089	0.2383
1910	16.397	0.1774	1957	39.737	0.2320
1911	14.445	0.1539	1958	39.938	0.2293
1912	15.528	0.1629	1959	41.474	0.2332
1913	16.525	0.1700	1960	42.999	0.2380
1914	15.340	0.1548	1961	43.568	0.2372
1915	17.559	0.1746	1962	45.557	0.2442
1916	17.551	0.1721	1963	47.312	0.2500
1917	19.342	0.1873	1964	49.097	0.2559
1918	20.177	0.1955	1965	51.122	0.2631
1919	17.300	0.1655	1966	53.936	0.2744
1920	21.098	0.1982	1967	55.552	0.2796
1921	16.144	0.1487	1968	58.788	0.2929
1922	16.917	0.1537	1969	61.504	0.3035
1923	21.344	0.1907	1970	63.596	0.3101
1924	20.159	0.1767	1971	65.192	0.3139
1925	22.023	0.1901	1972	68.188	0.3249
1926	22.058	0.1879	1973	70.711	0.3337

Table A3-5 *(continued)*

Year	Total	Per Capita	Year	Total	Per Capita
1974	69.260	0.3239	1980	73.887	0.3244
1975	70.913	0.3283	1981	72.126	0.3135
1976	71.080	0.3260	1982	69.539	0.2993
1977	72.807	0.3306	1983	69.516	0.2964
1978	75.226	0.3335	1984	69.428	0.2933
1979	76.283	0.3390			

Table A3-6 Energy Consumption in the United States by Source (quads)

Year	Coal	Oil	Gas	Nuclear	Wood	Water	Geotherm
1800	0.003	0	0	0	N.A.	N.A.	0
1805	0.004	0	0	0	N.A.	N.A.	0
1810	0.005	0	0	0	N.A.	N.A.	0
1815	0.007	0	0	0	N.A.	N.A.	0
1820	0.009	0	0	0	N.A.	N.A.	0
1825	0.012	0	0	0	N.A.	N.A.	0
1830	0.023	0	0	0	N.A.	N.A.	0
1835	0.047	0	0	0	N.A.	N.A.	0
1840	0.064	0	0	0	N.A.	N.A.	0
1845	0.121	0	0	0	N.A.	N.A.	0
1850	0.218	0	0	0	2.14	N.A.	0
1855	0.419	0	0	0	2.39	N.A.	0
1860	0.515	0.003	0	0	2.64	N.A.	0
1865	0.629	0.010	0	0	2.77	N.A.	0
1870	1.043	0.011	0	0	2.89	N.A.	0
1875	1.433	0.011	0	0	2.87	N.A.	0
1880	2.044	0.091	0	0	2.85	N.A.	0
1885	2.826	0.038	0.074	0	2.68	N.A.	0
1890	3.998	0.148	0.231	0	2.52	0.022	0
1895	4.925	0.160	0.132	0	2.31	0.090	0
1900	6.807	0.218	0.227	0	2.02	0.250	0
1901	7.338	0.238	0.253	0	N.A.	0.264	0
1902	7.724	0.346	0.269	0	N.A.	0.289	0
1903	9.112	0.427	0.285	0	N.A.	0.321	0
1904	8.907	0.507	0.297	0	N.A.	0.354	0
1905	9.951	0.580	0.335	0	1.84	0.386	0
1906	10.488	0.527	0.370	0	N.A.	0.414	0
1907	12.116	0.742	0.389	0	N.A.	0.441	0
1908	10.462	0.779	0.384	0	N.A.	0.476	0
1909	11.605	0.802	0.460	0	N.A.	0.513	0
1910	12.650	0.957	0.486	0	1.77	0.539	0
1911	12.380	0.989	0.490	0	N.A.	0.587	0
1912	13.373	1.006	0.535	0	N.A.	0.615	0
1913	14.170	1.152	0.558	0	N.A.	0.645	0
1914	12.836	1.259	0.569	0	N.A.	0.676	0
1915	13.228	1.347	0.606	0	1.69	0.691	0

(continued)

Table A3-6 Energy Consumption in the United States by Source (quads) (*continued*)

Year	Coal	Oil	Gas	Nuclear	Wood	Water	Geotherm
1916	14.663	1.433	0.726	0	N.A.	0.729	0
1917	16.132	1.690	0.765	0	N.A.	0.755	0
1918	16.888	1.845	0.694	0	N.A.	0.750	0
1919	13.732	2.088	0.714	0	N.A.	0.766	0
1920	15.426	2.542	0.744	0	1.61	0.775	0
1921	12.286	2.588	0.614	0	N.A.	0.656	0
1922	12.565	2.971	0.707	0	N.A.	0.675	0
1923	15.727	3.962	0.929	0	N.A.	0.727	0
1924	14.657	3.674	1.143	0	N.A.	0.685	0
1925	14.632	4.066	1.091	0	1.53	0.701	0
1926	15.835	4.256	1.202	0	N.A.	0.765	0
1927	14.917	4.328	1.319	0	N.A.	0.815	0
1928	14.865	4.715	1.429	0	N.A.	0.890	0
1929	15.350	5.263	1.748	0	N.A.	0.847	0
1930	13.571	5.600	1.772	0	1.46	0.785	0
1931	11.171	4.907	1.544	0	N.A.	0.692	0
1932	9.277	4.511	1.435	0	N.A.	0.726	0
1933	9.535	4.739	1.440	0	N.A.	0.729	0
1934	10.366	4.730	1.637	0	N.A.	0.721	0
1935	10.581	5.385	1.777	0	1.40	0.831	0
1936	11.988	5.993	1.999	0	N.A.	0.841	0
1937	12.503	6.471	2.221	0	N.A.	0.905	0
1938	9.909	6.340	2.113	0	N.A.	0.889	0
1939	11.060	6.726	2.285	0	N.A.	0.872	0
1940	12.472	7.344	2.453	0	1.36	0.917	0
1941	14.160	8.140	2.566	0	N.A.	0.975	0
1942	15.506	7.632	2.792	0	N.A.	1.177	0
1943	16.922	8.177	3.133	0	N.A.	1.347	0
1944	16.871	9.218	3.398	0	N.A.	1.387	0
1945	15.892	9.605	3.576	0	1.26	1.486	0
1946	14.407	9.956	3.680	0	1.24	1.446	0
1947	15.701	10.609	4.016	0	1.26	1.326	0
1948	14.816	11.744	4.535	0	1.27	1.393	0
1949	12.522	11.099	4.846	0	1.26	1.449	0
1950	12.874	12.432	5.725	0	1.17	1.440	0
1951	13.057	13.415	6.807	0	1.16	1.454	0
1952	11.715	13.923	7.379	0	1.13	1.496	0
1953	11.723	14.478	7.844	0	1.07	1.439	0
1954	10.047	14.708	8.198	0	1.04	1.388	0
1955	11.371	15.978	8.870	0	1.04	1.407	0
1956	11.574	16.567	9.448	0	1.01	1.487	0
1957	10.991	16.530	9.749	0	0.92	1.551	0
1958	9.713	17.078	10.617	0	0.89	1.636	0
1959	9.679	17.684	11.602	0	0.92	1.591	0
1960	9.995	18.178	12.327	0.01	0.83	1.657	0
1961	9.769	18.419	12.890	0.02	0.79	1.680	0
1962	10.048	19.116	13.810	0.03	0.73	1.821	0

Table A3-6 (*continued*)

Year	Coal	Oil	Gas	Nuclear	Wood	Water	Geotherm
1963	10.571	19.621	14.545	0.04	0.77	1.767	0
1964	11.120	20.053	15.351	0.04	0.63	1.907	0
1965	11.751	20.890	15.806	0.04	0.58	2.058	0
1966	12.345	21.835	17.098	0.06	0.53	2.073	0
1967	12.106	22.751	17.789	0.09	0.47	2.344	0
1968	12.509	24.217	19.125	0.14	0.46	2.342	0
1969	12.568	25.380	20.306	0.15	0.44	2.659	0
1970	12.516	26.423	21.342	0.24	0.43	2.650	0
1971	11.894	27.427	22.151	0.41	0.45	2.860	0
1972	12.328	29.522	22.368	0.58	0.45	2.940	0
1973	13.182	31.163	21.996	0.91	0.45	3.010	0
1974	12.765	30.012	21.303	1.27	0.60	3.310	0
1975	12.707	32.730	19.645	1.90	0.70	3.220	0.011
1976	13.634	31.301	19.953	2.11	1.00	3.070	0.012
1977	13.877	33.037	19.371	2.70	1.30	2.510	0.012
1978	13.781	33.793	19.681	3.02	1.80	3.140	0.01
1979	15.044	32.666	20.440	2.78	2.20	3.140	0.013
1980	15.364	29.754	20.409	2.74	2.48	3.120	0.017
1981	15.845	27.779	19.868	3.01	2.50	3.110	0.019
1982	15.277	26.600	18.477	3.13	2.48	3.560	0.017
1983	15.857	26.440	17.488	3.20	2.64	3.870	0.021
1984	17.157	26.970	17.945	3.55	N.A.	3.780	0.026

Table A3-7 Non-fuel and Total United States Consumption of Mineral Fuels (quads)

Year	Total Coal	Non-fuel Coal	Total Oil	Non-fuel Oil	Total Natural Gas	Non-fuel Natural Gas
1947	15.82	0.123	11.37	0.758	4.52	0.502
1948	14.94	0.121	12.56	0.813	5.03	0.498
1949	12.63	0.109	11.88	0.781	5.29	0.443
1950	13.00	0.129	13.32	0.888	6.15	0.425
1951	13.20	0.143	14.43	1.015	7.25	0.441
1952	11.84	0.125	14.96	1.037	7.76	0.381
1953	11.87	0.147	15.56	1.082	8.16	0.312
1954	10.17	0.123	15.84	1.132	8.55	0.350
1955	11.52	0.149	17.25	1.272	9.23	0.362
1956	11.72	0.146	17.94	1.373	9.83	0.386
1957	11.14	0.149	17.93	1.400	10.15	0.397
1958	9.83	0.117	18.53	1.452	11.00	0.378
1959	9.79	0.111	19.32	1.636	11.99	0.388
1960	10.12	0.125	19.92	1.742	12.70	0.372
1961	9.89	0.121	20.22	1.801	13.23	0.338
1962	10.17	0.122	21.05	1.934	14.12	0.311
1963	10.69	0.119	21.70	2.079	14.84	0.298
1964	11.25	0.13	22.30	2.247	15.65	0.297
1965	11.89	0.139	23.25	2.360	16.10	0.292
1966	12.48	0.135	24.40	2.565	17.39	0.295
1967	12.24	0.134	25.28	2.529	18.25	0.461
1968	12.66	0.151	26.98	2.763	19.58	0.455
1969	12.72	0.152	28.34	2.960	21.02	0.714
1970	12.66	0.144	29.52	3.097	22.03	0.687
1971	12.01	0.116	30.56	3.133	22.82	0.669
1972	12.45	0.122	32.95	3.428	23.04	0.672
1973	13.30	0.118	34.84	3.677	22.71	0.714
1974	12.88	0.115	33.45	3.438	22.03	0.727
1975	12.82	0.113	32.73	3.60	20.19	0.550
1976	13.73	0.096	35.17	3.870	20.57	0.620
1977	13.96	0.083	37.12	4.080	19.97	0.600
1978	13.85	0.069	37.97	4.170	20.29	0.610
1979	15.11	0.066	37.12	4.454	21.05	0.610
1980	15.43	0.066	34.20	4.446	21.04	0.630
1981	15.91	0.065	31.93	4.151	20.44	0.570
1982	15.32	0.043	30.23	3.630	18.99	0.510
1983	15.90	0.043	30.05	3.610	18.01	0.520
1984	17.20	0.043	31.00	4.030	18.50	0.560

Table A3-8 Disposition of U.S. Energy Consumption by End Use

	Total Quads			Fraction		
Year	Residential and Commercial	Industry	Transport	Residential and Commercial	Industry	Transport
1949	8.21	12.92	7.99	0.282	0.444	0.274
1950	8.86	14.29	8.49	0.280	0.452	0.268
1951	9.31	15.58	9.04	0.274	0.459	0.266
1952	9.54	15.22	9.00	0.283	0.451	0.267
1953	9.50	16.12	9.12	0.273	0.464	0.262
1954	9.79	14.98	8.90	0.291	0.445	0.264
1955	10.41	17.09	9.55	0.281	0.461	0.258
1956	10.97	17.68	9.86	0.285	0.459	0.256
1957	10.98	17.66	9.90	0.285	0.458	0.257
1958	11.65	16.73	10.00	0.304	0.436	0.261
1959	12.16	17.47	10.35	0.304	0.437	0.259
1960	13.05	17.94	10.60	0.314	0.431	0.255
1961	13.44	18.01	10.77	0.318	0.427	0.255
1962	14.27	18.68	11.22	0.323	0.423	0.254
1963	14.72	19.46	11.65	0.321	0.425	0.254
1964	15.23	20.61	11.99	0.318	0.431	0.251
1965	16.04	21.46	12.42	0.321	0.430	0.249
1966	17.06	22.53	13.09	0.324	0.428	0.248
1967	18.11	22.61	13.73	0.333	0.415	0.252
1968	19.24	23.58	14.84	0.334	0.409	0.257
1969	20.59	24.32	15.48	0.341	0.403	0.256
1970	21.72	24.72	16.07	0.347	0.395	0.257
1971	22.60	24.65	16.70	0.353	0.385	0.261
1972	23.69	25.64	17.70	0.353	0.382	0.264
1973	24.15	27.03	18.60	0.346	0.387	0.267
1974	23.73	26.41	18.11	0.348	0.387	0.265
1975	23.90	27.74	18.24	0.342	0.397	0.261
1976	25.02	25.65	19.09	0.359	0.368	0.274
1977	25.39	26.30	19.81	0.355	0.368	0.277
1978	26.09	26.53	20.59	0.356	0.362	0.281
1979	25.81	27.47	20.47	0.350	0.373	0.278
1980	25.65	25.47	19.69	0.362	0.360	0.278
1981	25.25	24.46	19.50	0.365	0.353	0.282
1982	25.63	21.93	19.07	0.385	0.329	0.286
1983	25.44	21.72	19.15	0.384	0.328	0.289
1984	25.98	23.25	19.86	0.376	0.336	0.287

Table A3-9 Energy from Mineral Sources and Total Energy Consumed in the United States (quads)

Year	Coal	Oil	Gas	Nuclear	Geotherm	Total
1850	0.22	0.00	0.00	0.00	0.00	2.3
1855	0.42	0.00	0.00	0.00	0.00	2.85
1860	0.52	0.00	0.00	0.00	0.00	3.20
1865	0.63	0.01	0.00	0.00	0.00	3.45
1870	1.04	0.01	0.00	0.00	0.00	3.99
1875	1.43	0.01	0.00	0.00	0.00	4.37
1880	2.04	0.09	0.00	0.00	0.00	5.05
1885	2.83	0.04	0.07	0.00	0.00	5.69
1890	4.00	0.15	0.23	0.00	0.00	7.00
1895	4.93	0.16	0.13	0.00	0.00	7.70
1900	6.81	0.22	0.23	0.00	0.00	9.61
1901	7.34	0.24	0.25	0.00	0.00	10.10
1902	7.72	0.35	0.27	0.00	0.00	10.59
1903	9.11	0.43	0.29	0.00	0.00	12.06
1904	8.91	0.51	0.30	0.00	0.00	11.93
1905	9.95	0.58	0.33	0.00	0.00	13.10
1906	10.49	0.53	0.37	0.00	0.00	13.61
1907	12.12	0.74	0.39	0.00	0.00	15.50
1908	10.46	0.78	0.38	0.00	0.00	13.91
1909	11.60	0.80	0.46	0.00	0.00	15.19
1910	12.65	0.96	0.49	0.00	0.00	16.51
1911	12.38	0.99	0.49	0.00	0.00	16.32
1912	13.37	1.01	0.53	0.00	0.00	17.38
1913	14.17	1.15	0.56	0.00	0.00	18.36
1914	12.84	1.26	0.57	0.00	0.00	17.15
1915	13.23	1.35	0.61	0.00	0.00	17.67
1916	14.66	1.43	0.73	0.00	0.00	19.34
1917	16.13	1.69	0.77	0.00	0.00	21.12
1918	16.89	1.84	0.69	0.00	0.00	21.94
1919	13.73	2.09	0.71	0.00	0.00	19.04
1920	15.43	2.54	0.74	0.00	0.00	21.21
1921	12.29	2.59	0.61	0.00	0.00	17.75
1922	12.56	2.97	0.71	0.00	0.00	18.52
1923	15.73	3.96	0.93	0.00	0.00	22.92
1924	14.66	3.67	1.14	0.00	0.00	21.72
1925	14.63	4.07	1.09	0.00	0.00	22.03
1926	15.84	4.26	1.20	0.00	0.00	23.58
1927	14.92	4.33	1.32	0.00	0.00	22.88
1928	14.87	4.71	1.43	0.00	0.00	23.39
1929	15.35	5.26	1.75	0.00	0.00	24.69
1930	13.57	5.60	1.77	0.00	0.00	23.27
1931	11.17	4.91	1.54	0.00	0.00	19.84
1932	9.28	4.51	1.43	0.00	0.00	17.47
1933	9.54	4.74	1.44	0.00	0.00	17.95
1934	10.37	4.73	1.64	0.00	0.00	18.95
1935	10.58	5.38	1.78	0.00	0.00	20.05
1936	11.99	5.99	2.00	0.00	0.00	22.28

Table A3-9 (*continued*)

Year	Coal	Oil	Gas	Nuclear	Geotherm	Total
1937	12.50	6.47	2.22	0.00	0.00	23.55
1938	9.91	6.34	2.11	0.00	0.00	20.69
1939	11.06	6.73	2.29	0.00	0.00	22.38
1940	12.47	7.34	2.45	0.00	0.00	24.61
1941	14.16	8.14	2.57	0.00	0.00	27.24
1942	15.51	7.63	2.79	0.00	0.00	28.49
1943	16.92	8.18	3.13	0.00	0.00	30.93
1944	16.87	9.22	3.40	0.00	0.00	32.20
1945	15.89	9.60	3.58	0.00	0.00	31.87
1946	14.41	9.96	3.68	0.00	0.00	30.78
1947	15.70	10.61	4.02	0.00	0.00	32.95
1948	14.82	11.74	4.54	0.00	0.00	33.79
1949	12.52	11.10	4.85	0.00	0.00	31.22
1950	12.87	12.43	5.73	0.00	0.00	33.68
1951	13.06	13.42	6.81	0.00	0.00	35.92
1952	11.72	13.92	7.38	0.00	0.00	35.67
1953	11.72	14.48	7.84	0.00	0.00	36.57
1954	10.05	14.71	8.20	0.00	0.00	35.40
1955	11.37	15.98	8.87	0.00	0.00	38.68
1956	11.57	16.57	9.45	0.00	0.00	40.11
1957	10.99	16.53	9.75	0.00	0.00	39.76
1958	9.71	17.08	10.62	0.00	0.00	39.95
1959	9.68	17.68	11.60	0.00	0.00	41.48
1960	10.00	18.18	12.33	0.01	0.00	43.01
1961	9.77	18.42	12.89	0.02	0.00	43.57
1962	10.05	19.12	13.81	0.03	0.00	45.56
1963	10.57	19.62	14.55	0.04	0.00	47.31
1964	11.12	20.05	15.35	0.04	0.00	49.10
1965	11.75	20.89	15.81	0.04	0.00	51.12
1966	12.35	21.84	17.10	0.06	0.00	53.94
1967	12.11	22.75	17.79	0.09	0.00	55.55
1968	12.51	24.22	19.13	0.14	0.00	58.79
1969	12.57	25.38	20.31	0.15	0.00	61.50
1970	12.52	26.42	21.34	0.24	0.00	63.60
1971	11.89	27.43	22.15	0.41	0.00	65.19
1972	12.33	29.52	22.37	0.58	0.00	68.19
1973	13.18	31.16	22.00	0.91	0.00	70.71
1974	12.77	30.01	21.30	1.27	0.00	69.26
1975	12.71	32.73	19.64	1.90	0.01	70.90
1976	13.63	31.30	19.95	2.11	0.01	71.07
1977	13.88	33.04	19.37	2.70	0.01	72.79
1978	13.78	33.79	19.68	3.02	0.01	75.22
1979	15.04	32.67	20.44	2.78	0.01	76.27
1980	15.36	29.75	20.41	2.74	0.02	73.87
1981	15.85	27.78	19.87	3.01	0.02	72.11
1982	15.28	26.60	18.48	3.13	0.02	69.52
1983	15.86	26.44	17.49	3.20	0.02	69.49
1984	17.16	26.97	17.95	3.55	0.03	72.00

Table A3-10 Energy from Nonmineral Sources Consumed in the United States (quads)

Year	Wood	Animal Energy	Water Power
1850	2.14	0.03	N.A.
1855	2.39	0.04	N.A.
1860	2.64	0.04	N.A.
1865	2.77	0.04	N.A.
1870	2.89	0.04	N.A.
1875	2.87	0.05	N.A.
1880	2.85	0.06	N.A.
1885	2.68	0.07	N.A.
1890	2.52	0.08	0.02
1895	2.31	0.09	0.09
1900	2.02	0.10	0.25
1901	2.00	0.10	0.26
1902	1.95	0.10	0.29
1903	1.90	0.10	0.32
1904	1.85	0.10	0.35
1905	1.84	0.10	0.39
1906	1.80	0.10	0.41
1907	1.80	0.10	0.44
1908	1.80	0.10	0.48
1909	1.80	0.10	0.51
1910	1.77	0.11	0.54
1911	1.76	0.11	0.59
1912	1.74	0.11	0.62
1913	1.72	0.11	0.65
1914	1.70	0.11	0.68
1915	1.69	0.11	0.69
1916	1.68	0.11	0.73
1917	1.67	0.11	0.76
1918	1.65	0.11	0.75
1919	1.63	0.11	0.77
1920	1.61	0.11	0.78
1921	1.60	0.10	0.66
1922	1.59	0.10	0.68
1923	1.57	0.10	0.73
1924	1.55	0.10	0.69
1925	1.53	0.10	0.70
1926	1.51	0.10	0.77
1927	1.49	0.10	0.82
1928	1.48	0.10	0.89
1929	1.47	0.10	0.85
1930	1.46	0.09	0.79
1931	1.45	0.08	0.69
1932	1.44	0.08	0.73
1933	1.43	0.08	0.73
1934	1.42	0.08	0.72
1935	1.40	0.08	0.83
1936	1.39	0.07	0.84

Table A3-10 (*continued*)

Year	Wood	Animal Energy	Water Power
1937	1.38	0.07	0.91
1938	1.37	0.07	0.89
1939	1.37	0.07	0.87
1940	1.36	0.06	0.92
1941	1.34	0.06	0.98
1942	1.32	0.06	1.18
1943	1.30	0.05	1.35
1944	1.28	0.05	1.39
1945	1.26	0.05	1.49
1946	1.24	0.05	1.45
1947	1.26	0.04	1.33
1948	1.27	0.04	1.39
1949	1.26	0.04	1.45
1950	1.17	0.04	1.44
1951	1.16	0.03	1.45
1952	1.13	0.03	1.50
1953	1.07	0.02	1.44
1954	1.04	0.02	1.39
1955	1.04	0.02	1.41
1956	1.01	0.02	1.49
1957	0.92	0.02	1.55
1958	0.89	0.01	1.64
1959	0.92	0.01	1.59
1960	0.83	0.01	1.66
1961	0.79	N.A.	1.68
1962	0.73	N.A.	1.82
1963	0.77	N.A.	1.77
1964	0.63	N.A.	1.91
1965	0.58	N.A.	2.06
1966	0.53	N.A.	2.07
1967	0.47	N.A.	2.34
1968	0.46	N.A.	2.34
1969	0.44	N.A.	2.66
1970	0.43	N.A.	2.65
1971	0.45	N.A.	2.86
1972	0.45	N.A.	2.94
1973	0.45	N.A.	3.01
1974	0.60	N.A.	3.31
1975	0.70	N.A.	3.22
1976	1.00	N.A.	3.07
1977	1.30	N.A.	2.51
1978	1.80	N.A.	3.14
1979	2.20	N.A.	3.14
1980	2.48	N.A.	3.12
1981	2.50	N.A.	3.11
1982	2.48	N.A.	3.56
1983	2.64	N.A.	3.87
1984	2.60	N.A.	3.78

Table A3-11 Fraction of Energy Consumed from Different Sources in the United States

Year	Animal	Wood	Hydro	Coal	Oil	Gas	Nuclear
1850	0.013	0.896	0.000	0.091	0.000	0.000	0.000
1855	0.014	0.839	0.000	0.147	0.000	0.000	0.000
1860	0.014	0.825	0.000	0.161	0.000	0.000	0.000
1865	0.012	0.803	0.000	0.183	0.003	0.000	0.000
1870	0.011	0.725	0.000	0.261	0.003	0.000	0.000
1875	0.011	0.658	0.000	0.328	0.002	0.000	0.000
1880	0.012	0.565	0.000	0.405	0.018	0.000	0.000
1885	0.012	0.471	0.000	0.497	0.007	0.013	0.000
1890	0.012	0.360	0.003	0.571	0.021	0.033	0.000
1895	0.012	0.299	0.012	0.639	0.021	0.017	0.000
1900	0.010	0.210	0.026	0.708	0.023	0.024	0.000
1901	0.010	0.198	0.026	0.726	0.024	0.025	0.000
1902	0.009	0.184	0.027	0.730	0.033	0.025	0.000
1903	0.008	0.158	0.027	0.756	0.035	0.024	0.000
1904	0.008	0.155	0.030	0.747	0.043	0.025	0.000
1905	0.008	0.141	0.029	0.759	0.044	0.026	0.000
1906	0.007	0.132	0.030	0.771	0.039	0.027	0.000
1907	0.007	0.116	0.028	0.782	0.048	0.025	0.000
1908	0.007	0.129	0.034	0.752	0.056	0.028	0.000
1909	0.007	0.119	0.034	0.764	0.053	0.030	0.000
1910	0.007	0.107	0.033	0.766	0.058	0.029	0.000
1911	0.007	0.108	0.036	0.759	0.061	0.030	0.000
1912	0.006	0.100	0.035	0.770	0.058	0.031	0.000
1913	0.006	0.094	0.035	0.772	0.063	0.030	0.000
1914	0.006	0.099	0.039	0.748	0.073	0.033	0.000
1915	0.006	0.096	0.039	0.749	0.076	0.034	0.000
1916	0.006	0.087	0.038	0.758	0.074	0.038	0.000
1917	0.005	0.079	0.036	0.764	0.080	0.036	0.000
1918	0.005	0.075	0.034	0.770	0.084	0.032	0.000
1919	0.006	0.086	0.040	0.721	0.110	0.037	0.000
1920	0.005	0.076	0.037	0.727	0.120	0.035	0.000
1921	0.001	0.090	0.037	0.692	0.146	0.035	0.000
1922	0.001	0.086	0.036	0.679	0.160	0.038	0.000
1923	0.000	0.068	0.032	0.686	0.173	0.041	0.000
1924	0.000	0.071	0.032	0.675	0.169	0.053	0.000
1925	0.000	0.070	0.032	0.664	0.185	0.050	0.000
1926	0.000	0.064	0.032	0.672	0.181	0.051	0.000
1927	0.000	0.065	0.036	0.652	0.189	0.058	0.000
1928	0.000	0.063	0.038	0.636	0.202	0.061	0.000
1929	0.000	0.060	0.034	0.622	0.213	0.071	0.000
1930	0.004	0.063	0.034	0.583	0.241	0.076	0.000
1931	0.004	0.073	0.035	0.563	0.247	0.078	0.000
1932	0.005	0.082	0.042	0.531	0.258	0.082	0.000
1933	0.004	0.080	0.041	0.531	0.264	0.080	0.000
1934	0.004	0.075	0.038	0.547	0.250	0.086	0.000
1935	0.004	0.070	0.041	0.528	0.269	0.089	0.000
1936	0.003	0.062	0.038	0.538	0.269	0.090	0.000

Table A3-11 (*continued*)

Year	Animal	Wood	Hydro	Coal	Oil	Gas	Nuclear
1937	0.003	0.059	0.038	0.531	0.275	0.094	0.000
1938	0.003	0.066	0.043	0.479	0.306	0.102	0.000
1939	0.003	0.061	0.039	0.494	0.300	0.102	0.000
1940	0.003	0.055	0.037	0.507	0.298	0.100	0.000
1941	0.002	0.049	0.036	0.520	0.299	0.094	0.000
1942	0.002	0.046	0.041	0.544	0.268	0.098	0.000
1943	0.002	0.042	0.044	0.547	0.264	0.101	0.000
1944	0.002	0.040	0.043	0.524	0.286	0.106	0.000
1945	0.002	0.040	0.047	0.499	0.301	0.112	0.000
1946	0.002	0.040	0.047	0.468	0.323	0.120	0.000
1947	0.001	0.038	0.040	0.477	0.322	0.122	0.000
1948	0.001	0.037	0.041	0.438	0.348	0.134	0.000
1949	0.001	0.040	0.046	0.401	0.356	0.155	0.000
1950	0.001	0.035	0.043	0.382	0.369	0.170	0.000
1951	0.001	0.032	0.040	0.364	0.373	0.190	0.000
1952	0.001	0.032	0.042	0.328	0.390	0.207	0.000
1953	0.001	0.029	0.039	0.321	0.396	0.214	0.000
1954	0.001	0.029	0.039	0.284	0.416	0.232	0.000
1955	0.001	0.027	0.036	0.294	0.413	0.229	0.000
1956	0.000	0.025	0.037	0.289	0.413	0.236	0.000
1957	0.001	0.023	0.039	0.276	0.416	0.245	0.000
1958	0.000	0.022	0.041	0.243	0.428	0.266	0.000
1959	0.000	0.022	0.038	0.233	0.426	0.280	0.000
1960	0.000	0.019	0.039	0.232	0.423	0.287	0.000
1961	0.000	0.018	0.039	0.224	0.423	0.296	0.000
1962	0.000	0.016	0.040	0.221	0.420	0.303	0.001
1963	0.000	0.016	0.037	0.223	0.415	0.307	0.001
1964	0.000	0.013	0.039	0.226	0.408	0.313	0.001
1965	0.000	0.011	0.040	0.230	0.409	0.309	0.001
1966	0.000	0.010	0.038	0.229	0.405	0.317	0.001
1967	0.000	0.008	0.042	0.218	0.410	0.320	0.002
1968	0.000	0.008	0.040	0.213	0.412	0.325	0.002
1969	0.000	0.007	0.043	0.204	0.413	0.330	0.002
1970	0.000	0.007	0.042	0.197	0.415	0.336	0.004
1971	0.000	0.007	0.044	0.182	0.421	0.340	0.006
1972	0.000	0.007	0.043	0.181	0.433	0.328	0.009
1973	0.000	0.006	0.043	0.186	0.441	0.311	0.013
1974	0.000	0.009	0.048	0.184	0.433	0.308	0.018
1975	0.000	0.010	0.045	0.179	0.462	0.277	0.027
1976	0.000	0.014	0.043	0.192	0.440	0.281	0.030
1977	0.000	0.018	0.034	0.191	0.454	0.266	0.037
1978	0.000	0.024	0.042	0.183	0.449	0.262	0.040
1979	0.000	0.029	0.041	0.197	0.428	0.268	0.036
1980	0.000	0.034	0.042	0.208	0.403	0.276	0.037
1981	0.000	0.035	0.043	0.220	0.385	0.276	0.042
1982	0.000	0.036	0.051	0.220	0.383	0.266	0.045
1983	0.000	0.038	0.056	0.228	0.380	0.252	0.046
1984	0.000	0.036	0.052	0.238	0.375	0.249	0.049

Table A3-12 Fraction of Energy from Mineral and Nonmineral Sources Consumed in the United States

Year	Nonmineral	Mineral	Year	Nonmineral	Mineral
1850	0.909	0.091	1938	0.113	0.887
1855	0.853	0.147	1939	0.103	0.897
1860	0.839	0.161	1940	0.095	0.905
1865	0.815	0.185	1941	0.087	0.913
1870	0.736	0.264	1942	0.090	0.910
1875	0.669	0.331	1943	0.087	0.913
1880	0.577	0.423	1944	0.084	0.916
1885	0.484	0.516	1945	0.088	0.912
1890	0.374	0.626	1946	0.089	0.911
1895	0.323	0.677	1947	0.080	0.920
1900	0.246	0.754	1948	0.080	0.920
1901	0.225	0.775	1949	0.088	0.912
1902	0.212	0.788	1950	0.079	0.921
1903	0.185	0.815	1951	0.073	0.927
1904	0.186	0.814	1952	0.074	0.926
1905	0.171	0.829	1953	0.069	0.931
1906	0.163	0.837	1954	0.069	0.931
1907	0.145	0.855	1955	0.064	0.936
1908	0.164	0.836	1956	0.063	0.937
1909	0.153	0.847	1957	0.063	0.937
1910	0.146	0.854	1958	0.064	0.936
1911	0.151	0.849	1959	0.061	0.939
1912	0.142	0.858	1960	0.058	0.942
1913	0.135	0.865	1961	0.057	0.943
1914	0.145	0.855	1962	0.056	0.944
1915	0.141	0.859	1963	0.054	0.946
1916	0.130	0.870	1964	0.052	0.948
1917	0.120	0.880	1965	0.052	0.948
1918	0.114	0.886	1966	0.048	0.952
1919	0.132	0.868	1967	0.051	0.949
1920	0.118	0.882	1968	0.048	0.952
1921	0.128	0.872	1969	0.050	0.950
1922	0.123	0.877	1970	0.048	0.952
1923	0.101	0.899	1971	0.051	0.949
1924	0.103	0.897	1972	0.050	0.950
1925	0.102	0.898	1973	0.049	0.951
1926	0.097	0.903	1974	0.056	0.944
1927	0.101	0.899	1975	0.055	0.945
1928	0.102	0.898	1976	0.057	0.943
1929	0.094	0.906	1977	0.052	0.948
1930	0.100	0.900	1978	0.066	0.934
1931	0.112	0.888	1979	0.070	0.930
1932	0.129	0.871	1980	0.076	0.924
1933	0.125	0.875	1981	0.078	0.922
1934	0.117	0.883	1982	0.087	0.913
1935	0.115	0.885	1983	0.094	0.906
1936	0.103	0.897	1984	0.089	0.911
1937	0.100	0.900			

Table A3-13 Total and per Capita Petroleum Consumption in the United States

Year	Total Consumption (quads)	Fuel Use (quads)	Non-fuel Use (quads)	Consumption per Capita (quads/million)
1860	0.003	0.00285	0.00015	0.000095
1865	0.01	0.0095	0.00050	0.00028
1870	0.011	0.0105	0.00055	0.00028
1875	0.011	0.0105	0.00055	0.00024
1880	0.096	0.091	0.0048	0.0019
1885	0.040	0.038	0.0020	0.0007
1890	0.156	0.148	0.0078	0.0025
1895	0.168	0.160	0.0084	0.0024
1900	0.229	0.218	0.0115	0.0030
1901	0.250	0.238	0.0125	0.0032
1902	0.364	0.346	0.0182	0.0046
1903	0.449	0.427	0.0225	0.0056
1904	0.534	0.507	0.0267	0.0065
1905	0.610	0.580	0.0305	0.0073
1906	0.555	0.527	0.0278	0.0065
1907	0.781	0.742	0.0391	0.0090
1908	0.820	0.779	0.0410	0.0092
1909	0.844	0.802	0.0422	0.0093
1910	1.007	0.957	0.0504	0.0109
1911	1.041	0.989	0.0521	0.0111
1912	1.059	1.006	0.0530	0.0111
1913	1.213	1.152	0.0607	0.0125
1914	1.325	1.259	0.0663	0.0134
1915	1.418	1.347	0.0709	0.0141
1916	1.508	1.433	0.0754	0.0148
1917	1.779	1.690	0.0890	0.0172
1918	1.942	1.845	0.0971	0.0188
1919	2.198	2.088	0.1099	0.0210
1920	2.676	2.542	0.1338	0.0251
1921	2.724	2.588	0.1362	0.0251
1922	3.127	2.971	0.1564	0.0284
1923	4.170	3.962	0.2085	0.0372
1924	3.867	3.674	0.1934	0.0339
1925	4.280	4.066	0.2140	0.0370
1926	4.480	4.256	0.2240	0.0382
1927	4.556	4.328	0.2278	0.0383
1928	4.963	4.715	0.2482	0.0412
1929	5.540	5.263	0.2770	0.0455
1930	5.895	5.600	0.2948	0.0479
1931	5.165	4.907	0.2583	0.0416
1932	4.748	4.511	0.2374	0.0380
1933	4.988	4.739	0.2494	0.0397
1934	4.979	4.730	0.2490	0.0394
1935	5.668	5.385	0.2834	0.0445
1936	6.308	5.993	0.3154	0.0492
1937	6.812	6.471	0.3406	0.0528

(continued)

Table A3-13 Total and per Capita Petroleum Consumption in the United States (*continued*)

Year	Total Consumption (quads)	Fuel Use (quads)	Non-fuel Use (quads)	Consumption per Capita (quads/million)
1938	6.674	6.340	0.3337	0.0514
1939	7.080	6.726	0.3540	0.0540
1940	7.730	7.344	0.3865	0.0585
1941	8.568	8.140	0.4284	0.0642
1942	8.034	7.632	0.4017	0.0596
1943	8.607	8.177	0.4304	0.0629
1944	9.703	9.218	0.4852	0.0701
1945	10.110	9.605	0.5055	0.0723
1946	10.480	9.956	0.5240	0.0741
1947	11.367	10.609	0.7580	0.0789
1948	12.557	11.744	0.8130	0.0856
1949	11.880	11.099	0.7810	0.0796
1950	13.320	12.432	0.8880	0.0878
1951	14.430	13.415	1.015	0.0935
1952	14.960	13.923	1.037	0.0953
1953	15.560	14.478	1.082	0.0975
1954	15.840	14.708	1.132	0.0975
1955	17.250	15.978	1.272	0.1044
1956	17.940	16.567	1.373	0.1066
1957	17.930	16.530	1.400	0.1047
1958	18.530	17.078	1.452	0.1064
1959	19.320	17.684	1.636	0.1086
1960	19.920	18.178	1.742	0.1103
1961	20.220	18.419	1.801	0.1101
1962	21.050	19.116	1.934	0.1128
1963	21.700	19.621	2.079	0.1147
1964	22.300	20.053	2.247	0.1162
1965	23.250	20.890	2.360	0.1197
1966	24.400	21.835	2.565	0.1241
1967	25.280	22.751	2.529	0.1272
1968	26.980	24.217	2.763	0.1344
1969	28.340	25.380	2.960	0.1398
1970	29.520	26.423	3.097	0.1440
1971	30.560	27.427	3.133	0.1472
1972	32.950	29.522	3.428	0.1570
1973	34.840	31.163	3.677	0.1644
1974	33.450	30.012	3.438	0.1564
1975	32.730	29.203	3.527	0.1515
1976	35.170	31.301	3.869	0.1613
1977	37.120	33.037	4.083	0.1685
1978	37.970	33.793	4.177	0.1683
1979	37.120	32.666	4.454	0.1649
1980	34.200	29.754	4.446	0.1502
1981	31.930	27.779	4.151	0.1388
1982	30.230	26.600	3.630	0.1301
1983	30.050	26.440	3.610	0.1281
1984	31.000	26.970	4.030	0.1310

Table A3-14 World Oil Production (billions of barrels)

Year	Bbbl/Year	Cumulative Production
1960	7.65	131.30
1961	8.19	138.95
1962	8.88	147.14
1963	9.54	156.02
1964	10.35	165.56
1965	11.06	175.91
1966	12.02	186.97
1967	12.91	198.99
1968	14.10	211.90
1969	15.22	226.00
1970	16.53	242.53
1971	17.46	259.99
1972	18.34	278.34
1973	20.32	298.66
1974	20.39	319.04
1975	19.30	338.34
1976	20.92	359.26
1977	21.79	381.05
1978	21.92	402.97
1979	22.83	425.80
1980	21.73	447.53
1981	20.40	467.93
1982	19.51	487.45
1983	19.34	506.78
1984	19.62	526.40

Table A3-15 United States Crude Oil Production (billions of barrels)

Year	Annual Production	Cumulated Production	Year	Annual Production	Cumulated Production
1860	0.0005	0.0005	1877	0.0134	0.0999
1861	0.0021	0.0026	1878	0.0154	0.1153
1862	0.0031	0.0057	1879	0.0199	0.1352
1863	0.0026	0.0083	1880	0.0263	0.1615
1864	0.0021	0.0104	1881	0.0277	0.1892
1865	0.0025	0.0129	1882	0.0303	0.2195
1866	0.0036	0.0165	1883	0.0235	0.2430
1867	0.0033	0.0198	1884	0.0242	0.2672
1868	0.0036	0.0234	1885	0.0219	0.2891
1869	0.0042	0.0276	1886	0.0281	0.3172
1870	0.0053	0.0329	1887	0.0282	0.3454
1871	0.0052	0.0381	1888	0.0276	0.3730
1872	0.0063	0.0444	1889	0.0351	0.4081
1873	0.0099	0.0543	1890	0.0458	0.4539
1874	0.0109	0.0652	1891	0.0543	0.5082
1875	0.0122	0.0774	1892	0.0505	0.5587
1876	0.0091	0.0865	1893	0.0484	0.6071

(continued)

Table A3-15 United States Crude Oil Production (billions of barrels) (*continued*)

Year	Annual Production	Cumulated Production	Year	Annual Production	Cumulated Production
1894	0.0493	0.6564	1941	1.402	25.212
1895	0.0529	0.7093	1942	1.387	26.599
1896	0.0610	0.7703	1943	1.506	28.104
1897	0.0605	0.8308	1944	1.678	29.782
1898	0.0554	0.8862	1945	1.714	31.496
1899	0.0571	0.9433	1946	1.734	33.230
1900	0.0636	1.007	1947	1.857	35.087
1901	0.0694	1.076	1948	2.020	37.107
1902	0.0888	1.165	1949	1.842	38.949
1903	0.1005	1.266	1950	1.974	40.923
1904	0.1171	1.383	1951	2.248	43.170
1905	0.1347	1.517	1952	2.390	45.560
1906	0.1265	1.644	1953	2.357	47.917
1907	0.1661	1.810	1954	2.315	50.232
1908	0.1785	1.989	1955	2.484	52.717
1909	0.1832	2.172	1956	2.617	55.334
1910	0.2096	2.381	1957	2.617	57.951
1911	0.2204	2.602	1958	2.449	60.400
1912	0.2229	2.825	1959	2.575	62.974
1913	0.2484	3.073	1960	2.575	65.549
1914	0.2658	3.339	1961	2.622	68.171
1915	0.2811	3.620	1962	2.676	70.847
1916	0.3008	3.921	1963	2.753	73.600
1917	0.3353	4.256	1964	2.787	76.387
1918	0.3559	4.612	1965	2.849	79.235
1919	0.3784	4.990	1966	3.028	82.263
1920	0.4429	5.433	1967	3.217	85.480
1921	0.4722	5.905	1968	3.329	88.809
1922	0.5575	6.463	1969	3.372	92.180
1923	0.7324	7.195	1970	3.518	95.698
1924	0.7139	7.909	1971	3.374	99.072
1925	0.7637	8.673	1972	3.373	102.445
1926	0.7709	9.444	1973	3.289	105.734
1927	0.9011	10.345	1974	3.132	108.866
1928	0.9015	11.246	1975	3.057	111.923
1929	1.007	12.254	1976	2.904	114.827
1930	0.898	13.152	1977	2.840	117.667
1931	0.851	14.003	1978	2.729	120.396
1932	0.785	14.788	1979	3.121	123.517
1933	0.906	15.694	1980	3.146	126.663
1934	0.908	16.602	1981	3.129	129.792
1935	0.997	17.598	1982	3.157	132.949
1936	1.100	18.698	1983	3.171	136.120
1937	1.279	19.977	1984	3.241	139.361
1938	1.214	21.192	1985	3.274	142.635
1939	1.265	22.457	1986	3.164	145.799
1940	1.353	23.810			

Table A3-16 United States Petroleum Proven Reserves (billions of barrels)

Year	Reserves	New Reserves	Cumulated Reserves
1899	2.5		2.50
1900	2.9	0.46	2.96
1901	3.0	0.16	3.12
1902	3.2	0.27	3.39
1903	3.4	0.29	3.68
1904	3.6	0.30	3.98
1905	3.8	0.32	4.30
1906	3.8	0.13	4.43
1907	3.9	0.23	4.66
1908	4.0	0.27	4.92
1909	4.2	0.38	5.30
1910	4.5	0.48	5.79
1911	5.0	0.71	6.50
1912	5.4	0.62	7.12
1913	5.5	0.32	7.44
1914	5.4	0.15	7.59
1915	5.5	0.37	7.95
1916	5.9	0.68	8.63
1917	5.9	0.30	8.93
1918	6.2	0.64	9.57
1919	6.7	0.86	10.43
1920	7.2	0.88	11.30
1921	7.8	1.04	12.35
1922	7.6	0.27	12.62
1923	7.6	0.56	13.18
1924	7.5	0.63	13.81
1925	8.5	1.71	15.52
1926	8.8	1.06	16.59
1927	10.5	2.47	19.06
1928	11.0	1.40	20.46
1929	13.2	3.10	23.56
1930	13.6	1.41	24.97
1931	13.0	0.30	25.27
1932	12.3	0.15	25.42
1933	12.0	0.49	25.90
1934	12.2	1.08	26.98
1935	12.4	1.13	28.12
1936	13.1	1.66	29.78
1937	15.5	3.54	33.32
1938	17.3	3.12	36.44
1939	18.5	2.35	38.79
1940	19.0	1.81	40.59
1941	19.6	1.89	42.48
1942	20.1	1.93	44.41
1943	20.1	1.37	45.78
1944	20.5	1.89	47.67

(continued)

Table A3-16 United States Petroleum Proven Reserves (billions of barrels) (*continued*)

Year	Reserves	New Reserves	Cumulated Reserves
1945	20.8	2.05	49.72
1946	20.9	1.76	51.48
1947	21.5	2.35	53.83
1948	23.3	3.65	57.48
1949	24.6	3.39	60.87
1950	25.3	2.46	63.33
1951	27.5	4.17	67.50
1952	28.0	2.74	70.25
1953	28.9	3.37	73.62
1954	29.6	2.97	76.59
1955	30.0	2.77	79.36
1956	30.4	2.91	82.27
1957	30.3	2.48	84.75
1958	30.5	2.85	87.60
1959	31.7	3.63	91.23
1960	31.6	2.47	93.70
1961	31.8	2.72	96.42
1962	31.4	2.25	98.67
1963	31.0	2.26	100.93
1964	31.0	2.77	103.70
1965	31.4	3.15	106.85
1966	31.5	2.95	109.80
1967	31.4	2.95	112.75
1968	30.7	2.55	115.30
1969	29.6	2.25	117.55
1970	39.0	12.74	130.30
1971	48.1	12.62	142.91
1972	36.3	−8.43	134.49
1973	35.3	2.37	136.86
1974	34.2	2.19	139.05
1975	32.7	1.63	140.68
1976	30.9	1.26	141.94
1977	30.7	2.70	144.64
1978	29.6	1.74	146.38
1979	28.5	1.63	148.01
1980	29.8	4.42	152.43
1981	29.4	2.75	155.18
1982	27.9	1.63	156.81
1983	27.7	2.96	159.76
1984	28.4	3.87	163.63
1985	28.4	3.24	166.87
1986	24.6	−0.53	166.35

Table A3-17 Ratio of Energy Consumed to Gross National Product for the United States (energy in quads and GNP in constant 1958 dollars)

Year	Quads per Billion GNP	Year	Quads per Billion GNP	Year	Quads per Billion GNP
1890	0.1327	1927	0.1206	1956	0.0899
1895	0.1230	1928	0.1225	1957	0.0879
1900	0.1249	1929	0.1213	1958	0.0893
1901	0.1178	1930	0.1268	1959	0.0868
1902	0.1224	1931	0.1172	1960	0.0882
1903	0.1328	1932	0.1211	1961	0.0877
1904	0.1329	1933	0.1268	1962	0.0860
1905	0.1361	1934	0.1228	1963	0.0859
1906	0.1266	1935	0.1183	1964	0.0845
1907	0.1419	1936	0.1154	1965	0.0827
1908	0.1388	1937	0.1159	1966	0.0820
1909	0.1301	1938	0.1073	1967	0.0823
1910	0.1375	1939	0.1069	1968	0.0832
1911	0.1324	1940	0.1083	1969	0.0848
1912	0.1334	1941	0.1033	1970	0.0872
1913	0.1397	1942	0.0957	1971	0.0871
1914	0.1366	1943	0.0918	1972	0.0862
1915	0.1419	1944	0.0891	1973	0.0846
1916	0.1440	1945	0.0897	1974	0.0834
1917	0.1561	1946	0.0985	1975	0.0864
1918	0.1445	1947	0.1063	1976	0.0821
1919	0.1300	1948	0.1044	1977	0.0797
1920	0.1515	1949	0.0963	1978	0.0784
1921	0.1390	1950	0.0948	1979	0.0773
1922	0.1251	1951	0.0937	1980	0.0751
1923	0.1382	1952	0.0903	1981	0.0715
1924	0.1313	1953	0.0886	1982	0.0705
1925	0.1228	1954	0.0870	1983	0.0679
1926	0.1241	1955	0.0883	1984	0.0659

Table A3-18 U.S. Cost of Fossil Fuels as a Fraction of Gross National Product

Year	Fuel Cost Fraction of GNP	Year	Fuel Cost Fraction of GNP	Year	Fuel Cost Fraction of GNP
1890	0.0166	1922	0.0381	1954	0.0268
1891	0.0165	1923	0.0381	1955	0.0270
1892	0.0163	1924	0.0331	1956	0.0275
1893	0.0172	1925	0.0301	1957	0.0282
1894	0.0176	1926	0.0328	1958	0.0272
1895	0.0181	1927	0.0288	1959	0.0262
1896	0.0189	1928	0.0262	1960	0.0257
1897	0.0164	1929	0.0274	1961	0.0252
1898	0.0164	1930	0.0266	1962	0.0247
1899	0.0184	1931	0.0206	1963	0.0242
1900	0.0203	1932	0.0244	1964	0.0232
1901	0.0200	1933	0.0242	1965	0.0222
1902	0.0206	1934	0.0285	1966	0.0217
1903	0.0265	1935	0.0265	1967	0.0218
1904	0.0242	1936	0.0278	1968	0.0211
1905	0.0227	1937	0.0291	1969	0.0211
1906	0.0215	1938	0.0265	1970	0.0217
1907	0.0246	1939	0.0249	1971	0.0203
1908	0.0242	1940	0.0259	1972	0.0201
1909	0.0209	1941	0.0255	1973	0.0223
1910	0.0222	1942	0.0216	1974	0.0431
1911	0.0218	1943	0.0200	1975	0.0433
1912	0.0225	1944	0.0207	1976	0.0460
1913	0.0257	1945	0.0209	1977	0.0498
1914	0.0242	1946	0.0237	1978	0.0461
1915	0.0228	1947	0.0299	1979	0.0565
1916	0.0258	1948	0.0365	1980	0.0728
1917	0.0350	1949	0.0292	1981	0.0784
1918	0.0341	1950	0.0342	1982	0.0711
1919	0.0288	1951	0.0274	1983	0.0586
1920	0.0446	1952	0.0260	1984	0.0555
1921	0.0392	1953	0.0274		

Table A3-19 Fraction of United States Gross National Product Spent on Fuels and Electricity

Year	Fraction of GNP for Energy	Year	Fraction of GNP for Energy
1970	0.083	1978	0.1101
1971	0.083	1979	0.1221
1972	0.0821	1980	0.1171
1973	0.0841	1981	0.1451
1974	0.1071	1982	0.1391
1975	0.1111	1983	0.1261
1976	0.1121	1984	0.1201
1977	0.1141		

Table A3-20 Production of Electricity in the United States by Utilities and Firms (billions of kilowatt hours per year)

Year	Utilities	Firms	Total
1950	329	59.5	388.7
1951	371	62.7	433.4
1952	399	63.8	463.1
1953	443	71.5	514.2
1954	472	73.0	544.6
1955	547	82.0	629.0
1956	601	84.1	684.8
1957	632	84.8	716.4
1958	645	79.7	724.8
1959	710	85.2	797.6
1960	753	88.3	844.2
1961	792	86.7	881.5
1962	852	91.2	946.5
1963	917	94.6	1011.4
1964	984	99.8	1083.7
1965	1055	102.3	1157.6
1966	1144	105.1	1249.4
1967	1214	102.9	1317.3
1968	1329	106.6	1436.0
1969	1442	110.6	1552.8
1970	1532	108.2	1639.8
1971	1613	103.2	1716.2
1972	1750	104.5	1854.5
1973	1861	105.2	1966.2
1974	1867	101.3	1968.3
1975	1918	85.4	2003.4
1976	2038	87.1	2125.1
1977	2124	87.3	2211.3
1978	2206	78.6	2284.6
1979	2247	72.9	2319.9
1980	2286	70.0	2356.0
1981	2295	70.0	2365.0
1982	2241	70.0	2311.0
1983	2310	70.0	2380.0
1984	2413	70.0	2483.0

Table A3-21 Total and per Capita Production of Electricity in the United States

Year	Net Production (billion kw-hours)	Kw-hours per Person		Net Production (billion kw-hours)	Kw-hours per Person
1902	5.97	75.4	1951	433.7	2811
1907	14.12	162.3	1952	462.8	2949
1912	14.75	154.7	1953	514.5	3224
1917	43.43	420.5	1954	545.0	3356
1920	56.56	531.3	1955	629.0	3806
1921	53.13	489.5	1956	685.1	4073
1922	61.20	556.2	1957	716.8	4185
1923	71.39	637.7	1958	724.7	4162
1924	75.89	665.1	1959	795.2	4472
1925	84.67	731.0	1960	841.3	4657
1926	94.22	802.6	1961	878.7	4784
1927	101.4	851.8	1962	943.2	5056
1928	108.1	896.8	1963	1012	5346
1929	116.7	958.8	1964	1084	5648
1930	114.6	930.6	1965	1157	5956
1931	109.4	881.0	1966	1249	6355
1932	99.4	795.2	1967	1317	6627
1933	102.7	816.7	1968	1436	7153
1934	110.4	872.8	1969	1553	7660
1935	118.9	933.8	1970	1640	7999
1936	136.0	1061	1971	1716	8264
1937	146.5	1136	1972	1855	8835
1938	142.0	1092	1973	1966	9279
1939	161.3	1231	1974	1968	9204
1940	179.9	1362	1975	2003	9276
1941	208.3	1561	1976	2125	9747
1942	233.1	1729	1977	2211	10040
1943	267.5	1957	1978	2285	10127
1944	279.5	2020	1979	2320	10308
1945	271.3	1939	1980	2356	10345
1946	269.6	1907	1981	2365	10281
1947	307.4	2133	1982	2311	9946
1948	336.8	2297	1983	2380	10148
1949	345.0	2313	1984	2483	10491
1950	388.7	2562			

Table A3-22 Fraction of U.S. Consumption of Primary Energy Used to Generate Electricity (quads)

Year	Fraction	Year	Fraction
1925	0.097	1955	0.193
1926	0.101	1956	0.199
1927	0.107	1957	0.207
1928	0.106	1958	0.203
1929	0.103	1959	0.211
1930	0.098	1960	0.213
1931	0.112	1961	0.216
1932	0.115	1962	0.220
1933	0.113	1963	0.225
1934	0.114	1964	0.232
1935	0.106	1965	0.237
1936	0.116	1966	0.243
1937	0.118	1967	0.248
1938	0.129	1968	0.255
1939	0.129	1969	0.266
1940	0.120	1970	0.274
1941	0.133	1971	0.280
1942	0.138	1972	0.288
1943	0.145	1973	0.296
1944	0.144	1974	0.304
1945	0.135	1975	0.300
1946	0.138	1976	0.316
1947	0.146	1977	0.324
1948	0.157	1978	0.326
1949	0.166	1979	0.326
1950	0.165	1980	0.342
1951	0.166	1981	0.354
1952	0.175	1982	0.360
1953	0.183	1983	0.369
1954	0.190	1984	0.386

Table A3-23 Generation of Electricity by Nuclear Power in the United States (fraction of all electricity generated and total in billions of kilowatt-hours)

Year	Total	Fraction
1960	1	0.0012
1961	2	0.0023
1962	2	0.0021
1963	3	0.0030
1964	3	0.0028
1965	4	0.0035
1966	6	0.0048
1967	8	0.0061
1968	13	0.0091
1969	14	0.0090
1970	22	0.0134
1971	38	0.0221
1972	54	0.0291
1973	83	0.0422
1974	114	0.0579
1975	173	0.0864
1976	191	0.0899
1977	251	0.1135
1978	276	0.1208
1979	255	0.1099
1980	251	0.1065
1981	273	0.1154
1982	283	0.1225
1983	294	0.1235
1984	328	0.1319
1985	384	0.1512
1986	414	0.1618

Table A3-24 Maximum Addition Rate for Computers Introduced from 1951 to 1979 (kilo-OPS per second)

Computer	Year	Max Add Rate	Ln(max add rate)
UNIVAC U1	1951	1.9	0.64185
IBM 704	1955	13.89	2.63117
IBM 7090	1959	76.3	4.33467
IBM 7030	1961	285.7	5.65494
CDC 6600	1964	3333.3	8.11172
BGH 6700	1969	5000	8.51719
IBM 370/1	1975	6250	8.74034
IBM 3033	1978	12500	9.43348

Table A3-25 Successive Maximum Performance Indices for U.S. Computers

Computer	Year	Speed (KOPS/s)	Cost (KOPS/$)	Memory (Kbytes)	NSF Linear Index[a]	Ln(NSF Linear Index)[a]	LAG'S Log Index[a]
UNIVAC UI	1951	0.27	7	8	0.011	−4.50986	5.699368
IBM 650	1954	0.29	45	20	0.027	−3.61191	6.320643
IBM 704	1955	3.79	50	192	0.246	−1.40242	6.356104
IBM 709	1958	10.23	91	192	0.265	−1.32802	6.556739
IBM 7090	1959	45.47	443	197	0.378	−0.97286	7.084649
CDC 1604	1960	20.4	374	256	0.381	−0.96495	7.028551
IBM 7094	1962	95.9	842	197	0.526	−0.64245	7.299739
CDC 3600	1963	74.9	849	2048	2.731	1.004667	7.302839
CDC 6600	1964	4090	33988	1280	13.694	2.616957	8.533079
BGH 6700	1969	8886	81540	6144	34.17	3.531348	8.82647
CDC CYB/7	1972	10220	38632	5770	34.479	3.54035	8.578481
IBM 3033	1978	19019	65932	16384	72.675	4.285997	8.758687

[a]The NSF index is a linear combination of speed, cost, and memory.

[b]The LAG index is a logarithmic combination of speed and memory with speed weighted twice as much as memory.

Table A3-26 Successive Most Powerful Computers Using Knight's Data and Definition of Computer Power for Scientific and Commercial Computation

Year	Scientific Power (OPS/sec)	Scientific Power per Unit Cost (OPS/cent)	Year	Commercial Power (OPS/sec)
1944	0.04			
1946	7.4	2.4	1946	44.65
1949	21.8	27.7	1950.4	253.8
1950.4	102.8	52.4	1951.3	324.7
1950.9	682.5	347.7	1953.8	666.2
1953.6	1002	318.7	1955.2	1063
1954.9	1925	178.7	1956.3	2087
1956.3	10670	1472.5	1956.3	3785
1957.9	36700	1040.1	1957.7	7473
1958.7	82050	6960.3	1957.9	15560
1959.9	97350	9483.8	1958.7	34000
1960.3	105844	15712.5	1959.9	45470
1960.5	142600	6586.7	1960.3	55740
1960.8	172200	10736.7	1961.4	631200
1961.4	371700	7723.9	1964.8	4091293
1963	459065	57245.4	1966	4307061
1964	7021619	583496.5		
1974	13389300			
1976	36275000			
1979	403165000			

Table A3-27 Diffusion of Computer Capacity

Year	MOPS[a]	OPS/cap
1955	0.12	0.00073
1956	0.447	0.00266
1957	0.882	0.00515
1958	1.785	0.01025
1959	3.31	0.01861
1960	7.48	0.04140
1961	14.8	0.08057
1962	25.1	0.1346
1963	47.4	0.2505
1964	108.6	0.5660
1965	213	1.096
1966	542	2.757
1967	1023	5.148
1968	1798	8.958
1969	2000	9.868
1970	2194	10.70
1971	3200	15.41
1972	4657	22.19

Table A3-27 (*continued*)

Year	MOPS[a]	OPS/cap
1973	5600	26.43
1974	6925	32.38
1975	8500	39.36
1976	10258	47.05
1977	13000	59.03
1978	26064	115.54

[a]Total diffusion in MOPS per second and per capita diffusion in OPS per second per person.

Table A3-28 Worldwide Cumulated Production of Tractors (millions)

Year	Production	Year	Production
1920	0.66	1947	5.22
1921	0.73	1948	5.75
1922	0.82	1949	6.30
1923	0.95	1950	6.80
1924	1.06	1951	7.37
1925	1.22	1952	7.78
1926	1.39	1953	8.17
1927	1.58	1954	8.42
1928	1.73	1955	8.75
1929	1.92	1956	8.97
1930	2.10	1957	9.19
1931	2.17	1958	9.44
1932	2.18	1959	9.70
1933	2.21	1960	9.85
1934	2.27	1961	10.02
1935	2.40	1962	10.21
1936	2.60	1963	10.41
1937	2.84	1964	10.62
1938	3.01	1965	10.87
1939	3.19	1966	11.14
1940	3.44	1967	11.38
1941	3.76	1968	11.59

Table A3-29 Tractor Efficiencies

Year	Fuel Efficiency HP − hr/gal	Mechanical Efficiency Ratio Draw HP to Belt HP	Total Fuel Efficiency = Belt HP − hr/gal
1920	4.49	0.522	2.34
1921	4.84	0.510	2.47
1922	5.54	0.542	3.00
1923	5.53	0.523	2.89
1924	6.04	0.540	3.26
1925	5.83	0.531	3.10
1926	5.77	0.480	2.77
1927	5.82	0.486	2.83
1928	5.78	0.550	3.18
1929	6.01	0.561	3.37
1930	6.31	0.580	3.66
1931	5.98	0.606	3.63
1932	7.57	0.685	5.18
1933	8.25	0.656	5.41
1934	7.36	0.640	4.71
1935	8.42	0.639	5.38
1936	8.16	0.641	5.23
1937	7.67	0.622	4.77
1938	8.95	0.670	6.00
1939	9.28	0.686	6.37
1940	10.51	0.694	7.29
1941	10.03	0.708	7.10
1947	8.77	0.703	6.16
1948	9.19	0.715	6.57
1949	9.93	0.704	6.99
1950	10.05	0.690	6.93
1951	10.45	0.696	7.27
1952	9.47	0.725	6.87
1953	9.99	0.721	7.20
1954	10.99	0.696	7.65
1955	10.9	0.718	7.82
1956	10.22	0.725	7.41
1957	10.62	0.742	7.88
1958	10.59	0.741	7.85
1959	9.54	0.731	6.98
1960	10.62	0.746	7.92
1961	10.49	0.796	8.34
1962	10.95	0.754	8.26
1963	10.15	0.760	7.72
1964	11.15	0.823	9.17
1965	11.03	0.831	9.16
1966	10.22	0.753	7.70
1967	11.01	0.661	7.27
1968	11.52	0.740	8.52

Source: From data compiled by Sahal (1981, page 341): Data are averages each year for worldwide tractor production.

Table A3-30 Successive Maximum Efficiencies of Farm Tractors as a Function of Cumulative Production (in horsepower per gallon)

Year	Total Efficiency[a]	Production	Year	Fuel Efficiency[a]	Production
1920	2.34	0.66	1920	4.49	0.66
1921	2.47	0.73	1921	4.84	0.73
1922	3.00	0.82	1922	5.54	0.82
1924	3.26	1.06	1924	6.04	1.06
1929	3.37	1.92	1930	6.31	2.10
1930	3.66	2.10	1932	7.57	2.18
1932	5.18	2.18	1933	8.25	2.21
1933	5.41	2.21	1935	8.42	2.40
1938	6.00	3.01	1938	8.95	3.01
1939	6.37	3.19	1939	9.28	3.19
1940	7.29	3.44	1940	10.51	3.44
1954	7.65	8.42	1954	10.99	8.42
1955	7.82	8.75	1964	11.15	10.62
1957	7.88	9.19	1968	11.52	11.59
1960	7.92	9.85			
1961	8.34	10.02			
1964	9.17	10.62			

[a]Total efficiency is fuel efficiency times efficiency of conversion to mechanical work at the draw bar. Tractor production is in millions of tractors.

Table A3-31 Tractor Efficiencies

Year	Efficiency
1920	2.34
1921	2.47
1922	3.00
1924	3.26
1929	3.37
1930	3.66
1932	5.18
1933	5.41
1938	6.00
1939	6.37
1940	7.29
1954	7.65
1955	7.82
1957	7.88
1960	7.92
1961	8.34
1964	9.17

Source: From data compiled by Sahal (1981, page 341): Data are successive maxima of averages each year in belt horesepower per gallon for world tractor production.

Table A3-32 Labor Productivity of Underground Coal Mining for the United States (tons per man-hour)

Year	Productivity	Year	Productivity	Year	Productivity
1900	0.344	1929	0.563	1958	1.340
1901	0.339	1930	0.588	1959	1.440
1902	0.354	1931	0.612	1960	1.520
1903	0.349	1932	0.624	1961	1.630
1904	0.365	1933	0.575	1962	1.710
1905	0.377	1934	0.604	1963	1.826
1906	0.392	1935	0.617	1964	1.960
1907	0.383	1936	0.631	1965	2.000
1908	0.389	1937	N.A.	1966	2.090
1909	0.389	1938	0.657	1967	2.150
1910	0.404	1939	0.703	1968	2.210
1911	0.409	1940	0.694	1969	2.230
1912	0.432	1941	0.690	1970	1.970
1913	0.423	1942	0.677	1971	1.710
1914	0.435	1943	0.699	1972	1.700
1915	0.459	1944	0.720	1973	1.660
1916	0.457	1945	0.720	1974	1.610
1917	0.440	1946	0.776	1975	1.360
1918	0.442	1947	0.784	1976	1.300
1919	0.449	1948	0.759	1977	1.250
1920	0.468	1949	0.774	1978	1.190
1921	0.494	1950	0.821	1979	1.290
1922	0.502	1951	0.869	1980	1.380
1923	0.525	1952	0.910	1981	1.470
1924	0.534	1953	1.001	1982	1.570
1925	0.528	1954	1.141	1983	1.850
1926	0.524	1955	1.183	1984	1.970
1927	0.530	1956	1.231	1985	2.050
1928	0.548	1957	1.273		

Table A3-33 Diffusion of the Use of Continuous Mining Machines for the Underground Mining of Coal in the United States (thousands of tons)

Year	Total Mined	Mined by Continuous Mining Machines	Fraction Mined by Continuous Mining Machines
1948	460.0	0.45	0.00098
1949	331.8	2.60	0.00784
1950	392.8	4.85	0.0123
1951	415.8	6.06	0.0146
1952	356.4	8.22	0.0230
1953	349.6	11.83	0.0338
1954	289.1	16.34	0.0565
1955	343.5	27.46	0.0799
1956	365.8	39.91	0.109
1957	360.6	53.78	0.149
1958	286.9	56.37	0.197
1959	283.4	65.79	0.232
1960	284.9	77.93	0.274
1961	272.8	84.32	0.309
1962	281.3	90.17	0.321
1963	302.3	104.35	0.345

Table A3-34 Cruising Speed, Passenger Capacity, and Passenger-Miles per Hour of Commercial Aircraft

Aircraft	Year	Miles/hr	Passengers	P-M/hr
Ford 4-AT-B	1926	111	11	1221
Ford 5-AT-B	1928	148	13	1924
Boeing 247D	1933	200	10	2000
Lockheed 10 Electra	1934	192	10	1920
Douglas DC-2	1934	213	14	2982
Vultee V-1A	1934	225	8	1800
Douglas DC-3	1935	220	21	4620
Boeing 307B	1938	228	33	7524
Curtis C-46	1940	241	36	8676
Lockheed Lodestar	1940	272	14	3808
Douglas DC-4r	1940	275	42	11550
Lockheed 649	1946	329	64	21056
Douglas DC-6	1947	315	68	21420
Convair CV-240	1947	347	40	13880
Boeing C-97A	1948	375	134	50250
Boeing 377	1949	375	86	32250
Martin 404	1950	312	40	12480
Lockheed 1049	1950	370	92	34040
Convair CV-340	1951	314	44	13816
Douglas C-124C	1952	265	200	53000
Douglas DC-7	1954	409	95	38885
Lockheed C-130	1955	360	92	33120
Douglas DC-7C	1955	400	95	38000
Lockheed 1649A	1956	377	92	34684
Douglas DC-8	1958	579	189	109431
Lockheed 188 Electra	1958	450	98	44100
Boeing 707-320B	1959	597	189	112833
Convair 880	1959	615	110	67650
Boeing 720B	1959	622	140	87080
C-135	1961	600	126	75600
Lockheed C141A	1963	571	154	87934
Boeing 727	1963	632	114	72048
Tupolev TU-144	1968	1550	140	217000
Boeing 747	1969	640	490	313600
Concorde	1969	1354	100	135400
Douglas DC-10	1970	575	270	155250
Lockheed 1011	1970	610	256	156160
Airbus	1972	570	336	191520

Table A3-35 Maximum Speeds of U.S. Military Aircraft Compared with Official World Aircraft Speed Records (miles per hour)

Year	Maximum Speed	Year	Official Records
1909	42	1906	26
1916	80	1907	33
1918	110	1909	48
1918	135	1910	68
1921	141	1911	83
1924	161	1912	108
1925	163	1913	127
1927	158	1920	194
1929	180	1921	206
1929	171	1922	223
1933	234	1923	267
1934	212	1924	279
1937	256	1932	294
1937	281	1933	305
1938	300	1934	314
1939	357	1935	352
1940	322	1937	380
1940	379	1939	469
1941	315	1945	606
1941	350	1946	616
1942	420	1947	651
1942	390	1948	671
1943	436	1952	699
1945	578	1953	755
1946	619	1955	822
1948	671	1956	1132
1950	600	1957	1208
1953	860	1958	1404
1954	1200	1959	1526
1956	1330	1961	1606
1958	1404	1962	1666
1961	1800	1965	2070
1965	1800		
1967	1900		

Table A3-36 Horsepower to Weight Ratio of Reciprocating Aircraft Engines

Year	Horsepower	Weight (pounds)	HP/lb
1902	12	179	0.0670
1902	52	151	0.3444
1912	80	312	0.2564
1913	100	270	0.3704
1914	350	880	0.3977
1914	90	390	0.2308
1916	150	467	0.3212
1917	420	857	0.4901
1921	485	775	0.6258
1922	435	680	0.6397
1926	600	685	0.8759
1929	1525	1469	1.0381
1934	1730	1450	1.1931
1936	1980	2115	0.9362
1937	2050	2390	0.8577
1944	1700	1940	0.8763
1945	3250	3670	0.8856
1948	3400	3675	0.9252

Table A3-37 Number of Locomotives in Service in the United States[a]

Year	Steam	Electric	Diesel	Other	Total
1890.5	N.A.	N.A.	N.A.	N.A.	30140
1891.5	N.A.	N.A.	N.A.	N.A.	32139
1892.5	N.A.	N.A.	N.A.	N.A.	33136
1893.5	N.A.	N.A.	N.A.	N.A.	34788
1894.5	N.A.	N.A.	N.A.	N.A.	35492
1895.5	N.A.	N.A.	N.A.	N.A.	35699
1896.5	N.A.	N.A.	N.A.	N.A.	35950
1897.5	N.A.	N.A.	N.A.	N.A.	35986
1898.5	N.A.	N.A.	N.A.	N.A.	36234
1899.5	N.A.	N.A.	N.A.	N.A.	36703
1900.5	N.A.	N.A.	N.A.	N.A.	37633
1901.5	N.A.	N.A.	N.A.	N.A.	39584
1902.5	N.A.	N.A.	N.A.	N.A.	41225
1903.5	N.A.	N.A.	N.A.	N.A.	43871
1904.5	N.A.	N.A.	N.A.	N.A.	46743
1905.5	N.A.	N.A.	N.A.	N.A.	48357
1906.5	N.A.	N.A.	N.A.	N.A.	51672
1907.5	N.A.	N.A.	N.A.	N.A.	55388
1908.5	N.A.	N.A.	N.A.	N.A.	57698
1909.5	N.A.	N.A.	N.A.	N.A.	58219
1910.5	N.A.	N.A.	N.A.	N.A.	60019
1911.5	N.A.	N.A.	N.A.	N.A.	62463
1912.5	N.A.	N.A.	N.A.	N.A.	63463
1913.5	N.A.	N.A.	N.A.	N.A.	65597
1914.5	N.A.	N.A.	N.A.	N.A.	67012
1915.5	66229	273	N.A.	N.A.	66502
1916.5	65021	293	N.A.	N.A.	65314
1917	65253	342	N.A.	N.A.	65595
1918	65699	371	N.A.	N.A.	66070
1919	67563	373	N.A.	N.A.	67936
1920	68592	385	N.A.	N.A.	68977
1921	68554	388	N.A.	N.A.	68942
1922	68733	389	N.A.	N.A.	69122
1923	68121	397	N.A.	N.A.	68518
1924	69005	409	N.A.	N.A.	69414
1925	69144	372	N.A.	N.A.	69486
1926	67713	379	1	5	68098
1927	66381	435	11	20	66847
1928	64843	467	N.A.	38	65348
1929	62642	617	N.A.	52	63311
1930	60572	621	25	39	61257
1931	59406	663	77	43	60189
1932	57820	709	80	43	58652
1933	55831	764	80	57	56732
1934	55302	789	85	52	54228
1935	50465	805	104	49	51423
1936	48477	884	130	50	49541
1937	46923	858	175	53	48009

(continued)

Table A3-37 Number of Locomotives in Service in the United States (*continued*)

Year	Steam	Electric	Diesel	Other	Total
1938	46342	872	293	48	47555
1939	45210	882	403	49	46544
1940	43604	879	639	50	45172
1941	42410	900	967	56	44333
1942	41911	895	1517	52	44375
1943	41755	892	1978	46	44671
1944	41983	907	2476	40	45406
1945	41921	902	3432	50	46305
1946	41018	885	4301	49	46253
1947	39592	867	5008	44	45511
1948	36942	864	6495	43	44344
1949	34581	867	8981	45	44474
1950	30344	856	12025	47	43272
1951	26680	827	15396	48	42951
1952	22590	817	19014	52	42473
1953	16737	790	22118	52	39697
1954	12274	713	24209	55	47251
1955	9041	669	25256	67	35033
1956	6266	639	26563	65	33533
1957	3918	616	28001	58	32593
1958	2608	597	29137	49	32391
1959	1488	562	29515	51	31616
1960	871	517	30097	54	31539
1961	374	498	30240	66	31178
1962	210	484	30123	72	30889
1963	136	441	30057	67	30701
1964	112	438	29898	58	30506
1965	93	402	29745	66	30296
1966	89	365	29552	55	30061
1967	76	347	29644	57	30124
1968	67	324	29428	55	29874
1969	N.A.	307	29031	110	29448
1970	N.A.	278	28711	101	29090
1971	N.A.	270	28773	79	29122

[a]Until 1915 the total number of locomotives were almost all steam engines with only a few electric locomotives in service.

Table A3-38 Fraction of Diesel and Steam Locomotives in Service in the United States

Year	Total	Steam	Diesel
1926	67714	1.0000	0.0000
1927	66392	0.9998	0.0002
1928	64843	1.0000	0.0000
1929	62642	1.0000	0.0000
1930	60597	0.9996	0.0004
1931	59483	0.9987	0.0013
1932	57900	0.9986	0.0014
1933	55911	0.9986	0.0014
1934	55387	0.9985	0.0015
1935	50569	0.9979	0.0021
1936	48607	0.9973	0.0027
1937	47098	0.9963	0.0037
1938	46635	0.9937	0.0063
1939	45613	0.9912	0.0088
1940	44243	0.9856	0.0144
1941	43377	0.9777	0.0223
1942	43428	0.9651	0.0349
1943	43733	0.9548	0.0452
1944	44459	0.9443	0.0557
1945	45353	0.9243	0.0757
1946	45319	0.9051	0.0949
1947	44600	0.8877	0.1123
1948	43437	0.8505	0.1495
1949	43562	0.7938	0.2062
1950	42369	0.7162	0.2838
1951	42076	0.6341	0.3659
1952	41604	0.5430	0.4570
1953	38855	0.4308	0.5692
1954	36483	0.3364	0.6636
1955	34297	0.2636	0.7364
1956	32829	0.1909	0.8091
1957	31919	0.1227	0.8773
1958	31745	0.0822	0.9178
1959	31003	0.0480	0.9520
1960	30968	0.0281	0.9719
1961	30614	0.0122	0.9878
1962	30333	0.0069	0.9931
1963	30193	0.0045	0.9955
1964	30010	0.0037	0.9963
1965	29838	0.0031	0.9969
1966	29641	0.0030	0.9970
1967	29720	0.0026	0.9974
1968	29495	0.0023	0.9977
1969	29031	0.0000	1.0000
1970	28711	0.0000	1.0000
1971	28773	0.0000	1.0000

Table A3-39 Average Traction of Locomotives in Service in the United States

Year	Lb Traction	Year	Lb Traction
1903.5	21781	1936	48367
1904.5	22804	1937	48972
1905.5	23666	1938	49412
1906.5	24741	1939	49803
1907.5	25781	1940	50395
1908.5	26356	1941	50905
1909.5	26601	1942	51217
1910.5	27282	1943	51811
1911.5	28291	1944	52451
1912.5	29049	1945	52822
1913.5	30258	1946	53217
1914.5	31006	1947	53735
1915.5	31501	1948	54506
1916.5	32380	1949	55170
1917	32840	1950	56333
1918	33932	1951	57075
1919	34995	1952	58476
1920	35789	1953	59966
1921	36365	1954	61339
1922	36935	1955	63152
1923	37441	1956	65005
1924	39177	1957	68745
1925	39891	1958	61515
1926	40666	1959	61312
1927	41886	1960	61408
1928	42798	1961	61314
1929	43838	1962	61969
1930	44801	1963	61415
1931	45225	1964	61533
1932	45764	1965	62311
1933	46299	1966	63096
1934	46916	1967	70900
1935	47712	1968	65267

Table A3-40 Fuel Efficiency of Railroad Freight Transport (ton-miles per 1000 Btus of fuel expended)

Year	Efficiency	Year	Efficiency
1921	0.10	1951	0.32
1922	0.11	1952	0.39
1923	0.10	1953	0.47
1924	0.11	1954	0.58
1925	0.11	1955	0.65
1926	0.11	1956	0.73
1927	0.12	1957	0.81
1928	0.13	1958	0.89
1929	0.13	1959	0.94
1930	0.12	1960	0.96
1931	0.12	1961	0.98
1932	0.11	1962	1.01
1933	0.12	1963	1.06
1934	0.12	1964	1.26
1935	0.12	1965	1.29
1936	0.13	1966	1.31
1937	0.13	1967	1.29
1938	0.13	1968	1.33
1939	0.14	1969	1.37
1940	0.14	1970	1.42
1941	0.15	1971	1.43
1942	0.18	1972	1.41
1943	0.18	1973	1.52
1944	0.18	1974	1.49
1945	0.18	1975	1.47
1946	0.18	1976	1.47
1947	0.19	1977	1.49
1948	0.20	1978	1.56
1949	0.24	1979	1.61
1950	0.27	1980	1.69

Table A3-41 First Adoptions of Diesel Locomotives by American Railroads

Year	Number[a]	Percent[b]
1925	1	4.2
1927	8	33.3
1929	9	37.5
1931	10	41.7
1933	10	41.7
1935	12	50.0
1937	14	58.3
1939	19	79.2
1941	21	87.5
1943	23	95.8
1945	23	95.8
1947	23	95.8
1949	23	95.8
1951	23	95.8
1953	23	95.8
1955	24	100
1957	24	100
1959	24	100

[a]Number of firms that had adopted their first diesel by the year given in the first column.
[b]Percentage of all firms that had adopted diesels. The total number of firms included was 24.

Table A3-42 Total and per Capita Miles of Railroad Track Operated and Miles of Track Built in the United States

Year	Miles Operated	Miles (per million people)	Miles Built
1830	23	1.8	40
1831	95	7.2	99
1832	229	16.9	191
1833	380	27.2	116
1834	633	44.0	214
1835	1098	73.2	138
1836	1273	83.6	280
1837	1497	95.5	348
1838	1913	118.6	453
1839	2302	138.7	386
1840	2818	164.6	491
1841	3535	201.2	606
1842	4026	222.8	505
1843	4185	225.2	288
1844	4377	229.1	180
1845	4633	229.6	277
1846	4930	244.1	333
1847	5598	269.7	263
1848	5996	281.0	1056

Table A3-42 (*continued*)

Year	Miles Operated	Miles (per million people)	Miles Built
1849	7365	335.9	1048
1850	9021	387.8	1261
1851	10982	474.4	1274
1852	12908	542.8	2288
1853	15360	628.8	2170
1854	16720	666.4	3442
1855	18374	670.9	2453
1856	22076	834.4	1471
1857	24503	902.0	2077
1858	26968	967.0	1966
1859	28789	1005.7	1707
1860	30626	971.9	1500
1861	31286	1037.6	1016
1862	32120	1038.2	720
1863	33170	1045.0	574
1864	33908	1041.3	947
1865	35085	928.7	819
1866	36801	1074.2	1404
1867	39050	1111.5	2541
1868	42229	1172.2	2468
1869	46844	1268.3	4103
1870	52922	1326.2	5658
1871	N.A.	N.A.	6660
1872	N.A.	N.A.	7439
1873	N.A.	N.A.	5217
1874	N.A.	N.A.	2584
1875	N.A.	N.A.	1606
1876	94665	2159.4	2575
1877	97308	2167.1	2280
1878	103649	2253.8	2428
1879	104756	2224.3	5006
1880	115647	2300.9	N.A.
1881	130455	2642.3	N.A.
1882	140878	2787.4	N.A.
1883	149101	2882.2	N.A.
1884	156414	2954.2	N.A.
1885	169506	2991.7	N.A.
1886	167952	3029.5	N.A.
1887	184935	3260.6	N.A.
1888	191376	3298.4	N.A.
1889	202088	3405.2	N.A.
1890	208152	3301.1	N.A.
1891	207446	3342.2	N.A.
1892	211051	3325.5	N.A.
1893	221864	3419.4	3024
1894	229796	3464.6	1760

(*continued*)

Table A3-42 Total and per Capita Miles of Railroad Track Operated and Miles of Track Built in the United States (*continued*)

Year	Miles Operated	Miles (per million people)	Miles Built
1895	233276	3352.6	1420
1896	239140	3451.4	1692
1897	242013	3418.1	2109
1898	245334	3391.2	3265
1899	250143	3384.4	4569
1900	258784	3400.8	4894
1901	265352	3420.2	5368
1902	274196	3463.7	6026
1903	283882	3520.7	5652
1904	297073	3615.5	3832
1905	306797	3660.1	4388
1906	317083	3710.7	5623
1907	327975	3769.5	5212
1908	333646	3761.1	3214
1909	342351	3783.3	3478
1910	351767	3806.7	4122
1911	362824	3865.5	3066
1912	371238	3894.0	2997
1913	379508	3903.4	3071
1914	387208	3906.8	1532
1915	391142	3890.2	933
1916	394944	3873.5	1098
1917	400353	3876.8	979
1918	402343	3898.4	721
1919	403891	3864.5	686
1920	406580	3819.1	314
1921	407531	3754.7	475
1922	409359	3719.8	324
1923	412993	3689.2	427
1924	415028	3637.1	579
1925	417954	3608.4	644
1926	421341	3589.0	
1927	424737	3568.3	
1928	427750	3549.5	
1929	429054	3523.5	
1930	429883	3489.6	
1931	429823	3462.2	
1932	428402	3428.6	
1933	425664	3386.6	
1934	422401	3339.4	
1935	419228	3291.6	
1936	416381	3248.4	
1937	414572	3214.7	
1938	411324	3164.8	
1939	408350	3116.5	
1940	405975	3072.7	

Table A3-42 (*continued*)

Year	Miles Operated	Miles (per million people)	Miles Built
1941	403625	3025.6	
1942	399627	2963.3	
1943	398730	2916.0	
1944	398437	2878.9	
1945	398054	2844.7	
1946	398037	2815.2	
1947	397355	2757.0	
1948	397203	2708.9	
1949	397232	2662.6	
1950	396380	2613.2	
1951	395831	2565.5	
1952	394631	2514.3	
1953	393736	2467.6	
1954	392580	2417.5	
1955	390965	2365.5	
1956	389668	2316.4	
1957	386978	2259.4	
1958	385264	2212.4	
1959	383912	2158.9	
1960	381745	2112.9	
1961	379415	2065.5	
1962	376290	2017.2	
1963	374522	1979.1	
1964	372300	1940.2	
1965	370636	1907.5	
1966	370104	1882.9	
1967	368030	1852.1	
1968	366238	1824.7	
1969	364915	1800.5	
1970	360330	1757.3	

Table A3-43 Percentage of U.S. Dwellings with Electric Service

Year	Farm	Non-farm	All Dwellings
1907	N.A.	N.A.	8.0
1912	N.A.	N.A.	15.9
1917	N.A.	N.A.	24.3
1920	1.6	47.4	34.7
1921	2	N.A.	37.8
1922	2.5	N.A.	40.0
1923	3	N.A.	44.2
1924	3.5	N.A.	48.6
1925	3.9	69.4	53.2
1926	4.8	N.A.	57.9
1927	5.9	N.A.	63.1
1928	7.3	N.A.	65.0
1929	9.2	N.A.	67.9
1930	10.4	84.8	68.2
1931	10.7	N.A.	67.4
1932	11.2	N.A.	67.0
1933	11.8	N.A.	66.7
1934	12.1	N.A.	67.1
1935	12.3	83.9	68.0
1936	14.5	N.A.	70.3
1937	18.3	N.A.	73.1
1938	23.4	N.A.	74.9
1939	27.4	N.A.	77.3
1940	32.6	90.8	78.7
1941	35.0	N.A.	80.0
1942	37.8	N.A.	81.2
1943	40.0	N.A.	81.3
1944	42.2	N.A.	84.0
1945	48.0	93.0	85.0
1946	53.3	N.A.	85.5
1947	60.2	N.A.	86.2
1948	66.8	N.A.	89.6
1949	72.9	N.A.	93.0
1950	77.7	96.6	94.0
1951	82.2	N.A.	95.2
1952	86.9	N.A.	96.1
1953	91.4	N.A.	97.2
1954	93.0	N.A.	97.9
1955	94.4	98.8	98.4
1956	95.9	99.2	98.8

Table A3-44 United States Households with Radios

Year	Households with Radios (millions)	Millions of Households	Percent with Radios
1922	0.06	24.85	0.24
1923	0.40	25.33	1.58
1924	1.25	25.81	4.84
1925	2.75	26.29	10.46
1926	4.50	26.79	16.80
1927	6.75	27.29	24.73
1928	8.00	27.80	28.77
1929	10.25	28.33	36.19
1930	13.75	29.90	45.98
1931	16.70	29.40	56.81
1932	18.45	29.95	61.60
1933	19.25	30.51	63.09
1934	20.40	31.08	65.63
1935	21.46	31.67	67.75
1936	22.87	32.27	70.88
1937	24.50	32.88	74.52
1938	26.67	33.50	79.61
1939	27.50	34.13	80.56
1940	28.50	34.95	81.55
1941	29.30	35.45	82.65
1942	30.60	36.13	84.69
1943	30.80	36.83	83.63
1944	32.50	37.54	86.57
1945	33.10	38.27	86.49
1946	34.00	39.02	87.13
1947	34.90	39.11	89.24
1948	37.62	40.58	92.72
1949	39.30	41.39	94.95
1950	40.70	43.55	93.45
1951	41.90	43.07	97.27
1952	42.80	43.95	97.38
1953	44.80	44.86	99.88
1954	45.10	45.79	98.50
1955	45.90	47.87	95.88
1956	46.80	48.90	95.70
1957	47.60	49.67	95.83
1958	48.50	50.47	96.09
1959	49.45	51.44	96.14
1960	50.19	52.80	95.06
1961	50.70	53.56	94.66
1962	51.31	54.76	93.68
1963	52.30	55.27	94.63
1964	54.00	56.15	96.17
1965	55.20	57.44	96.11
1966	57.00	58.41	97.59
1967	57.50	59.24	97.07
1968	58.50	60.81	96.20
1969	60.60	62.21	97.41
1970	62.00	63.40	97.79

Table A3-45 Aluminum Production Capacity (thousands of tons and production capacity per capita in pounds per person in the United States)

Year	Capacity	Capacity per Capita	Year	Capacity	Capacity per Capita
1910	17	0.37	1948	676	9.22
1915	45	0.90	1949	679	9.10
1919	63	1.21	1950	633	8.35
1927	82	1.38	1951	750	9.72
1928	83	1.38	1952	846	10.78
1929	100	1.64	1953	1142	14.31
1930	113	1.83	1954	1311	16.15
1931	125	2.01	1955	1388	16.80
1932	132	2.11	1956	1589	18.89
1933	134	2.13	1957	1776	20.74
1934	132	2.09	1958	2230	25.61
1935	130	2.04	1959	2403	27.03
1936	130	2.03	1960	2464	27.28
1937	133	2.06	1961	2484	27.05
1938	144	2.22	1962	2489	26.69
1939	131	2.00	1963	2509	26.52
1940	188	2.85	1964	2553	26.61
1941	145	2.17	1965	2795	28.77
1942	391	5.80	1966	2968	30.20
1943	771	11.28	1967	3319	33.41
1944	1164	16.82	1968	3668	36.55
1945	704	10.06	1969	3863	38.12
1946	785	11.10	1970	4121	40.19
1947	762	10.57			

Table A3-46 Number of Incandescent Lamps Produced in the United States

Year	Production (millions)	Production (per capita)	Year	Production (millions)	Production (per capita)
1899	25	0.34	1951	1070	6.94
1905	113	1.35	1952	864	5.50
1910	67	0.73	1953	1028	6.44
1914	89	0.90	1954	960	5.91
1919	225	2.15	1955	1057	6.40
1921	155	1.43	1956	1132	6.73
1923	233	2.08	1957	1112	6.49
1925	267	2.31	1958	1052	6.04
1927	335	2.81	1959	1212	6.82
1929	352	2.89	1960	1142	6.32
1931	320	2.58	1961	1155	6.29
1933	306	2.43	1962	1238	6.64
1935	388	3.05	1963	1254	6.63
1937	501	3.88	1964	1264	6.59
1939	517	3.95	1965	1320	6.79
1945	787	5.62	1966	1394	7.09
1946	774	5.47	1967	1391	7.00
1947	999	6.93	1968	1467	7.31
1948	1030	7.02	1969	1476	7.28
1949	975	6.54	1970	1582	7.72
1950	1200	7.91			

Table A3-47 Total and per Capita Number of Registered Automobiles in the United States

Year	Autos Registered (millions)	Registered Autos per Person	Year	Autos Registered (millions)	Registered Autos per Person
1900	0.008	0.00010	1944	25.57	0.1847
1901	0.014	0.00018	1945	25.80	0.1844
1902	0.023	0.00029	1946	28.22	0.1996
1903	0.033	0.00040	1947	30.85	0.2140
1904	0.055	0.00066	1948	33.36	0.2275
1905	0.077	0.00092	1949	36.46	0.2444
1906	0.106	0.00123	1950	40.34	0.2659
1907	0.140	0.00161	1951	42.69	0.2767
1908	0.194	0.00219	1952	43.83	0.2793
1909	0.306	0.00338	1953	46.43	0.2910
1910	0.458	0.00496	1954	48.47	0.2985
1911	0.619	0.00659	1955	52.14	0.3155
1912	0.902	0.00945	1956	54.21	0.3223
1913	1.190	0.01224	1957	55.92	0.3265
1914	1.664	0.01679	1958	56.89	0.3267
1915	2.332	0.02320	1959	59.45	0.3343
1916	3.368	0.03303	1960	61.68	0.3414
1917	4.727	0.04578	1961	63.42	0.3452
1918	5.555	0.05382	1962	66.08	0.3543
1919	6.679	0.06391	1963	69.06	0.3649
1920	8.132	0.07638	1964	71.98	0.3751
1921	9.212	0.08487	1965	75.26	0.3873
1922	10.70	0.09727	1966	78.12	0.3975
1923	13.25	0.1184	1967	80.41	0.4047
1924	15.44	0.1353	1968	83.59	0.4165
1925	17.48	0.1509	1969	86.86	0.4286
1926	19.27	0.1641	1970	89.28	0.4354
1927	20.19	0.1696	1971	92.70	0.4464
1928	21.36	0.1773	1972	97.10	0.4626
1929	23.12	0.1899	1973	102.00	0.4813
1930	23.03	0.1870	1974	104.80	0.4901
1931	22.40	0.1804	1975	106.70	0.4940
1932	20.90	0.1673	1976	110.20	0.5054
1933	20.66	0.1644	1977	112.30	0.5099
1934	21.54	0.1703	1978	116.60	0.5169
1935	22.57	0.1772	1979	118.50	0.5265
1936	24.18	0.1887	1980	121.60	0.5339
1937	25.47	0.1975	1981	123.10	0.5351
1938	25.25	0.1943	1982	123.70	0.5324
1939	26.23	0.2002	1983	126.60	0.5398
1940	27.47	0.2079	1984	128.10	0.5405
1941	29.26	0.2194	1985	131.90	0.5512
1942	27.97	0.2074	1986	135.40	0.5604
1943	36.01	0.2633	1987	139.00	0.5699

Table A3-48 Telephones in Service in the United States (per thousand people)

Year	Phones (per 1000)	Year	Phones (per 1000)	Year	Phones (per 1000)
1876	0.1	1912	90.7	1948	258.1
1877	0.2	1913	97.2	1949	270.4
1878	0.6	1914	100.6	1950	280.9
1879	0.9	1915	103.9	1951	292.9
1880	1.1	1916	109.5	1952	303.3
1881	1.4	1917	112.7	1953	312.7
1882	1.8	1918	115.2	1954	322.1
1883	2.3	1919	119.7	1955	337.2
1884	2.6	1920	123.4	1956	354.5
1885	2.7	1921	126.4	1957	368.2
1886	2.9	1922	128.7	1958	379.3
1887	3.0	1923	135.4	1959	394.8
1888	3.2	1924	139.2	1960	407.8
1889	3.4	1925	144.6	1961	418.0
1890	3.6	1926	149.5	1962	430.7
1891	3.7	1927	153.9	1963	442.9
1892	3.9	1928	158.9	1964	459.5
1893	3.9	1929	163.1	1965	479.0
1894	4.1	1930	162.6	1966	499.6
1895	4.8	1931	157.5	1967	519.3
1896	5.7	1932	138.5	1968	541.5
1897	7.1	1933	132.0	1969	565.2
1898	9.2	1934	133.0	1970	583.4
1899	13.3	1935	136.4	1971	N.A.
1900	17.6	1936	143.5	1972	N.A.
1901	23.0	1937	150.4	1973	N.A.
1902	29.0	1938	153.0	1974	N.A.
1903	34.5	1939	158.3	1975	661
1904	40.4	1940	165.1	1976	N.A.
1905	48.8	1941	175.3	1977	N.A.
1906	57.2	1942	183.4	1978	N.A.
1907	69.6	1943	191.6	1979	747
1908	72.4	1944	192.9	1980	758
1909	76.5	1945	198.1	1981	755
1910	82.0	1946	221.3	1982	715
1911	88.3	1947	239.7		

Table A3-49 Amount of Rayon and Acetate Yarns Available to United States Mills, Including Domestic Shipments and Imports

Year	Millions of Lb	Lb per Capita	Year	Millions of Lb	Lb per Capita
1911	2.1	0.022	1941	452.5	3.392
1912	2.9	0.030	1942	468.8	3.476
1913	4.0	0.041	1943	494.2	3.614
1914	5.1	0.051	1944	539.1	3.895
1915	6.6	0.066	1945	602.4	4.305
1916	6.6	0.065	1946	666.5	4.714
1917	6.8	0.066	1947	729.3	5.060
1918	6.0	0.058	1948	946.7	6.456
1919	9.3	0.089	1949	782.8	5.247
1920	8.7	0.082	1950	955.6	6.300
1921	19.8	0.182	1951	865.5	5.610
1922	25.0	0.227	1952	845.2	5.385
1923	32.8	0.293	1953	865.6	5.425
1924	42.4	0.372	1954	721.2	4.441
1925	58.4	0.504	1955	857.7	5.190
1926	60.9	0.519	1956	727.0	4.322
1927	100.1	0.841	1957	685.8	4.004
1928	100.3	0.832	1958	643.5	3.695
1929	131.8	1.082	1959	722.2	4.061
1930	118.4	0.961	1960	624.9	3.459
1931	157.8	1.271	1961	627.8	3.418
1932	152.1	1.217	1962	668.9	3.586
1933	211.9	1.686	1963	687.4	3.632
1934	194.8	1.540	1964	768.6	4.005
1935	252.8	1.985	1965	782.5	4.027
1936	297.6	2.322	1966	780.8	3.972
1937	267.2	2.072	1967	739.8	3.723
1938	274.1	2.109	1968	794.5	3.959
1939	359.9	2.747	1969	743.5	3.668
1940	388.8	2.943	1970	699.6	3.412

Table A3-50 Steel-making Capacity in the United States (thousands of short tons and per capita capacity in tons per person)

Year	Capacity	Tons/Person	Year	Capacity	Tons/Person
1887	5852	0.0991	1915	44454	0.4421
1889	7192	0.1168	1916	49266	0.4832
1892	8332	0.1272	1917	53914	0.5221
1894	10780	0.1579	1918	57083	0.5531
1896	13236	0.1860	1919	59174	0.5662
1898	15639	0.2109	1920	60220	0.5657
1901	23276	0.3000	1921	61928	0.5706
1904	26919	0.3276	1922	63135	0.5737
1908	36545	0.4120	1923	63383	0.5662
1914	42678	0.4306	1924	64137	0.5621

Table A3-50 (*continued*)

Year	Capacity	Tons/Person	Year	Capacity	Tons/Person
1925	65962	0.5695	1943	90589	0.6625
1926	63925	0.5445	1944	93854	0.6782
1927	65344	0.5490	1945	95505	0.6825
1928	66960	0.5556	1946	91891	0.6499
1929	69584	0.5714	1947	91241	0.6331
1930	71042	0.5767	1948	94233	0.6427
1931	75328	0.6068	1949	96121	0.6443
1932	76898	0.6154	1950	99983	0.6592
1933	76767	0.6108	1951	104230	0.6756
1934	78128	0.6177	1952	108588	0.6918
1935	78452	0.6160	1953	117547	0.7367
1936	78164	0.6098	1954	124330	0.7656
1937	78148	0.6060	1955	125828	0.7613
1938	80186	0.6170	1956	128363	0.7631
1939	81829	0.6245	1957	133459	0.7792
1940	81619	0.6178	1958	140473	0.8067
1941	85158	0.6384	1959	147634	0.8302
1942	88887	0.6591	1960	148571	0.8223

Table A3-51 Diffusion of Telegraphy in the United States (total messages in thousands and per capita in messages per person)

Year	Messages	Per Capita	Year	Messages	Per Capita
1867	5879	0.167	1889	54108	0.912
1868	6405	0.178	1890	55879	0.886
1869	7935	0.215	1891	59148	0.953
1870	9158	0.229	1892	62387	0.983
1871	10646	0.274	1893	66592	1.026
1872	12444	0.313	1894	58632	0.884
1873	14457	0.355	1895	58307	0.838
1874	16239	0.389	1896	58760	0.848
1875	17154	0.381	1897	58152	0.821
1876	18730	0.427	1898	62174	0.859
1877	21159	0.471	1899	61398	0.831
1878	23919	0.520	1900	63168	0.830
1879	25070	0.532	1901	65657	0.846
1880	29216	0.581	1902	69375	0.876
1881	32500	0.658	1903	69791	0.866
1882	38842	0.769	1904	67904	0.826
1883	41181	0.796	1905	67477	0.805
1884	42076	0.795	1906	71847	0.841
1885	42097	0.743	1907	74805	0.860
1886	43290	0.781	1908	62371	0.703
1887	47395	0.836	1909	68053	0.752
1888	51464	0.887	1910	75135	0.813

Table A3-52 Diffusion of Internal Combustion Farm Tractors in the United States (thousands of tractors and tractors per million people)

Year	Tractors (thousands)	Tractors per Million	Year	Tractors (thousands)	Tractors per Million
1910	1	0.01	1941	1665	12.48
1911	4	0.04	1942	1860	13.79
1912	8	0.08	1943	2055	15.03
1913	14	0.14	1944	2160	15.61
1914	17	0.17	1945	2354	16.82
1915	25	0.25	1946	2480	17.54
1916	37	0.36	1947	2613	18.13
1917	51	0.49	1948	2821	19.24
1918	85	0.82	1949	3123	20.93
1919	158	1.51	1950	3394	22.38
1920	246	2.31	1951	3678	23.84
1921	343	3.16	1952	3907	24.89
1922	372	3.38	1953	4100	25.69
1923	428	3.82	1954	4243	26.13
1924	496	4.35	1955	4345	26.29
1925	549	4.74	1956	4480	26.63
1926	621	5.29	1957	4570	26.68
1927	693	5.82	1958	4620	26.53
1928	782	6.49	1959	4673	26.28
1929	827	6.79	1960	4685	25.93
1930	920	7.47	1961	4695	25.56
1931	997	8.03	1962	4710	25.25
1932	1022	8.18	1963	4730	24.99
1933	1019	8.11	1964	4755	24.78
1934	1016	8.03	1965	4783	24.62
1935	1048	8.23	1966	4800	24.42
1936	1125	8.78	1967	4815	24.23
1937	1230	9.54	1968	4822	24.03
1938	1370	10.54	1969	4810	23.73
1939	1445	11.03	1970	4790	23.36
1940	1567	11.86			

Table A3-53 Diffusion of Television in United States Households

Year	Thousands of Households with TV	Percent Households with TV
1946	8	0.02
1947	14	0.04
1948	172	0.42
1949	940	2.27
1950	3875	8.90
1951	10320	23.96
1952	15300	34.81
1953	20400	45.48
1954	26000	56.79
1955	30700	64.13
1956	34900	71.37
1957	38900	78.31
1958	42924	85.04
1959	43950	85.45
1960	45750	86.65
1961	47200	88.13
1962	48855	89.21
1963	50300	91.01
1964	51600	91.90
1965	52700	91.75
1966	53850	92.20
1967	55130	93.07
1968	56670	93.19
1969	58250	93.63
1970	59550	93.93
1975	68500	96.32
1980	76300	94.46
1981	79900	N.A.
1982	81500	97.57
1983	83300	99.26
1984	83800	98.12
1985	84900	97.82

Table A3-54 Diffusion of Color Television in the United States

Year	Percent Households	Year	Percent Households
1960	0.57	1979	85.41
1965	4.87	1980	78.49
1970	32.96	1981	N.A.
1973	63.75	1982	85.48
1975	65.94	1983	88.06
1976	73.82	1984	88.75
1977	77.24	1985	89.53
1978	80.94		

Table A3-55 The Substitution of Power for Sail in the United States Merchant Fleet

Year	Thousands of Tons Power	Thousands of Tons Sail	Fraction Power
1810	1	1424	0.00070
1811	1	1231	0.00081
1812	2	1268	0.00157
1813	3	1164	0.00257
1814	3	1156	0.0026
1815	3	1365	0.0022
1816	6	1366	0.0044
1817	9	1391	0.0064
1818	13	1213	0.0106
1819	17	1243	0.0135
1820	22	1258	0.0172
1821	23	1276	0.0177
1822	23	1304	0.0173
1823	25	1312	0.0187
1824	22	1368	0.0158
1825	23	1400	0.0162
1826	34	1500	0.0222
1827	40	1590	0.0245
1828	39	1702	0.0224
1829	54	1207	0.0428
1830	64	1127	0.0537
1831	69	1198	0.0545
1832	91	1349	0.0632
1833	102	1504	0.0635
1834	123	1636	0.0699
1835	123	1702	0.0674
1836	146	1737	0.0775
1837	155	1742	0.0817
1838	193	1802	0.0967
1839	195	1901	0.0930
1840	202	1978	0.0927
1841	175	1956	0.0821
1842	230	1863	0.110
1843	237	1922	0.110
1844	272	2008	0.119
1845	386	2091	0.156
1846	348	2214	0.136
1847	405	2434	0.143
1848	428	2726	0.136
1849	462	2872	0.139
1850	526	3010	0.149
1851	584	3189	0.155
1852	643	3495	0.155
1853	605	3802	0.137
1854	677	4126	0.141

Table A3-55 (*continued*)

Year	Thousands of Tons Power	Thousands of Tons Sail	Fraction Power
1855	770	4442	0.148
1856	673	4199	0.138
1857	706	4235	0.143
1858	729	4320	0.144
1859	769	4376	0.149
1860	868	4486	0.162
1861	877	4663	0.158
1862	710	4402	0.139
1863	576	4580	0.112
1864	978	4008	0.196
1865	1067	4030	0.209
1866	1084	3227	0.251
1867	1192	3113	0.277
1868	1199	2509	0.323
1869	1104	2400	0.315
1870	1075	2363	0.313
1871	1088	2286	0.322
1872	1112	2325	0.324
1873	1156	2383	0.327
1874	1186	2474	0.324
1875	1169	2585	0.311
1876	1172	2609	0.310
1877	1171	2580	0.312
1878	1168	2521	0.317
1879	1176	2423	0.327
1880	1212	2366	0.339
1881	1265	2350	0.350
1882	1356	2361	0.365
1883	1413	2387	0.372
1884	1466	2414	0.378
1885	1495	2374	0.386
1886	1523	2210	0.408
1887	1543	2170	0.416
1888	1648	2124	0.437
1889	1766	2099	0.457
1890	1859	2109	0.468
1891	2016	2172	0.481
1892	2074	2178	0.488
1893	2183	2118	0.508
1894	2189	2023	0.520
1895	2213	1965	0.530
1896	2307	1928	0.545
1897	2359	1904	0.553
1898	2372	1836	0.564
1899	2476	1825	0.576
1900	2658	1885	0.585

(*continued*)

Table A3-55 The Substitution of Power for Sail in the United States Merchant Fleet (*continued*)

Year	Thousands of Tons Power	Thousands of Tons Sail	Fraction Power
1901	2921	1933	0.602
1902	3177	1942	0.621
1903	3408	1966	0.634
1904	3595	1945	0.649
1905	3741	1962	0.656
1906	3975	1899	0.677
1907	4279	1814	0.702
1908	4711	1761	0.728
1909	4749	1711	0.735
1910	4900	1655	0.748
1911	5074	1598	0.760
1912	5180	1539	0.771
1913	5333	1508	0.780
1914	5428	1433	0.791
1915	5944	1384	0.811
1916	6070	1311	0.822
1917	6433	1278	0.834
1918	7471	1210	0.861
1919	10416	1200	0.897
1920	13823	1272	0.916
1921	15745	1294	0.924
1922	15982	1288	0.925
1923	15821	1254	0.927
1924	15315	1185	0.928
1925	14976	1125	0.930
1926	14848	1092	0.931
1927	14507	989	0.936
1928	14344	915	0.940
1929	14162	825	0.945
1930	13757	757	0.948
1931	13528	673	0.953
1932	13568	625	0.956
1933	12862	563	0.958
1934	12687	500	0.962
1935	12535	441	0.966
1936	12267	379	0.970
1937	12170	312	0.975
1938	12007	261	0.979
1939	11952	221	0.982
1940	11353	200	0.983
1941	11047	182	0.984
1942	11072	166	0.985
1943	14052	142	0.990
1944	23217	129	0.994
1945	30247	115	0.996

Table A3-55 (*continued*)

Year	Thousands of Tons Power	Thousands of Tons Sail	Fraction Power
1946	35928	98	0.997
1947	35149	95	0.997
1948	30469	87	0.997
1949	28323	87	0.997
1950	28327	82	0.997
1951	27424	71	0.997
1952	27459	66	0.998
1953	27507	55	0.998
1954	27631	46	0.998
1955	26792	40	0.999
1956	26251	34	0.999
1957	25785	24	0.999
1958	24599	23	0.999
1959	24333	23	0.999
1960	23553	23	0.999
1961	21175	18	0.999
1962	20076	18	0.999
1963	20079	18	0.999
1964	20018	17	0.999
1965	19730	8	1.000

Table A3-56 Learning Curve Data for a Pressman on a New Two-color Press

Cumulative Output (thousands of impressions)	Impressions per Hour
101	2700
204	2840
286	2920
376	2910
460	3005
582	3100
682	3090
767	3235
852	3460
956	3480
1046	3490
1120	3560

Table A3-57 Efficiency of Incandescent and Fluorescent Lamps (lumens per watt)

Year	Incandescent Efficiency	Year	Lumens from 40-watt Fluorescent	Fluorescent Efficiency
1888	2	1938	1400	35.00
1906	4	1939	1850	46.25
1907	5	1941	2100	52.50
1908	8	1946	2300	57.50
1915	10	1949	2400	60.00
1916	12	1952	2450	61.25
1932	14	1953	2500	62.50
1940	16	1954	2550	63.75
1959	17	1954.5	2600	65.00
		1958	2750	68.75
		1959	2900	72.50
		1962	3000	75.00
		1963	3300	82.50
		1967	3350	83.75

Table A3-58 Efficiency of Artificial Light Sources (lumens per watt)

Source	Year	Efficiency
Candle	1850	0.1
Edison's first lamp	1879	1.6
Cellulose filament lamp	1894	2.6
Mercury arc	1901	12.7
Inert gas filled lamp	1913	19.8
Sodium lamp	1928	20
Mercury lamp	1935	40
Fluorescent lamp	1942	55
Gallium arsenide diode	1962	180

Table A3-59 Efficiency of U.S. Generation of Electricity by Fossil Fuel Generating Plants

Year	Percent Efficiency	Year	Percent Efficiency
1925	13.6	1955	29.2
1926	14.5	1956	29.8
1927	15.1	1957	30.0
1928	15.8	1958	30.8
1929	16.6	1959	31.4
1930	17.2	1960	31.9
1931	18.1	1961	32.3
1932	18.5	1962	32.5
1933	18.8	1963	32.7
1934	19.0	1964	32.8
1935	19.1	1965	32.9
1936	19.2	1966	32.8
1937	19.1	1967	32.8
1938	19.6	1968	32.9
1939	20.4	1969	32.6
1940	20.8	1970	32.5
1941	20.6	1971	32.6
1942	21.2	1972	32.9
1943	21.3	1973	32.9
1944	21.5	1974	32.7
1945	21.6	1975	32.8
1946	21.7	1976	32.9
1947	21.9	1977	32.7
1948	21.7	1978	32.9
1949	22.7	1979	32.9
1950	24.3	1980	32.8
1951	25.0	1981	32.6
1952	25.5	1982	32.7
1953	26.5	1983	32.7
1954	28.0	1984	33.4

Table A3-60 Steam Engine Efficiencies in Mechanical Energy Output (as a percent of thermal energy of fuel)

Engine	Year	Efficiency
Savery	1698	0.5
Newcomen	1712	1.3
Watt	1770	2.8
Watt	1796	4.1
Cornish	1830	10
Cornish	1846	15
Triple expansion	1890	18
Parson's turbine	1910	20
Steam turbine	1950	30
Steam turbine	1955	38

Table A3-61 Average Mileage of Passenger Autos in the United States

Year	Miles/Gal	Year	Miles/Gal
1936	15.04	1960	14.11
1937	15.05	1961	14.23
1938	15.04	1962	14.20
1939	15.05	1963	14.10
1940	15.05	1964	14.09
1941	15.06	1965	13.95
1942	14.85	1966	13.96
1943	14.58	1967	13.91
1944	14.48	1968	13.77
1945	14.56	1969	13.63
1946	14.60	1970	13.58
1947	14.60	1971	13.73
1948	14.63	1972	13.67
1949	14.66	1973	13.29
1950	14.69	1974	13.65
1951	14.72	1975	13.74
1952	14.43	1976	13.93
1953	14.47	1977	14.15
1954	14.73	1978	14.26
1955	14.45	1979	14.49
1956	14.36	1980	15.32
1957	14.44	1981	15.68
1958	14.12	1982	16.25
1959	14.12	1983	16.70

Table A3-62 Distribution of Research Funds at 91 United States Universities in 1985[a]

Start of Range (in millions)	Center of Range (in millions)	Number of Universities	Probability (per million)
30	—	—	—
40	35	11	0.1209
50	45	19	0.2088
60	55	11	0.1209
70	65	9	0.0989
80	75	6	0.0659
90	85	6	0.0659
100	95	6	0.0659
110	105	3	0.0330
120	115	2	0.0220
130	125	1	0.0110
140	135	5	0.0549
150	145	4	0.0440
200	175	4	0.0440
250	225	3	0.0330
400	325	1	0.0110

[a]Data are from a National Science Foundation fact sheet.

Table A3-63 Income Distribution in England in 1688

Number of Families	Income in Millions	Pounds per Family
19626	8.81	449
42960	4.46	104
94000	2.13	22.7
128025	10.9	85.1
256866	9.58	37.3
227440	12.21	53.7
284997	4.27	15.0
313183	2.04	6.5
23489	0.04	1.7

Table A3-64 Size Distribution of U.S. Cities in 1984[a]

Population (in millions)	Number of Cities	Percent of Cities	Probability in Range
0.1	174	100	
0.2	71	40.80	0.5920
0.3	51	29.31	0.1149
0.4	31	17.82	0.1149
0.5	21	12.07	0.0575
0.6	14	8.05	0.0402
0.7	12	6.90	0.0115
0.8	8	4.60	0.0230
0.9	7	4.02	0.0057
1	6	3.45	0.0057
2	3	1.72	0.0172
3	2	1.15	0.0057
4	1	0.57	0.0057

[a]See appendix table listing for description of column headings.

Table A3-65 National and World Levels of Industrialization from 1750 to 1980[a]

Country	1750	1800	1830	1860	1880	1900	1913	1928	1938	1953	1963	1973	1980
Austria-Hungary	7	7	8	11	15	23	32						
Belgium	9	10	14	28	43	56	88	116	89	117	183	291	316
France	9	9	12	20	28	39	59	82	73	95	167	259	277
Germany	8	8	9	15	25	52	85	101	128	144	244	366	395
Italy	8	8	8	10	12	17	26	39	44	61	121	194	231
Russia/U.S.S.R.	6	6	7	8	10	15	20	20	38	73	139	222	252
Spain	7	7	8	11	14	19	22	28	23	31	56	144	159
Sweden	7	8	9	15	24	41	67	84	135	163	262	405	409
Switzerland	7	10	16	26	39	67	87	90	88	167	259	366	354
U.K.	10	16	25	64	87	100	115	122	157	210	253	341	325
Canada	N.A.	5	6	7	10	24	46	82	84	185	237	370	379
U.S.	4	9	14	21	38	69	126	182	167	354	393	604	629
Japan	7	7	7	7	9	12	20	30	51	40	113	310	353
China	8	6	4	4	4	3	3	4	4	5	10	18	24
India	7	6	6	3	2	1	2	3	4	6	8	14	16
Brazil	N.A.	N.A.	N.A.	4	4	5	7	10	10	13	23	42	55
Mexico	N.A.	N.A.	N.A.	5	4	5	7	9	8	12	22	36	41
Developing countries	8	8	11	16	24	35	55	71	81	135	194	315	344
Third World	7	6	6	4	3	2	2	3	4	5	8	14	17
World	7	6	7	7	9	14	21	28	31	48	66	100	103

[a]The industrialization index is defined relative to the United Kingdom taken as 100 in 1900.

Table A3-66 Annual Percentage Growth Rates of Levels of Industrialization (exponential averages)

Time Period	Mid-range Year	U.K.	U.S.	Germany	France	Russia U.S.S.R.	Japan	World	Developing Countries	Third World
1750–1800	1775	0.94	1.62	0.00	0.00	0.00	0.00	−0.31	0.00	−0.3
1800–1830	1815	1.49	1.47	0.39	0.96	0.51	0.00	0.51	1.06	0.00
1830–1860	1845	3.13	1.35	1.70	1.70	0.45	0.00	0.00	1.25	−1.35
1860–1880	1870	1.54	2.97	2.55	1.68	1.12	1.26	1.26	2.03	−1.44
1880–1900	1890	0.70	2.98	3.66	1.66	2.03	1.44	2.21	1.89	−2.03
1900–1913	1906.5	1.08	4.63	3.78	3.18	2.21	3.93	3.12	3.48	0.00
1913–1928	1920.5	0.39	2.45	1.15	2.19	0.00	2.70	1.92	1.70	2.70
1928–1938	1933	2.52	−0.86	2.37	−1.16	6.42	5.31	1.02	1.32	2.88
1938–1953	1945.5	1.94	5.01	0.79	1.76	4.35	−1.62	2.91	3.41	1.49
1952–1963	1958	1.86	1.05	5.27	5.64	6.44	10.39	3.18	3.63	4.70
1963–1973	1968	2.98	4.30	4.05	4.39	4.68	10.09	4.16	4.85	5.60
1973–1980	1976.5	−0.69	0.58	1.09	0.96	1.81	1.86	0.42	1.26	2.77

Table A3-67 Ratios of Levels of Industrialization of Rich to Poor Countries

Year	Developed to Third World	U.K. to World	U.S. to World	U.K. to Third World	U.S. to Third World
1750	1.14	1.43	0.57	1.43	0.57
1800	1.33	2.67	1.50	2.67	1.50
1830	1.83	3.57	2.00	4.17	2.33
1860	4.00	9.14	3.00	16.00	5.25
1880	8.00	9.67	4.22	29.00	12.67
1900	17.50	7.14	4.93	50.00	34.50
1913	27.50	5.48	6.00	57.50	63.00
1928	23.67	4.36	6.50	40.67	60.67
1938	20.25	5.06	5.39	39.25	41.75
1953	27.00	4.38	7.38	42.00	70.80
1963	24.25	3.83	5.95	31.63	49.13
1973	22.50	3.41	6.04	24.36	43.14
1980	20.24	3.16	6.11	19.12	37.00

Table A3-68 The Takeoff to Exponential Growth of Industrialization[a]

Country	Take-off Date (curve)	Take-off Date (L.A.G.)	A (growth rate)	B	Correlation Coefficient	Standard Deviation
Austria-Hungary	1830	1815	0.0166	28.449	0.9995	0.0875
Belgium	1800	1775	0.0188	31.713	0.9992	0.1710
France	1830	1775	0.0206	35.432	0.9992	0.1735
Germany	1830	1815	0.0258	45.138	0.9995	0.1383
Italy	1860	1845	0.0270	48.238	0.9983	0.2296
Russia	1830	1815	0.0175	30.442	0.9998	0.0509
U.S.S.R.	1928	1920	0.0499	93.123	0.9998	0.0918
Spain	1830	1815	0.0128	21.418	0.9999	0.0304
Spain	1938	1932	0.0510	95.895	0.9988	0.2037
Sweden	1830	1815	0.0267	46.972	0.9996	0.1316
Switzerland	1800	1775	0.0201	34.008	0.9992	0.1840
U.K.	1750	1775	0.0158	25.405	0.9995	0.1476
Canada	1860	1830	0.0354	64.084	0.9996	0.1339
U.S.	1750	1775	0.0234	39.942	0.9986	0.2475
Japan	1860	1870	0.0249	44.616	0.9983	0.1702
Japan	1953	1950	0.0843	160.800	0.9994	0.1776
China	1953	1953	0.0586	112.775	0.9997	0.0576
India	1913	1913	0.0315	59.574	0.9991	0.0805
Brazil	1900	1900	0.0285	52.686	0.9970	0.2193
Mexico	1900	1905	0.0261	48.068	0.9967	0.2196

[a]For each country the data were fitted to an exponential equation of the form industrial level $= \exp [A^* (\text{Year}) - B]$.

References

Abernathay, William J. and James M. Utterback; (1978); "Patterns of Industrial Innovation"; *Technology Review*, Vol. 80, No. 7, pp. 1–9.

Adams, W. and J. Dirlam; (1966); "Big Steel, Invention and Innovation"; *Quart. J. of Econ.*, Vol. LXXX, No. 20, p. 175.

Alchian, A.; (1963); "Reliability of Progress Curves in Aircraft Production"; *Econometrica*, Oct.

Bairoch, Paul; (1982); "International Industrialization Levels from 1750 to 1980"; *Jour. European Economic History*, Vol. 11, pp. 269–333.

Baker, Norman R., Jack Siegman, and Jon Larson; (1971); "The Relation Between Certain Characteristics of Industrial Proposals and Their Subsequent Disposition"; *IEEE Trans. on Eng. Manag.*, Vol. EM-18, No. 4, pp. 118–124.

Baloff, Nicholas; (1966*a*); "The Learning Curve: Some Controversial Issues"; *J. Indust. Economics*, Vol. 14, pp. 275–282.

Baloff, Nicholas; (1966*b*); "Startups in Machine Intensive Production Systems"; *J. Ind. Eng.*, Vol. XVII, pp. 25–32.

Baloff, Nicholas; (1967); "Estimating the Parameters of the Startup Model—An Empirical Approach"; *J. Ind. Eng.*, Vol. XVIII, pp. 248–253.

Baloff, Nicholas; (1971); "Extension of the Learning Curve—Some Empirical Results"; *Operations Research*, Vol. 22, pp. 329–340.

Barnouw, Erik; (1966); *A Tower of Babel*, Oxford University Press, New York.

Blackman, A. Wade, Jr.; (1974); "The Market Dynamics of Technological Substitution"; *Tech. Forecast. and Soc. Change*, Vol. 6, pp. 41–63.

Brobst, D. A. and W. P. Pratt; (1973); "United States Mineral Resources"; Geological Survey Professional Paper 820; U.S. Department of the Interior; Government Printing Office, Washington, D.C.

Casetti, E.; (1969); "Why Do Diffusion Processes Conform to Logistic Trends?"; *Geog. Analysis*, Vol. 1. pp. 101–105.

Casetti, E. and R. K. Kemple; (1969); "Concerning the Testing of Spatial Diffusion Hypothesis"; *Geog. Analysis*, Vol. 1, pp. 254–259.

Cerveny, R. P. and K. E. Knight; (1983); "Performance of Computers," in *Encyclopedia of Computer Science and Engineering*, 2nd edition, pp. 1127–1131.

Coale, A. J.; (1983); "Fertility in Less Developed Countries"; *Science*, Vol. 221, pp. 828–832.

Cole, R. R.; (1958); "Increasing Utilization of the Cost-Quantity Relationship in Manufacturing"; *J. Ind. Eng.*, Vol. 9, pp. 122–128.

Conway, R. and A. Schultz, Jr.; (1959); "The Manufacturing Progress Function"; *J. Ind. Eng.*, Vol. 10, pp. 39–53.

Cook, Earl; (1976); "Limits to Exploitation of Nonrenewable Resources"; *Science*; Vol. 191, pp. 677–682.

Cooper, Arnold and Dan Schendel; (1976); "Strategic Responses to Technological Threats"; *Business Horizons*, Feb.

Daniels, Guy and J. M. Snyder; (1984); *Translation of N. Kondratieff's Papers*; Richardson and Snyder, New York.

Davies, Stephen; (1979); *The Diffusion of Process Innovations*; Cambridge University Press, Cambridge, England.

Duijn, J. J. van; (1983); *The Long Wave in Economic Life*; George Allen & Unwin, London.

Enos, J. L.; (1962); *Petroleum Progress and Profits*; M.I.T. Press, Cambridge, Mass.

Erickson, Don V.; (1973); *Armstrong's Fight for FM Broadcasting*; University of Alabama Press, University, Ala.

Fenichel, Allen H.; (1964) "Quantitative Analysis of the Growth and Diffusion of Steam Power in Manufacturing in the United States, 1839–1919"; Ph.D. Dissertation, University of Pennsylvania.

Fisher, J. C. and R. H. Pry; (1971); "A Simple Substitution Model of Technological Change"; *Tech. Forecast. and Soc. Change*, Vol. 3, No. 1, pp. 75–88.

Flamm, Kenneth; (1988*a*); *Creating the Computer: Government, Industry and High Technology*; Brookings Institution, Washington, D.C.

Flamm, Kenneth; (1988*b*) *Targeting the Computer: Government Support and International Competition*; Brookings Institution, Washington, D.C.

Fliegel, F. C. and J. E. Klein; (1966); "Attributes of Innovations as Factors in Diffusion"; *Amer. J. Soc.*, Vol. 72, pp. 235–248.

Floyd, A. (1968); "Trend Forecasting: A Methodology for Figure of Merit"; *Proc. First Annual Tech. and Manag. Conf.*; J. Bright, Ed.; Prentice-Hall, Englewood Cliffs, N.J.

Foster, Richard N.; (1979); "Strategic Management of Technology"; speech delivered at an Exposition of Specialized Chemical Companies.

Fowler, John M.; (1977); "Energy and the Environment" in *Perspectives on Energy*; L. C. Ruedisili and M. W. Firebaugh, Eds.; Oxford University Press, New York (1978), p. 16.

Freeman C., J. Clark and Luc Soete; (1982); *Unemployment and Technical Innovation*; Greenwood Press, Westport, Conn.

Galbraith, J. K.; (1952); *American Capitalism*; Houghton Mifflin, Boston, Mass.

Girifalco, L. A.: (1983); "Malthus Modified"; *J. Resource Management and Technology*, Vol. 12, No. 3, pp. 191–203.

Girifalco, L. A. and S. Simozar: (1985); "Multilevel Technological Substitution"; *J. Resource Management and Tech.*, Vol. 14, No. 1, pp. 45–54.

Goeller, H. E. and A. M. Weinberg; (1976); "The Age of Substitutability: What Do We Do When the Mercury Runs Out?"; *Science*, Vol. 191, pp. 691–699.

Gold, Bela; (1981); "Technological Diffusion in Industry: Research Needs and Shortcomings"; *J. Ind. Econo.*, Vol. XXIX, No. 3, pp. 247–269.

Gold, Bela, W. S. Pierce, G. Rosegger and M. Perlman: (1984); *Technological Progress and Industrial Leadership: The Growth of the U.S. Steel Industry*; Lexington Books, D. C. Heath & Co., Lexington, Mass.

Griliches, Z.; (1957); "Hybrid Corn: An Exploration in the Economics of Technological Change"; *Econometrica*, Vol. 25, pp. 501–522.

Hamberg, Daniel; (1966); *R&D: Essays on the Economics of Research and Development*; Random House, New York.

Hewett, D. F.; (1929); "Cycles in Metal Production"; *Transaction AIME*, pp. 65–93, and *AIME Tech. Publication 183*, 31 pages.

Hirsch, W. Z.; (1952); "Manufacturing Progress Functions"; *Rev. of Econ. and Statist.*; Vol. 34, May, pp. 143–145.

Hubbert, M. K.; (1978); "Survey of World Energy Resources" in *Perspectives on Energy*; L. C. Ruedisili and M. W. Firebaugh, Eds.; Oxford University Press, New York (1978).

Hughes, Thomas P.: (1983); *Networks of Power: Electrification in Western Society*; Johns Hopkins University Press, Baltimore.

Jewkes, J., D. Sawyers and R. Stillerman; (1969); *The Sources of Invention*, 2nd edition, Macmillan, New York.

Kakela, Peter J.; (1981); "Iron Ore: From Depletion to Abundance"; *Science*, Vol. 212, pp. 132–136.

Keyfitz, N.; (1982); "The Demographic State of the World" in *Outlook for Science and Technology*, a report of The National Research Council; W. H. Freeman, San Francisco.

Knight, K. E.; (1966); "Changes in Computer Performance"; *Datamation*, Vol. 12, No. 9, pp. 40–54.

Knight, K. E.; (1968); "Evolving Computer Performance: 1963–1967"; *Datamation*, Vol. 14, No. 1, pp. 31–35.

Knight, K. E.; (1973); "An Application of Technological Forecasting to the Computer Industry" in *A Guide to Practical Technological Forecasting*; J. R. Bright and M. F. Schoeman, Eds.; Prentice-Hall, Englewood Cliffs, N.J.

Koeppl, Gerald W.; (1982); *Putnam's Power from the Wind*, 2nd edition; Van Nostrand Reinhold, New York.

Kravis, Irving, Alan Heston and Robert Summers; (1982); *World Product and Income: International Comparisons of Real Gross Product*; Johns Hopkins University Press, Baltimore.

Kravis, Irving B. and Robert E. Lipsey; (1984); "The Diffusion of Growth in the World Economy" in *International Comparisons of Productivity and Causes of Slowdown*; John W. Kendrick, Ed.; Ballinger Publishing Company, Cambridge, Mass., for the American Enterprise Institute.

Kuznets, Simon; (1930); *Secular Movements in Production and Prices*; Houghton Mifflin, New York.

Landes, David S.; (1969); *The Unbound Prometheus: Technological Change and Industrial Development in Western Europe from 1750 to the Present*; Cambridge University Press, Cambridge, England.

Lehfeldt, R. A.; (1916); "The Normal Law of Progress." *J. Royal Statistic. Soc.*, May, pp. 329–332.

Levy, F. K.; (1965); "Adaptation in the Production Process"; *Management Science*, April, pp. 136–154.

Lipsey, Robert E. and Irving B. Kravis; (1987); "Saving and Economic Growth: Is the United States Really Falling Behind?"; Report No. 901; The Conference Board

Inc.; (jointly sponsored by the American Council of Life Insurance and the Conference Board).

Maddala, G. S. and P. T. Knight; (1967); "International Diffusion of Technical Change: A Case Study of the Oxygen Steel Making Process"; *Econ. Journ.*, Vol. 77, No. 307, pp. 531–538.

Maddison, Angus; (1982); *Phases of Capitalist Development*; Oxford University Press, Oxford, England.

Mansfield, Edwin; (1961); "Technical Change and the Rate of Imitation"; *Econometrica*, Vol. 29, No. 4, pp. 741–766.

Mansfield, Edwin; (1964); "Industrial Research and Development Expenditures: Determinants, Prospects and Relation to Size of Firm and Inventive Output"; *J. Pol. Economy*, Vol. 72, pp. 334–336.

Mansfield, Edwin; (1968*a*); *The Economics of Technological Change*; W. W. Norton & Co., New York.

Mansfield, Edwin; (1968*b*); *Industrial Research and Technological Innovation*; W. W. Norton & Co., Inc., New York.

Mansfield, Edwin; (1972); "Contribution of R&D to Economic Growth in the U.S."; *Science*, Vol. 175 (Feb.), pp. 477–486.

Mansfield, Edwin; (1980); "Research and Development, Productivity and Inflation"; *Science*, Vol. 209, No. 4, pp. 1091–1093.

Mansfield, Edwin, John Rapoport, Anthony Romeo, Edmond Villani, Samuel Wagner and Frank Husic; (1977); *The Production and Application of New Technology*; W. W. Norton & Co., New York.

Mansfield, Edwin, Anthony Romeo, Mark Schwartz, David Teece, Samuel Wagner and Peter Brach; (1982); *Technology Transfer, Productivity and Economic Policy*; W. W. Norton & Co., New York.

Martino, J. P.; (1983); *Technological Forecasting for Decision Making*, 2nd edition; Elsevier Science Publishing Co., New York.

Marx, Karl; The *Communist Manifesto* was written with F. Engels and appeared in 1848. *Das Kapital* appeared in 1867. Nathan Rosenberg gives an illuminating treatment of Marx's view of technology in his *Inside the Black Box: Technology and Economics*; Cambridge University Press, Cambridge, England (1982).

McKelvey, V. E.; (1972); "Mineral Resources and Public Policy"; *Amer. Scientist*, Vol. 60, No. 1, pp. 32–40.

Meadows, D. H., D. L. Meadows, J. Randers and W. W. Behrens III; (1972); *The Limits to Growth*, 2nd edition; New American Library, New York.

Mensch, Gerhard; (1979); *Stalemate in Technology*; Ballinger Publishing, Cambridge, Mass.

Mueller, W. F.; (1962); "The Origins of the Basic Inventions Underlying Du Pont's Major Product and Process Innovations, 1920 to 1950" in *The Rate and Direction of Inventive Activity*; National Bureau of Economic Research, pp. 323–346.

Nasbeth, L. and G. F. Ray; (1974); *The Diffusion of New Industrial Processes*; Cambridge University Press, Cambridge, England.

Parmenter, Robert H.; (1964); "Research Project Selection (An Industrial Researcher's View); *Research Management*, Vol. VII, No. 3, pp. 225–233.

Pearl, R.; (1925); *The Biology of Population Growth*; Alfred A. Knopf, New York.

Peck, M. J. (1962); "Invention in the Postwar Aluminum Industry" in *The Rate and*

Direction of Inventive Activity; R. R. Nelson, Ed.; pp. 279–298; Princeton University Press, Princeton, N.J.

Pemberton, H. E.; (1936); "The Curve of Culture Diffusion Rate"; *Amer. Sociology Rev.*, Vol. 1, No. 4, pp. 547–556.

Penner, S. S. and L. Icerman; (1981); *Energy, Volume I: Demands, Resources, Impact, Technology and Policy*, 2nd edition; Addison Wesley Publishing Co., Reading, Mass.

Petrov, B.; (1982); "The Advent of the Business Portfolio"; *J. of Bus. Strategy*, Fall.

Phister, Montgomery Jr.; (1979); *Data Processing Technology and Economics*; Digital Press and Santa Monica Publishing Co., Bedford, Mass.

Rose, David J.; (1986); *Learning About Energy*; Plenum Press, New York and London.

Rostow, W. W.; (1980); *Why the Poor get Richer and the Rich Slow Down*; University of Texas Press, Austin.

Sahal, Devendra; (1981); *Patterns of Technological Innovation*; Addison-Wesley Publishing Co., Reading, Mass.

Scherer, Frederick; (1965); "Firm Size, Opportunity and the Output of Patented Inventions"; *Amer. Econ. Rev.*, pp. 1097–1125.

Schultz, T. Paul; (1981); *Economics of Population*; Addison-Wesley Publishing Co., Reading, Mass.

Schumpeter, J. A.: (1934); *The Theory of Economic Development*; Harvard University Press, Cambridge, Mass.

Sharif M. N. and C. Kabir; (1976); "A Generalized Model for Forecasting Technological Substitution"; *Tech. Forecast. and Soc. Change*; Vol. 8, pp. 353–364.

Singer, Charles, E. J. Holmyard, A. R. Hall and Trevor I. Williams, Eds. (1958); *A History of Technology*; Oxford University Press, New York and London.

Sorokin, Pitirim; (1957); *Social and Cultural Dynamics*; Porter Sargent, Boston, Mass. {a one-volume abridgement of his four-volume work, 1937–1941}.

Spengler, Oswald; (1926); *The Decline of the West*, a translation by Charles Francis of the 1923 original published in Munich; Alfred A. Knopf, New York.

Summers, Robert and Alan Heston; (1984) "Improved International Comparison of Real Product and its Composition: 1950–1980"; *Review of Income and Wealth*, June, pp. 207–262.

Summers, Robert and Alan Heston; (1988); "A New Set of International Comparisons of Real Products and Price Levels Estimates for 130 Countries, 1950–1985"; *The Rev. of Income and Wealth*, Series 34, No. 1, pp. 1–25. (Accompanied by three data diskettes; compatible with spreadsheet programs.)

Toynbee, Arnold J.; (1946); *A Study of History*, abridgement of volumes I–VI by D. C. Somervell; Oxford University Press, Oxford, England.

Usher, A.; (1954); *A History of Mechanical Inventions*; Harvard University Press, Cambridge, Mass.

Zacharia, K. C. and My T. Vu; (1988); *World Populations Projections*, 1987–88 edition; Johns Hopkins University Press, Baltimore.

Additional Readings

Agriculture

Gray, R. B. and E. M. Dieffenbach; (1957); "Fifty Years of Tractor Development in the U.S.A."; *Agricultural Engineering*, Vol. 38, pp. 388–397.

Zink, W. L.; (1931); "The Agricultural Power Takeoff"; *Agricultural Engineering*, Vol. 12, pp. 209–210.

Automation

Amber, G. H.; (1962); *Anatomy of Automation;* Prentice-Hall, Englewood Cliffs, N.J.

Asfall, C. Ray; (1985); *Robots and Manufacturing Automation;* John Wiley & Sons, New York.

Baranson, Jack; (1983); *Robots in Manufacturing;* Lomond Publishing Co., Mt. Airy, Md.

Diebold, J.; (1952); *The Advent of the Automated Factory;* Van Nostrand, New York.

Dorf, Richard; (1984); *Robotics and Automated Manufacturing;* Reston Publishing Co., Reston, Va.

Jerome, Harry: (1934); *Mechanization in Industry;* National Bureau of Economic Research, New York.

Automobiles

Abernathy, William, J.; (1982); *The Productivity Dilemma: Roadblock to Innovation in the Automobile Industry;* Johns Hopkins University Press, Baltimore.

White, L. J.; (1971); *The Automobile Industry Since 1945;* Harvard University Press, Cambridge, Mass.

Aviation

Archer, R. D.; (1967); "Evolution of the Supersonic Shape"; *Space/Aeronautics,* Vol. 48, pp. 89–104.

Constant, Edward W.; (1980); *The Origin of the Turbojet Revolution;* Johns Hopkins University Press, Baltimore.

Gibbs-Smith, Charles H.; (1965); *The Invention of the Airplane;* Toplinger Press, New York.

Miller, Ronald and David Sawyers: (1970); *The Technical Development of Modern Aviation;* Praeger, New York.
Phillips, Almarin; (1971); *Technology and Market Structure: A Study of the Aircraft Industry;* Heath–Lexington Books, Lexington, Mass.
Reguero, M. A.; (1957); *An Economic Study of the Military Airframe Industry;* Wright Patterson Air Force Base, Ohio.
Sahal, D.; (1976*a*); "On the Conception and Measurement of Trade-Offs in Engineering Systems: A Case Study of the Aircraft Design Process"; *J. Tech. Forecast. and Soc. Change,* Vol. 8, pp. 371–384.
Sahal, D.; (1976*b*); "On the Conception and Measurement of Technology: A Case Study of the Aircraft Design Process"; *J. Tech. Forecast. and Soc. Change,* Vol. 8, pp. 385–400.
Schlaifer, R. and S. D. Heron; (1950); *The Development of Aircraft Engines and Fuels;* Harvard Business School, Cambridge, Mass.
Wright, T. P.; (1936); "Factors Affecting the Cost of Airplanes"; *J. Aeronautical Science,* Vol. 3, pp. 122–128.

Chemicals

Beer, John J.; (1959); *The Emergence of the German Dye Industry;* University of Illinois Press, Urbana, Ill.
Enos, J. L.; (1962); *Petroleum Progress and Profits;* MIT. Press, Cambridge, Mass.
Multauf, R. P.; (1966); *The Origins of Chemistry;* Oldburne, London.

Communications

Aitken, J. G.; (1985); *The Continuous Wave: Technology and American Radio 1900–32;* Princeton University Press, Princeton, N.J.
Blade, G. G.; (1926); *History of Radio, Telegraphy and Telephony;* Radio Press Ltd., London.
Hybels, Saundra and Dana Ulloth; (1978); *Broadcasting: An Introduction to Radio and Television;* Van Nostrand Reinhold, New York.
Lichty, Lawrence W. and Malachi C. Topping; (1975); *American Broadcasting: A Source Book on the History of Radio and Television;* Hastings House, New York.
MacLaurin, W. R. (1949); *Invention and Innovation in the Radio Industry;* Macmillan, New York.
Pierce, J. P.; (1968); *The Beginning of Satellite Communication;* San Francisco Press, San Francisco, Ca.
Stearn, Gerald; (1977); *The Telephone Book;* Riverwood Publishers, New York.

Computers

Andrews, E. G.; (1963); "Telephone Switching and the Early Bell Laboratories Computers"; *Bell System Technical Journal,* March, pp. 341–353; reprinted in *Annals of Hist. Comp.;* (1982); Vol. 4, pp. 13–19.
Arden, Bruce W., Ed.; (1980); *What Can Be Automated?;* MIT Press, Cambridge, Mass.

Becker, Joseph and R. M. Hayes; (1965); *Information Storage and Retrieval;* John Wiley & Sons, New York.

Bernstein, Jeremy; (1964); *The Analytical Engine: Computers—Past, Present and Future;* Random House, New York.

Brainerd, J. G. and T. K. Sharples; (1948); "The ENIAC"; *Elect. Eng.*, Vol. 67, pp. 163–172.

Burks, Arthur W. and Alice R. Burks; (1981); "The ENIAC: First General Purpose Computer"; *Ann. Hist. Comp.*, Vol. 3, pp. 4–12.

Buzbee, B. L., R. H. Edwald and W. J. Worlton; (1982); "Japanese Supercomputer Technology"; *Science*, Vol. 218, pp. 1189–1193.

Case, John; (1985); *Digital Future: The Personal Computer Industry;* William Morrow and Co., New York.

Chase, G. C.; (1980); "History of Mechanical Computing Machinery"; *Ann. Hist. Comp.*, Vol. 2, pp. 198–226.

Chow, G. C.; (1967); "Technological Change and the Demand for Computers"; *American Economic Review;* Vol. 57, pp. 1117–1130.

Connolly, J.; (1967); *History of Computing in Europe;* IBM World Trade Corp.

Dertouzos, M. I. and J. Moses, Eds.; (1980); *The Computer Age—A Twenty Year View;* MIT Press, Cambridge, Mass.

Eames, Charles and Ray Eames; (1973); *A Computer Perspective;* Harvard University Press, Cambridge, Mass.

Evans, Susan H. and Peter Clarke; (1981); *The Making of the Micro;* Van Nostrand Reinhold, New York.

Fishman, Katherine Davis; (1981); *The Computer Establishment;* McGraw-Hill, New York.

Flamm, Kenneth; (1988*a*); *Creating the Computer: Government, Industry and High Technology;* Brookings Institution, Washington, D.C.

Flamm, Kenneth; (1988*b*) *Targeting the Computer: Government Support and International Competition;* The Brookings Institution, Washington, D.C.

Fleck, G., Ed.; (1973); *A Computer Perspective;* (by the Office of Charles and Ray Eames); Harvard University Press, Cambridge, Mass.

Freiberger, Paul and Michael Swaine; (1984); *Fire in the Valley: The Making of the Personal Computer;* Osborne/McGraw-Hill, New York.

Goldstine, Herman H.; (1971); *The Computer: From Pascal to von Neumann;* Princeton University Press, Princeton, N.J.

Graham, Alan K.; (1982); "Software Design: Breaking the Bottleneck"; *IEEE Spectrum*, pp. 43–50.

Hawkes, Nigel; (1972); *The Computer Revolution;* E. P. Dutton, New York.

Hollingdale, S. H. and G. C. Tootill; (1967); *Electronic Computers;* Penguin Books, Baltimore.

Hyman, Anthony; (1982); *Charles Babbage, Pioneer of the Computer;* Princeton University Press, Princeton, N.J.

Joseph, E. C.; (1972); "Future Computer Architectures—Polysystems"; *COMPCOM–72 Digest of Papers, IEEE Computer Society*, pp. 149–154.

Kidder, Tracy; (1981); *The Soul of a New Machine;* Little, Brown and Co., Boston.

Knight, K. E.; (1966); "Changes in Computer Performance" *Datamation*, Vol. 12, No. 9, pp. 40–54.

Knight, K. E.; (1968); "Evolving Computer Performance: 1963–1967"; *Datamation*, Vol. 14, No. 1, pp. 31–35.

Lavington, Simon; (1980); *Early Computers: The Story of Vintage Computers and the People Who Built Them;* Manchester University Press, Manchester (Digital Press, U.S.).

Luckoff, Herman; (1979); *From Dits to Bits: A Personal History of the Electronic Computer;* Robotics Press, Portland, OR.

McCorduck, Pamela; (1979); *Machines Who Think: A Personal Inquiry into the History and Prospects of Artificial Intelligence;* W. H. Freeman, San Francisco.

McCorduck, Pamela; (1985); *The Universal Machine;* McGraw-Hill, New York.

McKie, W. and Richard B. Mnacke; (1983); *IBM and the U.S. Data Processing Industry: An Economic History;* Praeger, New York.

Metropolis, N., J. Howelett and Gian-Carlo Rota, Eds.; (1980); *A History of Computing in the Twentieth Century: A Collection of Essays;* Academic Press, New York.

Moreau, R.; (1985); *The Computer Comes of Age: The People, the Hardware and the Software;* MIT Press, Cambridge, Mass.

Osborne, A.; (1979); *Running Wild—The Next Industrial Revolution;* Osborne/McGraw-Hill, Berkeley, Calif.

Phister, Montgomery Jr.; (1979); *Data Processing Technology and Economics;* Digitial Press and Santa Monica Publishing Co., Bedford, Mass.

Ralston, Anthony, Ed.; (1983); *Encyclopedia of Computer Science and Engineering,* 2nd edition; Van Nostrand Reinhold, New York.

Rosen, S.; (1969); "Electronic Computers; A Historical Survey"; *Computing Surveys ACM,* Vol. 1, No. 1, pp. 7–36.

Serrell, R., M. M. Astrahan, G. W. Patterson and I. B. Payne; (1962); "The Evolution of Computing Machines and Systems"; *Proc. IRE,* Vol. 50, pp. 1039–1058.

Shurkin, Julian, L.; (1981); *Engines of the Mind: A History of The Computer;* W. W. Norton & Co., New York.

Spinrad, R. J.; "Office Automation"; *Science,* Vol. 215, pp. 808–813.

Stern, Nancy; (1981); *From ENIAC to UNIVAC: An Appraisal of the Eckart-Mauchly Computers;* Digital Press, Bedford, Mass.

Stoneman, Paul; (1976); *Technological Diffusion and the Computer Revolution: The U.K. Experience;* Cambridge University Press, Cambridge, England.

Tropp, H.; (1974); "The Effervescent Years: A Retrospective"; *IEEE Spectrum;* Vol. 11, No. 2, pp. 70–79.

Turn, R.; (1974); *Computers in the 1980s;* Columbia University Press, New York.

Wildes, Carl C. and Nilo A. Lindgren; (1985); *A Century of Electrical Engineering and Computer Science at MIT;* MIT Press, Cambridge, Mass.

Williams, Michael R.; (1985); *A History of Computing Technology;* Prentice-Hall, Englewood Cliffs, N.J.

Wulforst, Harry; (1982); *Breakthrough to the Computer Age;* Scribners & Sons, New York.

Corporations

Drucker, Peter; (1983); *Concept of the Corporation,* 2nd edition; (1st edition in 1946); New American Library (Harper & Row), New York.

Hart, P. E.; (1962); "The Size and Growth of Firms"; *Economica;* Vol. 39, No. 113, pp. 29–39.

Ijiri, Y. and H. A. Simon; (1977); *Skew Distributions and the Size of Business Firms;* North Holland, Amsterdam.
Magee, S.; (1977); "Multinational Corporations: the Industry Technology Cycle and Development"; *Journal of World Trade Law,* Vol. 11.
Penrose, E. T.; (1968); *The Theory of the Growth of the Firm;* Oxford University Press, Oxford.
Simon, Herbert A. and Charles P. Bonini; (1958); "The Size Distribution of Business Firms"; *Amer. Econ. Rev.,* Vol. 48, No. 4, pp. 607–617.

Diffusion and Substitution

Babcock, J. M.; (1962); "Adoption of Hybrid Corn: A Comment"; *Rural Sociology,* Vol. 27, pp. 332–338.
Bain, N. J. T.; (1962); "The Growth of Television Ownership in the United Kingdom"; *International Econ. Rev.,* Vol. 3, pp. 145–167.
Bain, N. J. T.; (1963); "The Growth of Demand for New Commodities; *J. Royal Statistic. Soc.,* Vol. 126, Ser. A.
Baker, M. J.; (1979); *Industrial Innovation: Technology Policy Diffusion;* Macmillan, London.
Bass, F. M. (1969); "A New Product Growth Model for Consumer Durables"; *Management Sci.,* Vol. 15, pp. 215–227.
Becker, M. H.; (1970); "Sociometric Location and Innovativeness: Reformulation and Extension of the Diffusion Model"; *Amer. Sociological Rev.,* Vol. 35, pp. 267–283.
Belcher, J. "Acceptance of the Salk Polio Vaccine"; *Rural Sociology,* Vol. 23, pp. 158–170.
Bernhardt, I.; (1970); "Diffusion of Catalytic Techniques Through a Population of Medium Sized Petroleum Refining Firms"; *J. Ind. Econ.,* Vol. 19, pp. 59–64.
Bernhardt, I. and K. E. Mackenzie; (1972); "Some Problems in Using Diffusion Models for New Products"; *Manag. Sci.,* Vol. 19, pp. 187–200.
Blackman, A. Wade, Jr.; (1974); "The Market Dynamics of Technological Substitution"; *Tech. Forecast. and Soc. Change,* Vol. 6, pp. 41–63.
Bonus, H.; (1973); "Quasi-Engel Curves, Diffusion and the Ownership of Major Consumer Durables"; *J. Pol. Economy,* Vol. 81, No. 3, pp. 655–677.
Brittain, J. E.; (1976); "The International Diffusion of Electrical Power Technology, 1870–1920" *J. Econ. Hist.,* Vol. 36, pp. 108–130.
Brown, L. A.; (1968); "Diffusion Dynamics: A Review and Revision of the Quantitative Theory of the Spatial Diffusion of Innovations"; C. W. H. Gleerup, Lund, Sweden.
Brown, L. A.; (1969); "Diffusion of Innovation: A Macro View"; *Econ. Devel. and Cultural Change,* Vol. 17, pp. 189–209.
Brown, L. A.; (1975); "The Market and Infrastructure Context of Adoption: A Spatial Perspective on the Diffusion of Innovation"; *Econ. Geog.,* Vol 51, pp. 185–216.
Brown, L. A. and K. R. Cox; (1971); "Empirical Regularities in the Diffusion of Innovation"; *Annals Assoc. Amer. Geog.,* Vol. 61.
Brown, L. A. and B. Lentek; (1973); "Innovation Diffusion in a Developing Economy"; *Econ. Dev. and Cultural Change,* Vol. 21, pp. 274–292.
Carter, C. F. and B. R. Williams; (1957); *Industry and Technical Progress: Factors*

Governing the Speed of Application of Science; Oxford University Press, Oxford, England.
Casetti, E.; (1969); "Why Do Diffusion Processes Conform to Logistic Trends?"; *Geog. Analysis,* Vol. 1, pp. 101–105.
Cassetti, E. and R. K. Kemple; (1969); "Concerning the Testing of Spatial Diffusion Hypothesis"; *Geog. Analysis,* Vol. 1, pp. 254–259.
Coleman, J., E. Katz and H. Menzel; (1957); "The Diffusion of Innovation Among Physicians"; *Sociometry,* Vol. 20, pp. 253–270.
Crain, R. L.; (1966); "Fluoridation: The Diffusion of an Innovation Among Cities"; *Social Forces,* Vol. 44, pp. 467–476.
Davies, Stephen; (1979); *The Diffusion of Process Innovations;* Cambridge University Press, Cambridge, England.
Dodd, S. C.; (1956); "Testing Message Diffusion via Harmonic Logistic Curves"; *Psychometrica,* Vol. 21, pp. 191–201.
DuBoff, R. B.; (1964); "A Note on the Substitution of Inanimate for Animal Power"; *J. Pol. Economy,* Vol. LXXII, pp. 196–197.
Edmonson, M. S.; (1961); "Neolithic Diffusion Rates"; *Current Anthropology,* Vol. 1, pp. 71–102.
Evensen, R.; (1974); "International Diffusion of Agrarian Technology"; *J. Econ. Hist.,* Vol. 34, pp. 51–90.
Ferry, A. G., A. Sullivan, R. J. Dolan and G. F. Marsh; (1967); "The Adoption Process: S-Curve or J-Curve?"; *Rural Sociology,* Vol. 32, No. 2, pp. 220–222.
Fisher, J. C. and Pry, R. H.; (1971); "A Simple Substitution Model of Technological Change"; *Tech. Forecast. and Soc. Change,* Vol. 3, No. 1, pp. 75–88.
Fliegel, F. C. and J. E. Klein; (1966); "Attributes of Innovations as Factors in Diffusion"; *Amer. J. Soc.,* Vol. 72, pp. 235–248.
Gersho, A. and D. Mitra; (1975); "A Simple Growth Model for the Diffusion of a New Communication Service"; *IEEE Trans. on Systems, Man and Cybernetics,* Vol. 5, pp. 209–216.
Gold, Bela; (1981); "Technological Diffusion in Industry: Research Needs and Shortcomings"; *J. Ind. Econ.,* Vol. XXIX, No. 3, pp. 247–269.
Gold, B., G. Rosseger and W. S. Pierce: (1970); "Diffusion of Major Technological Innovations in U.S. Steel Industry"; *J. Ind. Econ.,* Vol. XVIII, No. 3, pp. 218–241.
Griliches, Z.; (1957); "Hybrid Corn: An Exploration in the Economics of Technological Change"; *Econometra,* Vol. 25, pp. 501–522.
Hagerstrand, T.; (1965); "A Monte Carlo Approach to Diffusion"; *European Jour. Soc.,* Vol. 6, pp. 43–67.
Hagerstrand, T.: (1967); *Innovation as a Spatial Diffusion Process;* University of Chicago Press, Chicago.
Hurter, Arthur. P., Jr. and Albert Rubinstein: (1978); "Market Penetration by New Innovations: The Technological Literature"; *Tech. Forecast. and Soc. Change,* Vol. 11, pp. 197–221.
Katz, E., M. L. Levin and H. Hamilton; (1963); "Traditions of Research on the Diffusion of Innovations"; *Amer. Sociological Rev.,* Vol. 28, pp. 237–252.
Lakhani, Hyder; (1975); "Diffusion of Environment-Saving Technological Change: A Petroleum Refining Case Study"; *Tech. Forecast. and Soc. Change,* Vol. 7, pp. 33–53.

Lenz, R. C. and H. W. Lanford; (1972); "The Substitution Phenomena"; *Bus. Horizons*, Vol. 15, No. 1, pp. 63–68.

Maddala, G. S. and P. T. Knight; (1967); "International Diffusion of Technical Change: A Case Study of the Oxygen Steel Making Process"; *Econ. Jour.*, Vol. 77, No. 307, pp. 531–538.

Mahajan, V. and R. A. Peterson; (1979*a*); "First Purchase Diffusion Model of New Product Acceptance"; *Tech. Forecast. and Soc. Change*, Vol. 15, pp. 127–146.

Mahajan, V. and R. A. Peterson; (1979*b*); "Integrating Time and Space in Technological Diffusion Models"; *Tech. Forecast. and Soc. Change*, Vol. 14, pp. 231–241.

Mahajan, V. and M. E. F. Schoeman; (1977); "Generalized Model for the Time Pattern of the Diffusion Process"; *IEEE Trans. Eng. Manag.*, Vol. 24, pp. 12–19.

Mansfield, Edwin; (1961); "Technical Change and the Rate of Imitation"; *Econometrica*, Vol. 29, No. 4, pp. 741–766.

Martino, J. P.; (1972); "The Effect of Errors in Estimating the Upper Limit of a Growth Curve"; *Tech. Forecast. and Soc. Change*, Vol. 4, pp. 77–84.

Mason, R. and A. Halter; (1968); "The Application of a System of Simultaneous Equations to an Innovation Diffusion Model"; *Soc. Forces*, Vol. 47, pp. 183–195.

Nasbeth, L. and G. F. Ray; (1974); *The Diffusion of New Industrial Processes;* Cambridge University Press, Cambridge, England.

Nevers, J. V.; (1972); "Extensions of a New Product Growth Model"; *Sloan Manag. Rev.*, Vol. 13, No. 2, pp. 77–91.

Pederson, P. O.; (1970); "Innovation Diffusion Within and Between National Urban Systems"; *Geog. Analysis*, Vol. 2, pp. 203–254.

Pemberton, H. E.; (1937); "The Curve of Culture Diffusion Rate"; *Amer. Sociology Rev.*, Vol. 1, No. 4, pp. 547–556.

Poznanski, K. Z.; (1983); "International Diffusion of Steel Technologies"; *Tech. Forecast. and Soc. Change*, Vol. 22, p. 23.

Rapoport, A.; (1956); "The Diffusion Problem in Mass Behaviour"; *General Systems Yearbook*, Vol. 1, pp. 48–55.

Rapoport, A. and L. I. Rubhuhn; (1952); "On the Mathematical Theory of Rumor Spread"; *Bulletin of Mathematical Biophysics*, Vol. 14, p. 375.

Ray, G. F.; (1983); "The Diffusion of New Technology: A Study of Ten Processes in Nine Industries"; *National Institute Economic Review*, Vol. 48.

Renshaw, E. F.; (1963); "The Substitution of Inanimate Energy for Animal Power"; *J. Pol. Econ.*, Vol. LXXI, p. 284.

Robertson, T. S.; (1967); "The Process of Innovation and the Diffusion of Innovation"; *J. of Marketing*, Vol. 31, pp. 13–18.

Robinson, A. E. H.; (1974); "The Early Diffusion of Steam Power"; *J. of Econ. Hist.*, Vol. 34, pp. 91–107.

Rogers, E. M.; (1962); *Diffusion of Innovations;* The Free Press, New York.

Rosegger, C.; (1979); "Diffusion and Technological Specificity: The Case of Continuous Casting"; *J. of Ind. Econ.*, Vol. XXVIII, No. 1, pp. 39–53.

Rosenberg, Nathan; (1972); "Factors Affecting the Diffusion of Technology"; *Explorations in Economic History;* Fall.

Sharif, M. N. and A. K. M. Abdul Haq; (1979); "Determinants for Forecasting Technological Substitution" *Tech. Forecast. and Soc. Change*, Vol. 14, pp. 59–81.

Sharif, M. N. and C. Kabir; (1976); "A Generalized Model for Forecasting Technological Substitution"; *Tech. Forecast. and Soc. Change,* Vol. 8, pp. 353–364.

Sharif, M. N. and K. Ramanathan; (1982); "Binomial Innovation Diffusion Model with Dynamic Potential Adopter Population"; *Tech. Forecast. and Soc. Change,* Vol. 20, pp. 63–97.

Smith, Sir Grafton E.; (1971); *The Diffusion of Culture;* Kennikat Press, Fort Washington, N.Y.

Stapleton, Earl; (1976); "The Normal Distribution as a Model of Technological Substitution"; *Tech. Forecast. and Soc. Change,* Vol. 8, pp. 325–334.

Stern, M. O., R. U. Ayres and A. Shapanka; (1975); "A Model for Forecasting the Substitution of One Technology for Another"; *Tech. Forecast. and Soc. Change,* Vol. 7, pp. 37–79.

Tilton, J. E.; (1971); *International Diffusion of Technology: The Case of Semiconductors;* Brookings Institution, Washington, D.C.

Utterback, J.; (1974); "Innovation in Industry and the Diffusion of Technology"; *Science,* Vol. 183, pp. 620–626.

Wilson, L. O.; (1972); "An Epidemic Model Involving a Threshold"; *Math. Bioscience,* Vol. 15, pp. 109–121.

Yupa, L. S.; (1977); "The Green Revolution: A Diffusion Model"; *Annals Assoc. Amer. Geog.,* Vol. 67, pp. 350–359.

Drugs

Wardell, William M., Maureen S. May and Gene Trimble; (1982); "New Drug Development by U.S. Pharmaceutical Firms: . . . Analysis of Trends in Acquisition . . . of Drug Candidates"; *Clinical Pharmacology and Therapeutics,* Vol. 32, pp. 407–417.

Economics

Ayres, C. E.; (1961); *Towards a Reasonable Society;* University of Texas Press, Austin.

Ayres, C. E.; (1962); *The Theory of Economic Progress;* 2nd ed.; Schocken Books, New York.

Barnett, H. and C. Morse; (1963); *Scarcity and Growth;* Johns Hopkins University Press, Baltimore.

Conrad, G. R. and R. H. Plotkin; (1968); "Risk/Return: United States Industry Pattern"; *Harv. Bus. Rev.,* Vol. 46, pp. 90–99.

David, P. A.; (1975); *Technical Choice: Innovation and Economic Growth;* Cambridge University Press, Cambridge, England.

Eltis, W. A.; (1963); "Investment, Technical Progress and Economic Growth"; *Oxford Economic Papers,* Vol. 15, pp. 32–52.

Floud, Roderick and Donald McCloskey, Eds.; (1981); *The Economic History of Britain Since 1700, Vol. 1: 1700–1860;* Cambridge University Press, Cambridge, England.

Freeman, C.; (1974); *The Economics of Industrial Innovation;* Penguin Books, Middlesex, England.

Griliches, Zve; (1958); "The Demand for Fertilizer: An Economic Interpretation of a Technological Change"; *J. Farm Economics,* Aug., pp. 591–606.

Hufbauer, G.; (1966); *Synthetic Materials and the Theory of International Trade;* Duckworth, London.

Jacobs, Jane; (1984); *Cities and the Wealth of Nations;* Random House, New York.

Kravis, Irving B. and Robert E. A. Lipsey; (1984); "The Diffusion of Growth in the World Economy" in *International Comparisons of Productivity and Causes of Slowdown;* John W. Kendrick, Ed.; Ballinger Publishing Company, Cambridge Mass. (for the American Enterprise Institute).

Kuznets, Simon; (1930); *Secular Movements in Production and Prices;* Houghton Mifflin, New York.

Kuznets, Simon; (1966); *Modern Economic Growth; Rate, Structure and Spread;* Yale University Press, New Haven.

Kuznets, Simon; (1968); *Toward a Theory of Economic Growth;* W. W. Norton & Co., New York.

Kuznets, Simon; (1971); *Economic Growth of Nations: Total Output and Production Structure;* The Belknap Press of Harvard University Press, Cambridge, Mass.

Lipsey, Robert E. and Irving B. Kravis; (1987); "Saving and Economic Growth: Is the United States Really Falling Behind?" Report No. 901; The Conference Board, Inc.; (jointly sponsored by the American Council of Life Insurance and the Conference Board).

Maddison, Angus; (1982); *Phases of Capitalist Development;* Oxford University Press, Oxford, England.

Nelson, R. R.; (1959); "The Economics of Invention: A Survey of the Literature"; *The J. of Bus.,* Vol. 32, pp. 113–114.

North, Douglas C.; *The Economic Growth of the United States, 1790–1860";* Prentice-Hall, Englewood Cliffs, N.J.

Rosenberg, Nathan; (1971); *The Economics of Technological Change;* Penguin Books, London.

Rosenberg, Nathan; (1972); *Technology and American Economic Growth;* Harper & Row, New York.

Rosenberg, Nathan; (1976); *Perspectives on Technology;* Cambridge University Press, Cambridge, England.

Rosenberg, Nathan; (1979); "Technological Interdependence in the American Economy"; *Technology and Culture;* Jan.

Rosenberg, Nathan; (1982); *Inside the Black Box: Technology and Economics"*; Cambridge University Press, Cambridge, England.

Rostow, W. W.; (1975); *How It All Began: Origins of the Modern Economy;* McGraw-Hill, New York.

Rostow, W. W.; (1980); *Why the Poor Get Richer and the Rich Slow Down;* University of Texas Press, Austin.

Schmookler, Jacob; (1952); "The Changing Efficiency of the American Economy"; *The Rev. of Econ. and Statistics,* Vol. XXXIV, pp. 214–231.

Schmookler, Jacob; (1968); *Invention and Economic Growth;* Harvard University Press, Cambridge, Mass.

Schumpeter, J. A.; (1934); *The Theory of Economic Development;* Harvard University Press, Cambridge, Mass.

Schumpeter, J. A.; (1939); *Business Cycles: A Theoretical, Historical and Statistical Analysis of Capitalistic Processes;* McGraw-Hill, New York.

Summers, Robert and Alan Heston; (1988); "A New Set of International Comparisons of Real Products and Price Levels Estimates for 130 Countries, 1950–1985"; *The*

Rev. of Income and Wealth, Series 34, No. 1, pp. 1–25. (Accompanied by three data diskettes; compatible with spreadsheet programs.)

Thurow, Lester; (1980); *The Zero Sum Society;* Basic Books, New York.

Vatter, Harold G.; (1975); *The Drive to Industrial Maturity;* Greenwood Press, Westport, Conn.

Electric Light and Power

Anderson, J. M.; (1974); "The Electric Lamp; 100 Years of Applied Physics"; *Physics Today,* Vol. 32, No. 10, pp. 33–40.

Bright, Arthur H., Jr.; (1949); *The Electric Lamp Industry;* McGraw-Hill, New York.

DuBoff, R. C.; (1967); "The Introduction of Electric Power in American Manufacturing"; *Economic History Review,* Dec., pp. 513.

Hughes, Thomas P.: (1983); *Networks of Power: Electrification in Western Society;* Johns Hopkins University Press, Baltimore.

Hughes, W. R.; (1971); "Scale Frontiers in Electric Power"; in *Technological Change in Regulated Industries;* W. M. Caoron, Ed.; pp. 44–85; Brookings Institution, Washington, D.C.

Krick, N.; (1969); "Turbogenerators Today"; *Brown Bovari Review,* Vol. 10, pp. 368–379.

MacLaren, Malcolm; (1943); *The Rise of the Electrical Industry During the Nineteenth Century;* Princeton University Press, Princeton, N.J.

Martino, Joseph P.; (1972); "The Stepwise Growth of Electric Generators"; *Tech. Forecast. and Soc. Change,* Vol. 3, pp. 465–471.

Passer, Harold; (1953); *The Electrical Manufacturers;* Harvard University Press, Cambridge, Mass.

Energy

Banks, F. E.; (1977); *Scarcity, Energy and Economic Progress;* Lexington Books, D. C. Heath & Co., Lexington, Mass.

Cook, E.; (1976); *Man, Energy and Society;* W. H. Freeman, San Francisco.

Dorf, Richard C.; (1981); *The Energy Factbook;* McGraw-Hill, New York.

Fischer, J. C.; (1974); *Energy Crisis in Perspective;* John Wiley & Sons, New York.

Rose, David J.; (1986); *Learning About Energy;* Plenum Press, New York and London.

Stobaugh, Robert and Daniel Yergen; (1979); *Energy Future;* Random House, New York.

Entrepreneurship

Jacobs, Michael; (1982); "The Physicist as Entrepreneur"; *Physics Today,* pp. 35–40.

Madoque, Modesto A.; (1980); "Entrepreneurs, Champions and Technological Innovation" *Sloan Management Rev.*, pp. 59–76.

Roberts, Edward B. and Alan Frohman; (1972); "Internal Entrepreneurship: Strategy for Growth"; *Bus. Quart.*, Vol. 37, No. 1, pp. 71–78.

Firearms

Greener, W. W.; (1967); *The Gun and Its Development;* Bonanza Books, New York.

Forecasting

Ayres, R. V.; (1969); *Technological Forecasting and Long Range Planning;* McGraw-Hill, New York.

Blackman, A. Wade Jr.; (1972); "A Mathematical Model for Trend Forecasts"; *Tech. Forecast. and Soc. Change.*, Vol. 3, pp. 441–452.

Blackman, A. Wade Jr.; (1973*a*); "A Cross Impact Model Applicable to Forecasts for Long Range Planning"; *Tech. Forecast. and Soc. Change*, Vol. 5, pp. 233–242.

Blackman, A. Wade Jr.; (1973*b*); "New Venture Planning: The Role of Technological Forecasting"; *Tech. Forecast. and Soc. Change*, Vol. 5, pp. 25–49.

Blackman, A. Wade Jr.; (1970); "Normex Forecasting of Jet Engine Characteristics"; *Tech. Forecast. and Soc. Change*, Vol. 2, pp. 61–76.

Boorstin, D. J.; (1978); *The Republic of Technology: Reflections on Our Future Community;* Harper & Row, New York.

Bright, J.; (1974); *A Brief Introduction to Technological Forecasting*, 3rd edition; Pemaquid Press, Texas.

Bright, J.; (1974); *Technological Forecasting for Industry and Government;* Prentice-Hall, Englewood Cliffs, N.J.

Bright J. and M. E. F. Schoeman, Eds.; (1973); *A Guide to Practical Technological Forecasting;* McGraw-Hill, New York.

Cetron, M. J.; (1969); *Technological Forecasting;* Gordon and Breach, New York.

Cetron, M. J. and C. A. Ralph; (1971); *Industrial Applications of Technological Forecasting;* Wiley Interscience, New York.

Delaney, Charles L.; (1973); "Technology Forecasting: Aircraft Engine Design"; *Tech. Forecast. and Soc. Change.*, Vol. 5, pp. 249–252.

Faurt, L. A. and J. W. Woodcock; (1960); "Early Prediction of Market Success for New Grocery Products"; *J. Marketing*, Vol. 25, No. 2, pp. 31–38.

Fusfeld, A. R.; (1970); "Technological Progress Functions: A New Technique for Forecasting"; *Tech. Forecast. and Soc. Change*, Vol. 1, pp. 301–312.

Heppenheimer, T. A.; (1983); *The Real Future;* Doubleday & Co., Garden City, N.Y.

Jones, H. and B. Twiss; (1978); "Forecasting Technology for Planning Decisions"; Macmillan Press, London.

Martino, Joseph P., (1970); "Correlation of Technological Trends"; *Tech. Forecast. and Soc. Change*, Vol. 1, pp. 347–354.

Martino, Joseph P.; (1971); "Examples of Technological Trend Forecasting for Research and Development Planning"; *Tech. Forecast. and Soc. Change*, Vol. 2, pp. 247–260.

Martino, J. P.; (1983); *Technological Forecasting for Decision Making*, 2nd edition; Elsevier Science Publishing Co., New York.

Sahal, D.; (1975); "A Generalized Logistic Model for Technological Forecasting"; *Tech. Forecast. and Soc. Change*, Vol. 7, pp. 81–97.

Sharif, M. and G. Uddin; (1975); "A Procedure for Adapting Technological Forecasting Models"; *Tech. Forecast. and Soc. Change,* Vol. 7, pp. 99–106.

Simmonds, W. H. C.; (1972); "The Analysis of Industrial Behaviour and its Use in Forecasting"; *Tech. Forecast. and Soc. Change,* Vol. 3, pp. 205–224.

Growth

Nelson, R. R. and S. G. Winter; (1975); "Growth Theory From an Evolutionary Perspective: The Differential Productivity Puzzle"; *Amer. Econ. Rev.,* Vol. 65, pp. 338–344.

Reeve, E. C. R. and J. S. Huxley; "Some Problems in the Study of Allometric Growth", in *Essays on Growth and Form;* W. E. Le Gros and P. B. Medawar, Eds.; Clarendon Press, Oxford.

Steindl, J.; (1965); *Random Processes and the Growth of Firms;* Griffin, London.

Turner, M. E., R. A. Blumenstein and J. L. Sebaugh; (1969); "Generalization of the Logistic Law of Growth"; *Biometrics,* Vol. 25, pp. 620–626.

Willis, R. E.; (1979); "Statistical Considerations in the Fitting of Growth Curves"; *Tech. Forecast. and Soc. Change,* Vol. 15, pp. 107–125.

History

Canfield, L. H. and H. R. Anderson; (1964); *The Making of Modern America;* Houghton Mifflin, New York.

Cardwell, D. S. L.; (1972); *Turning Points in Western Technology;* Neale Watson Academic Publications, Inc., New York.

Clark, Victor: (1929); *History of Manufactures in the United States;* McGraw-Hill, New York.

Cochran, Thomas C. and William Miller; (1942); *The Age of Enterprise: A Social History of Industrial America;* Harper & Row, New York.

Daumas, Maurice, Ed.; (1979); (English translation from the French, first published in 1968); *A History of Technology and Invention;* (3 vols.); Crown Publishers, New York.

De Bono, Ed. (1974); *Eureka! An Illustrated History of Inventions from the Wheel to the Computer;* Holt, Rinehart and Winston, New York.

de Camp, L. Sprague; (1961); *The Heroic Age of American Invention;* Doubleday and Company, New York.

Derry, T. K. and Trevor I. Williams; (1960); *A Short History of Technology from the Earliest Times to A. D. 1900;* Clarendon Press, Oxford.

Douglas, R. W. and S. Frank; (1972); *A History of Glassmaking;* G. T. Foulis & Co, Henly-on-Thames, England.

Eco, Umberto and G. B. Zorzoli; (1962); *A Pictorial History of Inventions;* Wiedenfield and Nicolson, London; translated from the Italian published by Casa Editrice Valentino Bompiani & C. Milano (1961).

Gillespie, C.; (1957); "The Natural History of Industry"; *Isis,* p. 48.

Habakkuk, H. J.; (1962); *American and British Technology in the 19th Century: The Search for Labour Saving Inventions;* Cambridge University Press, Cambridge, England.

Holmyard, E. J.; (1954); *A History of Technology;* Clarendon Press, Oxford, England.

Klemm, Friedrich; (1954); *A History of Western Technology;* 1964 translation by D. W. Singer; MIT Press, Cambridge, Mass.

Kranzberg, Melvin and Carrol W. Pursell; (1967); *Technology in Western Civilization, Vol. I., 1750–1900; Vol. II, 20th Century;* Oxford University Press, London.

Landes, David S.; (1969); *The Unbound Prometheus: Technological Change and Industrial Development in Western Europe from 1750 to the Present;* Cambridge University Press, Cambridge, England.

Lilley, Samuel; (1966); *Men, Machines and History;* International Publishers, New York.

Musson, A., and E. Robinson; (1960); "Science and Industry in the Late 18th Century"; *Economic History Review,* Dec.

Noble, David F.; (1979); *America By Design; Science, Technology and the Rise of Corporate Capitalism;* Alfred A. Knopf, New York.

Oliver, John; (1956); *History of American Technology;* The Ronald Press Co., New York.

Singer, C. and E. J. Holmyard; (1950); *A History of Technology;* Oxford University Press, London.

Singer, Charles, E. J. Holmyard, A. R. Hall and Trevor I. Williams, Eds.; (1958); *A History of Technology;* Oxford University Press, New York and London.

Usher, A.; (1954); *A History of Mechanical Inventions;* Harvard University Press, Cambridge, Mass.

Williams, Trevor I.; (1982); *A Short History of Twentieth-Century Technology, c. 1900–c. 1950;* Clarendon Press, Oxford, England.

Williams, Trevor I.; (1987); *The History of Invention: From Stone Axes to Silicon Chips;* Facts on File Publications, New York.

Industrial Revolution

Chapman, S. D. and J. D. Chambers; (1970); *The Beginnings of Industrial Britain;* University Tutorial Press, London.

Crafts, N. F. R.; (1985); *British Economic Growth During the Industrial Revolution;* Clarendon Press, Oxford, England.

Deane, Phyllis; (1965); *The First Industrial Revolution;* Cambridge University Press, Cambridge, England.

Hills, R. L.; (1970); *Power in the Industrial Revolution;* Manchester University Press, Manchester.

Lane, Peter; (1978); *The Industrial Revolution: The Birth of the Modern Age;* Barnes & Noble, New York.

Mantoux, Paul; (1928); *The Industrial Revolution in the Eighteenth Century;* University of Chicago Press, Chicago; (1983 publication of 1961 revision of 1928 translation).

Musson, A. and E. Robinson; (1969); *Science and Technology in the Industrial Revolution;* Manchester University Press, Manchester, England.

Smelser, N. J.; (1959); *Social Change in the Industrial Revolution: An Application to the British Cotton Industry;* University of Chicago Press, Chicago.

Innovation and Invention

Argyris, C.; (1965); *Organization and Innovation;* Irwin-Dorsey Press, Homewood, Ill.

Baker, R.; (1976); *New and Improved: Inventors and Inventions That Have Changed the Modern World;* British Museum Publications, Ltd.

Barnett, H. G.; (1953); *Innovation: The Basis of Cultural Change;* McGraw-Hill, New York.

Burnes, Tom and G. M. Stalker; (1961); *The Management of Innovation;* Tavistock, London.

Drucker, Peter; (1985); *Innovation andEntrepreneurship;* Harper & Row, New York.

Dummer, G. W. A.; (1983); *Electronic Inventions and Discoveries,* 3rd edition; Pergamon Press, New York.

Gilfillan, S. C.; (1935*a*); *Inventing the Ship;* Follett, Chicago.

Gilfillan, S. C. (1935*b*); *The Sociology of Invention;* Follet, Chicago.

Hill, Christoper T. and James M. Utterback, Eds.; (1979); *Technological Innovation for a Dynamic Economy;* Pergamon Press, New York.

Jewkes, J., D. Sawyers and R. Stillerman; (1969); *The Sources of Invention;* Macmillan, New York.

Langrish, J., M. Gibbons, W. C. Evans and R. D. Jevons; (1972); *Wealth from Knowledge: Studies of Innovation in Industry;* Macmillan Press, London.

Li, Yao Tzu, D. G. Jannson and E. G. Cravalo; (1980); *Technological Innovation in Education and Industry;* Van Nostrand Reinhold, New York.

Mahdavi, K. B.; (1972); *Technological Innovation: An Efficiency Investigation;* Beckman's, Stockholm.

Mansfield, Edwin; (1968); *Industrial Research and Technological Innovation;* W. W. Norton & Co., New York.

Mansfield, Edwin, John Rapoport, Anthony Romeo, Edmond Villani, Samuel Wagner and Frank Husic; (1977); *The Production and Application of New Technology;* W. W. Norton & Co., New York.

Morse, Dean and A. W. Warner; (1966); *Technological Innovation and Society;* Columbia University Press, New York.

Nelson, R. R.; (1962); *The Rate and Direction of Inventive Activity;* Princeton University Press, Princeton, N.J.

Rogers, E. M. and F. F. Shoemaker; (1971); *Communications and Innovation: A Cross Cultural Approach;* The Free Press, New York.

Roman, Daniel D.; (1980); *Science, Technology and Innovation: A Systems Approach;* Grid Publishing Co., Columbus, Ohio.

Sahal, Devendra; (1981); *Patterns of Technological Innovation;* Addison-Wesley Publishing Co., Reading, Mass.

Strassman, W.; (1959); *Risk and Technological Innovation;* MIT Press, Cambridge, Mass.

The Textile Institute; (1983); *World Textiles: Investment, Innovation, Invention;* Manchester.

Zwicky, Fritz; (1969); *Discovery, Invention, Research;* Macmillan, New York.

Iron and Steel

Carr, J. S. and W. Taplin; (1962); *History of the British Steel Industry;* Harvard University Press, Cambridge, Mass.

Davis, E. W.; (1964); *Pioneering with Taconite;* Minnesota Historical Society, St. Paul, Minn.

Gold, Bela, W. S. Pierce, G. Rosegger and M. Perlman: (1984); *Technological Progress and Industrial Leadership: The Growth of the U.S. Steel Industry;* Lexington Books, D. C. Heath & Co., Lexington, Mass.

Hogan, William; (1972); *The 1970's: Critical Years for Steel;* Lexington Books, D. C. Heath & Co., Lexington, Mass.

Hyde, Charles; (1977); *Technological Change and the British Iron Industry;* Princeton University Press, Princeton, N.J.

Temin, Peter; (1964); *Iron and Steel in 19th Century America;* MIT Press, Cambridge, Mass.

Wertime, Theodore A.; (1962); *The Coming of the Age of Steel;* University of Chicago Press, Chicago.

Long Waves

Daniels, Guy and J. M. Snyder; (1984); *Translation of N. Kondratieff's Papers;* Richardson and Snyder, New York.

Dewey, Edward R. and Edwin F. Dakin: (1947); *Cycles: The Science of Prediction;* Henry Holt and Co., New York.

Dujin, J. J. van; (1983); *The Long Wave in Economic Life;* George Allen & Unwin, London.

Freeman C., J. Clark and Luc Soete; (1982); *Unemployment and Technical Innovation;* Greenwood Press, Westport, Conn.

Mensch, Gerhard; (1979); *Stalemate in Technology;* Ballinger Publishing, Cambridge, Mass.

Manufacturing

Pratten, C. F.; (1971); *Economies of Scale in Manufacturing Industry;* Cambridge University Press, Cambridge, England.

Tann, J.; (1970); *The Development of the Factory;* Cornmarket Press, London.

Metals

Atchison, Leslie; (1960); *A History of Metals;* (two volumes); McDonald and Evans, London.

Hadfield, Robert; (1925); *Metallurgy and Its Influence on Modern Progress;* Chapman and Hall, London.

Smith, C. S.; (1960); *A History of Metallography;* University of Chicago Press, Chicago.

Microelectronics

Braun, E. and S. McDonald; (1978); *Revolution in Miniature—The History and Impact of Semiconductor Electronics;* Cambridge University Press, Cambridge, England.

Friedrichs, Guenther and Adam Schaff; (1982); *Microelectronics and Society: A Report to the Club of Rome;* New American Library (Pergamon Press), New York.

Reid, T. R.; (1985); *The Chip: The Microelectronics Revolution and the Men Who Made It;* Simon and Schuster, New York.

Queisser, Hans; (1988); *The Conquest of the Microchip;* Harvard University Press, Cambridge, Mass.

Mining

Lord, Eliot; (1883); *Comstock Mining and Miners;* Howell North, Berkeley, Calif; (1959 reprint of the 1883 edition).

Plastics

Friedal, Robert; (1983); *Pioneer Plastic: Making and Selling of Celluloid;* University of Wisconsin Press, Madison, Wisc.

Kayfman, M.; (1963); *The First Century of Plastics;* The Plastics Institute, London.

Population

Keyfitz, N.; (1968); *Introduction to the Mathematics of Population;* Addison-Wesley Publishing Co., Reading, Mass.

Pearl, R.; (1925); *The Biology of Population Growth;* Alfred A. Knopf, New York.

Schultz, T. Paul; (1981); *Economics of Population;* Addison-Wesley Publishing Co.; Reading, Mass.

Post-Industry

Bell, Daniel; (1973); *The Coming of Post Industrial Society;* Basic Books, Inc., New York.

Stein, G. Harry; (1975); *The Third Industrial Revolution;* Putnam, New York.

Pre-History

Carter, Thomas F.; (1955); *The Origin of Printing and Its Spread Westward,* 2nd edition; revised by L. Carrington Goodrich; Ronald Press, New York.

Cohen, Martin Nathan; (1977); *The Food Crisis in Prehistory;* Yale University Press, New Haven.

Productivity

Burns, Arthur F.; (1934); *Production Trends in the United States Since 1872;* National Bureau of Economic Research.

Fabricant, S.; (1959); *Basic Facts on Productivity;* Columbia University Press, New York.

Hollander, S.; (1965); *The Sources of Increased Efficiency;* MIT Press, Cambridge, Mass.

Kendrick, J. W.; (1961); *Productivity Trends in the United States;* Princeton University Press, Princeton, N.J.

Kendrick, J. W.; (1973); *Postwar Productivity Trends in the United States, 1948–1969;* Princeton University Press, Princeton, N.J.

Salter, W. E. G.; (1969); *Productivity and Technical Change;* Cambridge University Press, Cambridge, England.

Public Policy

Nelson, R., M. Peck and E. Kalachek; (1967); *Technology, Economics and Public Policy;* Brookings Institution, Washington, D.C.

Nelson, Richard, Ed. (1982); *Government and Technical Progress: A Cross Industry Analysis;* Pergamon Press, New York.

Railroads

Fishlow, Albert; (1965); *American Railroads and the Transformation of the Ante-Bellum Economy;* Harvard University Press, Cambridge, Mass.

Fogel, R.; (1964); *Railroads andEconomic Growth;* Johns Hopkins University Press, Baltimore.

Research and Development

Archibald, R. D.; (1976); *Managing High Technology Programs and Projects;* John Wiley (Interscience), New York.

Blake, S. P.; (1978); *Managing Responsive Research and Development;* W. H. Freeman, San Francisco.

Bright, J. R.; (1964); *Research, Development and Technological Change;* Prentice-Hall, Englewood Cliffs, N.J.

Hamberg, Daniel; (1966); *R&D: Essays on the Economics of Research and Development;* Random House, New York.

Kornhauser, William; (1962); *Scientists in Industry; Conflict and Accommodation;* University of California Press, Berkeley, Cal.

Mansfield, Edwin; (1968); *Industrial Research and Technological Innovation;* W. W. Norton & Co., New York.

Marcson, Simon; (1960); *The Scientist in American Industry;* Industrial Relations Section, Department of Economics, Princeton University.

Marschak, Thomas, T. Glennan and R. Summers; (1967); *Strategy for R&D;* Springer-Verlag, New York.

Orlans, H.; (1973); *The Non-Profit Research Institute;* McGraw-Hill, New York.

Pelz, Donald C. and Frank M. Andrews; (1966); *Scientists in Organizations;* John Wiley & Sons, New York.

Quinn, J. B.; (1967); *Yardsticks for Industrial Research;* Ronald Press, New York.

Roberts, E. B.; (1964); *The Dynamics of Research and Development;* Harper & Row, New York.

Sahal, D.; (1980); *Research, Development and Technological Innovation: Recent Perspectives in Management;* D. C. Heath, Lexington, Mass.

Seiler, R.; (1965); *Improving the Effectiveness of Research and Development;* McGraw-Hill, New York.

Stanley, A. and K. White; (1965); *Organizing the R&D Function;* American Manufacturing Association.

Terleckyi, N.; (1974); *The Effect of R&D on Productivity Growth in Industries;* National Planning Association, Washington, D.C.

Tybout, R., Ed.; (1965); *Economics of Research and Development;* Ohio University Press, Columbus, Ohio.

White, F.; (1961); *American Industrial Research Laboratories;* Public Affairs Press, Washington, D.C.

Resources

Altenpohl, D.; (1980); *Materials in World Perspective;* Springer-Verlag, Berlin.
Banks, F. E.; (1983); *Resources and Energy; An Econometric Analysis;* D. C. Heath & Co., Lexington, Mass.
Barnett, H. and C. Morse; (1963); *Scarcity and Growth;* Johns Hopkins University Press, Baltimore.
Fischman, L. L.; (1980); *World Mineral Trends and U.S. Supply Problems;* Resources for the Future, Washington, D.C.
Peach, W. N. and J. A. Constantine; (1972); *Zimmerman's World Resources and Industries;* Harper & Row, New York.
Simon, Julian L.; (1981); *The Ultimate Resource;* Princeton University Press, Princeton, N.J.

Rocketry

Bainbridge, William; (1976); *The Spaceflight Revolution;* John Wiley & Sons, New York.
Von Braun, Wernher, and Frederick Ordway; (1987); *History of Rocketry and Space Travel;* Thomas Crowell Co., New York.

Sociology

Allen, Francis R.; (1971); *Socio-Cultural Dynamics;* Macmillan, New York.
Duncan, Greg J.; (1984); *Years of Poverty, Years of Plenty;* Institute of Social Research, Ann Arbor, Mich.
Foster, Davis; (1980); *Innovation and Employment;* Pergamon Press, London.
Gilfillan, S. C.; (1935); *The Sociology of Invention;* Follett Publishing Co., Chicago.
Hamblin, R. L., R. B. Jacobson and J. L. L. Miller; (1973); *A Mathematical Theory of Social Change;* John Wiley & Sons, New York.
Smelser, N. J.; (1959); *Social Change in the Industrial Revolution: An Application of Theory to the British Cotton Industry;* University of Chicago Press, Chicago.
Sorokin, Pitirim; (1957); *Social and Cultural Dynamics;* Porter Sargent, Boston, Mass.; {a one-volume abridgement of his four-volume work, 1937–1941}.

Steam Power

Bruce, A. W. (1952); *The Steam Locomotive in America;* W. W. Norton & Co., New York.
Dickenson, H. W.; (1939); *A Short History of the Steam Engine;* Cambridge University Press, Cambridge, England.
Hunter, L. C.; (1949); *Steamboats on the Western Rivers: An Economic and Technological History;* Harvard University Press, Cambridge, Mass.
Kolin, I.; (1974); *The Evolution of the Heat Engine;* Longman's, London.

Pursell, C. W.; (1969); *Early Stationary Steam Engines in America: A Study in the Migration of a Technology;* Smithsonian Institution Press, Washington, D.C.

Tunzelman, G. N. von; (1978); *Steam Power and British Industrialization to 1860;* Clarendon Press, Oxford, England.

Wik, R. M.; (1953); *Steam Power on the American Farm;* University of Pennsylvania Press, Philadelphia.

Technological Change

Brown, Murray; (1966); *On the Theory and Measurement of Technological Change;* Cambridge University Press, Cambridge, England.

Dosi, G.; (1984); *Technical Change and Industrial Transformation;* St. Martin's Press, New York.

Lave, Lester B.; (1966); *Technological Change: Its Conception and Measurement;* Prentice-Hall, Englewood Cliffs, N.J.

Mansfield, Edwin; (1968); *The Economics of Technological Change;* W. W. Norton & Co., New York.

Nordhaus, W. D. (1969); *Invention, Growth and Welfare: A Theoretical Treatment of Technological Change;* MIT Press, Cambridge, Mass.

Schoen, D. A.; (1967); *Technology and Change;* Delacorte Press, New York.

Walker, C. R.; (1968); *Technology, Industry and Man: The Age of Acceleration;* McGraw-Hill, New York.

Technology Transfer

Allen, Thomas; (1977); *Managing the Flow of Technology: Technology Transfer and the Dissemination of Information Within the R&D Organization;* MIT Press, Cambridge, Mass.

Bradbury, F., P. Jervis, R. Johnson and A. Pearson, Eds.; (1978); *Transfer Processes in Technical Change;* Sijthoff and Noordhoff, Alphen van den Rijn.

Carter, C. F. and B. R. Williams; (1957); *Industry and Technical Progress: Factors Governing the Application of Science;* Oxford University Press, Oxford, England.

Davidson, H. F., M. J. Cetron and J. D. Goldhar, Eds.; (1974); *Technology Transfer;* Noordhoff, Leiden.

Gruber, W. H. and D. G. Marquis, Eds.; (1969); *Factors in the Transfer of Technology;* MIT Press, Cambridge, Mass.

Hayden, Eric W.; (1976); *Technology Transfer to Europe;* Praeger, New York.

Mansfield, Edwin, Anthony Romeo, Mark Schwartz, David Teece, Samuel Wagner and Peter Brach; (1982); *Technology Transfer, Productivity and Economic Policy;* W. W. Norton & Co., New York.

Nau, Henry; (1976); *Technology Transfer and U.S. Foreign Policy;* Praeger, New York.

Rogers, E. M. and F. Shoemaker; (1971); *Communication of Innovation: A Cross Cultural Approach;* Free Press, New York.

Sovel, M. T.; (1968); *Technology Transfer—A Selected Bibliography;* Denver Research Institute, Denver, Colo.

Spencer, D. L. and A. Woroniak; (1967); *The Transfer of Technology to Developing Countries;* Praeger, New York.

Textiles

Copeland, M. T.; (1912); *The Cotton Manufacturing Industry of the United States;* Harvard University Press, Cambridge, Mass.

Miles, C.; (1966); *Lancashire Textiles: A Case Study of Industrial Change;* Cambridge University Press, Cambridge, England.

Index

Abbans, Marquis Jouffry d', 381
Abel, F., 390, 396
Aberdeen Ballistic Research Laboratory, 242–243, 245
Acetate, 398–399, 474
AC generator, 393, 394
AC motor, 395
Ada, Augusta (Countess of Lovelace), 235
AEG–Telefunken, 263–264
Agriculture, 285, 294, 305, 315, 318, 327, 374, 378, 379, 382, 384, 385, 396, 402
Aiken, Howard, 237, 238–239
Airplanes, 399, 405, 455–458
Aldrin, Edwin, 409
Alkali, 308, 379
Allometric equation, 357, 366–370
Aluminum, 395, 470
Amdahl, Gene M., 265
American Telephone and Telegraph (AT&T), 156, 157
Ammonia, 401
AM radio, 157–159
Analogue computer, 232, 236–237, 257
Analytical Engine, 235, 236, 247
Andersen, Harlan, 261
AN/FSQ7 computer, 258
Aniline purple, 388
Anschutz-Kaempfe, Hermann, 400
Anthracite, 384
Apple Computer, 410
Applied research, 101–102
Archer, Frederick Scott, 387
Arc lamp, 388
Arithmetic, 231, 232, 241
Arithmetic-logic unit (ALU), 266
Arkwright, Richard, 376
Armstrong, Edwin, F., 154, 157, 158, 406
Armstrong, Neil, 409
Army (U.S.), 252
Aspdin, John, 383
Astrolabe, 236
Atanasoff, John V., 239, 244
AT&T. *See* American Telephone and Telegraph
ATLAS computer, 256, 268
Atmospheric engine, 373, 377
Atomic bomb, 407
Atomic power, 406, 408
Attitudes, 21–22
Automatic Computing Engine (ACE), 248–249
Automation, 399
Automobiles, 8, 204–205, 316–317, 397, 472, 484
Aviation, 55–60, 112–115
Axiomatic core, 13–14

Babbage, Charles, 233–235, 238, 247, 248
Babbage, Henry, 235
Backus, John, 260
Baekeland, L. H., 400
Baird, J. L., 404, 408
Bairoch index, 288, 290, 293
Bakelite, 400
Baldwin, Stephen, 233
Banting, Frederick G., 402
Bardeen, John, 407
Barnack, Oskar, 402
Basic research, 101–102, 127
Battery, 389, 393
Bell, Alexander Graham, 392, 398
Bell, Chichester, 398
Bell Labs, 237–238, 240, 252, 407
Bensley, Thomas, 382
Bentham, Samuel, 381
Benz, Karl, 397
Berger, Hans, 403
Berliner, Emile, 393, 398
Berry, Clifford, 239, 244
Berry, Thomas S., 318–320
Berthollet, C. L., 379, 380
Bessemer, Henry, 388
Bessemer process, 388
Bevan, E. J., 396, 399
Bigelow, Julian, 250
Billings, John Shaw, 258
BINAC computer, 256, 280
Binary adoption model, 343
Binary substitution, 145–146, 148–154, 359–360
Binary system, 237, 245, 264
Biomass, 221, 224, 225
Bissel, George, 389
Black, Joseph, 377
Blanc, Honore, 382
Blast furnace, 373, 377, 384
Blathy, Otto, 394
Bleaching, 379, 380
Bolton, Edward K., 403

Bolton, W. B., 392
Boring mill, 377
Boulton, Matthew, 377
Bourne, D., 377
Brainerd, John Grist, 243–244, 245, 251, 252
Brandenberger, Jacques E., 402
Brattain, Walter, 407
Brearley, Harry, 401
Breeder reactor, 220–221
Britain. *See* Great Britain
Brown, Joseph, 312, 389
Brunel, Marc I., 381
Bullard, W. H. G., 156
Burks, Arthur, 245, 250, 251, 253
Burroughs, William S., 233
Burroughs (company), 262
Burton, William M., 401
Bush, Vannevar, 236–237, 240
Business cycles, 17, 91–95

C (computer language), 268
Calculators, 236, 237
Cambridge University, 254, 255
Camera, 395, 402
Campagnie International pour l'Informatique, 263
Carbon-zinc cell, 391
Carbutt, John, 395
Carding machine, 374, 376–377
Carlson, Chester, 408
Carothers, Wallace C., 403
Carpathia, 154
Cartwright, Edmund, 97, 379
Casetti hypothesis, 138, 139
Catalytic cracking, 405
Catch-up phenomenon, 302–305
Cathode ray tubes, 250, 254, 260, 268
Caventou, Joseph B., 382
Cawley, John, 373
CBS News, 257
Cellophane, 402
Celluloid, 392
Cellulose acetate, 398–399
Cement, 383
Census Bureau, 256, 257, 258, 280
Central processing unit (CPU), 266
Chain, E. B., 406
Challenge and response, 15
Chalmers, William, 404
Chambers, Carl, 253
Charch, William Hale, 402
Chardonnet, Count Hillaire de, 396
Chedaker, Joseph, 245
Chemicals, 375, 379, 380, 386, 388, 390, 392, 401
Cities (U.S.), 486
Clermont, 381
Clustering, 31–35, 40, 41, 80–81, 86–88, 170–173
Coal, 73–79, 180, 181, 184, 213, 214, 227–228, 307, 413, 415, 454–455
Coal gas, 381
Codes, 239–242, 256
Coke, 373, 377
Collins, Arthur M., 403
Collodion process, 387, 392
Color television, 408, 477
COLOSSUS computer, 241–242, 248, 254
Colt, Samuel, 385
Competition, 24–25, 42, 356–359
Computers, 28–29, 230–284, 448–450
  companies, 258–266
  diffusion of, 279–282
  evolution of modern technology, 266–269
  performance, 269–279
  personal, 410
  post-ENIAC, 253–258
  precursors, 233–241
  significance of, 230–233
  see also specific computers
Computer-Tabulating-Recording Company, 259
Comrie, L. J., 236, 254
Comstock lode, 229, 323–325
Concorde aircraft, 59
Conditional resources, 217
Conrad, Frank, 156
Construction, 383
Consumption, 312–314, 327–328, 331–332, 346–347
  see also specific items consumed, e.g., Energy
Control Data Corporation, 261, 263
Converter steel, 387–388, 393
Cook, Earl, 323
Cooke, James, 378
Cooke, William F., 210, 385
Copper, 323, 332–333
Copper plate printing, 375–376, 378
Cordite, 396
Corporations, 25, 42, 99–101, 102, 109–112, 123–124, 173–174, 336, 337
Cotton, 309
Cotton gin, 379–380
Couffignal, Louis, 263
Cray, Seymour, 261
Crease resistant fabrics, 403
Crompton, Samuel, 378
Crooke, William, 394
Cross, C. F., 396, 399
Cross-sectional analysis, 366
Crucible process, 375
Crude oil, 183–184, 229, 439–440
Currency, 299
Cycles, 81
Cylinder press, 382
Cylinders, 373–374, 377

Daguerre, L. J. M., 385
Daguerreotype, 385
Daimler, Gottlieb, 396
Daniell, John F., 385

Darby, Abraham, 373
Data collection, 340
Data smoothing, 370–371
Data tables, 411–489
Davis, E. W., 325
DC motor, 391
Decaying exponential rise, 351–353
DEC (company), 262
Decimal system, 245
Deere, John, 384
Defensive technology, 6–7
De Forest, Lee, 154, 156, 400
Demographic transition, 329–331
Denmark, 303
Depression, economic, 94
Deri, Max, 394
Desch, Joseph, 240
Despeissis, Louis Henri, 396
Development. *See* Research and development
Dewar, James, 396
Dialectic materialism, 16
Diesel, Rudolf, 398, 403–404
Diesel engine, 398
Diesel locomotive, 148–154, 461, 464
Difference Engine, 234–236
Differential Analyzer, 236, 243, 254
Diffusion, 3–4
  analysis, 136–143, 144
  causal factors, 173–175
  and clustering of technologies, 170–173
  of computers, 279–282
  definition, 3, 144
  of radio broadcasting, 154–160
  studies, 165–169
  technological, 35–45, 130–136
Digital electronic computers. *See* Computers
Digital Equipment Corporation, 261
Diode, 400
Dollis Hill (Great Britain), 242
Double acting engine, 378
Drake, Edwin L., 389
Drugs. *See* Pharmaceuticals; specific drugs
Dry plate process, 392
Dummer, W. A., 267, 409
Dunlop, John Boyd, 395
Du Pont (company), 402, 403, 405
Durfee, B. M., 238
Duryea, J. Frank, 397
Dynamite, 390–391
Dynamo, 390

Eastman Company, 395
Eastman Kodak, 240
Echo I, 409
Eckert, J. Presper, 243–247, 249–253, 254, 256, 257, 261, 268, 280
Economic growth, 294–302
Edison, Thomas, 97–98, 210, 312, 393, 394, 397
EDSAC computer, 254–255
EDVAC computer, 249, 250, 253–255, 257
Efficiency, 47–48, 195–205
Einthoven, Willem, 399
Electricity, 186, 200–201, 203, 208–214, 221, 225, 315–317, 385, 414–415, 444–448, 468, 483
Electric motor, 391
Electrocardiograph, 399
Electroencephalograph, 403
Electrolytic process, 395
Electronic Computer Corporation, 265
Electronic Control Company, 256
Electrotechnical Laboratory (ETL), 264
Elevator, 387–388
Ellehammer, J. C., 406
Energy
  consumption, 205–215, 366, 412–415, 423–427, 429–436, 443
  dynamics of usage, 176–229
  in the economy, 185–195
  efficiency of utilization, 195–205
  Hewett-Hubbert model, 225–229
  and industrialization, 313
  inventory, 215–225
  sources, 183–185, 306–307
  transition, 176–183
  see also specific sources, e.g., Oil; Coal; Natural gas
Engine
  airplane, 55, 57
  internal combustion, 393, 396, 398, 476
  jet, 405
  liquid fueled, 402–403
  steam. *See* Steam engine
Engineering, 125
Engineering Research Associates, 256
English Electric, 263
Engstrom, Harold, 256
ENIAC computer, 28, 233, 239, 242–248, 251, 252, 253, 259, 267, 268, 274, 283, 407
ENIGMA (decoding machine), 241–242
Environmental issues, 77, 79
ERA 1101 computer, 256, 261
Ericsson, John, 386
Essential parameters, 51
Ether, 386
Europe, 298
Evans, Oliver, 310, 381
Explosives, 390–391, 396
Exponential equation, 342, 343–348

Farm equipment, 83, 84
  see also Tractor
Farmer, Moses, 391
FCC. *See* Federal Communications Commission
Federal Coal Mine Health and Safety Act, 77, 79
Federal Communications Commission (FCC), 157–159
Felt, Dorr E., 233

Ferranti (company), 262, 263
Fessenden, Reginald A., 154, 155
Figures of merit, 49–55, 60, 269
Filament lamps, 393–394
Film, 395
Firearms, 383, 385, 389, 394
Fitch, John, 381
Fitch, Stephen, 386
Fleming, Alexander, 406
Fleming, John Ambrose, 400
Flint, Charles R., 259
Florey, H. W., 406
Flowers, T. H., 242
Flying shuttle, 374
FM radio, 157–159, 406
Focke, Heinrich, 406
Ford, Henry, 397
Forrester, Jay, 257
FORTRAN, 260
Fossil fuels, 185–187, 205, 208, 219–221, 226, 444, 483
Foster, R. N., 107
France, 263, 264, 290, 382
Frankel, Stanley P., 246
Frequency modulation. *See* FM radio
Froelich, John, 68
Fujitsu (company), 264, 265
Fulton, Robert, 381
Functional figure of merit, 53–54
Furnaces, 373, 377, 384, 389

Gagarin, Yuri, 409
Gale, Professor, 386
Garbett, Samuel, 375
Gas, natural. *See* Natural gas
Gas lighting, 381
Gasoline engine, 396
Gasoline tractor, 396, 402
GDP. *See* Gross Domestic Product
Gellman Research Associates, 98, 99, 100
General Electric, 156, 262
General Motors, 149, 150, 403–404
Generator, 388, 390, 391, 393, 394
Generic research, 102
Geothermal energy, 223
Germanium diodes, 255, 267
Germany, 263, 264, 290
Gilchrist, Percy, 393
Glass, 399
GNP. *See* Gross National Product
Goddard, Robert H., 402
Godowski, Leo, Jr., 404
Goeller-Weinberg inventory, 328, 332–334
Goldmark, Peter, 407, 408
Goldstine, Adele, 246
Goldstine, Herman, 243–245, 249, 250, 251, 253, 255
Gompertz curve, 354, 355
Gompertz rate equation, 343, 353–354
Goodwin, Hannibal, 395
Goodyear, Charles, 386
Government, 23–24
Governor, 378
Gramme, Zenobe T., 390, 391
Grant, George B., 235
Great Britain, 262–263, 288–291, 298, 305–306, 309, 485
Great Depression, 318–319, 326
Gribeauval, Jean-Baptiste de, 311, 382
Gross Domestic Product (GDP), 192–195, 299–304
Gross National Product (GNP), 186–192, 286–287, 293, 295–299, 302, 318, 319–321, 330, 339, 340, 366, 412, 422, 443–444
Growth laws, 339–343
Gyrocompass, 400
Gyrostabilizer, 400

Haber, Fritz, 401
Haber process, 401
Hall, Albert, 404
Hall, Charles Martin, 395
Hamilton, F. E., 238
Hargreaves, James, 376
Hartree, Douglas, 246
Havens, Brian, 259
Haynes, Elwood, 401
Heat, 196–197
Heath Robinson machine, 242
Helicopter, 406
Hematite, 325–326
Henry, B. Tyler, 389
Henry, Christopher, 389
Heroult, Paul Louis Touissaint, 395
Hewett-Hubbert model, 225–229, 322, 323, 325, 326
Hierarchies, 28–30
Highs, Thomas, 376
Hitachi (company), 265
Hjorth, Soren, 388, 391
Hoff, Ted, 409
Hoffman, August Wilhelm von, 388
Holburn Viaduct station, 210
Hollerith, Herman, 258–259
Holmes, Frederick Hale, 388, 391
Holmes, Hogden, 380
Honeywell-Bull (company), 263
Hooper, S. C., 156
Horizontal analysis, 366
Horse-Hoing Husbandry, 374
Horsepower, 314–317
Houdry, Eugene J., 405
Howe, Elias, 387
Hubbert-Hewett model. *See* Hewett-Hubbert model
Huntsman, Benjamin, 375
Huskey, Harry, 249, 256
Hussey, Obed, 384
Hyatt, John Wesley, 392

Hydrocarbons, 183–184, 208, 333
Hydroelectric power. *See* Water power
Hydrogen bomb, 408
Hypothetical resources, 217

IAS. *See* Institute for Advanced Study
IBM (International Business Machines), 236–240, 245, 251, 257–265, 267, 280
Idealistic view, 14
Ideational view, 14
Idiosyncratic factors, 10–11, 30
Incandescent lighting, 393–394, 471, 482
Incremental technology, 5–7
India, 309
Industrialization, 487–489
Industrial research, 98, 101
Industrial Revolution, 17, 18, 21, 120, 176, 295, 305–306, 308, 309, 321, 327, 330, 375
Industrial transition, 285–334
  American, 309–321
  catch-up phenomenon, 302–305
  economic growth and, 295–302
  first, 305–309
  material resources, 321–326
  measures of industrialization, 285–288
  world, 288–298, 326–334
Innovation, 4–6
  and business cycles, 91–95
  causes of, 87
  chronology of, 372–410
  clustering of, 31–35, 40, 41, 80–81, 86–88
  and corporate size, 99–101
  definition, 3
  and economic growth, 18
  frequency of, 80–86
  as heroic individual achievement, 19
  and incremental progress, 19–20
  and inherent technological possibilities, 26–27
  and learning curves, 41
  and long-wave cycle, 17, 81, 92–95
  low-tech vs. high-tech, 135
  origins of, 95–99
  profitability to corporation, 173–174
  time lag between invention and, 88–91
  see also Technological change
Institute for Advanced Study (IAS), 250, 251, 255, 257
Institutional structures, 25–26
Insulation, 196
Insulin, 402
Integrated circuits, 267–268, 409
Intel (company), 268
Interchangeable parts, 382, 383
Internal combustion engine, 393, 396, 398, 476
International Business Machines. *See* IBM
International Computers Limited, 263
Invention, 98–99
  definition, 3
  "golden age" of, 96
  and time lags, 88–91
  see also specific inventions and inventors
Iron, 308, 312, 325–326, 373, 375, 377, 378, 384, 387
Italy, 303
Ives, H. E., 408

Jacquard, J. M., 380
Japan, 262, 264–265, 290, 291, 298–299
Javel water, 379
Jefferson, Thomas, 382
Jet engine, 57, 405
Jet propulsion, 57
J. Lyons and Co. Ltd., 255
Jobs, Steve, 410
Johnson, Eldridge, 398

Kay, John, 374
Kelly, William, 387
Kelvin, Lord, 236
Kennedy, John, 380
Kilburn, T., 250
Kilby, Jack, 267, 409
Kimball, Fred M., 397
Knowledge, 22–23
Kodachrome, 404
Kodak camera, 395
Kondratieff, Nikolai, 17, 81
Konig, Friedrich, 382
Kravis-Lipsey analysis, 299–300, 302, 303
Kroll, W. J., 407

Labeling, 4
Labor, 16
Lake, C. D., 238
Lake Superior, 325–326
Land, Edwin, 407
Lane, John, 384
Langley, Samuel Pierpont, 399
Lardner, D., 235
Laser, 409
Lathe, 375, 376, 380, 386
Lead acid battery, 389
Lead chamber, 375
Learning, 48
Learning curves, 37, 41, 42, 47, 60–67, 77–78, 366–370, 481
Leblanc, Nicholas, 379, 390
Lebon, Phillipe, 381
Leclanche, Georges, 391
Leibnitz, Gottfried Wilhelm, 233
Lennarts, E. H., 255
Lenoir, Ettiene, 393
LEO computer, 255
Lighting, 32–35, 52–53, 381, 388, 393–394, 471, 482

Lipsey-Kravis data. *See* Kravis-Lipsey analysis
Liquid fueled engine, 402–403
Liston, Robert, 386
Little, Arthur D., 398
Locomotives, 148–154, 310, 383, 403–404, 459–462, 464
Lodge, Sir Oliver, 154
Logarithms, 234
Logic, 231, 232, 237, 241, 247
Logistic equation, 343, 348–351, 352, 357
Lognormal distribution, 360–366
Long, Crawford W., 386
Long playing record, 407–408
Long-wave cycle, 17, 81, 92–95
Looms, 379, 380
Lumiere, Auguste, 397
Lumiere, Louis, 397

Machine Bull (company), 263
Machine gun, 394
Machine tools, 375–377, 380, 381, 382, 386, 389
Macroeconomics, 25
Maiman, T. H., 409
Malthusian principle, 295, 313, 327
Manchester University, 254, 255
Manley, Charles, 399
Mannes, Leopold, 404
Marconi, Guglielmo, 154, 398
Mark I computer, 238, 239, 254, 262
Mark II computer, 238
Mark III computer, 239
Mark IV computer, 239
Market index, 117–119
Markets, 26
Martin, Emil, 389
Martin, Pierre, 389
Marx, Karl, 15–17
Mass production, 311
Mass storage (computer), 266, 268
Mathematical tables, 234
Mathematics, 231, 247
Mauchly, John, 243–247, 249, 251, 253, 254, 256, 257, 261, 280
Maudslay, Henry, 380, 381
Maurer, Edward, 401
Maxim, Hiram, 394
Mazer, Stan, 409
McCormick, Cyrus, 384
McKelvey diagram, 216–218, 322
Meader, Ralph, 256
Measurement, 340
Mechanization, 16, 75–76, 309, 314–315, 384
Meiji Restoration, 292
Meikle, Andrew, 379, 385
Memory, computer, 266, 268
Mensch. *See* Schumpeter-Mensch theory
Meritens, Baron M. de, 393
Metals, 226, 310–311, 382
Meteorology, 252
Methyl methacrylate, 404
Metropolis, Nicholos, 246
Microcomputer, 268
Microeconomics, 25
Microphone, 393
Microprocessor, 268, 409–410
Middle Ages, 295
Miles, G. W., 399
Milling machine, 389
Minerals, 178, 179, 216–218, 306, 308, 321–326, 328, 428, 430, 436
Minicomputers, 261–262, 264
Mining, 73–79, 226, 307, 323, 415, 454–455
Monitor, 386
Moore School, 242–247, 249–253, 254, 257, 259, 283
Mork, Harry S., 398
Morrison, William, 397
Morse, Samuel, 210, 386
Mortar, 383
Morton, W. T. G., 386
Motion pictures, 397
Multilevel substitution, 145–148, 160–165, 359–360
Multivariable measures, 55
Mumma, Robert, 240
Murdock, William, 381

Nadin, Lauren, 398
Napier, John, 233
National Bureau of Standards, 255, 256
National Cash Register Corporation, 240
National Defense Research Council, 237
National objectives, 23
National Physical Laboratory, 248–249, 254
National Science Foundation, 98, 99, 101, 274–279
Natural gas, 181, 184, 213, 214, 227–228, 413
Natural resources, 344, 346–347
Naval Research Laboratory, 238
Navy (U.S.), 238, 239–240, 248, 256, 260
Neoprene, 403
Newcomen, Thomas, 373
Newcomen engine, 44, 202, 373–374
Newman, M. H. A., 242, 254
Nielson, James Beaumont, 384
Niepce, Joseph N., 385
Nieuwland, Julius A., 403
Nipkow, P., 404
Nippon Electric (NEC), 264, 265
Nitrocellulose, 396
Nitroglycerine, 390–391
Nixdorf (company), 264
Nixon, Francis, 375
Nobel, Alfred, 390–391

Nollet, F., 388
Nonessential parameters, 51
Nonpolarizing cell, 385
Nonrenewable resources, 215–216
NORC computer, 260
Norris, William, 256, 261
North, Simeon, 311, 382
Northrop Aviation, 280
Noyce, Robert, 268, 409
Nuclear power, 169–170, 180, 185, 220–221
Nuclear reactor, 408
Nylon, 405

Oberth, Hermann, 403
Ohain, Hans von, 405
Oil, 180, 183–184, 187, 209, 213, 214, 218, 227–229, 303, 389, 401, 405, 413, 437–442
Olsen, Kenneth, 261
Open hearth process, 389–390
Optical fibers, 410
Otis, Elisha G., 387–388
Otto, Nickolaus August, 393
Oughtred, William, 233
Owens, M. J., 399

Paddle wheel steamer, 381
Parallel bar mechanism, 378
Parker, James, 383
Parker, John E., 256
Parkes, Alexander, 392
Parsons, Charles, 97, 394
Pascal (computer language), 268
Patent Arms Manufacturing Company, 385
Patents, 252, 253
Patterns, 30–40, 205–215
Paul, Lewis, 374
PDP minicomputers, 261
Pearl Street station, 210
Pelletier, Pierre Joseph, 382
Pelletization, 218, 229, 326
Pender, Dean, 252, 253
Penicillin, 406
Performance, 37, 46–55
Perkin, W. H., 388
Personal computer, 410
Petroleum. *See* Oil
Pharmaceuticals, 382–383
  see also specific drugs
Phonograph, 397–398
Phosphorus, 333
Photography, 385, 387, 392, 395, 397, 402, 404, 407
Physical resources, 22
Pig iron, 387, 390
Pilot ACE, 249, 254
Pitts, Hiram, 385
Pitts, John, 385
Pixii, Antoine Hippolyte, 391
Planimeter, 236
Planing machine, 382
Plankalkul, 241
Plante, Raimond Louis Gaston, 389, 393
Plastics, 400, 402, 404
Plough, 382, 384
Pneumatic tire, 395
Polaroid photography, 407
Politics, 24
Polynomials, 234
Population, 295, 327, 329–331, 341, 344–345, 347, 420–421
Portland cement, 383
Power, Harold J., 154, 156
Power loom, 379
Priestley, Joseph, 97
Prindle, Karl Edwin, 402
Printing, 382, 386–387
Production costs, 63–67
Productivity, 37, 60–63, 73–79, 293, 318
Programming, 254
Progress function, 61
Propeller, 56–57
Puddling process, 378
Pulley blocks, 381
Pumps, 307
Punched cards, 258–259, 380
Pyroscaphe, 381

Quinine, 382

Radar, 404
Radical technological change, 5–7, 135
Radio, 154–160, 398, 406, 408, 469
Radio Corporation of America. *See* RCA
Railroads, 310, 311, 384
  freight transport, 463
  track, 464–467
  see also Locomotives
Raistrick, H., 406
Rajchman, Jan, 250
Ramsden, Jesse, 376
Ransome, Robert, 382
Rationalism, 18
Rayon, 396, 398, 474
RCA (Radio Corporation of America), 155–159, 250, 264, 267, 406, 408
Reaper, 384
Registers (memory units), 266
Regulations, government, 24
Reichenbach, H. M., 395
Rejewski, Marian, 242
Relative technological position (RTP), 116–117
Relay Interpolator, 237–238
Remington Rand, 256–257, 261
Renewable resources, 215–216, 218, 221–225
Repeating rifle, 389

Research
  basic, 101–102, 127
  funds, 485
  generic, 102
Research and development
  aggressive, 103, 104, 127, 128
  corporate level, 123–124
  defensive, 103
  functions of, 101–104
  incremental, 103, 104, 126, 128
  industrial, 119–122
  laboratory level, 124–125
  portfolio, 115–119
  project level, 125–128
  radical, 103, 104, 128
  successful, 104–105, 123–129
  and technological cycles, 106–109
  and technology-driven corporate cycle, 109–112
Reserves, 216
Revolver, 385
Rifle, 383, 389
Roberts, Richard, 380, 382
Rocket, 384
Rockets, 402–403, 406
Roe, Richard, 386–387
Roebling, Ferdinand W., 259
Roebuck, John, 375
Rostow analysis, 295–302, 318–319
Rotary press, 386–387
Rubber, 386, 403
Rumsey, James, 381
Russia. *See* Soviet Union
Rutishauser, Heinz, 241

SAGE Air Defense System, 258, 260, 268
Saltpeter, 375
Sarnoff, David, 158
Satellites, 409
Saturation effect, 315
Savery, Thomas, 373
Savery engine, 201–202, 373, 374
Sayce, B. J., 392
Schawlow, Arthur, 409
Scheutz, Edvard, 235
Scheutz, George, 235
Schonbein, C. F., 390, 396
Schreyer, Helmut, 240
Schuler, Max, 400
Schumpeter-Mensch theory, 17, 81–82, 92–93
Schutzenberger, Paul, 398
Science, 97–98, 251
Scientific Data Systems, 261–262
Screw-cutting lathe, 376, 380
Screw-driven lathe, 375
Screw-driven steamship, 386
Screw propeller, 386
S-curve, 60, 137–138, 142, 144–145, 171, 294
SEAC computer, 255–256, 267
Seaton, Charles W., 258, 259
Seed drill, 374, 378
Selective Sequence Electronic Calculator, 239
Selectron, 250
Self-sustaining growth, 285, 295, 342–343
Semiconductors, 83, 85
Sensate view, 14
701 computer, 260
702 computer, 260
Sewing machine, 387
Shale oil, 218
Sharpless, Kite, 245, 253
Shaw, Arthur, 245, 253
Ships, 386, 478–481
Shockley, William, 407
Sholes, C. S., 392
Slide lathe, 375
Siemens, E. W. von, 390, 391
Siemens, Friedrich, 390
Siemens, William, 390
Siemens (company), 263–264, 389–390
Sikorsky, Igor, 406
Silliman, Benjamin, 389
Silver, 324–325
Singer, Isaac M., 387
Single variable measures, 49
650 computer, 260, 263
Skewed size distribution, 38–39, 41
Skills, 22–23
Skin, airplane, 56
Slater, Samuel, 309
Smith, Horace, 389
Smokeless powder, 396
Sobrero, Ascanio, 390
Societe d'Electronique et d'Automatisme, 263
Soda ash, 379
Sodium carbonate, 390
Software, 260, 268
Solar energy, 223–224
Solvay, Alfred, 390
Solvay, Ernest, 390
Solvay process, 390
Sorokin, Pitirim A., 14–15
Sound reproduction, 397–398, 407–408
Soviet Union, 290, 291
Speculative resources, 217
Spengler, Oswald, 13–14
Sperry, E. A., 400
Spinning, 376, 379
Spinning jenny, 376
Spinning mule, 378, 380
Sprague, Frank J., 211
Sprengel, Hermann, 394
Springfield Armory, 383
Sputnik, 409
Stainless steel, 401
Starley, J. K., 397

Steamboat/steamship, 381, 386
Steam engine, 44–45, 148, 201–202, 307–308, 309, 310, 373–374, 377–381, 383, 384, 461, 484
Steam-powered blast, 377
Steam turbine, 394
Steel industry, 10–11, 130, 132–133, 160–165, 308, 312, 313, 373, 375, 377, 378, 387–388, 393, 474–475
Stephenson, George, 310, 383, 384
Stephenson engine, 310
Stevens, John, 381
Stibitz, George R., 237, 238
Stone, John S., 154
Straus, Henry, 256
Strauss, Benno, 401
STRETCH supercomputer, 267
Sturgeon, William, 391
Substitution
  analysis, 134, 143–145
  causal factors, 173–175
  interrupted, 169–170
  multilevel, 145–148, 160–165
  studies, 165–169
  technological, 35
Sulfuric acid, 375
Superacceleration equation, 343, 354–356
Supersonic transports, 58
SWAC computer, 255–256
Swan, Joseph, 210, 393, 394
Swinburne, James, 400
System 360 computers, 260, 265, 267
System 370 computers, 265
Systemic factors, 10–11
Systemic trends, 40–43
Systems, 28–30

Tabulating Machine Company, 259
Taconite, 218, 229, 325–326
Tainter, Charles, 398
Targeted research, 102
Technical figure of merit, 53–54
Technical performance, 37, 48–55
Technological attractiveness (TA), 116–117
Technological change
  basic concepts of, 1–45
  defining, 1–8
  determinants of, 20–28
  and function, 2
  mathematical analysis of, 335–371
  measuring, 8–12
  radical, 5–7, 135
  self-sustaining growth, 20
  systemic model of, 43–45
  theories of, 12–20
  time sequence, 2–3
  see also Innovation
Technological diffusion. *See* Diffusion
Technological discontinuities, 112–115
Technological dynamics, 1, 27, 117
Technological enterprise, 29, 337–338
Technological progress, 46–48
Technology
  and corporate cycle, 109–112
  defensive, 6–7
  definition, 1, 7
  history of, 96
  incremental, 5–7, 42
  and industrialization, 286–287, 295–296
  inherent possibilities in, 26–28
  interdependence of parts, 7
  substitution, 134, 143–145, 359–360
Telegraph, 210, 385, 386, 389, 390, 475
Telephone, 392, 473
Television, 36, 86, 404, 408, 477
Telstar 1, 409
Tennant, Charles, 380
Tesla, Nikola, 395
Texas Instruments, 268
Textile industry, 308–310, 374–380, 396, 398, 403, 405
Theile, Edmund, 396
Thermal cracking, 401
Thiout, Antoine, 375
Thomas, Sidney Gilchrist, 393
Thomas J. Watson Astronomical Computing Bureau (Columbia), 238
Thompson, T. R., 255
Thomson, R. W., 395
Thomson, William, 390
Thornton, William, 383
Three Mile Island, 170
Threshing machine, 379, 385
Tidal power, 223
Time-sharing, 268
Tires, 395
Titanic, 154
Titanium, 407
Total resource potential, 217
Tousard, Louis, 382
Townes, C. H., 409
Toynbee, Arnold, 15, 22
Tractor, 67–73, 396, 402, 451–453, 476
Transatlantic cable, 390, 409
Transformer, 394
Transistor, 267, 268, 407, 408
Transportation, 380–381, 383, 384
  see also specific means of transportation
Travis, Irven, 252, 253
Trevithick, Richard, 380–381, 383
Triode, 400
Tull, Jethro, 374
Tungsten lamps, 394
Turbine, 394
Turbojet, 57, 113–114, 405
Turing, Alan, 242, 247–251, 254
Turret lathe, 386
Two-color press, 481
Typewriter, 392

Underground coal mining, 73–79
Unemployment, 287
Unidata (company), 263
Unisys (company), 262
United Fruit Company, 156
United Kingdom. *See* Great Britain
United Nations International Comparison Project, 299, 302
United States, 289–291, 302–303, 309–321
UNIVAC computer, 255–257, 261, 262, 267, 268, 280
University of Tokyo, 264
UNIX, 268

Vacuum tubes, 239–242, 244, 267
Vail, Alfred, 386
Values, 21–22
Varley, S. A., 390, 391
Vaucanson, Jacques de, 375
Veblen, Oswald, 243
Viscose process, 396
Voltaic cell, 385
Von Braun, Wernher, 403
Von Neumann, John, 246–255
Vulcanized rubber, 386

Walker, William H., 398
Walters press, 387
Ward, Joshua, 375
Warren, Reid, 252
Water frame, 376
Water power, 180, 218, 222, 307, 414
Water-powered printing, 380
Watson, Thomas, 259
Watson, Thomas, Jr., 261
Watson-Watt, Robert, 404
Watt, Gregory, 381
Watt, James, 44, 45, 97, 307, 308, 309, 377, 378, 379
Waves, 81
Wax cylinder, 398
Weapons, 311, 382, 383, 385, 389
Weaving, 374, 379, 380
Wells, Horace, 386
Wesson, Daniel B., 389
Westinghouse Company, 156
Wet collodion process, 387
Weygandt, Cornelius, 243
Wheatstone, Charles, 210, 385, 390, 391
Wheeler, S. S., 211
Whirlwind computer, 257
Whitney, Eli, 97, 379
Whittle, Frank, 405
Wilde, Henry, 391, 394
Wilkes, Maurice, 254–255
Wilkinson, John, 377
Williams, F. C., 250, 254
Williams, S. B., 237
Wilson, Allen B., 387
Wind power, 222
Woodworking, 381
Woolrich, John Stephen, 391
Worden, 400
World competition, 24–25
Wozniak, Steve, 410
Wright brothers, 55, 399
Wrought iron, 378
Wyatt, John W., 312

Xerography, 408
Xerox Corporation, 262

Yamashita, Hideo, 264
Young, Owen D., 156

Zipernowski, Karl, 394
Zuse, Konrad, 240–241, 247
Zworykin, Vladimir, 250, 252, 405